Ph Rider

Robert Kastenbaum

University of Massachusetts, Boston
Cushing Hospital

HUMANS DEVELOPING
A Lifespan Perspective

Allyn and Bacon, Inc.
Boston London Sydney Toronto

Library of Congress Cataloging in Publication Data

Kastenbaum, Robert.
Humans developing.

Bibliography: p.
Includes index.
1. Developmental psychology. I. Title.
BF713.K386 155 78-21103
ISBN 0-205-06513-9

Printed in the United States of America.

Drawings within the artwork by Deborah Johnson.

Chapter opening photos: page 4 by Tania Mychajlyshyn-D'Avignon; page 28 by The Psychological Corporation, 304 East 45th St., New York; page 54 courtesy of C. Paul Hodgkinson, M.D., Henry Ford Hospital; page 98 by Sam Sweezy, Stock, Boston; page 134 © 1978 by Thomas S. Wolfe; page 206 by Charles Gatewood, Stock, Boston; page 338 by David Kelley; page 374 by Bobbi Carrey; page 510 by Peter Karas; page 540 by Paul S. Conklin; page 576 courtesy of CBS Columbia Records; pages 170, 242, 292, 416, 438, 466 by Talbot Lovering.

Production Editor: Nancy Doherty
Interior Designer: Nancy McJennett
Manufacturing Buyer: Karen Mason

Contents

Preface ix

About This Book 1

1 **Beginnings . . . Ends . . . and Middles 4**

Chapter Outline 5
Beginnings? 6
Ends? 9
Middles? 19
Summary 23
Your Turn 24

2 **Science and Scenario 28**

Chapter Outline 29
Scenario 1: The Blind Infant as Protagonist 31
The Developmental Mission and Its Methodologies 40
Summary 50
Your Turn 51

3 **The Prenatal Scene 54**

Chapter Outline 55
Scenario 2: The Prehistory of Michael 56
Psychological Influences on the Prenatal Scene 62
Genetic Influences on the Prenatal Scene 71
Life Before Birth: The Pattern of Prenatal Development 82
Placenta Determinism: From Mother to Child 87
Hazard: Abortion 90
Summary 93
Your Turn 92

4 **Birth and the First Thirty Days** 98
Chapter Outline 99
Birth Has Many Meanings 100
Does the Newborn's World Bloom, Buzz, and Confuse? 103
Unlearned Behavior: The Reflexes 106
Perceiving, Learning, and Preferring 112
Scenario 3: Michael-the-New 122
Hazard: Postpartum Psychosis 124
Summary 128
Your Turn 130

5 **The First Two Candles: Growing Toward People** 134
Chapter Outline 135
The Infant as Wary Adventurer: Exploratory and Attachment Behavior 136
Scenario 4: Michael Attaches 141
Parenting: The Effects of Parental Mood and Behavior 144
Hazard: Parental Assault 157
Summary 166
Your Turn 162

6 **The First Two Candles: Moving and Experiencing** 170
Chapter Outline 171
On the Move: The Child Becomes Mobile 172
Experiencing and Expressing 180
The Child Speaks: The Early Development of Language 182
Mental Development in the Very Young Child 190
Summary 199
Your Turn 200

7 **Inside Out or Outside In? A Choice of Perspectives on Humans Developing** 206
Chapter Outline 207
The Perspectives in Overview 208
From the Inside: Development Is Instinctual 209
From the Outside: We Develop Through Experience 221
Becoming a Person Through Interaction 230
Summary 236
Your Turn 237

8 **The Preschool Scene** 242
Chapter Outline 243
Scenario 5: Out-of-the-Body Experiences 244
Mental Development in the Preschool Child 249
Child's Play 262
Part of the Family: Development of Social Relationships 271
Day Care 277

Hazard: A Parent Dies *280*
Summary *285*
Your Turn *286*

9 **From Home to School: The Child Steps Out 292**
Chapter Outline *293*
Growth and Stability *294*
The Child as Philosopher *301*
Hazard: The Slow Learner *320*
Scenario 6: School Bells for Michael *325*
Scenario 7: School Bells in Peyrane *329*
Three Michaels *331*
Summary *331*
Your Turn *332*

10 **The Child Puts It All Together 338**
Chapter Outline *339*
Celebrating the Ten Year Old *340*
"Paying" Attention *342*
How Does the Ten Year Old Learn and Think? *347*
Playing Fair: Development of Moral Judgment *356*
Scenario 8: "Please Like Me" *362*
Hazard: Fatal Accidents *365*
Summary *369*
Your Turn *370*

11 **Childhood in Perspective 374**
Chapter Outline *375*
What Is Becoming of Childhood? *376*
Girls, Boys, and Children: Sex Differences in Development *386*
Identity and Interaction *397*
Recent Studies that Illuminate and Challenge Our Understanding of Childhood *403*
Summary *412*
Your Turn *413*

12 **Ripening: The Body Comes of Age 416**
Chapter Outline *417*
The Pre-years: Biological Changes on the Verge of Puberty *419*
Into Puberty *425*
Why Adolescence? A Biological Perspective *434*
Summary *435*
Your Turn *436*

13 **Who Is the Adolescent? 438**
Chapter Outline *439*

Interpretations of Adolescence *442*
Interpreting the Interpretations *460*
Summary *463*
Your Turn *464*

14 **Intimate Thoughts and Relationships in Adolescence 466**
Chapter Outline *467*
Moving Through Adolescence with the "Normal" Boy *468*
Adolescent Girls in Self-portrait *480*
The Mature and the Less Mature Adolescent *483*
The Continuity of Value Orientations Between Parents and Child in Adolescence *487*
Sex Stereotypes and Competitive Behavior *489*
Loving Intimacy in Adolescence *491*
Mutual Reconstruction of Reality: Is This What Friends Are For? *498*
Death as Thought and Reality *500*
Summary *505*
Your Turn *506*

15 **Prime Time: The Early Adult Years 510**
Chapter Outline *511*
Life Satisfaction: Anticipated and Actual *512*
The Able Young Adult: Physical, Mental, and Social Competence *520*
Scenario 10: Ch'eng-jen *533*
Hazard: The One-way Ride *534*
Summary *535*
Your Turn *536*

16 **Neither Young nor Old 540**
Chapter Outline *541*
Conceptions of Middle Age *544*
What We Do and What We Become: Occupation and Mid-life Development *551*
What We Have Been and What We Become: The Continuity of Self *557*
Scenario 11: Turning the Corner *565*
Hazard: Alcoholism *567*
Toward Understanding the Middle Years of Life *568*
Summary *571*
Your Turn *572*

17 **The Old Person: In Stereotype, Reality, and Potentiality 576**
Chapter Outline *577*
What Is Aging? *578*
The Competent Old Person *587*
Scenario 12: Toward Sanyāsa *602*
Hazard: Victimization by Crime *605*

The Old Person in Our Society *606*
Intimate Thoughts, Feelings, and Relationships *613*
Death, Dying, and Grief *617*
Loving and Sexuality *620*
Summary *624*
Your Turn *625*

If There Were Another Chapter 631

Glossary 634
Name Index 639
Subject Index 642

Preface

Perhaps you have had experiences like this yourself. I boarded the rollercoaster in a rickety amusement park to the general disapproval of my parents and a cluster of other grown-ups. They would have nothing to do with this contrivance themselves, yet they didn't want to stand in the way of my big thrill. Excited, confident, apprehensive, and downright scared—all these words described my state of mind. The first few rises and falls were gentle enough to bolster my fading courage. But then came the real twistings, the slow ascensions, and the steep, screaming plunges. As the coaster (and my innards) rose for still another climb, I saw the jovial sign posted ahead: "Under construction. Proceed at your own risk." This, I hoped, was a joke. Somehow the ride did come to its appointed end with my body, if not my mind, still in one piece. There was not much time to dwell on how I *really* felt because now the grown-ups were approaching and it was obviously incumbent on me to tell them what fun it had been and how, no, really, it hadn't been scary at all.

I did not know just then that I would be writing a text on lifespan human development. For that matter, I knew very little about the rises and falls, lurches, twistings, and turnings that are so characteristic of human lives. Many years later I completed a course on human development (not quite as exciting as the rollercoaster ride!). From lecture and text I learned quite a bit. I was still not well prepared, however. What might I expect in my own life? How was I to understand other people who were already in their middle and later years? In those days, courses on human development almost invariably ran out of track early. The typical student was given the opportunity to look at childhood through a sort of rear-view mirror while hurtling toward his or her own future with little in the way of guide posts.

This situation has started to change in recent years. More and more psychologists, teachers, nurses, and others in the "people professions" now realize that humans developing is a phenomenon not limited to the first few years of life. Those of us who write and teach are not fulfilling our entire responsibility if we transport the learner to a hairpin curve in the road of life and then give him or her a hearty slap on the back and say, "Proceed at your own risk!" Life, indeed, is always "under construction" and each individual has personal tracks to make. Yet we have the right to expect a little more help from each other. Fortunately, research on humans developing throughout the entire lifespan has also increased markedly in recent years. It is now possible to draw on more extensive and sophisticated inquiries into human behavior and experience beyond childhood. Attention to the full scope of the human lifespan does not detract from intensive concern with infancy and childhood. On the contrary, we can now put childhood into a more adequate perspective. We can examine the relationship between the earliest development of social, personality, and cognitive characteristics of the individual and who this person is likely to be some seventy or eighty years hence. The challenges to research skill and theory construction are immense, but at least many of the problems are now being seriously addressed.

Precisely what will you find in this book? You will notice an overall plan that traverses the lifespan from conception (actually, from *before* conception) through old age. It just makes sense to follow life's characteristic contours. Interwoven there is a context-setting, and reflective theme that occasionally steps outside the chronological sequence. The first two chapters are examples of this approach, as is the later exploration of the "inside-outside" dialectic (Chapter 7) and the attempt to put childhood itself into perspective (Chapter 11). Each chapter has organizational features that will make it easier for you to find and return to topics of particular interest: the chapter outline at the beginning and a carefully prepared summary at the end. Most chapters have other features that will help you bridge the gap between a book (*any* book) and "real life." You will find *scenarios* in which people are treated as people. In other words, there are miniature slices-of-life dramas that illuminate and bring together topics that often have to be presented separately and analytically from the research standpoint. These scenarios usually are drawn from research and provide specific information as well as the opportunity to see lives in action. You will also find illustrations of *hazards* that confront people at particular points in the lifespan. These are, again, drawn from research. Attention to hazards is one way of discouraging writer and reader from treating humans developing as a sort of fairy tale or railroad schedule. Not everybody moves effortlessly from Point A to Point B as we are sometimes led to believe.

You will also find sections that invite you to take an immediate and direct role in developmental inquiry yourself. These are the *Your Turn* sections. Even the most overburdened (and, dare I say, laziest?) reader will have something to gain simply by thinking over these suggested enterprises. A reader as enthusiastic, perceptive, and energetic as yourself will, of course, gain immeasurably more by actually following through with selected *Your Turns.*

The *Learner's Companion,* a separate publication, was prepared to help you "process information" (one of the buzz phrases of our time) presented in this text. It may also be useful to you in approaching other courses as well. The *Instructor's Companion* is intended to make life just a little easier for your professor who is faced each semester with the challenge of shaping and reshaping the course.

Why is this book called *Humans Developing* instead of *Human Development*? Glad you asked! Life is a process or set of processes. When we set words in type and enclose them between covers it is hard to avoid giving the impression that we have somehow captured and fixed our subject matter. Here it is, friends: all the questions, all the answers! I have done my best to present significant questions and answers, but without lulling us into a precarious sense of contentment or misrepresenting the dynamic and complex nature of humans developing. There are many places in this book where material is laid out in good and proper order. The material itself is up to date, as you have the right to expect, and you will also find most of the topics that most developmentalists think are important. But you will also find controversy and criticism, alternative views of the same phenomena, and excursions into areas of human behavior and experience that are too important to neglect even if they are not customarily given central attention in the book next door. You will see, for example, that death is not merely tacked onto the last few pages of a chapter on old age. Similarly, old age itself is not merely tacked on because the subject has become slightly fashionable of late. Our thoughts, feelings, and confrontations with mortality accompany us throughout the entire lifespan; compartmentalizing this topic would not be quite facing up to the realities of humans developing. Anticipations of old age and questions bearing on the processes of aging are also threaded through the book at various appropriate points before the topic is treated in detail. Other themes important to the full scope of humans developing appear in various forms at various places throughout the book to facilitate a true lifespan perspective.

Much of the material in this book is drawn from psychology and closely neighboring disciplines. But you will also come across material from the health-related fields, the humanities, and virtually any source that seemed useful in promoting a well-rounded view. The book is intended for learners of any adult age who have a serious interest in humans developing: psychologists, yes, but also future nurses, social workers, teachers, and the people who make and enforce our laws.

Did you know that a book such as this is read in manuscript form by many people, including experts who rip into the text to discover weaknesses, omissions, errors, and failures to communicate clearly? Their reviews are sent to the author anonymously (an antihomicidal measure, I believe). He then ignores all complimentary comments, but goes straight to the criticisms and is not worth living with for the next few days. Eventually, however, he makes constructive use of the criticisms and the result is a better book. I thank all the phantom reviewers for their perceptive, informed, cranky, and often entertaining comments. I hope my number one nonphantom reviewer appreciates how much I have valued her responses to the manuscript and to its author (both on his good days and bad weeks). Cynthia and David *and* David have helped me so much just by being their own luxuriantly developing selves while the book was coming along. By now they are accustomed to finding some of their comments and exploits recorded on the pages of my articles and books and will have fun searching for the examples in this one. It is customary to thank the manuscript typist, but there's no point in wasting a compliment on him; he and the author have to share the same head and hands whether they get along or not. I hope you enjoy the characterful artwork of Deborah Johnson and the physical design of a book that the people at Allyn and Bacon have treated rather kindly. They did insist that I write a Preface before going to press, and now we're both through with it!

Robert Kastenbaum

About This Book

Nobody comes to this book cold. Everybody knows something about human development. All of us have been younger; all of us are becoming older. Just the experience of being alive guarantees that we are not strangers to the topic of humans developing. Why, then, do we need this or any other book?

Part of the answer is obvious. We could all use more information. But what is "more"? Information is not always a blessing. Facts may fly at us in such a confusing swarm that we feel pressured and distracted rather than enlightened. Facts can also be unrelated to the questions that are uppermost in our minds. If we are wise, we will be selective about the "more." It is worthwhile to have information that is more up to date, more closely based upon careful research studies, and more comprehensive than most of us can obtain from our own particular experiences. Preferably, this information should be selected for its potential value to us—information that can really make a difference in what we think and do. This book attempts to provide you with information of the type I have been describing.

Part of the answer is less obvious. Some of what we "know" about human development before cracking a book is only half-right (which means, of course, that it is half-wrong). Studying human development does not just add bits and pieces to our store of information, it also gives us the opportunity to revise or discard notions that cannot stand the light of scientific inquiry. And who would be satisfied with bits and pieces, anyway? More useful and exciting is a frame of reference, a way of thinking about human development. People, after all, are not bits and pieces, facts and statistics. They are not put together according to simple mechanical principles. Without a frame of reference, even the most alert observer will find himself or herself adrift among his or her factual bits and pieces. This book aims to convey a way of looking at human development as well as a careful selection of factual material. This statement can be taken a step further. This book aims to encourage your own development as a keen, lifelong observer of human growth and transformation.

Perhaps you have not seriously thought of yourself as a psychological observer. And perhaps you have not bargained for a book that wants to become a part of your life. But consider for a moment where you have been and where you are going. Nobody else has experienced life in exactly your way. Although your future is likely to involve many people—including some you have yet to meet—it will also be your own distinctively personal life. No book can adequately represent your own trajectory, even though in some respects your life resembles those of other people. A book devoted to your own development is not a bad idea. But if that book were completed today, by tomorrow it would be on its way to becoming out-dated. Keeping a perspective on your own development is a perpetual challenge.

Is this challenge worth the effort? There are three points to keep in mind here:

1. Everything we observe and conclude about other people must filter its way through the kind of person we are. Self-knowledge is crucial. To be accurate and wise in our relations with others requires a knack for monitoring our own thoughts and actions. If we leave ourselves out of the picture, our knowledge of developmental psychology is apt to be empty and artificial.
2. There is pleasure in observing ourselves and others with an informed eye. Some people hear more in music, taste more in food, and see more in gardens and forests than other people. We apply the term *connoisseur* to a person who has an unusually keen ability to understand and appreciate some aspect of life. Have you thought of becoming a connoisseur of life itself? Developmental psychology is only one of the many pathways to quickening the appreciation of human life. Unfortunately, however, this topic has often been approached either with the resigned attitude that "well, I guess I should know something about it" or the strictly practical attitude that "there may be a payoff here." These approaches have their validity. Being well acquainted with human development will prove helpful in decision-making, time and again. Yet the pleasure factor is there as well. The more we know, *really know* about development, the more we are able to savor, to enjoy each other. An all-out dedication to observing human development rewards us with a heightened capacity for enjoyment and a deepened capacity to feel for others in their times of crisis and suffering. Is appreciation a goal of science? We seldom if ever hear this said. But this book contends that appreciation of human development—apart from its practical consequences—is a significant goal.
3. Facts become outdated, circumstances change, unprecedented events occur. Today's best knowledge could become quaint or misleading in the years ahead. Keeping the general pattern of human development in perspective requires as much continuing effort as does the monitoring of our own personal lives. Lifelong students of human development

> evolve a knack for identifying and analyzing new events and influences. They recognize that closing their minds along with their books is an invitation to obsolescence. They also recognize that to allow their knowledge to become out-moded is not so very different from allowing themselves to become out-moded.

It is for these reasons that I have not presented you with this book as a passive object. I have not written a head of cabbage. You can automatic your way through this book, dutifully scanning page after page, only to discard them like so many used-up leaves. You can, of course [illegible] your privilege. But you cannot blame me for hoping that you will bring your best to this topic. I have attempted to come as close to you as the form and requirements of a book will allow. I have attempted to communicate not only the world of facts and research, but also something of the relationship between science and daily life. In many instances I have relied on you to bring the topic to life. You are asked to "localize" and adjust material that is presented. You are asked to make your own observations. You are asked to become an insider in the search for understanding, not the polite recipient of "established truths." I have tried not to forget that you probably are coming at the systematic study of human development as a near-beginner; I did not assume a lot of previous knowledge on your part. Technical terms and concepts will be introduced when they are needed. I have not crowded you with impressive-sounding material intended more for the eye of other professors than for the student. On the other hand, I haven't insulted your intelligence. Being new to a particular technical field is not the same thing as being stupid.

I will be slipping definitions and notes in the margin now and then

It is also my hope that this book will be especially useful to people whose lives and careers will be intimately related with other people. This is not just the psychologist, but the nurse, the educator, the social worker, and the minister, to name a few of the people-professions, all of whom have responsibilities that require a good working knowledge of human development. Those of you who may be on your way to such careers will find samplings of future problems and solutions. But whether or not you will ever have professional responsibilities for the care of other people, it is likely that people will depend upon you in various ways. As a parent, neighbor, and colleague you will have to make decisions that can gain from an appreciation of the developmental dynamics involved.

There is more that could be said about this book, including some of the other ways in which it differs from traditional presentations of developmental psychology. But I do not care much for long introductions. We are ready to begin. Welcome!

BEGINNINGS . . . ENDS . . . AND MIDDLES

1

CHAPTER OUTLINE

I
There are alternative views about the beginning of human existence and development.
Viability, survivability / Eastern versus Western assumptions / Characteristics of the preconceived and preborn

II
Choices also exist as to the end of development.
When the destination is reached / Types of development: height and creativity compared / Directional change / Reversibility / Developmental progression / Death / Social immortality

III
Developmental phenomena are usually observed when they are already in process.
Cross-sectional, longitudinal and cross-sequential research designs / Cohort effects / Age effects

BEGINNINGS?

Begin at the beginning and end at the end. This reasonable advice is not easy to follow in the study of humans developing. Consider beginnings for a moment. Precisely when does a *human* life start? At conception? At birth? Somewhere between conception and birth? According to some viewpoints, a truly human existence does not develop until months or even years after birth. The newborn, for example, may not be accepted as a person until he or she has demonstrated **viability** (survival potential). This attitude is often found where the infant mortality rate is very high. It is as though society has decided not to invest itself completely in what might prove to be a "bad risk." Imagine yourself, for example, to be a young man or woman living with your spouse in a blighted area of a large city, unable to rise beyond poverty despite hard work. Your first baby is soon dead. The second suffers the same fate . . . and the third, fourth, fifth. Assume that it is still very important to both of you to raise a family despite the severe infant mortality rate. How would you regard the sixth child while it is still within the high-risk age range? Burdened by the past deaths and fearful that this child also will fail to survive, any of us might suspend some of our emotional investment in the new baby. The infant has to prove itself viable before we permit ourselves to accept it as an individual human personality. One of the most famous individual personalities in your grandparents' youth belonged to a man who came out of the circumstances that have just been described. Seventeen children were born and seventeen were swiftly carried away by the ravages of disease before the birth of Enrico Caruso, a man who became one of the most celebrated voices and personalities in the history of opera.

Viability is the ability to survive

Perhaps this example will help to alert us that when others do not share our personal views of human nature and **development,** the reasons for these differences are worth understanding rather than condemning. We might have the conviction that a neonate is human from the moment of conception or from the first breath outside its mother's body. Nevertheless, there are many other perspectives from which all aspects of humans developing can be viewed and these begin, as we are starting to see, right from the beginning. You might come across the belief, for example, that the young one cannot properly be regarded as a person until he or she has demonstrated certain "distinctively human" qualities. The person must have reached the "age of reason," "developed a sense of moral responsibility," or "made a name for himself or herself."

Development is, literally, an unfolding of potential

An even more extreme position can be taken. Adults sometimes are not treated as though they were persons in the full sense of the term. Historically, some societies have regarded people who dwell on the other side of their city walls as something less than human. In our own time we have witnessed everything from subtle discrimination to extreme torture and murder of people who were not accorded fully human status by their persecutors. Political enemies (and even police officers) have been called *pigs;* ethnic and racial slurs continue to portray other people as something less than human (it was psychologically easier to "waste a gook" than kill a person); individuals opposed to a particular regime have been labelled as *vermin* who are thereby candidates for *extermination;* the massacre of millions of Jews by Nazi Germany was accompanied by a descriptive language in which they were degraded and by physical markings (the infamous yellow stars) intended to set them apart from other people. Current demands for equality by various sets of people in our own society reflect their perception that, although physically and mentally mature, they are not regarded as fully enfranchised members of the human race. Individual maturation does not always guarantee that one will be accepted as a human being. The consequences of being defined as *human* as contrasted with *something-other-than-fully-human* are far from trivial: both the length and the quality of life may be affected.

Just as one might suspend the judgment that "here is a human being" until birth or long afterward, it is also possible to maintain that life begins *before* conception. For centuries a variety of beliefs in the transmission of souls or identities through time and space have been accepted. The child who is yet to be conceived may already pre-exist in another form. Whether or not we share this belief, we must reckon with it as an influential frame of reference. We may be accustomed to regarding the **neonate** as brand new. Others, however, might see a continuation of life forces that have already enjoyed a long history. Richard A. Kalish, one of the few psychologists to concern himself with this question, has observed:

A **neonate** is a newly born individual

> Although it is common in Asia, psychological pre-existence is rarely assumed in Western countries. Sporadic stories of the Bridey Murphy variety do appear from time to time, implying that there is within the presently living person the possibility of gaining awareness of previous existence. By and large, however, existence in the Occident seems to begin at or following conception and extends—according to traditional theology—infinitely into the future. Young children may ask where the souls are kept before they are allowed to accompany their bodies, but adults are likely to ignore the question or consider it nonsensical.
>
> Buddhism and Hinduism, however, take it for granted that existence is continuous. Any given existence begins infinitely far back in time and will go through innumerable forms of life until Nirvana or, failing that, infinitely far into the future. Although the traditional Hindu or Buddhist does not deem it possible to gain an awareness of what life was like during his prior existence, he assumes that his present life is related to his earlier lives. Eastern life, like Eastern music, has no beginning, middle, or end; it is continuous, like an end-

> less merry-go-round with an ever-changing variety of moving animals. Western life, like Western music, has a distinguishable beginning, one or several climaxes, and a type of end, sometimes with a bang and sometimes with a whimper, after which it changes form and may no longer be heard by living human beings. (Kalish, 1968, p. 252).

Religious belief and cultural tradition influence our interpretation of human development in its earliest stages. And, as we will see throughout this book, cultural factors continue to influence both an individual's development and our perception of an individual's development throughout the entire life cycle.

There is still another sense in which it might be said that the development of a particular human being precedes its conception. Many of the circumstances that will influence his or her development are on the scene before conception. The parents are of a certain age and socioeconomic status. Individually and as a couple, they have a certain genetic lineage. The population subgroup to which the child will belong has its own profile: certain diseases and disorders occur at a relatively high rate, others are rare; certain activities and interests are typical, others are harder to find; families may tend to be large or small; organized religion may be a dominant, unifying influence, or may scarcely touch one's life. The subgroup itself may be highly integrated with other subgroups or, at the other extreme, function as a self-contained in-group that discourages intimate dealings with outsiders.

Information of the kind mentioned above can be used to predict characteristics of the preborn and preconceived as well as those already on their feet. These predictions will often be wrong or partially wrong. Nevertheless, we can learn much about the most likely characteristics of a preconceived individual: the diseases most likely to afflict him or her, the occupations most

This—or any other family configuration—establishes a climate of expectation, demand, and opportunity that influences the course of development. (Photo by Talbot Lovering.)

likely to appeal to him or her, the age at which he or she is most likely to marry (as well as height and weight at birth). If we are resourceful enough, it might also be possible to predict his or her opinions on many topics, voting preferences, and some aspects of his or her personality make-up.

What I am suggesting here is that the neonate is likely to take on many characteristics of the very particular part of the world he or she enters. In this sense, the neonate's biography begins before conception: what he or she will become can be anticipated, although imperfectly, by what is on the scene before he or she appears in the flesh (*see* Chapter Three).

ENDS?

Consider now the other extreme. When does development *end?* There are many choices here as well.

WHEN THE OBJECTIVE OR DESTINATION IS REACHED

Perhaps development ends when it reaches its objective or destination. This is an appealing choice. We develop until we get to where we were supposed to go. Height is one of the simpler examples. We grow taller and taller. Eventually we reach a certain altitude, and that's it—at least for awhile. Because this is such a familiar fact to all of us, it makes a useful test case. But is this process of vertical expansion a good example of human development in general? Let us first review some of the things we know about this aspect of physical development:

1. Height can be measured with accuracy and ease.
2. Growth along the height dimension is usually accompanied by other indices of maturation (weight being perhaps the most obvious one).
3. The peak height for an individual is determined by considering only his or her own history; similarly, the termination of height growth is also documented by his or her personal measurements.
4. Although height growth may seem to have come to an end at a certain time, we know that another spurt or two might still occur.
5. In the United States and some other nations, the "average" person has been emerging as taller than his or her parents who, in turn, were taller than *their* parents, etc.
6. There are some national-ethnic-racial-geographical differences in average height, as well as sex differences.
7. After growth (i.e., height increase) terminates, there is a period of time during which no appreciable change is observed. Eventually, however, a *decrease* in height can be documented.

Differences *within* groups are greater than those *between* groups

There is something else about height that should be made explicit. We have had so much experience with height growth, both our own and that of others', that firm expectations have been established. We know what passes for normal height at various ages, and what range of variation can be found. We know that while increase in height eventually will be followed by a decrease, the opposite ordinarily does not occur. We know enough to be surprised and curious if, for example, an eighty-year-old began to show a growth spurt.

Now let's see how the attainment of height serves as an example or model for development in general. If everything else about human development closely resembled height, then we might be able to rely upon the formula that development ends when it gets where it is going. A big *if!* There are many aspects of the total developmental process that refuse to fit into this mold. Think, for example, of *creativity.* When, or under what circumstances, does the development of creativity end?

Immediately we are faced with the problem that creativity cannot be measured with accuracy and ease. There is no measuring instrument available that can compare with the ruler and its reliable markings. This means

Creative behavior is as much a part of human development as physical growth. Creativity has no set limits of fixed destination. (Photo by Talbot Lovering.)

that we cannot move with assurance from the origins of creativity through its peaking and, therefore, cannot be confident about our assessment of its decline either. As a matter of fact, we cannot be certain that creativity does peak or decline—or that it might not surge back after a period of decline, quite unlike height growth.

The basic rules that seem to govern the development and termination of height growth just may not apply to creativity or to some other aspects of human development. Apart from reasons that have already been mentioned, we should recognize that height and creativity seem to differ in another crucial respect: growth in height shows up in more of the same—more inches. An inch is an inch to the ruler. Growth in creativity, however, may show up as something *different* from what had been seen before. The change is likely to be a matter of quality rather than quantity. And if it is not appropriate to speak of a person as gaining an "inch of creativity," neither does it seem right to speak of a person "losing an inch."

Maturation is the process by which the young attain adult structure and function

More differences come to mind. Creativity is as difficult to define as height is easy to define. Creativity may or may not be accompanied by other indices of **maturation.** Is each generation more or less "creative" than the preceding one? Is one ethnic group more creative than another, or creative along a different dimension? Perhaps it does not even make sense to ask the same type of question in talking about height and creativity, yet both are part of the process of humans developing. Do we really evaluate an individual's

Figure 1-1. *Height and creativity: two models of human development.*

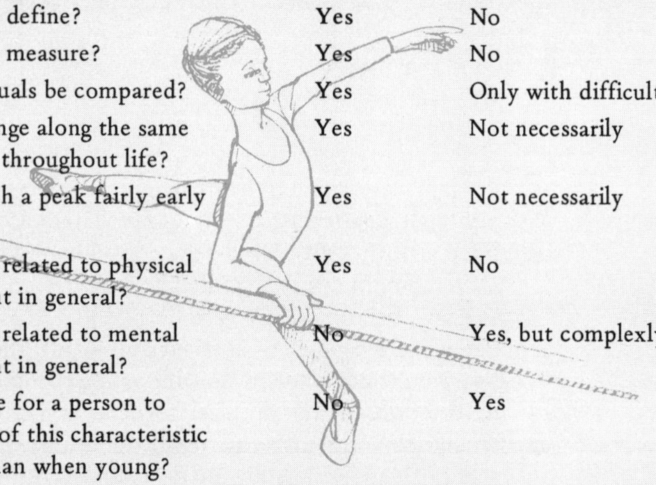

Question	Height	Creativity
Is it easy to define?	Yes	No
Is it easy to measure?	Yes	No
Can individuals be compared?	Yes	Only with difficulty
Does it change along the same dimensions throughout life?	Yes	Not necessarily
Does it reach a peak fairly early in life?	Yes	Not necessarily
Is it closely related to physical development in general?	Yes	No
Is it closely related to mental development in general?	No	Yes, but complexly
Is it possible for a person to show more of this characteristic when old than when young?	No	Yes

creativity in terms of his or her own previous behavior? Or do we tend to make comparative ratings, sizing him or her up with others of the same age or others who have engaged in the same form of creativity? All in all, our approach to the development of creativity differs markedly from our approach to height growth. If we insisted upon treating height and creativity in an identical manner, we would probably be doing damage to at least one of these dimensions of human growth, if not both.

The height-creativity comparison challenges the basic notion that development ends when it reaches its objective or destination. Does creativity even have a "final destination"? Or is it, in principle, capable of perpetual development? Could we even agree on what creativity is? On what it should be? The objective or developmental destination of creativity is elusive. The standards we use to define and judge creativity are influenced by many personal and social factors. What one person or subgroup regards as a creative innovation may be seen as an act of destruction by others. Many aspects of human development are ambiguous or controversial. It might be comforting and convenient to apply the same terms and criteria to all manifestations of development, but it would also be highly misleading.

Perhaps we should not speak too glibly about development having a specific goal or outcome. This practice leads itself not only to unintentional confusion, but on occasion to deliberate distortions and propagandizing. In each instance we must identify precisely what aspect of humans developing we have in mind and discover, if we can, the "rules" that apply to this phenomenon in particular.

This discussion does not force us to scrap the idea that development may have certain natural or built-in limits. But it does suggest that we think critically in every situation where this possibility arises. Let's move on to another possible answer to the question, "When does development end?"

WHEN AGING BEGINS

Perhaps development ends when aging begins. This is an interesting answer although, again, not a simple one. There are similarities with the answer we have already considered (that development ends when it achieves its goal). Development is still regarded as *directional change*. We do not just change in size: we become taller. We do not just change in the way we think: we become more capable of abstract and logical thought. But with this new alternative we do not have to depend so much upon the idea of a final objective or destination. In its simplest form, this approach only requires us to notice when development starts to go the other way. Showing signs of shrinking in size or thinking less abstractly (more "concretely") could be described as *reverse development*. More often, it is known as *aging*. Whether or not aging and reverse development should really be considered equivalent is a question that will be explored later in this book. For now, it is enough to recognize that

Physical development has concluded; signs of aging can be detected—yet the adult may be reaching new depths of insight and new heights of achievement, all well sustained by continuing vigor. (Photo by Jean Boughton.)

a process of undoing seems to be at work. The physical and psychological achievements that took years to reach their peak begin to give way.

The same phrase can capture both the end of development and the beginning of aging. "I am slowing down" is a good example. This statement might refer to a specific area of functioning: the athlete who is losing his speed afoot, or the businessman finding it difficult to keep up with the tempo of competitive pressures. The speaker might also be reporting his or her general self-assessment: "It is not just my legs, not just my capacity for work under pressure. It is me. *I* am slowing down." This individual might still be highly competent when compared with many others. In terms of his own life trajectory, however, he has gone past the point of continued improvement—at least on those characteristics that are under consideration. "Slowing down" represents the end of development at the same time that it proclaims the onset of aging.

If we accept this approach, we are then able to work consistently with many other phenomena in the study of human development. What goes up we call *development*, what goes down we label *aging*. Whenever a human function or characteristic declines we conclude that we are in the presence of aging. The field of vision shrinks. Aging! Bones become thinner and more brittle. Aging! Skin wrinkles, loses its elasticity. It is aging, of course! Sexual activity diminishes. What else—aging! Ability to learn new material is not what it used to be. Right again—aging!

This approach is popular and influential. It has the advantages already mentioned, and one could find further arguments in its favor. Let's look at some of the problems instead.

The first problem is not so much a defect of the approach itself as it is a temptation placed in our path. We can agree, if we like, to use *aging* when we want to refer to reversals and negative changes. However, it is easy to slip into a misleading usage. *Why* has his field of vision decreased? Aging! *Why* are her bones thinner and more brittle? Ditto! Skin wrinkled, sexuality diminished, learning impaired—why? Ditto, ditto, ditto! Designation and explanation are not necessarily equivalent. Even scientists, educators, and practitioners sometimes stumble here. It is one thing to apply the term *aging* to certain kinds of changes; it is quite another thing to imply that aging *causes* the changes. This, of course, amounts to saying that aging causes itself, a proposition that hardly advances our knowledge. But, as we have said, this is a temptation rather than an actual defect in the approach that is now being considered. If we are alert enough to avoid this lure—a tempting one because it leads us to believe that we know more than we really do—then we only have to face the direct limitations and complications of the approach that maintains development ends when aging begins.

One of these limitations has to do with *reversibility.* Can we still speak of *aging* if the negative changes prove to be temporary? Suppose that a decline in learning ability is shown by two fifty-year-old men. Five years later, one of these men has remained at the diminished level, even lost a little more. The other one, however, has bounced back. Should we judge that one of these men was "really" aging, and the other not? Or that aging can be temporary and reversible? Whichever alternative we choose, we will have given up much of the simplicity and self-evident validity that makes this approach so attractive at first. As we continue our study of human development throughout this book, it will be seen that: (1) not all negative changes should be interpreted as manifestations of aging, (2) not all negative changes are permanent, and (3) aging does not have to be regarded as an entirely negative process. Let us linger a moment on that last point. If we commit ourselves to considering the aging process as change in a negative direction, then we will have a hard time recognizing and interpreting anything that might *improve* as we live longer. This does not mean that we would completely miss such phenomena, but the bias would be there.

Wouldn't it be nice to have something to blame for all our problems?

Another complication. It is not always possible to place a positive or negative sign next to the changes that occur in our physical, mental, or social status. As a child gains increasing coordination of his or her body, it seems appropriate to consider this a plus or *developmental progression.* As coordination begins to falter many years later, it seems appropriate to apply the minus sign. However, many changes cannot be labeled *positive* or *negative,* at least, not without an argument. You may have heard it said that people tend to become more conservative politically as they grow older. This is not always true (and "conservatism" is subject to a variety of definitions). But suppose that the statement were accurate, would it be a progression or a regression to adopt a more conservative political philosophy? The judgment on this question is more likely to depend upon the judge's own political philosophy than upon objective research. Take another example. A man shifts his favorite participant sport from tackle football to golf. Is this a negative shift, and thus a sign of aging? Or is it part of a change in life-style that really cannot be judged on a simple positive-negative basis?

I am suggesting that many significant aspects of human development would either be ignored or misinterpreted if we relied entirely upon the guideline: development ends when aging starts, and aging is to be understood as a shift to the negative side (loss, decrease, weakening).

There is another complication which may prove a blessing in disguise. Close examination reveals that we do not usually develop or age at a uniform rate. This applies to changes taking place within a single individual as well as differences among individuals. A particular child may quickly become agile physically, for example, while taking her time in the development of language. Similarly, a particular adult may show signs of physical aging, but no signs at all of impairment in his thought and language. If we insist on taking a very simple approach to development–aging, then we are likely to be disturbed by the possibility that a person may be maturing and de-maturing at the same time. But why insist on such simplicity? The notion that aging picks up where development leaves off can encourage us to look carefully at differences and variations—even though it also lends itself to sloppy use by those who want their answers quickly, easily, and painlessly.

The approaches we have been considering may or may not jibe with your own thinking. If you have been reserving judgment on when development ends, then you will be interested in the third and final alternative that will be offered here.

WHEN WE DIE

Perhaps development does not end until death. As long as life goes on, development continues. This alternative may seem peculiar to those who have accepted without question the traditional emphasis on early development. Many still regard *child* development to be virtually identical with *human* development. If we were fully committed to this view, then it would be difficult to imagine that personal growth might be found throughout the entire life cycle.

Another bias enters here as well. The social-behavioral sciences and the people professions have given surprisingly little attention to dying and death. One of the consequences (or is it one of the causes?) of this neglect is the assumption that as a person approaches death all changes that occur are negative, **regressive,** and for the worse. The possibility that development might persist throughout the second half of life and even into the dying process has

Regression, in personality, is a shift toward an earlier, simpler way of functioning

There are people who seem always to find something new in life and in themselves. For such men and women, old age is not necessarily the end of development. (Photo by Frank Siteman, Stock, Boston.)

seldom been taken seriously. We will look into this bias—and why it is changing—at various points throughout the book. It has been brought to your attention at this time so that you will know I am dealing with a somewhat innovative view.

What is to be said for this idea? First of all, it has a hopeful and democratic sort of appeal. *Development*, as you probably have noticed, is a word we tend to reserve for something that we like; it is a nod of approval as well as a scientific term. (*Should* it be used in both ways? What do you think?) When we say that development continues throughout life, we are speaking in positive and hopeful terms. It is easier to face the future if we believe that development (the "good stuff") will accompany us on our personal journey to the limits of our lives. The idea is democratic in the sense that no zone within our total life cycle is slighted. Development can be found whether we are concerned with the young, the old, or the inbetween.

The idea is useful because it encourages us to raise questions and make observations that we might fail to do otherwise. Who knows what we might discover if we took seriously the position that development continues throughout the entire life cycle! What changes have we overlooked or misinterpreted in the past? How can our perceptions and definitions of development be sharpened by examining the new ways in which growth may reveal itself after childhood and adolescence? Once we *expect* to find signs of further maturation, we are more likely to recognize them when they do occur (although the over-eager observer then runs the risk of seeing development when it is not there).

The attractiveness of this alternative, however, does not guarantee its validity. Do we *want* to believe that development goes on as long as life endures? The burden is then upon us to demonstrate this proposition if, indeed, it can be demonstrated. Wishful thinking is hardly enough; sensitive research and careful reasoning are required. Although research on this topic has not been abundant, we will keep our eyes open for material bearing upon the possible persistence of development throughout the entire life cycle as we move along in this book.

Another type of question confronts us if we wish to pursue this alternative. Assume that development does continue until death. Precisely what does this mean? Defining death is far from simple; in fact, our definition is now in the process of being changed or "updated" (Cutler, 1969). Life processes may persist after the pronouncement of death has been made. An individual may be in a state of **clinical death** (circulation and respiration suspended, but irreversible organic damage not yet advanced) for a brief period of time, or he or she may be in an unresponsive, comatose condition for weeks, months, or even years. The *moment of death* is not always easy to determine. More significantly for our purposes, it is not always clear when the *person* has died.

In **clinical death,** vital functions have ceased, but nonreversible bodily damage has not occurred

There are circumstances (such as deep, prolonged coma) in which the observer has the strong impression that the person is no longer there, even though the body lingers. It then may appear sensible to judge that development as a human has terminated. And so we find ourselves coming back to the question of human status that engaged our attention when we first started to inquire about the beginnings of development. Two important variations should be mentioned at this point:

1. Development as a human being may appear to halt well in advance of physical death. An individual may be regarded as either psychologically or socially dead. **Psychological death** is a term that comes to mind when we encounter somebody who appears to have lost the capacity for experiencing life, somebody whose identity as a person has been virtually submerged. By contrast, when a person is *treated* as though he or she were not a living human being, we may speak of that person as **socially dead.** Consider for a moment an aged patient who is confined to bed and does not speak. He may be treated by those in his environment as though he were dead. In no way is he made a participant in conversation or action. Socially he is dead, but if he is, in fact, "out of it" mentally and emotionally, we might consider him psychologically dead as well. It has been observed, however, that many people who *look* psychologically dead and who are treated as though socially dead are still capable of experiencing and interpreting what is happening to them (Kastenbaum, 1969). The two conditions, psychological death and social death, do not necessarily occur together under all circumstances and their relationship to physical death is also variable.

In **psychological death,** mental life has been lost, but biological functioning remains

In **social death,** the individual is treated as nonexistent by others

Instances of psychological and social death could be cited to challenge the idea that development continues throughout life. "Here is an individual who is alive, but not developing." But we might also say: "Although the heart is still beating, the person we knew is gone. A death of the person has taken place and, without the human person, who is there to develop?"

The thoughtful reader will recognize that we have not delved into this topic as extensively as it deserves. Perhaps enough has been said, however, to suggest that development until death is a proposition that should be considered in light of *who* is developing and what we understand by *death*.

2. Human development may persist through or beyond death. Doctrines of immortality come first to mind here. Belief in some form of survival after death was cherished by many of our ancestors and is still, of course, very much with us today. Interestingly, some so-called primitive societies took a more developmental view of survival after death than our own familiar notion of heaven as a place or state of unending bliss. It is impressive and instructive to learn how humankind has envisioned the prospect of life beyond the grave (*See* Ducasse, 1961 for a useful survey and discussion). For our purposes it is enough to remember that there are many perspectives from which it could be maintained that human development, in some form, continues "on the other side." Whether or not you hold one of these positions, it is worth keeping in mind that belief in survival is a relevant attitude in the study of human development. It certainly has a bearing on when we believe development ends.

Social immortality is the continuation of the person as a memory or influence in others' lives

Social immortality is another facet of belief in continued development after death. You may be carrying forward the name of a parent, grandparent, or other forebearer. Perhaps you will bestow your own name upon an infant. Symbolically this can project your identity into future times after your physical death. With or without giving our names to our offspring, the begetting of children can support the desire to have something of ourselves persist. Social extension of the self after death (the other side of the coin, perhaps, from social death) can be sought through other means as well. A few people have become statues in the park. A few others have become endowed chairs or sofas in universities, scholarships, and ships of the sea. Artists, authors, scientists, and inventors may live on in their achievements. There may be even more subtle forms of remaining part of the social scene after death, for example, donating body parts for transplantation.

Are attempts at social immortality really developmental? Not by traditional standards. The living person is dead, therefore he or she cannot develop. Yet, it is not necessary to indulge in ghosts and migrant spirits to appreciate that certain aspects of the self or the human person are the property of society during one's lifetime (e.g., reputation) and are thereby also subject to change after death.

MIDDLES?

The beginning and end points of human development are difficult to specify, as we have seen. Attention is usually focused on what takes place between start and finish. Even the traditional emphasis on early development ordinarily does not probe into the very beginnings; still less does it explore the question of termination. We are in a better position to appreciate middles, however, when we recognize that there are many problems and alternatives associated with both extremes of the life cycle. How we interpret the broad span of human development depends to some extent upon our understanding of the extreme points. How can we hope to comprehend what development "is" if we do not have some idea about where it has been and where it is going?

Most of the time we are concerned with developmental phenomena that are already in process. This is just another way of saying that we come upon the scene some place in the middle. Both the infant and the aged person have a certain history before we make their acquaintance. Usually we are not in a position to follow up our observations indefinitely. People come into focus for us, and then disappear from view: they move on, or we do. This is true for many, although not necessarily all, of the people we meet in the course of our lives. It is even more true, perhaps, in the research encounters that provide the basic information for the field of human development. The specialist in child development may see boys and girls for only one set of observations and never set eyes on them again. The specialist in adult development may also limit himself or herself to a single set of observations with men and women he or she did not know as children. A study in which people are observed at one point in their lives is known as **cross-sectional.** Some of the conclusions that have been drawn about human development come largely from cross-sectional research. Suppose, for example, that we were interested in learning how moral development, the sense of right and wrong, develops from age five to age fifteen. The most convenient research strategy might involve studying different children who at this moment are age five, age six . . . age fourteen, age fifteen. All the ages would be covered, so conclusions might be drawn about changes from age five through age fifteen. In fact, we would have *no* information about the dynamics of moral development between the ages of five and fifteen in one individual child. Cross-sectional research, useful as it can be, creates a sort of composite person who is built from information obtained by a variety of people who represent different ages.

In **cross-sectional studies,** people of different ages are studied at the same time

When we are able to study the *same* person over a period of time, then we are in a better position to understand the actual developmental patterns involved. This is known as **longitudinal research.** It is much more costly and time-consuming than cross-sectional study. Even with longitudinal research designs, however, we seldom encompass more than a few years of the person's total development. Mostly, the researcher enters and departs in

In **longitudinal research,** the same people are studied over a period of time the midst of a person's developmental career, just as people often enter and depart our personal lives. Longitudinal studies that cover a large span of the individual's life are quite expensive and depend upon favorable conditions for research. Projects of this kind are highly valued by developmentalists. We will be drawing upon them throughout this book.

Cross-sequential research combines cross-sectional and longitudinal designs

There is still another general type of research. **Cross-sequential research** combines features of both the cross-sectional and the longitudinal approaches. Some individuals enter the study at one point in time and are followed up longitudinally. Others enter at a later time. These fresh participants begin at the same chronological age as the first wave—who have in the meantime, of course, grown older. Still other waves of participants may join the study as it progresses.

Cohort effects represent the influence of social conditions at a particular time

Through this type of research we can compare people who reach the same age at different moments in our social history. This helps us to distinguish between what may be called generational or **cohort effects** and differences that are more closely related to maturation (age effects). Girls who are ten years old today, for example, are growing up in a social context in which the role of women is different from that of their mothers and grandmothers. It is likely that there are some psychological, social, and biological characteristics of being a ten-year-old girl that are relatively constant across these generations, but other characteristics are markedly different for a vari-

If some members of this family have different values than the others, is it because of age differences or different patterns of life experience? (Photo by Talbot Lovering.)

ety of socio-cultural reasons. Cross-sequential research of various types provides one of the more powerful techniques for sorting out the relative significance of cohort and age effects (Schaie, 1967).

Our personal experience with "middle-ness" is worth pursuing a little further. Consider your mother and father. When did you enter their lives? Obviously, we have all missed knowing our parents as children and, usually, our grandparents as young adults, and our great-grandparents as middle-aged adults. Similarly, few of us are likely to know our children as aged adults (or our grandchildren as middle-aged adults, or our great-grandchildren as young adults). How rare it is to know a person from beginning to end on the time dimension (and this is apart from the question of the *depth* of our knowledge).

We generally encounter other people in the middle of their lives, when we are in the middle of our own lives as well. This has implications both for our personal knowledge and scientific knowledge of human development. As we first become acquainted with our parents, we see them through the eyes of infants and children. When we are a few years older, so are they. Each of us is a kind of *observing system*. How we observe the world—what we take in and what we make of it—is always relative. What we are at the moment cannot really be separated from what we perceive. And the what-we-are changes somewhat from moment to moment, year to year.

Where, then, is the fixed point? As humans observing other humans, we have no absolute fixed point. We continue to develop-age while observing and interacting with others who are also developing-aging. Both we and the objects of our observation are in motion at the same time.

One of the implications of experiencing middle-ness for our personal lives is different generations (our parents, ourselves, our children) are locked into particular angles of observation of each other. This is not necessarily good or bad; it just is. Recognition of these built-in differences can be helpful in understanding those who are younger and older than ourselves. Some of the material that is offered later in this book is intended to augment your knowledge of age-related angles of observation.

The absence of a fixed point has an important implication for the study of human development as well as for our own personal transactions. From researcher to you as reader, every step of the way, there is the opportunity for the observing system to influence what is observed. What is your own age? How might this particular age perspective influence your approach to human development? Many students of human development, for example, are in the late adolescent–young adult age range. Whatever is characteristic of perception, thought, needs, and moods of people of this age group is likely to influence what they look for and what they remember in their studies. Certain aspects of development may be of great interest and other aspects seen as dull or trivial, regardless of the value placed on these topics by the "experts." And those who have steeped themselves in human development, who are active contributors to knowledge, do they not also have a selective influence on

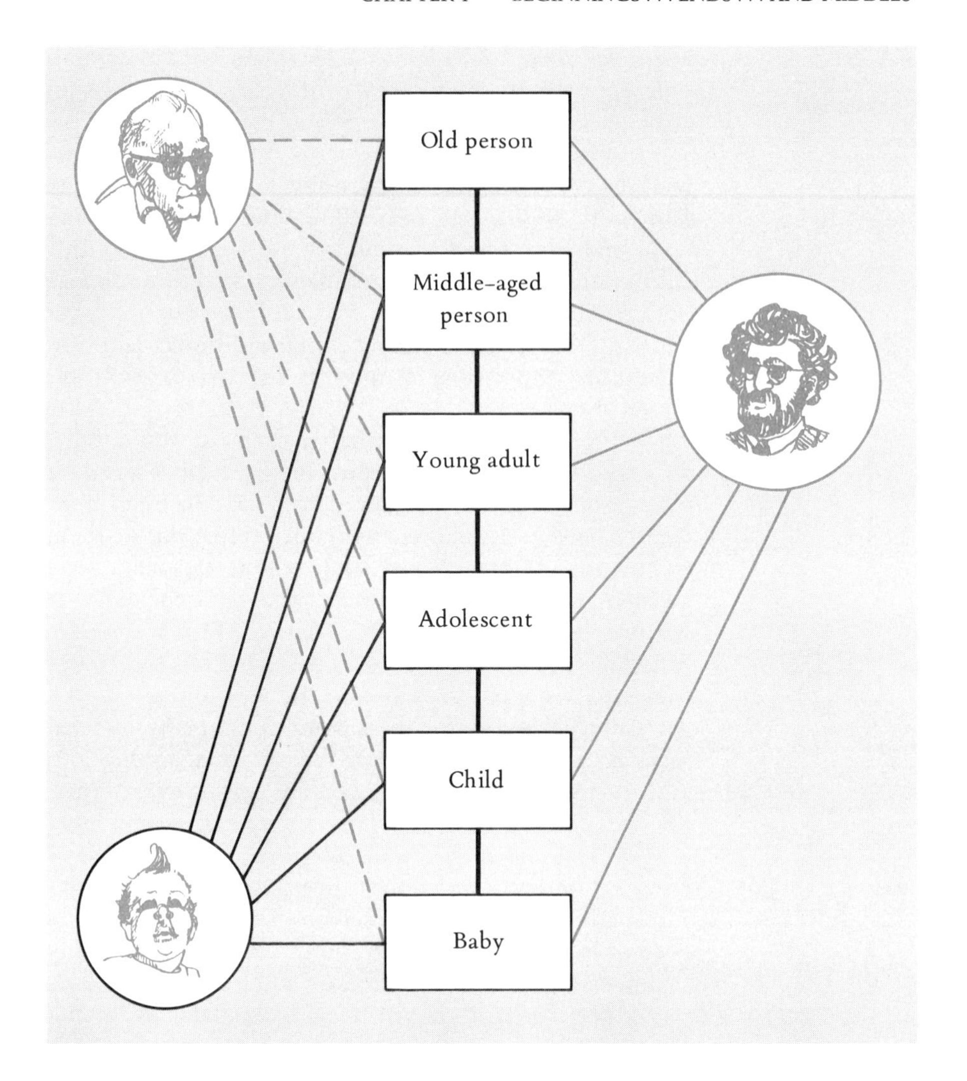

Figure 1-2. *The observer and the observed: the developmental angle. Differences in the developmental angle of observation are likely to result in different data and interpretations. Two researchers, a young researcher studying the aged and an aged researcher studying the young, may have the same width of angle, for example, but the different future-past directionality could influence the data each records and the interpretations each makes. In most developmental research the observer is young or middle-aged. Does this introduce a bias? Is it "better" to have observer and observed of the same age? What might we learn if babies and children could be the observers? Our data always reflect some configuration of the observer-observed developmental angle.*

what eventually comes forth as "fact"? An elderly psychologist studying children and a youthful psychologist studying elders are simple examples of the developmental match between the researcher and the researched. More subtle differences can also be found if we care to look.

Whatever our role in the study of human development, we can reduce error by remembering that observations always tell us something about the observer as well as the observed.

Being in the middle is a familiar enough experience for all of us. Yet there is much to learn about the process we have been designating as *development, growth,* or *maturation.* (These terms will continue to be used interchangeably until otherwise noted.) The middle, after all, is where some aspects of development rise up to take shiny precedence over everything else, only to be eclipsed themselves by newer developments. By any accounting, the middle itself is rich with beginnings and ends. Some of these possess a natural schedule, for example, "baby teeth" slipping from the gums of first or second graders. Other events are less predictable, more variable from person to person. Throughout this book, both the general and the individual aspects of development will engage our attention. We will also attempt to sort out the "more important" aspects from the "less important." This is a risky effort; perhaps it cannot really be done with much confidence. However, we do need to establish priorities, and to find a way of maneuvering through the vast realm known as human development. In short, we need method, and we need a frame of reference. Next chapter, please . . .

SUMMARY

When does human development begin and when does it end? Although difficult to answer, these questions must be raised if we are to think clearly about what comes between beginning and end.

People in our own society tend to believe that human development begins either at conception or at birth. There are alternative views, however:

1. We are not worth counting as human until an age of survivability has been reached.
2. We are not worth counting as human until we have displayed certain qualities or made a name for ourselves.
3. We precede our own conceptions because souls are in transit from one generation to the next.
4. We precede our own conceptions because what we will become as individuals is to some extent prefigured by the environment that awaits us.
5. We may never be regarded as human in the eyes of some people because of our color, sex, national origin, religion, or personal characteristics.

YOUR TURN

1. Where does development take us? In some instances the destination appears fairly clear, as in height growth. In other instances it is very difficult to gauge just how far a person might develop, as in creativity. I suggest you turn over in your own mind many different dimensions of human development. See if you can divide these into dimensions with relatively known or fixed points of final maturation and dimensions whose upper reaches seem to resist confident measurement. You might begin by simply ruling a page down the middle and heading one side *Relatively Fixed Dimensions* and the other *Relatively Open Dimensions.* The lists could begin with the already given examples of height and creativity. As you think of other dimensions do not concern yourself too much with the precise names you give to these, but rather just select terms that will identify the dimensions to your own satisfaction. You might, for example, think of *curiosity* as a dimension of human development. Is curiosity more like height, a characteristic whose growth and ultimate limit are probably well established—or is it more like creativity? How about *finger dexterity,* the ability to perform agile actions with fingers? Perhaps your own list, then, would start out like this:

Relatively Fixed Dimensions
Height
Finger dexterity (?)

Relatively Open Dimensions
Creativity
Curiosity

See if you can identify and classify at least a dozen possible dimensions of human development. If some of the classifications raise questions you cannot answer at the moment, then you might so indicate as I have done with finger dexterity. (It does have important physical components, but can I be sure that the upper range of finger dexterity has been or really can be established?) Reflect on the reasons you use for making these distinctions. This attempt to sort out different types of development, simple and rough as it is, is not a bad way to begin your thoughtful examination of the vast range of thoughts, feelings, and actions this field encompasses.

2. We have already come upon the bias that development implies positive or favorable changes over time, while aging implies negative, unfavorable changes. What changes can you think of that are part of "development" but strike you as either *un*favorable or neither better nor worse? What changes can you think of that are part of "aging" but strike you as being either favorable or neither better nor worse than the previously existing conditions? Do not be unduly troubled if the distinction between development and aging becomes more complicated or blurry as you give this task a try—or if the same thing happens with your ideas of what constitutes favorable and unfavorable change. Work at this question even if you have a hard time coming up with results that truly satisfy you. Is development always "good"? Is aging always "bad"? Throughout this book, and in many other sources, you will continue to come across information and ideas that are relevant to this question.

3. Waiting on the shelves of your college library are many journals that report studies in psychology and related fields. These periodicals are an important part of the stream of knowledge that freshens and enriches science. Here you will find detailed reports of specific experiments as well as theoretical essays, critical surveys of certain topics, book reviews, and the like. Go to the shelves that house journals having to do with human development. These include, for example, *Child Development*, the *Journal of Genetic Psychology*, the *Journal of Experimental Child Psychology*, the *International Journal of Aging & Human Development*, and quite a few others (not necessarily the same journals in all libraries). Select, let's say, six articles (brief ones will do just fine) that report specific studies, empirical research. Read the abstract first (the abstract is a short summary of the entire article that appears right under the title in many but not all journals) to get the general sense of the article and to make sure that it is a report of research and not an essay or critique. Now turn to the section that is headed *Method* or *Procedure* or something of that kind. Pay particular attention to what is said about the time framework of the observations that were made for each participant. (Research participants have traditionally been called *subjects*, abbreviated *Ss*. I think this is an insensitive and imprecise way to speak of people, and I have not employed this usage in this book—but go change the world!) We have arrived finally at the specific mind-work I am proposing here. See if you can determine the basic nature of the research design in this study. Is it cross-sectional? Longitudinal? Cross-sequential? Which type of design was most common across the several studies you read? Compare your findings on the count with others who also try this assignment. Although you are probably new to the reading of journal-reported research, you are entitled to your opinions: do you think the research design used for each of these studies was the most appropriate type? Would the conclusions perhaps have been more definitive if a different type of design had been used? In particular, do you find examples in which cross-sectional research was conducted where longitudinal or cross-sequential would have been more appropriate? It is not too early to be a critic—certainly, not too early in your own development to take journal in hand and read fiercely.

The end of human development has several alternatives to choose from as well. Development ends: (1) when it reaches its objective or destination, (2) when aging begins, or (3) at death. In examining these three major alternatives, we noticed that when we think development ends depends much upon the specific aspect of development we have in mind. An illustrative comparison was made between height growth and creativity. It is tempting to use height as the model for total development because it is such a familiar, easy-to-measure dimension. However, to do so would neglect creativity and other vital, if more challenging aspects of development. We also saw that to explain everything that takes place in the later years of life as *aging* is really not much of an explanation at all. The concept of aging must be examined as carefully as the concept of development.

The attractive idea that development continues throughout the entire lifespan was presented and will be pursued later in this book. But it was acknowledged that under some circumstances development as a human being seems to halt well in advance of physical death. It was also noted that various forms of development after death have their advocates in our society.

For most of our lives, we are neither at the beginning nor at the end; we are right in the middle of other people's lives and our own. Our angle of observation affects how we see others and how they see us, including our views on the developmental process. Researchers must also take into account their relationship to the people and phenomena they are studying. Seldom is it possible to study the development of one human being all the way through, from beginning to end. Developmental research sometimes provides a longitudinal perspective, but often we must depend on cross-sectional observations that do not provide definitive information on what came before and what will come after the researcher concludes his or her project.

Reference List

Cutler, D. R. (Ed.). *Updating life and death.* Boston: Beacon Press, 1969.

Ducasse, C. J. *The belief in a life after death.* Springfield, Ill.: Charles C Thomas, Publisher, 1961.

Kalish, R. A. Life and death: Dividing the invisible. *Social Science & Medicine,* 1968, 2, 249–259.

Kastenbaum, R. Psychological death. In Leonard Pearson (Ed.), *Death and dying: Current issues in the treatment of the dying person.* Cleveland, Ohio: Case Western Reserve University Press, 1969.

Schaie, K. W. Age changes and age differences. *The Gerontologist,* 1967, 7, 128–132.

SCIENCE AND SCENARIO

2

CHAPTER OUTLINE

I

The language of dramatic action can help us to integrate research bits and pieces to discover their larger pattern of interrelationships.

Scenario / Protagonist / Setting / Natural time / Developmental time / Phenomenological time

II

The developmental mission and its methodologies leads us to consideration of age and time as index variables. What are the criteria for developmental change?

Index variables / Developmental change / Evolutionary change / Phases or stages / Orthogenetic Principle (Werner)

III

Studies of human development require decisions about setting, timing, and management of variables.

Single-observer approach / Natural experiment / Observer biases / Cross-sectional research / Longitudinal research / Cross-sequential research / Single-variable research / Multi-variate research / Experimental interventions

A mother speaks to her three-month-old infant. Will she receive a smile in return? A smile is entirely possible at this early age (Emde, 1969; Fraiberg, 1971b). And the infant is already behaving selectively: he or she is more likely to smile for mother or father than for a stranger.

Suppose, however, that the infant is either totally blind or limited to perception of light and dark. Will he or she show recognition of the parental voice in the same way as a three month old who has no visual deficit? Almost. The sightless baby also tends to respond selectively to his or her parent's voice at three months of age. But the smile may not be forthcoming until the parent has spoken several times (Fraiberg, 1971a).

Here, then, are a few bits and pieces from the vast research literature on infant development. This material can serve as well as any to introduce the range of purpose, method, theory, and style that exists in the study of humans developing. We notice immediately that there is value in considering the child with an impairment at the same time that we inquire into the developmental pattern of the unimpaired child. Important principles of development are often revealed when individuals with different characteristics are compared. This is true not only with vision, but with many other physical, psychological, and social characteristics. In the present instance, we learn more about the significance of vision in early development as we learn how babies with little or no vision come to terms with life. Furthermore, a developmental psychology that was limited to describing only "normal" growth would exclude many people at all age levels. One of our aims is to encompass the full range of human experience within the developmental approach.

Let's continue with the bits and pieces that have already been sampled. Journals and books are full of such information. But must facts be collected, shared, and memorized just because they *are* facts? Or is there a way of working with the facts of human development that would make them of greater significance to us? What would you do, for example, with the information given at the beginning of this chapter? What might be worth knowing about this topic? How could additional information be acquired? And what might we do with that information? The developmental scientist must answer such questions. He or she must form a more integrated picture out of many small elements. Both the detailed information and the effort to draw valid inferences and construct larger views are essential to understanding humans developing.

Let's step behind the research scenes for a moment. Here is a research case history cast in the form of a scenario, the first of a number I will be presenting throughout the book to illustrate how bits and pieces of information come together in daily life.

Dimensions of Individual Difference	Definition
Activity level	Amount and vigor of body motion
Perceptual sensitivity	Degree of response to sensory stimulation
Mobility	Neuromuscular actions that lead to change of body position in space
Bodily self-stimulation	Actions on one's own body such as rubbing, rocking, thumb-sucking
Spontaneous activity	Behavior when awake and alert, not in discomfort, or not exposed to focused external stimulation
Body need states and gratifications	Manner in which hunger and thirst and other needs are expressed
Object-related behavior	All behavior that is responsive to objects or things in the environment
Social behavior	All behavior that is responsive to people

Figure 2-1. *We are individuals right from the start. These categories were defined by Escalona (1968) to guide research into observable individual differences from early infancy through childhood. Infants differ from each other on all of these dimensions. Tyler (1978) provided a more recent follow-up to the study of individual differences. It would be a mistake to overemphasize developmental changes in general and lose sight of distinctive patterns shown by individual infants and children.*

SCENARIO 1: THE BLIND INFANT AS PROTAGONIST

The **protagonist** is the hero or central character

The work of Selma Fraiberg and her University of Michigan colleagues might be regarded as an attempt to discover the unwritten scenario in which the blind infant finds himself or herself as **protagonist.** The researchers displayed an interest not only in describing, but also in attempting to revise, the scenario to the infant's advantage. The attention of these observers was caught first by problems they noticed in young children who had been blind since birth. Many of these children were not developing in the expected way. In Fraiberg's description:

> Typically, these children appear to have no significant human ties. Language, if present at all, is echolalic. There is no definition of body boundaries,

> of self and other. There are motor stereotypes of the trunk and the hands—such as rocking, lateral rotation of the head and trunk, empty fingering. Many of these children have not achieved mobility. (Fraiberg, 1971b).

At this point we seem to have the protagonist clearly identified: the young child with multiple impairments. However, for Fraiberg and her colleagues, it was necessary to focus in more carefully. Two types of blind children were excluded: those whose development did not show the deficits described and those who were known to be afflicted with brain damage or other significant handicaps in addition to blindness. Instead of waiting to study the child who already displayed conspicuous impairment, Fraiberg and her colleagues decided to begin with the very young infant.

Think for a moment about this decision; it is just one of many that the developmental scientist must make. What would be the advantages of centering the research around the blind child who already demonstrates difficulties in an obvious way? We would be certain that we had the "right" protagonist, the person whose situation aroused our interest in the first place. The study would also be relatively inexpensive, both with respect to time and to money. There are limits to the amount of money available to support research, and to the number of available qualified personnel. If we did not have to wait for the protagonist to grow up, we might be able to reduce expenses and also have our results in hand much sooner. These are among the possible advantages of restricting the **scenario** to a closely limited time span. Perhaps you can think of other advantages as well.

The **scenario** is the general situation from which a story develops

There is one major advantage to beginning with the very young infant: the observers have an opportunity to learn what actually does take place—and what fails to take place—during the process of development. This is an expensive kind of research, even if restricted to the infancy-childhood years alone. Yet it is doubtful that any satisfactory shortcut exists. Other methods can help to illuminate the process of development, but there seems to be no substitute for the patient observation of people growing up and growing old.

The when-to-begin decision is only one of the questions about *timing* that must be considered in studying human development. When should the study end? At how many different points between start and finish should research observations be made? And precisely when should these points of observation be introduced? There are no automatic answers available. How a research team will respond to these questions depends upon its specific goals, the available resources, and a variety of practical conditions.

And so we begin to recognize that the research project itself has a developmental history and a scenario of its own. There are participants, settings, situations, goals, tensions, possibilities, and limitations—as well as vulnerability to events that are external to the project. We will not dwell further on the scenario of the research project as such. It is worth remembering, however, that those who seek to understand humans developing are continuously moving through their own complex situations at the same time they are in-

Heroes, or protagonists, are not always six-footers wearing ten-gallon hats. We are all protagonists in our own life scenarios. (Courtesy of The March of Dimes.)

volved in studying others. Under these conditions, the experiences of the protagonist and of the researcher cannot be regarded as entirely independent. Every fact we have in developmental psychology and related fields comes from specific interactions between researchers, who have their own talents and limitations, goals and restraints, and those they have observed. The principle of relativity is not confined to theoretical physics by any means. It might be satisfying to pretend that developmental facts grow on trees and roll obligingly into textbooks, but that's just not so.

For Scenario 1, then, we have as protagonist a baby who was blind from birth (or could perceive only the difference between light and dark) but who had no other significant handicap. Fraiberg's longitudinal studies involved forty-three infants, of whom ten were selected for intensive observations. We will concentrate upon a single representative, recognizing that details of the scenario would vary somewhat from infant to infant if each developmental career were explored in depth.

What about the *setting*? There is a parallel here between the research process and the dramatic arts. A theatrical production provides us with a highly selective glimpse into the protagonist's life. We are free to assume, for example, that he or she sleeps at times and makes use of the bathroom. But bedroom and bathroom scenes are not represented on stage unless a particular dramatic point is to be made thereby. We accept the illusion that he or

Like the theatre, the experimental laboratory offers a selective view of human life in which only a few events actually occur before our eyes. (Photos by Daniel Bernstein for Tufts–New England Medical Center Hospitals and courtesy of Loeb Drama Center, Harvard University.)

she exists when not on stage, and that he or she inhabits various settings not brought to our attention. In most studies of human development there is a similar pairing of "realities." The blind infant in Scenario 1 exists continuously. He is always in one place or another, whether or not a researcher also happens to be on the scene. That is the point. Even a systematic and intensive research project merely comes and goes in the protagonist's lifespace, just as the on-stage scene only provides a temporary view, a sampling of action that had its origins in scenes unseen.

This pairing creates a curious and potentially confusing situation. Which is the "real reality?" The researcher could insist that he or she is communicating careful objective observations and analyses. Therefore, if we demand precise and dependable information we should rely upon the researcher's measures. But a contrary argument can be made that the infant's basic reality is the life he or she moves forward through in his or her total environment. Of this total, only a few, highly selected snippets are utilized as research settings. In other words, it is the off-stage sequence that should be considered fundamental for understanding. The researcher's contribution can be valued highly, but both the play and the research manuscript borrow their significance from the continuous and much larger pattern of human events that are *not* presented to us directly. There is a distinction, then, between the selected settings from which research data derive, and those in which people actually unfurl their lives. Keeping this distinction clear in our minds is likely to improve our perspective.

The primary setting in Fraiberg's research was the baby's home. While the home exists continuously, it became a research scene for approximately one and one half hours on a twice-monthly basis. Another approach might have been to utilize a special setting for research purposes, such as an observational laboratory. You might find it interesting to reflect upon the relative advantages and disadvantages of studying infants at home or under special laboratory conditions.

We now come to *situations*. What were people doing in the setting? This question is answered in part by describing readily observable actions: what baby was doing, what mother was doing. The question is answered more fully if we include some of the purposes, motives, or dynamics that seemed to be in operation. The typical situation here involved mother and baby at home behaving as both saw fit. The total observational period usually involved several subsituations: a feeding, a diapering, a bath, etc. It would also include time during which baby was left to his own devices. A few minutes were set aside in most sessions for the observer to conduct simple tests, such as handing baby an object. The scene, then, could be regarded as consisting of three types of situations:

- Mother caring for baby
- Baby with time on hands
- Baby responding to observer

Natural time is the standardized, public passage of time

The *movement* in this scenario occurs through what we might call real or **natural time.** The protagonist actually does grow older and is expected to mature. This imparts a sense of change and direction and provides the occasion for new action to unfold. In a theatrical presentation, the same effect is intended through the *illusion* of time passage. We may see the protagonist as a youth in Act 1 and as an elder in Act 3, although less than two hours of natural time have elapsed. Cross-sectional research resembles the illusion of theatrical representation in this respect. Fraiberg might have observed Newborn A and Toddler B, and from these observations on two different youngsters have offered a view of what takes place between infancy and age two in blind children. Instead, she decided in favor of a longitudinal approach in which the protagonist (baby) moves through natural time.

A **syndrome** is several symptoms that occur in the same person

We have sketched in the protagonist, a few other characters, the setting, the situation, and movement. Let's attend now to *plot*. What are likely to be the major themes and conflicts as baby moves toward the years of childhood? A *fate* theme hovers. Will baby, handicapped at first only by blindness, eventually be overtaken by the multiple-impairment **syndrome** that is known to befall some children who begin life as he did? There is also something of a *mystery* theme as well. If blindness is the villain and serious impairment the possible fate, then what clues can be discovered as to how the crime is committed? By what processes or mechanisms does visual limitation lead to multiple handicaps? An answer to this question would improve our understanding of the way in which normal development of visual functioning contributes to the overall maturation of a child's physical, psychological, and social skills. The scenario also embodies a variant of the classic Powers-of-Evil-versus-Powers-of-Good struggle. Can something be done—in time—to counteract the forces that threaten to deny the protagonist his full opportunity in life?

The actual child selected as the protagonist for this scenario was born three months premature. At birth he weighed only two pounds, three ounces. His blindness was attributed to a condition known as **retrolental fibroplasia.**

In **retrolental fibroplasia,** the retina detaches from the inner surface of the eye due to exposure to excess oxygen

The research observations indicated that baby was developing well. How well? Evaluation of his progress depends upon the criteria one decides to use. At 13 months of age, he could sit well without support, use his hands capably, and express himself verbally with "Dada," "dog," and utterances of his own invention that made his needs known. However, baby was not creeping about, and he was at a marked disadvantage in dealing with lost or displaced

Figure 2-2. *Development of a blind, premature baby: a choice of comparisons. The blind, premature baby in Scenario 1 showed less development at the one-year mark than either average sighted or blind full-term infants. But when we consider the fact that the protagonist was, in effect, three months younger at twelve months postnatal, some of the discrepancy disappears. This illustrates the way in which choice of comparisons can influence our interpretation of an individual's development.*

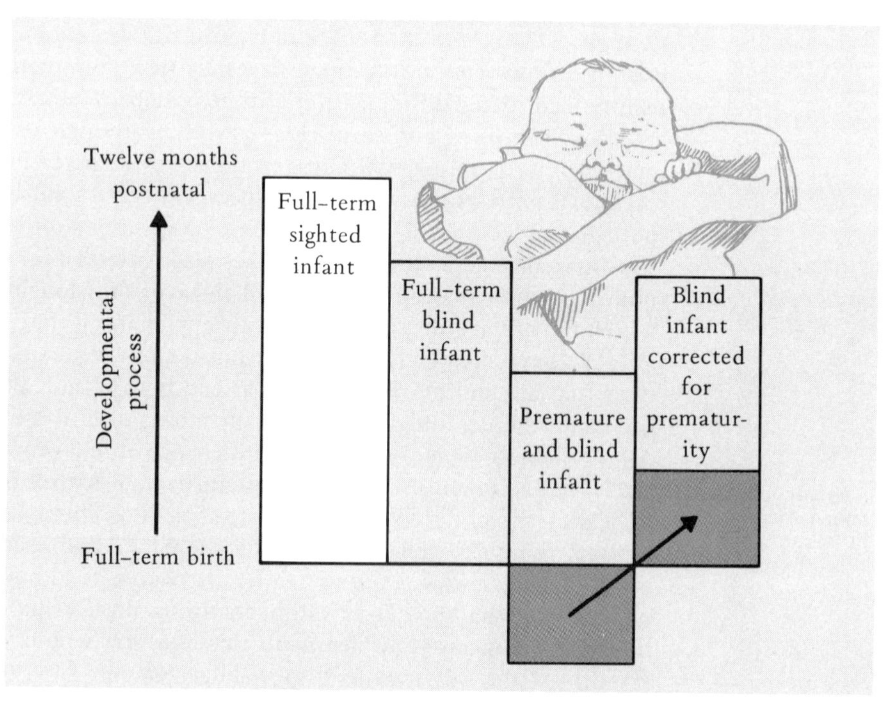

objects. On some important dimensions, then, his *developmental time* was slower than sighted children of his age. He was also a little behind the progress of most blind children. Remember, though, that he was born three months premature. When this fact is taken into account, baby was approximately on schedule when compared with other blind children, although not with same-age children in general.

Even in this very young age range there are multiple sets of expectations for interpreting the timing and pattern of human development. Baby could be considered *retarded* in some (but not all) ways if we insist upon comparing him with full-term sighted infants. Taking his handicaps into account, much of his retardation can be reinterpreted as a reasonable pattern of development. Furthermore, it would be difficult to interpret baby's development at all if he had been the only infant whose progress had been observed. Research on other youngsters, both sighted and blind, provides a context for viewing the behavior of this particular infant. Even if our core interest is the understanding of individual developmental careers, we are in a better position to do so when we are aware of the more general patterns and norms. In turn, the general patterns take on more significance when contrasted with individual pathways that depart from the usual.

At about this time a new situation arose. The problem at first had nothing to do with baby. His maternal grandfather died. Mother left immediately to visit her father's family and attend the funeral. Baby was placed in the care of various family members and friends. Mother returned three days later, and soon thereafter baby "showed alarming symptoms and began to regress in nearly all areas of development . . . [exhibiting] screaming fits which lasted for hours. Our own observers found them nearly indescribable. These were screams or shouts of a repetitious, chanting character, and were practically unceasing for most of his waking hours. During these attacks the child's face was curiously immobile and expressionless. . . ." (Fraiberg, 1971a, p. 391).

Baby abandoned his toys. He seemed to lose interest in playing—and also in eating. He would accept only a few mouthfuls of food, and then vomit. At night he slept very little. When mother held baby he would stop screaming for a moment or two, but he could hardly be said to calm down. He would "crawl desperately all over her body as if trying to get as much as possible of his own body surface in touch with his mother's body" (p. 391).

From an infant who had been progressing well despite his handicaps, baby became a highly disturbed person. Fraiberg described this change as "a pathological regression in nearly every sector of his development . . ." (p. 391). **Regression** is a fascinating term, one that will be encountered again. It implies a sort of going backward through time, a return to an earlier stage of functioning. The concept is more complex than that, however, and we will not try to do it justice here. We simply acknowledge that Fraiberg had to select a strong expression (pathological regression) to indicate the extent of baby's disturbed pattern of functioning.

Regression, more broadly, is any return to an earlier, simpler stage

Mother found she could not cope with the situation. It was clear to her that baby was reacting to their separation, but she did not know how to reassure him. If baby could *see*, then he would have visual proof that mother was with him. A sighted infant in a similar situation might follow mother around, keeping an eye on her. Additionally, the sighted infant could learn to anticipate mother's absences. "There goes mother, putting on her hat and coat and fishing about in her purse for the keys. I'd better start crying my head off!" As Fraiberg noted, the sighted infant at least has the opportunity to be active in the face of danger. Separation is a prime threat to the very young who depend so heavily on the nourishment and protection of significant others.

Baby in this scenario could neither track mother adequately when she was present nor anticipate her disappearances. How, then, could he know when she was really with him? Body contact seemed to be the only mode. He knew mother was there when he was clinging to her.

In attempting to explain baby's terror and disturbance, Fraiberg used a term that is often heard from those who employ a psychoanalytic approach (that is, an approach originating with Sigmund Freud and his disciples). Because baby had no means of anticipating mother's ordinary comings and goings throughout the day, each separation, no matter how fleeting, was

Some researchers prefer to observe developmental phenomena in its "natural" setting—home. These observations take parent-child interactions into account rather than just the infant's behavior. (Photo by Talbot Lovering.)

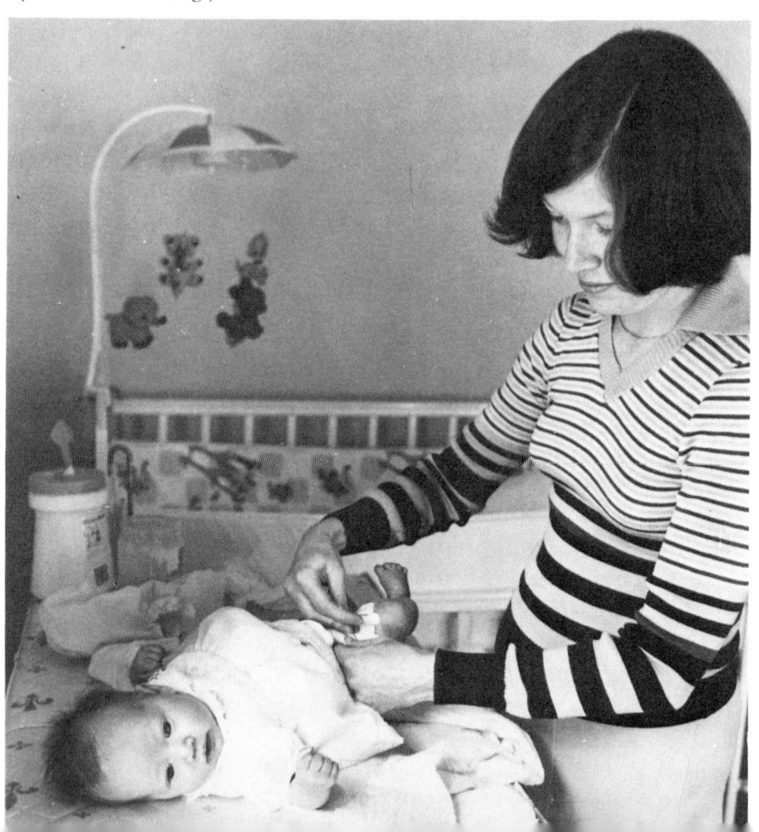

Trauma is a physical or emotional injury or both

experienced as a "repetition of the **trauma."** In other words, baby had just been through an extremely upsetting experience: the absence of mother for a prolonged period of time. This was the trauma. Now he was hypersensitive; *any* loss of contact might bring back the terrifying experience of abandonment.

Place the spotlight on mother for a moment. She was crucial to baby at this point in his life. But what of her own psychological state? This was a woman who was mourning the death of her father. She herself was a child who had suffered loss. Unlike baby, she could distinguish between losses that are temporary, and those that are extended or permanent. In this state of mind she could not easily give baby's needs total priority over her own. The screaming fits made mother furious. When she could not take the screaming any more, she would shut baby in his room—alone. Although this was an understandable action from mother's viewpoint, it had the effect of repeating the trauma baby was trying so desperately to avoid.

What is, or should be, the scholar's role in all of this? Fraiberg and her colleagues were conducting a developmental study. They had not entered this home with the promise or intention of helping. Actually, students of human development are often careful to avoid intermixing "help" with their scientific observations. Unless the helping techniques are included ahead of time as a systematic part of the study there is danger that the research findings might be "contaminated" by the investigators' double role. But there are dangers on the other side, too. Who can sit back and watch others suffer and enter into great jeopardy without trying to be of assistance? These decisions are not always easy to make.

Perhaps the decision was easier for Fraiberg because mother requested help; perhaps it was easier because the project staff were also skilled clinicians. In any event, the team did offer its assistance with what they saw as a crisis related to mother-baby separation. It is instructive to learn both what they suggested and why.

The researchers reasoned this way: baby was not only terrified, he was also fiercely *angry*. The observers had been sensitive enough to recognize the rage component of baby's hoarse screams. A sighted infant can express his anger vocally, as our protagonist did. But he or she can also fight back—kick out, strike with hands and fists. Blind infants, however, remain mouth-centered in their expressions of anger and aggression. Near the end of the first year of life, the blind infant is less likely to have shifted his or her aggressive energies to the arms and legs. Furthermore, in this baby's specific case, he did not have a clear *target* for his rage. He was not really mad *at* mother, because he did not yet understand cause-effect relations. What he probably felt was *helpless*. And his only method of overcoming that sense of helplessness was to swing into action, to fight back somehow. It is, of course, extremely difficult to fight back when one does not have the motor patterns for physical action and also lacks targets.

Fraiberg and her colleagues had a suggestion for mother: As soon as baby starts shouting, provide him with pots, pans, and banging toys. Encourage him to bang away with his hands and fists. Mother tried the suggestion, and it worked! Within a few days, the unbearable sound of shouting and screaming had stopped. To be sure, it was replaced by the din of pot banging. But mother could accept this type of noise more easily, and it was evident that baby was in much better spirits. Baby did not have to bang pots all day long to discharge tension; he just did it when he felt the need. A few days later, he was much improved in all areas of functioning. He played, he ate, he slept, and only occasionally did he shout or cling to mother. In short, the 14-month-old boy was his old self again.

I will end the scenario at this point. Age 14 months can hardly be said to constitute an ending, but we have at least seen a favorable outcome to a developmental episode that could have turned out quite badly. We have also seen an example of developmental research that combined direct observation, theoretical formulations, and practical advice that helped a family through a difficult time. The study concerned itself with many bits and pieces, only a few of which have been reported here. Yet each observation, no matter how fleeting, had its value in contributing to understanding and improving the total situation. Would the bits and pieces have been as interesting or instructive by themselves, treated as isolated examples of scientific data? On the other hand, would we have learned anything at all if there had not been attention to details and methodology? We really do not have to choose between the close observation and analysis of human behavior in small units (e.g., how long it takes for baby to smile at mother's voice) and the comprehension of situations that involve people interacting over a protracted period of time. The student of human development, whether beginner or expert, attempts both to master the details and to discover their larger pattern of interrelationships.

THE DEVELOPMENTAL MISSION AND ITS METHODOLOGIES

We have started to acquaint ourselves with the developmental mission and its methodologies. What is the developmental mission? The mission of developmental research and theory is to contribute precise, reliable, and useful information to everybody capable of using it. We all make decisions and take actions that assume some facts about human development. How much better it would be if those assumptions had been tested and the facts validated. The developmental mission is also to distinguish core developmental change from other phenomena which occur in our lives. For some people, the mission includes the construction of systems of thought (theories or models, to use these terms a little loosely) that integrate specific facts and small sets of facts into a larger view. Still again, for others the developmental mission is part of a more general effort to improve the human condition by helping to eliminate hazards and obstacles and stimulating optimal development.

Researchers approach the fulfillment of this developmental mission by many methods. Some of these have already been described. It is now time to introduce several other key considerations in the study of humans developing. These are:

1. Age, time, change, and development
2. Criteria for developmental change
3. Methods of studying human development
4. "Big" and "little" theories

AGE, TIME, CHANGE, AND DEVELOPMENT

The study of development is closely associated with age, time, and change. But precisely how? It would be simple if we could describe development as something that happens as we move from one age to another, or from one point in time to the next. But development is more complicated. It would also be simple if we could describe development as change. Again, it is more complicated.

Younger and older members of the same species may seem different from each other in various ways. There are two conclusions we might be tempted to draw: (1) The differences between younger and older organisms are the result of the developmental process at work; or (2) The differences between younger and older organisms are the result of the age differential.

These are two very different conclusions. The first proposition asserts that age differences necessarily represent developmental differences. Whenever a younger and an older organism look or act differently, then we have an instance of developmental change. The second proposition goes beyond this. It implies that age or the passage of time somehow causes the change. Neither of these propositions is accepted at face value by developmental scientists.

Index variables point to the influential factors in a situation

Chronological age and the passage of time are **index variables** rather than forces in themselves. Often it is useful to know the chronological age of an organism, or how much time elapses between the appearance of one characteristic and the appearance of another. But it would be careless to attribute causal properties to age or time. Age and time are very important, but essentially empty, categories. We still must discover what happens at a particular age, what processes are at work during an interval of passing time. Developmentalists often use age and time as shorthand or approximate ways of indexing the phenomena in which they are interested. They do not assume that age or time precisely locate an individual's developmental status or explain why changes occur.

Developmentalists pay particular attention to making a clear discrimination between *age changes* and *age differences.* Often it is not difficult to determine that two organisms differ in age. It is more difficult, however, to determine whether or not observable differences between the organisms represent maturation/development or the outcome of some other type of pro-

The garb and life-style favored by an elderly person may have little to do with age, but much to do with the nature of early experiences and upbringing. (Photo by Hugh Rogers, Monkmeyer Press Photo Service.)

cess. Compare, for example, the speech patterns of young and old adults within the same family constellation. I have often been in contact with hospitalized old men and women of Italian extraction and their family members. After awhile it became obvious that they often differed in speech patterns. One typical difference was that the hospitalized old people tended to speak in Italian accents and often raced off completely in that language when they became excited. Their children and grandchildren, by contrast, were strictly Boston-American in speech pattern. The obvious conclusion is we become more Italian with age and time. This is the type of conclusion we stumble toward when no distinction is made between age-related differences and age-related change. In this example one might speculate that it is not the empty variable of age, but the impact of illness that has led to "the Italian difference." This would be a peculiar speculation, but at least it would recognize the likelihood that differences in the behavior of people of different ages can be caused by factors other than maturation/development. A more promising speculation would be that the age differences are related to cohort factors. The old men and women are more likely to speak Italian now because they have been speaking Italian for years. We would conclude, then, that it is true that there are age differences, but these derive largely from life experience factors present in the entire generations from which these young and old people started out.

An elderly couple, e.g., may be more politically liberal than their grandchildren

We can all recognize obvious situations such as this, but often in developmental research it is much more difficult to separate age changes from age differences. The cautious investigator attempts to rule out other possible ways of accounting for differences between younger and older organisms before coming to the conclusion that their differences represent the kind of change that should be called maturational or developmental.

This brings us back to the term *change* again. We usually think of development as change, but not all change can usefully be thought of as development. Illness may result in change (e.g., impaired mobility and strength), but we are not apt to describe this as development. Other changes may seem too unsystematic or trivial to be regarded as developmental or may appear much more closely linked with other processes. In recent years, for example, many people have taken to wearing more casual garb, such as blue jeans. This is a change of some social consequence perhaps, but there would be little point in characterizing it as maturational or developmental. Certain kinds of change are more likely to be interpreted as developmental than others. It is possible to make errors and either be too inclusive or too exclusive in the type of changes we regard as developmental. When you turn off your light tonight, some developmental researchers will still be lying awake with their minds agitated by the challenge of distinguishing "true" developmental change from all the other conditions that lead to age differences in structure and function.

CRITERIA FOR DEVELOPMENTAL CHANGE

What distinguishes developmental change from change in general? Let's first use the process of elimination to achieve a closer focus. There are situations in which changes occur from Status A to Status B. These changes are "real" enough. But if we are patient we will see that the altered state returns to its original form. The menstrual cycle and the hibernation cycle in some animals are examples of such alternations in status within the same organism. Alternating status and periodicities ordinarily would not be considered as examples of developmental change—although the process by which an immature female achieves the ability to reproduce and concurrently enters into periodic changes would be an example of maturation or developmental change.

Another type of nondevelopmental change has already been mentioned. Sociocultural conditions are always in motion, always shifting in the mixture of opportunities and pressures, emphases and concerns. Individuals are strongly influenced both by their past histories of immersion in particular socio-cultural circumstances and by the current environmental dynamics. As we will see in Chapter 7, there is more than one way to interpret the relationship between socio-cultural influences and the unfolding of genetic characteristics within the individual. One might insist that only genetically programed changes constitute true development. At the other extreme, it could be argued that development, especially in humans, is largely the result of experiences that are mediated through the socio-physical environment. In the middle, there are many positions that try to account for both genetic and environmental contributions to change. The point for now is researchers are becoming increasingly cautious about concluding that any observed differences between people of different ages or generations are *intrinsic.* They do not want to mistake characteristics of a particular generation with its distinctive blend of socio-cultural experiences for basic, universal developmental change. This is one of the reasons why cross-sequential research methods that attempt to encompass both changes over time in the same person and differences among people who grow up at different moments in social time are becoming increasingly favored.

There is another very important kind of change that can be distinguished from development in theory, but which requires persistent and inventive research to do so in practice. *Evolution* has much in common with development. The term refers to a type of change that has a clear directionality. More "developed" or more "evolved" organisms are not only different, their structures and functions can also be classified as having reached a higher level. The brain of *Homo sapiens*, for example, is regarded (by *Homo sapiens*, it is true!) as more *evolved* than the most nearly parallel structures in the flatworm, anchovy, or armadillo. But the brain of both the adult armadillo and adult human being are more *developed* than the same organs in their newborns. Evolution usually refers to directional changes in the level of structure and function of a species. Evolutionary processes are thought to take place over long arches of time that far exceed the lifetimes of individuals within the species. By contrast, development refers to those basic changes within the life of a particular organism through which it reaches or approaches the potential of the species.

While this may seem clear enough in principle, it is difficult to establish the relationship between developmental and evolutionary change in practice. Should we assume that the evolutionary process has run its course, that Nature rests content since creating human beings? Or is it possible that the slow onward and upward progression is still at work? Consider the evolution of brain/mental dimensions alone. *If* there is a general evolutionary movement in the mental sphere, then this would introduce an error of unknown magnitude into our conclusions about the cognitive functioning of people who have entered human society at different points in time. The infant who is born today might have a very slight evolutionary edge over the one who was born twenty-five, fifty, or seventy-five years ago. We might think we are seeing the effects of a developmental or aging process alone in comparing people of different ages, when the actual process involves both evolutionary and developmental/aging processes. And, yes, we might also choose to believe that instead of a universal progression within the human species, there might be an involution or negative evolution. Perhaps we have past our crest, and succeeding generations will operate with slight reductions in brainpower.

The possible relationships between developmental and evolutionary change are fascinating to the person who takes a broad view of the past and future of the human race, indeed, of all life on this planet. The person who has but a casual or narrow interest in human development perhaps can ignore this topic. But until we know much more about the relationships between both types of change, we will remain limited and uncertain in drawing conclusions about human development.

The transition from girl to young woman occurs at different rates for different people, but the process of physical development is clearly directional and universal. (Photo by Talbot Lovering.)

I have mentioned in passing some of the major characteristics of the kind of change that is most likely to be viewed as developmental. A few more characteristics can now be added and a summary presented.

Observing just part of a cycle may give the mistaken impression of directional change

1. Developmental change has direction. In this respect it can be distinguished from cyclical or alternating changes on the one hand and random or unsystematic changes on the other.

2. In a lifespan perspective, development is movement in a direction from a relatively vulnerable and dependent status to one of self-sufficiency with the capacity to reproduce others of the same species.

3. Development is often portrayed as proceeding through phases or stages in which qualitative differences emerge. This has become perhaps the most common and influential view, and much supportive data has been offered. However, other developmentalists emphasize the *continuity* of the organism throughout its lifespan. They have other data and theory to offer that de-emphasizes qualitatively different phases or stages. The stages/phases approach, then, is often identified with development, but it is not without constructive competition today.

4. If we climb to a higher rung on the ladder of abstract thinking, we will discover still another view of development. This approach is well exemplified by the work of Heinz Werner (1948), a psychologist who brought an exceptionally rich intellectual background to his theory and research. Think of a condition in which what meets our observation is without much structure. Something is "there," but there are few separate components or functions. The "something" could be almost anything—an acorn, a microscopic organism at one point in its cycle, an orchestra tuning up, the way our own mind might seem to us as we slowly awaken from deep sleep. According to Werner

this is where development always begins. It is a global and undifferentiated state, to use his phraseology. Eventually more structure emerges. There are more parts and more distinct activities. Later, if development continues, the emerging structures and functions become more highly organized with respect to each other, more coordinated. The particular type of organization that Werner specifies is hierarchic or layered. Lower levels of organization are subsumed under higher layers of organization. The composer of a symphony, for example, charts a rich variety of sound structures from individual instruments and, through the skills of the conductor, achieves a performance in which main and subsidiary themes, rhythms, and tonal qualities are integrated and balanced. That same symphony may be a mess during rehearsal when the ability to integrate all the differentiated components of the symphony has not yet been perfected.

Take one of the other examples. In slowly waking up, we may not have completely differentiated our sleeping from our waking states. For a moment or two we may have only a vague awareness of our surroundings. Soon our familiar mental world starts to reassert itself, but it may take a little while before we have our thoughts, feelings, and perceptions in perspective. This is one illustration of Werner's **orthogenetic principle.** This term is not quite as fearsome as it sounds. In fact, we have already covered its essence: development is said to begin in a relatively global, undifferentiated state, it proceeds to intermediary states of increasing differentiation, and on (perhaps) to hierarchic organization of the component structures and functions.

The **orthogenetic principle** states development begins in a global state

This is a rather special view of development; it strikes some people as too abstract and far-reaching. Notice some of its interesting properties, however. The orthogenetic principle can be applied to events that occur in either very short or very long units of time—events ranging from microseconds to an entire lifespan or more. It is not dependent on any one particular subject matter. Anything that shows the processes specified by the orthogenetic principle might be regarded as a developmental phenomenon. This approach to development has enormous potential to unify our thinking across usual domains of knowledge, from the biological to the social. The serious student of human development will not have savored all the possibilities of his or her subject matter without examining the work of Werner and others who think they see fundamentally similar principles in operation at many levels of natural phenomena including, but not limited to, the familiar realm of personal maturation.

METHODS OF STUDYING HUMAN DEVELOPMENT

Research into human development requires many of the same basic skills and attitudes other forms of scientific investigation require. There are some aspects of developmental research, however, that deserve special mention.

The **naturalistic method** studies behavior in its ordinary setting

The naturalistic method. Through the centuries many observations have been made in naturalistic settings. In other words, people observed other people wherever they happened to be, without introducing special instrumentation or controls. The naturalistic method is still used today; we can observe children in school or at play, see how young adolescents arrange their own groups at dances, etc. Sooner or later, much of the more controlled and "sophisticated" research has to find its way back to application in naturalistic settings. It is likely, therefore, that there will always be a place in developmental studies for the careful observation of people in their familiar settings.

At times developmentalists are able to make use of opportunities to examine "natural experiments." Nature sometimes produces unusual or distinctive situations that show developmental phenomena in a rare light. A famous example of this sort is the so-called "Wild Boy of Aveyron" (Lane, 1976). Victor was a boy of about 12 years of age who had apparently grown up alone in a French forest, having only a few, sporadic human contacts. Jean-Marc-Gaspard Itard, a young physician with a strong sense of social consciousness, worked intensively with the boy. The results have had a lasting impact on educational methods, and have put into focus some problems of the relationship between nature and nurture. These problems have since occupied many researchers and theorists. This natural experiment is well worth our attention today.

The **descriptive method** is observation only

The descriptive method. Early contributions to the understanding of human development often relied upon the observations and impressions of individuals, and, therefore, upon their inadvertent biases as well. A parent might keep a lively and informative journal on the growth of his or her child. Later generations of researchers had to bear in mind that they were reading not just about the child but about the child-as-viewed-by-the-parent. The observer-observed unit of research is a legitimate one, but it is difficult to differentiate some aspects of the observed person's own development from the relational network.

More recent studies take care to minimize and account for the role of the observer. Direct observation remains important, but now there are many technical means for achieving greater precision and dependability. It is possible, for example, to technically monitor an infant's eye movements or to tape-record his or her verbalizations. This affords the researcher precision not available to earlier generations.

The **multi-variate method** studies many aspects at the same time

The multi-variate method. Developmental research has also become more complex and resourceful on the level of experimental design and statistical analysis. We have already mentioned the three basic types of research design with respect to the time line: cross-sectional, longitudinal, and cross-sequential. There is a whole art and science of designing researches of these types. Many developmentalists are also making extensive use of multivariate, as contrasted with single-variable, research designs. Instead of concentrating on the developmental career of a single aspect of a person's functioning, today's researcher is more likely to take many aspects into account at the same time. This is in keeping with the need to avoid over-isolating one function from the entire person or one stimulus from the entire environmental array. The facts about human development often require paying attention to complex patterns of information, some of which come from the individual, others from the environment. The continued growth of statistical programs, research designs, and computer resources for dealing with multivariate situations has opened new opportunities for the developmental researcher.

In **controlled experimentation,** formal research designs are used to ensure valid conclusions

Controlled experimentation. Controlled experimentation has also become of increasing importance. Many studies involve more than observing the individual. Experimental variations are introduced. We will see some examples of this approach throughout the book.

If you browse through journals that report new studies of human development you may notice one general type of experimental intervention that is especially popular. A number of contemporary researchers are curious about

Figure 2-3. *The stages of cognitive development as seen by Jean Piaget.*

Stage	Approximate Age	Some Characteristics of the Stage
Sensorimotor	Birth to 2 years	Relates and adapts to the world without the use of images and other internal representations. Gradually comes to develop the concept of object permanence, as seen in the ability to search for hidden and missing things.
Preoperational	2-7 years	Can form mental representations of people, objects, and events. Important advances in ability to receive and communicate through language. Still quite egocentric: can see the world only through his or her own perspective.
Concrete operations	7-11 years	Comprehends perspectives of other people. Shows marked development in logical thinking and understanding of rules, games, and some mathematical relationships. Essentially, can perform mental operations securely in practical or concrete situations.
Formal operations	11 years to adulthood	Abstract reasoning develops. Can test hypotheses, think about thought, imagine possibilities as well as cope with existing phenomena. Thought is more flexible.

the upper limits of behavior. What *might* an infant or child be able to do at this particular point in his or her life if given a special opportunity or special training? Experimentation can be used, then, not only to establish more adequately the usual course of development, but also to see how much range there is for accelerating, expanding, or modifying development. (The possibility that attempts to hasten or modify human development might be injurious in some way has not escaped notice. The American Psychological Association is one of a number of professional and scientific organizations that have established ethical standards and review practices.)

BIG AND LITTLE THEORIES

There are theories of human development, and there are theories of human development! As you become interested in a particular topic, you will find that there usually are several different views from which to choose, each supported by some evidence and speculation. "Little theories" are concerned with a fairly limited topic area. This is not to say that either the topics or the theories are trivial. In fact, some of the most ingenious studies and best

Figure 2-4. *The psychosocial stages of development as seen by Erik Erikson.*

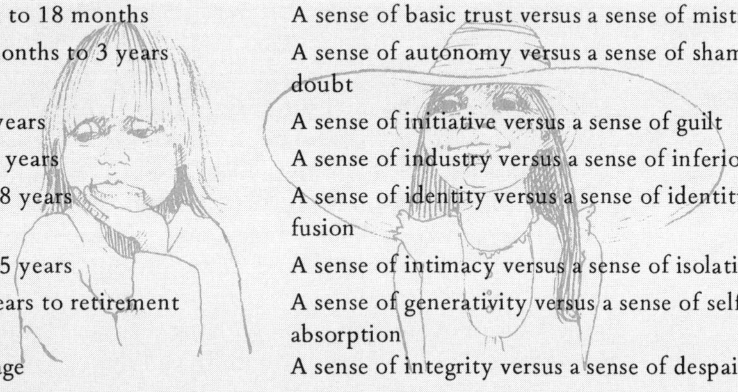

Phase	Approximate Age	The Developmental Challenge at This Time
1	Birth to 18 months	A sense of basic trust versus a sense of mistrust
2	18 months to 3 years	A sense of autonomy versus a sense of shame and doubt
3	3–6 years	A sense of initiative versus a sense of guilt
4	6–12 years	A sense of industry versus a sense of inferiority
5	12–18 years	A sense of identity versus a sense of identity confusion
6	18–35 years	A sense of intimacy versus a sense of isolation
7	35 years to retirement	A sense of generativity versus a sense of self-absorption
8	Old age	A sense of integrity versus a sense of despair

thought-out conceptions center around problems that can be closely defined. There are several theories regarding the function of play in childhood, for example (*see* Chapter 8), and about the question of intellectual decline in later life (*see* Chapter 15). Relatively few developmentalists offer "big theories," conceptions that try to cover an enormous range of phenomena. Even among the big theories, there are significant differences in scope and focus.

Figures 2–3 and 2–4 present two of the most influential big theories of development in schematic form. The theories themselves are explored in appropriate places throughout the book. They are outlined here chiefly to indicate how different they are in emphasis. Even though both are of the stage/phase type, the views elaborated by Jean Piaget (1954) and Erik Erikson (1963) address themselves to different issues and different data. Each has more to say about certain zones of the life span than others; each raises different problems and requires somewhat different research approaches for definitive testing.

We are likely to turn to Piaget, for example, when we want to understand the sweep of intellectual development from infancy through adolescence. He offers a broad view of intelligence as our primary mode of adaptation and tries to show how we progressively move toward a "construction" of reality. Piaget does have something to tell us about postadolescent development as well, and about social relationships in addition to intellectual functioning. However, mental development in the early years of life is that domain of Piaget's theory that has commanded the most attention.

Erikson, by contrast, delves more into personality development and an individual's relationships with other people. He believes there are certain "developmental tasks" that must be mastered at a particular stage in order to move on successfully to the next. Erikson also differs from Piaget in his more inclusive view of the total life span (although his coverage is richer in some developmental zones than others).

Theorists differ, then, in the level, scope, area, and detail they try to encompass. There is no single theory that even attempts to account for the full range of developmental phenomena both in broad outline and rich detail. We will have to work with bits and pieces of important facts that do not fit in clearly with established theory, with little theories that do their best to study limited sets of phenomena, and with big theories that take on larger realms of phenomena. Those who enjoy both the *analysis* of phenomena and the challenge of *synthesizing* elements into larger sets of relationships will probably have the best time in studying humans developing.

SUMMARY

Science involves two processes: analysis and synthesis. It may seem at first as though research produces only bits and pieces, isolated facts that are difficult to relate to life as we know it. But developmental scientists also attempt to integrate specific facts into networks of relationships and theories.

YOUR TURN

The scenario technique gives us the opportunity to perform experiments—in our own minds if not elsewhere—that can advance our understanding of humans developing. It is based on the language of dramatic action, which in turn is based on human lives in their various settings. I will suggest a few variations on Scenario 1 that you might be interested in working out for yourself.

- ***Variant A.*** Baby has normal vision, but is almost completely deaf. How would such an infant respond to the separation experience? What kind of information would you seek to deal with this problem?
- ***Variant B.*** The situation is as described in Scenario 1 except that mother is also blind.
- ***Variant C.*** The situation is as in Scenario 1, but father stays home from work to care for baby while mother is away.

Many other variants could be proposed. Perhaps you can devise one yourself that you would prefer to develop. But do try at least one variant on Scenario 1. Go back through the process that has been described here. Think of the many ways in which the variant condition might affect the development of baby and the situation of those around him. Summarize your thoughts on paper. Then, why not search for additional observations to support, modify, or extend your scenario? You might draw these observations from the library, from conversation, or from what you can discover with your own eyes and ears.

One of the ways this kind of exploration is worthwhile is the challenge it offers to your own potential as an observer of humans developing. What can you come up with? How well can you think through the complexities and possibilities? How will your approach and results compare with what others produce? Give yourself a whirl as a scenario writer, drawing your materials from: (1) the scenario given in this chapter, (2) what you can find in the world around you, and (3) what you can find in the world within you. I suggest that you be tolerant of your own first efforts. Once you get the knack of it, you will find much to enjoy in this approach to studying the developmental process.

In this chapter I introduced the language of dramatic action as an approach to integrating research data with real life. A *protagonist* was engaged in a *scenario* that involved *settings, situations,* and movement through *natural time, developmental time,* and *phenomenological time* as the plot unfolded. An extended scenario was presented based on the work of a developmental research team. We followed the early development of an infant who was born prematurely and without vision. Patterns of development for sighted and blind infants were compared. A crisis arose in the protagonist's life because of a family situation, but it was resolved with the sensitive intervention of the research team. We learned something about the relationship between the researcher and the researched and, in "Your Turn," you were invited to try out your own potential skills as one who can observe and think about human development.

I then turned more broadly to the question of the *developmental mission* and its *methodologies.* Development is closely related to *age, time,* and *change.* However, the relationship is not as simple as it might first appear. Some of the problems in distinguishing between age-related differences and age-related changes were noted. Age and time were more properly regarded as *index variables* rather than as explanations for devlopmental phenomena.

Criteria for developmental change were explored. We saw the difference between cyclical and alternating phenomena and socio-cultural shifts. Particular attention was given to the similarities and differences between developmental and *evolutionary* change, although this topic needs much more consideration.

Developmental change is usually regarded as having the characteristics of directionality and progression from a relatively dependent to a relatively self-sufficient status. Although often associated with a phase or stage approach, development can also be seen as more of a quantitative than a qualitative set of changes. The *orthogenetic principle* was introduced as an important example of a very broad theory of development whose potential has yet to be fully explored.

Methods of studying development include the naturalistic approach and the natural experiment, as well as single-observer approaches. Current methods emphasize degrees of control and precision not generally available before. There is also a more resourceful use of *multivariate* research designs and more *experimental* interventions to study both the normal and the upper limits of human capabilities at a particular point in development. Examples were given of two "big theories" (Piaget and Erikson), their relationship to each other and to developmental phenomena in general was touched upon.

Reference List

Emde, R. N., & Koenig, K. L. Neonatal smiling, frowning, and rapid eye movement states. II: Sleep-cycle study. *Journal of the American Academy of Child Psychiatry,* 1969, *4,* 214–222.

Erikson, E. H. *Childhood and society.* New York: W. W. Norton & Company, 1963.

Escalona, S. K. *The roots of individuality.* Chicago: Aldine Publishing Company, 1968.

Fraiberg, S. Intervention in infancy: A program for blind infants. *Journal of the American Academy of Child Psychiatry,* 1971, *10,* 381–405. (a)

Fraiberg, S. Smiling and stranger reaction in blind infants. In J. Hellmuth (Ed.), *Exceptional infant* (Vol. 2). New York: Brunner/Mazel, 1971, pp. 110–127. (b)

Lane, H. *The wild boy of Aveyron.* Cambridge, Mass.: Harvard University Press, 1976.

Piaget, J. *The construction of reality in the child.* New York: Basic Books, 1954.

Tyler, L. E. *Individuality.* San Francisco: Jossey-Bass, Inc., Publishers, 1978.

Werner, H. *Comparative psychology of mental development.* New York: International University Press, 1948.

Wolff, P. H. Observations on the early development of smiling. In B. M. Foss (Ed.), *Determinants of infant behavior.* New York: John Wiley & Sons, Inc., 1963.

THE PRENATAL SCENE

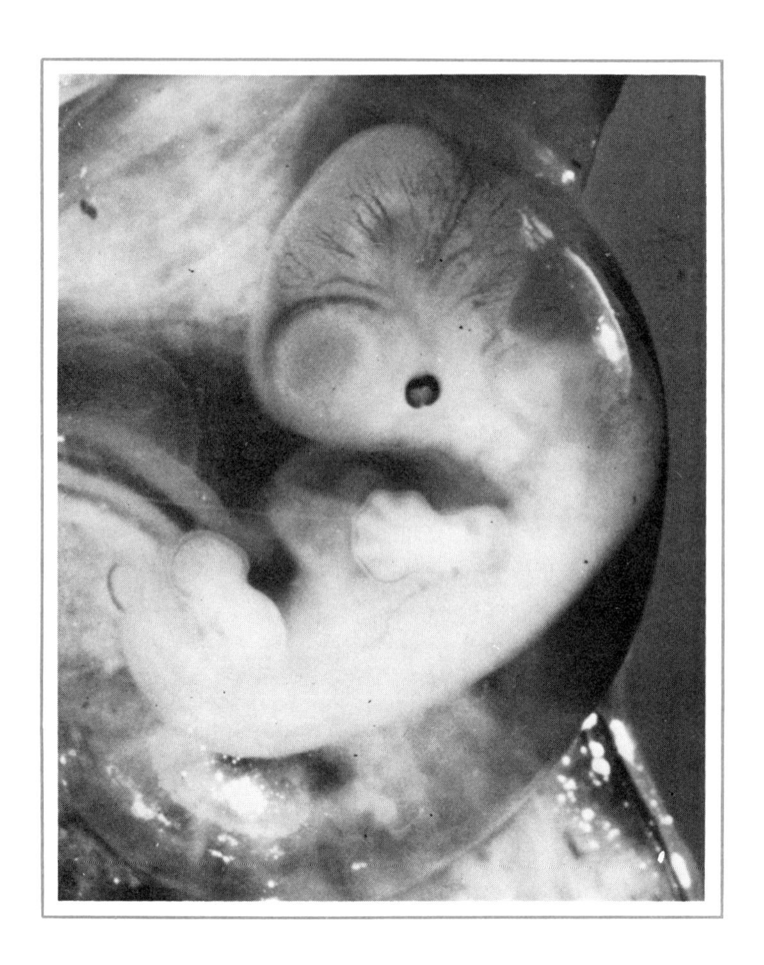

3

CHAPTER OUTLINE

I
The prenatal scene includes the larger human environment with its expectations, resources, and hazards.

Timing / Social prematurity / Spatial dimensions / Birth order / Family intellectual climate / Socioeconomic class

II
Physical and genetic components are an important part of the prenatal scene.

Somatic cells / Germ cells / Ova / Sperm / Chromosomes / Genes / DNA / Nucleotides / Zygote / Mitosis / Genotype / Phenotype / Preservation (Mendel) / Segregation (Mendel) / Dominant and recessive traits / Genocide / Eugenics / Mutation / Behavioral genetics / Twins / Phenylketonuria (PKU) / Schizophrenia

III
Three phases of development mark the sequence of events between conception and birth.

Zygote / Implantation / Embryo / Placenta / Differentiation / Ectoderm, mesoderm, endoderm / Cephalocaudal Principle / Fetus / Individual differences / Reflexes / Behavioral differentiation / Viability / Placenta determinism / Abortion

Precisely what, where, and when is the prenatal scene?

Human development does not begin with birth. We could not hope for a comprehensive understanding of development if we neglected the earliest phases of human growth. At this point it would be customary to introduce two very important topics: (1) genetic influences, and (2) the sequence of growth from conception to full term. Despite its significance, this focus is too narrow. Our curiosity should not be bounded by the womb.

Let's begin with a scenario. I will use only rather commonplace elements to set our thoughts in motion.

SCENARIO 2: THE PREHISTORY OF MICHAEL

The protagonist in Scenario 2 is a person who does not yet exist. We are concerned at this point only with his prehistory. Reduced to simple elements, Michael's prehistory might read as follows:

1. Gloria and Tom are serious about each other. Marriage seems likely. There is not much discussion about raising a family at this time, but Gloria *knows* that she wants to bear a child by Tom, sooner or later.

2. The couple does marry. They have an understanding with each other that they will not have children until both have completed college and made a reasonable start in their careers. This means that the possibility of a Michael is clearly acknowledged, but his existence is postponed to the not-so-immediate future.

3. Michael-the-Possible exercises an influence over Gloria and Tom. In particular, their sexual relations are conducted with awareness (mostly on Gloria's part) that pregnancy at this time would upset their plans and mutual agreement. There are times, however, when one or the other is tempted to suggest conception.

4. Eventually one partner or the other expresses readiness to add another generation to their family. For awhile, the foremost question for Gloria and Tom is the parenthood decision. The level of tension, uncertainty, and interaction in their relationship changes in this period. It can be resolved in many different ways. Let's say that after a few weeks of doubt and hesitation, Gloria and Tom decide to go ahead.

5. There is also a period of time between the decision and the achievement. This trying period provides another test of Gloria and Tom's relationship. Curious developments can occur at this time, and yet they seem quite

natural when they do. In this instance, Tom now becomes even more interested in producing a child than Gloria, whose motivation had been stronger at first. Tom suddenly finds himself thinking of his *son*. It will be a son, of course. He is a little surprised to find this prospect looming so important to him, for he had not given it much thought before.

6. Pregnancy begins. Another set of thoughts, feelings, and actions follows from this event. These include:

This is a way of helping the child have his or her own identity

- *The naming*. If "it" is a boy, then Michael. Gloria had suggested Thomas, Jr. This appealed to Tom, but they eventually decided in favor of Michael. It appears true of our current generation that relatively few young parents are inclined to bestow their given names upon their children (Kastenbaum, 1974).
- *The placing*. Serious thought is now given to establishing and equipping a room for Michael. This might involve moving to a larger apartment or buying a first home. Well before he is born, then, Michael has a place in the world waiting for him.
- *The funding*. Although Gloria and Tom had discussed the financial aspect before, now the planning (and worrying) becomes more specific.

Other responses to the pregnancy could also be listed: Gloria's fears about her own well-being and the health of the baby, for example. She decides to avoid any exposure to X rays since expert opinion declares that *no* dosage of radiation can be considered safe (Stevenson, 1973). Tom is concerned that he might become more tied down than he would prefer. Both find themselves receiving advice from relatives and friends that is sometimes helpful, sometimes not.

7. Gradually, Michael takes on an identity and reputation despite the fact that he has not yet made his postnatal appearance. Some of his prereputation is based upon isolated bits of information Michael himself has contributed to, for example, "The little devil is really kicking today." But much of the regard in which he is held depends upon the expectations and needs of his parents-to-be. His grandparents may already be sending Michael to medical school, while Tom is fantasizing an athlete who will have the career he did not, and Gloria is mostly concerned about how well he will mind her. Some components of Michael's prereputation may be held lightly and will be easy to shift; others may prove so tenacious that Michael is intently measured for size against them when he does arrive on the scene. Differences between Gloria's and Tom's expectations can be relatively small, or they can be so great that Gloria and Tom (not to mention various relatives) hold contradictory views concerning who Michael is and will be.

Let's review the components of this scenario. Consider the time dimensions (there are several). Notice that Michael pre-existed earlier in the mind of one parent-to-be than in the other. His pre-existence, therefore, followed

Are these people ready to have a baby? A thoughtful couple allows time to explore each other's ideas and feelings before conceiving. (Photo by Talbot Lovering.)

two or more time tracks. This seems to be fairly common, although there do not seem to be any clear data to support the proposition. Notice also that Michael's prehistory consisted of several stages:

- *Stage 1:* The possibility and desirability of a Michael is acknowledged. He exists in the minds of at least two people who have the potentiality for bringing about his physical actualization.
- *Stage 2:* This is a period of determined nondevelopment. In terms of his parent's plans, Michael would be a *socially premature* baby if conceived at this time, even if his pregnancy were full-term. Time continues to pass, of course, but **developmental time** is held back.
- *Stage 3:* Development time begins now. There is the resolve and effort to conceive. The developmental process is in motion, although one cannot yet speak of an actual physical being who is developing.
- *Stage 4:* Conception occurs, and the physical pattern of prenatal development is underway. At this point the official clock begins to tick, and Michael, for the first time, is both a physical and a socio-symbolic personage.

Developmental time is the tempo at which actual growth occurs

Consider now the *spatial* dimensions. It is only when we reach Stage 4 that Michael's prenatal scene actually includes the intrauterine environment. Either before or after conception, space in the external world may have been established for his use (i.e., a nursery room set aside) and thus be part of the prenatal scene. But the prenatal scene, in the larger sense, includes all the settings in which both Gloria and Tom function.

Take an example or two. Tom feels the need to earn more money to provide an adequate postnatal scene for Michael. He decides to moonlight, accepting a job that requires evening hours and which has a negative effect upon his energy level and mood. Part of Michael's prenatal scene, then, is the second job, because this will present him with a father who is more weary and irritable than he would otherwise have been. Gloria's share of the prenatal scene becomes increasingly lonely. She is seeing less of Tom, and eventually she leaves her own job. With more time alone or socializing with neighbors, Gloria's habits change. She finds herself eating more often and has gone back to smoking. As a result she puts on more weight than she intended, feels depressed about it, experiments with diets she and her friends improvise, switches back and forth from her diet to the one recommended by her physician, etc. Gloria is also unhappy about smoking again and worries somewhat that this might have an effect upon her child. She knows that heavy smoking has been correlated with a high incidence of premature birth, spontaneous abortion, and low birth weight (Timiras, 1972, p. 383). Michael thus inherits a mother who has some preoccupation with her body image and whose mood may fluctuate somewhat as she subsequently attempts to diet. Should her dietary habits during pregnancy prove unwise and extreme, this would be another instance in which the prenatal social scene might influence the child's early developmental career.

Although some psychological and social influences may be operative while Michael is within the womb, it is probable that more of the influences will take hold when he emerges. Does the pattern of coping and stress experienced by his parents result in an orderly or disorderly postnatal scene? Are Gloria and Tom feeling confident and ready for him, or are they distracted, beseiged, and subpar?

Not everybody will agree with the approach I have suggested here. Some will prefer to limit consideration of the prenatal scene to those events that take place within the intrauterine space after the time of conception. By contrast, I have proposed that Michael's prenatal scene occurs along several time tracks, with various stages discernible. Furthermore, it seems to me that the people who most intimately represent Michael take his future with them wherever they go. In this way, they expose Michael-the-Possible to both direct and indirect influences. And, by implication, those who would like to do right by Michael could begin by making life a little more comfortable and a little less stressful for Gloria and Tom before their child appears on the scene.

A VARIANT ON THE SCENARIO: MICHAEL-THE-LONG-AWAITED

Suppose that we are still concerned with Gloria and Tom. Everything that was said about them continues to hold true—up to a point. They have moved into Stage 3 of Michael's prehistory. The variant I now introduce is time,

time in the form of delay. Gloria and Tom consider themselves ready to conceive and care for their first child. But month after month goes by without results. A year passes, two years, more.

Reflect upon this situation. The most important thing that is happening is what is not happening. From a physical and objective standpoint the situation is essentially static. Gloria and Tom are not that much older than they were at the beginning of Stage 3. Their life pattern has not been affected by all the little changes that occur when a first child makes the scene—no baby-sitter problems, no special expenses, no getting up in the middle of the night to see why the baby is crying, etc. But there may indeed have been some changes, perhaps significant ones at that. Let's consider a few of the shifts that are likely to have occurred in the prenatal scene as natural time moves on while developmental time stands still.

1. Gloria and Tom wonder who is "at fault." Depending on their feelings about themselves and each other, this can either become a major source of tension and discontent in their relationship, or it can be put into perspective as a specific problem for which a specific solution should be sought.

2. The couple tries various ways of increasing the probability of conception. Each may undergo a physical examination and laboratory tests. Medical recommendations for diet, surgery, or psychotherapy may be made. Quite opposite kinds of advice may be given: plan sexual activities more carefully by the fertility calendar, try a different position during intercourse, stop thinking so much about conceiving, take a vacation together, etc. The opportunities for confusion and contradictory advice are numerous.

3. The delay in conception might not be a negative factor in itself. However, a subtle change could be taking place in the couple's sense of *futurity.* Gloria and Tom had some ideas or assumptions about how their life together would develop. Now either or both might be experiencing a sense of being stalled, or being behind schedule. Other plans may be delayed further because of Michael's nonappearance. For example, Gloria may have in mind a professional career that she had been willing to postpone until her children reached a certain age. In this situation, it would not be out of the question for Gloria, Tom, or both to be irked at Michael for interfering with their plans. Although this attitude might not be held at a conscious level, and certainly would not represent their only or most significant orientation toward their first child, it could nevertheless exert an influence when he does come along.

Artificial insemination introduces the sperm by laboratory procedures

4. As time goes on, Gloria and Tom may consider making other arrangements. **Artificial insemination** is still something out of the ordinary for childless couples, but it could be that this appears to be the only answer for them. Adoption is another, more likely prospect. Whether or not they follow through on either of these possibilities, the prenatal scene will have become one in which a number of important adjustments have been attempted.

5. It is also possible that after awhile Gloria and Tom have made so many adjustments to life without a baby that their original motivation is weakened. Perhaps they have raised a dog or cat as a sort of temporary substitute for Michael. The pet receives some of the affection that had been waiting for the baby and, in return, provides a measure of distraction and amusement. If Gloria and Tom are otherwise very active in their careers and outside pursuits, the pet may seem to be as much child as they absolutely need at the moment. Again, the longer they live together without a child, the more their patterns may become comfortable and fixed. It may now seem inconvenient to have a baby. Underneath the day-by-day functioning, Gloria and Tom may still have a deep desire to bring a child into the world. However, this desire can become increasingly covered over by the little adjustments of living that have turned into a fixed routine.

6. If Michael persists in not being conceived for many years, then the effect of his absence can make itself known in a variety of ways: where Gloria and Tom live, who they select for close friends (e.g., couples with or without children), etc. Wanting him makes them a different couple from the one they would have been had they never attempted to conceive. A sensitive observer should be able to see the difference between two couples who otherwise appear very similar. Neither couple has children. But one couple has planned and yearned for a child; the other has been decisive about its intention to remain childless. The physical situation for both couples might be quite comparable. But it is a quite different story to lack something (or someone) you *want,* than it is to simply be lacking something. Gloria and Tom *know* that their marriage is missing one of its most important components, although they may have discovered a number of ways to insulate themselves from the frustration and disappointment.

7. Suppose now that Michael finally does move to Stage 4: he is a fetus and on his way at last! Many implications of his late arrival can be anticipated. One possibility is that Michael will have waiting for him a pair of not-so-young parents who are psychologically ready to spoil him with long held-back affection. Their original desire for a child has been magnified by the long delay, and this could influence the style in which Michael will be raised. The fact that his parents have been around a little longer can also result in a different method of childrearing than would have been used a few years earlier. The financial situation might also be appreciably different and have an influence on his development. Furthermore, there is the possibility that Gloria's pregnancy might be a little more difficult than it would have been at first. Based upon statistical expectations, Michael would now run more risk of being born with some kind of problem that is related to the relatively older age of his mother (Jayant, 1964). These last two points are not intended to alarm those who are planning to have children a few years later than the average. Maternal age is only one of many factors that are related to ease of pregnancy and constitution of the newborn. Many women have per-

fectly normal pregnancies and offspring in the later years of their reproductive cycle. But there remains a slightly greater risk factor, on the average, if we compare, say, a twenty-five-year-old woman and a thirty-five-year-old woman in their respective first pregnancies.

All the factors mentioned above, in various possible combinations, suggest that Michael's prenatal scene could be quite different if the time dimension is varied appreciably from the pattern given in Scenario 2. The total result could be "better," "worse," or just "different" for Michael and his parents. Even if the genetic transmission of characteristics to Michael turned out to be precisely the same in both situations, we would be well advised to expect some differences in his pattern of development based upon the varying configurations of the total prenatal scene.

PSYCHOLOGICAL INFLUENCES ON THE PRENATAL SCENE

Scenario 2 and its variant have introduced some psychosocial factors that can often be observed on the prenatal scene. Most of these factors do not exercise a direct influence over the fetus; some of them, in fact, exist long before conception. Nevertheless, important components of the world in which the baby will find himself or herself have already become established on the prenatal scene. The baby's reputation, for example, may have preceded him or her, along with a name, and an expectation to live up to certain parental hopes and aspirations. We will now examine two possible psychosocial influences in more detail.

BIRTH ORDER

Over the years, much speculation and controversy have developed over the effect of birth order on personality development. More than a century ago, Sir Francis Galton (1874) came up with a relationship between intellectual distinction and the position of the firstborn among British scientists. A number of American psychologists, including the trailblazing developmentalist Lewis Terman (1925), made similar discoveries. Special attention should be given to the observations of Alfred Adler, whose work stimulated much of what has followed. Adler was one of the brilliant pioneers of psychoanalysis who, after close association with Freud, broke away to cultivate his own approach.

Adler was well ahead of his time in his belief that the socio-physical environment was integral to human development. He told us: "It is a common fallacy to imagine that children of the same family are formed in the same environment. Of course there is much which is the same for all children in the same home, but the psychological situation of each child is individual and differs from that of others, because of the order of their succession" (cited by Ansbacher & Ansbacher, 1956, p. 376 in their comprehensive selection from his writings).

Adler did not intend for us to regard birth order in a rigid, mechanistic manner. For him, birth order was significant because of the *situation* it implied. "Thus, if the eldest child is feeble-minded or suppressed, the second child may acquire a style of life similar to that of an eldest child; and in a large family, if two are born much later than the rest, and grow up together separated from the older children, the elder of these may develop like a first child. Such differences also happen sometimes in the case of twins" (Ansbacher & Ansbacher, 1956, p. 377). Unfortunately, Adler's ideas seem to have been misunderstood or oversimplified. Some of the research undertaken to prove or disprove his propositions was not designed well for this purpose.

More specifically, Adler suggested that the firstborn usually receives more attention. He or she is most likely to be "spoiled," if the parents and other relatives are inclined to spoil anybody. Yet the firstborn's place in the sun of parental affection is perilous. When child 2 arrives, the little king or queen finds the throne pulled from under him or her. He or she is upset by the alteration in household power dynamics—everybody now seems too interested in the new arrival. Dethroned, the firstborn may carry this experience for the rest of his or her life, with profound effects upon adult personality. Adler also offered further observations that take into account the period of

The firstborn child is often a teacher, protector, and junior parent for younger siblings. This might be why firstborns tend to score higher on intelligence tests. (Photo by Talbot Lovering.)

time the firstborn has in which to enjoy only-child status and other significant variables.

Dethronement is the loss of special status when a sibling is born

Adler's main point was that we should expect the firstborn to be different from other children in the same family because of: (1) the focal position he or she enjoyed at first, and (2) the shock of **dethronement** (a term introduced by Adler to describe this experience). He believed that expression of the eldest child's anxieties could be found in dream life. The firstborn often dreams of falling because ". . . they are on top, but are not sure they can keep their superiority. [By contrast] Second children often picture themselves running after trains and riding in bicycle races. Sometimes this *hurry* in his dreams is sufficient by itself to allow us to guess that the individual is a second child" (Ansbacher & Ansbacher, 1965). Adler saw the motives and lifestyle of the secondborn (and of later children) to be influenced by their birth order situation, although the nature of the influence is different from that of the firstborn.

Much research was stimulated by the proposition that birth order has significant implications for lifespan development. As early as 1931, H. E. Jones could find approximately two hundred studies bearing upon this topic. He concluded his critical review by declaring that the effects of birth order on personality had not been clearly established despite all the efforts (Jones, 1931). This conclusion dampened some of the scientific enthusiasm for birth order effects, although popularized treatments and a number of unsystematic observations that appeared consistent with Adler's views kept the ideas in circulation.

The status of birth order effects was elevated again in the behavioral sciences when Schacter (1959) found clear differences between firstborn and only children on the one hand, and laterborn children on the other. The "firsts" and "onlys" seemed to have a greater need or preference to be with others. This finding stimulated another wave of birth order research, including approaches that were more relevant and sophisticated than most of the earlier inquiries. A review of the new research confirmed that birth order did seem to play some importance after all (Schooler, 1972). Firsts and onlys, for example, appeared more likely to become intellectually eminent, especially in the sciences. Schacter's finding that they tended to seek out the company of others also seemed to be confirmed. It was not so much that the firstborn was sociable and outgoing in the usual sense of these terms; rather, he or she felt more need to affiliate with others during times of anxiety and stress. Presumably, this stronger need for affiliation has something to do with the firstborn's early relationship with mother, a relationship that might involve a greater level of dependency (Hilton, 1967).

Motivational and achievement differences have since been found among elementary school children who occupy various positions in birth order (Adams & Phillips, 1972). The researchers noted that most previous studies found intellectual supremacy for the firstborn, but that there was conflicting data about motivational differences.

The Adams and Phillips study included approximately equal numbers of Mexican–, Anglo–, and Black–American students. This was a commendable feature that we are now beginning to find in a variety of developmental research studies. As it turned out, there were no differences in birth order effects among the three racial-ethnic groupings, nor did socioeconomic status seem to make a difference. But this would never have been known if the psychologists had not taken the trouble to raise and answer the questions.

Adams and Phillips's most important findings were that: (1) firstborn students perform higher on measures of intellectual and academic functioning, but (2) the difference can probably be explained on the basis of motivation rather than on the basis of basic intellectual ability. Adams and Phillips suggest that the cumulative results of their own and previous research studies indicate: ". . . the beginning of an empirical 'solution' to the birth order puzzle. Not only do parents have a higher expectation for . . . the firstborn child in comparison with the laterborn, but, as evidenced by their higher school motivation, firstborns appear to be living up to this expectation" (Adams & Phillips, 1972, p. 162).

Today many developmentalists are trying to evaluate an even newer set of challenging data on birth order effects. Lillian Belmont and Francis A. Marolla (1973) were able to study the relationship between birth order and intelligence test scores for almost all the nineteen-year-old men in the Netherlands who were born between 1944 and 1947. These military examination data provided a population of 386,144 individuals—quite a colossal sample for any developmental study. The intelligence test was the Raven Progressive Matrices, a measure that attempts to reduce cultural bias through a visual-reasoning format rather than a verbal format. Belmont and Marolla not only had a large sample available, but they were able to examine the relationship between place in the birth order and family size. It was found that the brightest young men had come from the smallest families. Furthermore, whatever the family size, those who came on the scene earlier tended to be brighter.

You may be thinking that the socioeconomic level of the family might have been making the real difference, with family size and birth order being secondary. Belmont and Marolla were able to control statistically for these factors. It still turned out that within a particular social class the most intelligent children were from the smaller families. The lastborn children also showed lower intelligence test scores than the firstborns. One could speculate on biological reasons for these findings, but Belmont and Marolla found no sound basis for invoking biological explanations.

The next step in birth order research was taken by Robert Zajonc and Gregory Markus (Zajonc, 1976). They developed the concept of a *family intellectual climate.* This concept purports that other things being equal, a child's mental development will benefit if there is a high level of quality mental life going on around him or her. This idea was fashioned into a mathematical model by which the intellectual climate of a particular household could be estimated. Assume, for example, a childless couple whose own intellectual level is average. Let the average be represented by the number 100. The home at this point has a total mental saturation of 200, but since this must be divided by the number of members (2), we are back to 100 again. When the first child is born into this household, his or her intellect is estimated to be zero or thereabouts. The family's mental score is still a combined 200, but when divided now by three individuals, the index drops to 67. When a second child comes along, the index will drop again, etc., etc.

A technical point must be introduced here. The IQ score with which we are all so familiar is not a reflection of "absolute" intellectual performance on the particular measure used. It is age-adjusted. An IQ score of 100 means average-for-this-person's-age. This procedure has its advantages, but it does not clearly convey the absolute "correctness" of the individual's test-measured intellect. Zajonc and Markus bypassed the IQ score to focus instead on the actual intellectual performance level independent of age. This leads them to reason as follows: "After two years the first child has about four percent of his adult intellect, so the second child enters an environment of 100 plus 100 plus four plus zero, for an average level of 51. The second child enters a less intelligent atmosphere" (Zajonc, 1975, p. 39).

The complete mathematical model developed by Zajonc and Markus to determine intellectual performance (family intellectual climate) is more complex than has been stated here. Zajonc has now applied it not only to the Belmont and Marolla data but to a variety of other data from around the world. In what seems to be the most systematic analysis of the most data ever performed in the area of birth order and family size, Zajonc has offered the following conclusions:

This data trend involves large numbers of people. Use it on siblings at your own risk!

1. The smaller the family, the brighter the children.
2. When intervals between birth of siblings are relatively short, the earlier-born children do better on intelligence tests.
3. When the intervals between birth of siblings are large, the effects of birth order are cancelled.
4. Long intervals between sibling births enhance intellectual growth.
5. The relatively lower intellectual performance of children of multiple births is related to the adverse effects of short intervals between births.
6. Only children do not have the chance to teach younger siblings, so they lose some of the benefits of smaller family size.
7. Point 6 applies to last children as well. Whatever the family size, they do not perform on intelligence tests as well as might be otherwise expected because they do not have younger siblings to teach.
8. When a parent is missing from the family, the children's intellectual performance is lower.
9. Differences in birthrates, average order of births, intervals between children, and family size show up in measures of intellectual performance when large populations are compared with each other.

Figure 3–1. *Birth order and spacing and IQ. Smaller families with long intervals between births show higher IQ's on the average than larger families with shorter intervals (Zajonc, 1975). Firstborns in both configurations tend to score higher than later-borns. Further research may help to explain these differences and put them into better perspective.*

10. Family pattern differences between different nations, ethnic or racial groups, and geographic regions of the same country are also related to differences in aggregate intellectual performance.
11. The average birth order of males and females is different (i.e., the number of firstborn males is different from the number of firstborn females, the number of second-born males is different from the number of second-born females, etc.). These differences are also reflected in intellectual performance scores (Zajonc, 1976).

This is quite a cluster of findings. Although the research and the conceptual and mathematical model make no reference to the earlier observations of Adler, it is clear that his sensitivity to the family environment and its impact on subsequent development is taken into account here.

How are these findings to be evaluated? Zajonc himself cautions that there is much more to the family environment than the variables he has concentrated upon so far. In other words, the role of birth order and family size might be exaggerated temporarily until similar in-depth examination can be pursued in other dimensions of psychosocial and biological influence.

Quantifiable information can be converted to numbers for further analysis

From a critical standpoint, I am particularly concerned about the concept of individual and family intelligence that was applied here. The basic concept of family intellectual climate, and the attempt to translate it into **quantifiable** terms, is stimulating and potentially of much value. But to say that a two-year-old, for example, has only 4 percent of his or her adult intellect is such a narrow view of mental functioning that it cannot be allowed to pass unchallenged. You will see for yourself in many of the following chapters that the infant and child have high degrees of that kind of adaptive functioning we know as *intelligence.* Much of the reality of infant and child development is overlooked when we simply classify a youngster as having a certain small percentage of adult intelligence. It ignores, for example, possible qualitative differences in the very nature of intelligence from infancy to adulthood. This kind of simplification is useful for purposes of statistical analysis, but it is not adequate for those who want their conclusions to be grounded in a firm understanding of the nature of intellectual development. Other types of criticism are to be expected as well—partially because few studies are flawless, and mostly because Zajonc's work does have many implications for social planning as well as the understanding of developmental phenomena.

Among the many positive contributions of this approach, attention may be given particularly to the recognition of the sibling as *teacher.* This deserves further exploration in its own right. Zajonc, himself an only child, has given us a lesson in the vigorous and innovative analysis of data relevant to early human development, but we will still have to draw our own conclusions.

SOCIOECONOMIC CLASS

In some of the studies we have reviewed, socioeconomic differences seemed less important than birth order and family size. This is worth keeping in mind. At times it is tempting to attribute almost all individual differences in development to socioeconomic background; at other times, one is tempted to forget background factors altogether. A balanced approach will always raise the question of the contribution socioeconomic situation makes toward a child's developmental career, but not expect the same answer on all occasions. Consider a few of the ways in which the socioecomonic aspects of the prenatal scene might influence development.

Vernon L. Allen has noted that, "Most . . . psychological theories are woefully inadequate for dealing with the types of problems presented by poverty" (1970, p. 149). We do not know enough about the psychological dynamics of poverty to be sure we are raising the right questions and attempting to answer them in the most appropriate way. Socioeconomic influences on personality development undoubtedly exist, but we do not yet have a systematic approach to this subject.

This does not mean that we have learned nothing about the effects of deprivation upon the young. There is quite a substantial amount of research literature on the effects of deprivation (of food and water, for example) on the behavior of young animals when they subsequently reach maturity. Some of these studies are of considerable interest, but we will stick to human research here.

One of the most thorough reports on deprivation in human beings was prepared by Birch and Gussow (1970). After reviewing the available research, they concluded that: " . . . poor women viewed as a group, and especially poor nonwhite women (of whom the great majority in the childbearing population are Negro), begin childbearing younger, repeat it more rapidly than the better-off and white women, and are more likely to continue it into the older ages and higher birth orders where the rates of complication are strikingly high" (pp. 90–91). Low socioeconomic status thus seems clearly related to a configuration of higher risk factors for children. The odds are indeed more against Michael if his parents are impoverished. Furthermore, "Given the tendency among the poor to an earlier initiation of childbearing, it is inevitable that groups of women bearing babies when they are teen-agers will include a high proportion of lower-class mothers, among whom there is an excess of illegitimacy, untreated disease, poor health practices, and other factors associated with poor reproductive outcome" (p. 91). The unborn child brought into a grossly unstable socio-physical environment might be expected to develop under a variety of handicaps.

Although our focus here is upon the developmental career of the infant-to-be, it is important to acknowledge the additional risk factor deprivation creates for the mother herself. The pattern of risk is more complex than I can describe here. I will simply pass along Birch and Gussow's well-warranted conclusion that, "the reproductive habits associated with poverty interact with the biological risk associated with certain age and parity* relationships in such a way as to produce an increased risk in childbearing for women in the lower social classes" (p. 93).

The interrelated risk to both mother and child can be illustrated from another perspective if we pursue the earlier reference to health maintenance. Adequate medical care is less often provided to the pregnant woman of lower

* *Parity*, in this context, refers to the number of children, if any, a woman has already borne.

The socioeconomic context into which a child is born tends to preselect some of his or her experiences. (Photo by Talbot Lovering.)

socioeconomic background. If this results in problems for the mother, then the child is likely to feel the effects in various ways. An ailing, suffering parent can hardly be expected to provide as much care, attention, etc. to her latest infant, and may also be forced to deprive her other children of the attention they need. The infant may also experience his or her mother's medical problems even more directly if the opportunity to be born in sound health is jeopardized. The prenatal scene thus includes the health service delivery system as well as the household.

Birch and Gussow and a number of others have documented the grim situation that too often prevails when a low-income woman does seek help during her pregnancy. Birch and Gussow conclude that, "So long as clinic care is provided reluctantly, patronizingly, inconsiderately, impersonally, and at inconvenient times and locations, poor women will continue to seek an inadequate amount of prenatal attention" (1970, p. 163). This does not tell the whole story but, by and large, it is a story of heightened risk for both mother and fetus.

Birth order and socioeconomic class were selected as two significant dimensions of the prenatal scene that continue to be the subject of scientific inquiry and social concern. Many questions remain on both topics. Precisely how, for example, does socioeconomic status affect a particular aspect of the child's developmental career? Poverty-level income cannot of itself *cause* psychological and physical phenomena. Students of human development are interested in identifying the specific ways in which a condition leads to or influences a certain outcome. We do not really explain anything by saying, "He is that way *because* he is a firstborn," or "She is experiencing these difficulties *because* she comes from a low socioeconomic background." There are plenty of individual exceptions to the statistical regularities, and even when the expectations are confirmed for a particular person, we should remember that low income does not directly "cause" a particular form of behavior.

You wouldn't fall for such shaky explanations, would you?

I would not like to leave the impression that there is anything intrinsically better or worse about being born into a particular position in the birth order or on the socioeconomic ladder. Certain problems and hazards are more common in one position than in another: so are certain opportunities and enriching experiences. It is not the purpose of developmental psychology to sort people out as more or less valuable, especially on the basis of such background factors as birth order and socioeconomic class. But it is part of our purpose to identify significant situations and their likely influence on the course of lives.

GENETIC INFLUENCES ON THE PRENATAL SCENE

We have been exploring some of the psychosocial processes that are likely to operate during the prenatal period, although they may operate in different ways for different people. Now it is time to address ourselves to the physical and genetic components of the prenatal scene. This involves moving to a different level of data. We will find, however, that it does not mean a complete neglect of psychosocial influences, even when our focus comes down to genes, chromosomes, and the like.

THE LANGUAGE OF HEREDITY

The most basic facts about human heredity are easy to present and understand. However, first it will be useful to brush up on our grasp of the most frequently used terms.

The cells in our bodies can be divided into the *somatic* and the *germ*. The somatic or body cells comprise most of the structure of the adult—our muscles, nerves, bones, and organs. The germ cells undergo a developmental career all their own, resulting in the end products we know as *ova* in females and *sperm* in males.

Chromosomes transmit genetic information from parent to child

Both the ova and sperm and all other cells possess a set of minute structures known as **chromosomes.** There are twenty-three chromosomes contributed by each partner when fertilization takes place. The fertilized egg thus contains forty-six chromosomes: its total endowment from the parents (and from their own ancestors).

Genes are tiny protein units that make up the larger chromosomes

Each of these tiny chromosomes is made up of structures that are even smaller, the **genes.** But what is the gene itself? Most scientists now accept the proposition put forth by J. D. Watson and F. H. Crick (1953) who suggested that it was comprised of a substance known as *DNA* (deoxyribonucleic acid). This is a high order or complex protein found in the nucleus of the cell. Its most unique and vital talent is the ability to duplicate itself, create more DNA. This property has led some scientists to attribute life itself to DNA. Obviously, it is at least one of the most fascinating links between the inorganic and those forms that none of us would hesitate to describe as living organisms.

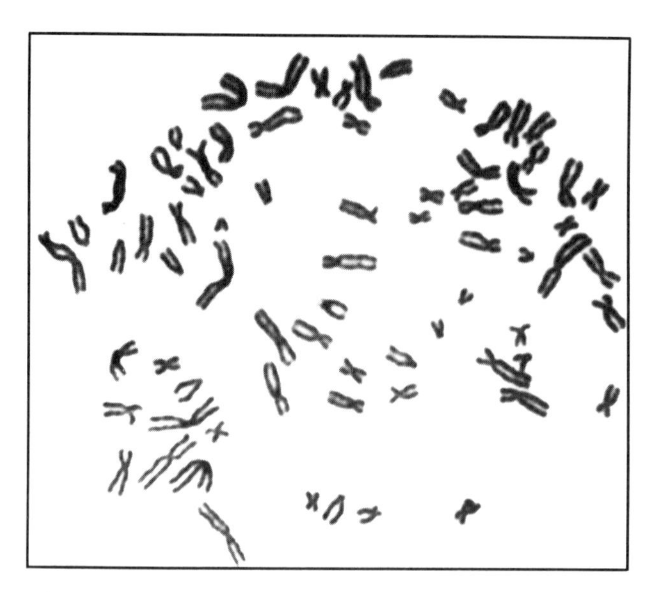

This is a close-up of the building blocks of our own genetic structure—the chromosomes. Each carries many genetic messages from one generation to the next. (Courtesy of The March of Dimes.)

DNA is regarded as that functional unit that transmits the instructions or coded messages from one generation (gene-ration) to the next. Kornberg (1968) reported that each genetic unit of DNA includes approximately one thousand smaller components known as *nucleotides.* These incredibly small components are lined up in a precise manner that encodes specific genetic messages. From such unbelievably small microstructures come the plan that results eventually in the adult human being.

Watson and Crick's image of DNA structure—the so-called double helix—has become famous. We might picture this for ourselves as something along the lines of a ladder. It is not a proper ladder that goes straight up and down. Its two sides are more like chains; they coil around each other in serpentine fashion. The ladder therefore sprirals along the vertical dimension. The "sides" of this ladder are made up of phosphate and sugar molecules, while the "crossbars" are a combination of chemicals from the group known as bases. While the double helix may sound like a prop left over from a fairy tale theater, it is in their specification of this structure that Watson and Crick made their most fundamental contribution.

A **zygote** is a fertilized egg

Mitosis is the process of cell division and subdivision

The fertilized egg itself is called a **zygote.** "Identical twins" are known as monozygotic (i.e., they shared the same fertilized egg).

Mitosis is an action term. It is applied to the process of division and subdivision by which the zygote accomplishes its growth.

One other set of terms should be introduced. There can be important differences between the genetic code that a person carries and the way in which that code has expressed itself in his or her own constitution. Each of

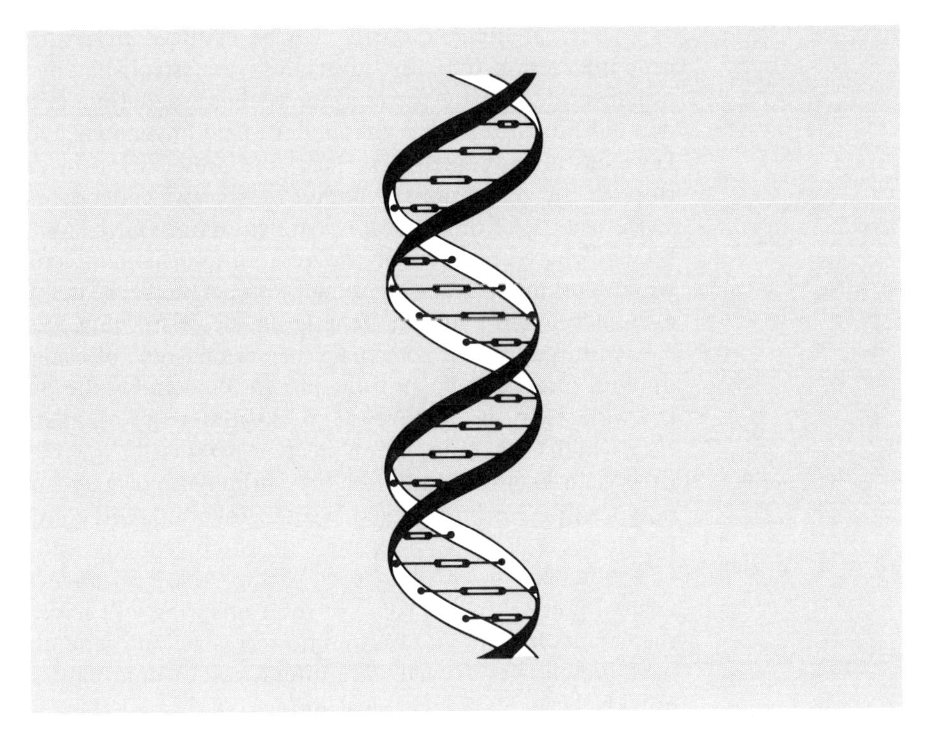

Figure 3-2. *Watson and Crick's double helix.*

Genotype is the inherited *disposition* toward a certain characteristic

Phenotype is the inherited characteristic actually expressed

us has a certain **genotype,** or set of inherited materials. We know each other, however, by our particular **phenotypes.**

We now have enough hold on the language of heredity to formulate a more systematic account of how the total process works (within the limits of current scientific knowledge, of course).

HEREDITY IN ACTION

The pooled transmission of heritable characteristics from parents to child involves a remarkable contrast between predictable and unpredictable outcomes. It is highly predictable that certain principles of gene transmission and interaction will operate. We know, for example, that the zygote will receive genetic components that have been conveyed through the centuries with little if any alteration. The newest human is sent on his or her way with most ancient "instructions" from his or her ancestors. This is related to a critical difference between somatic and germ cells. As we mature and age, our somatic cells change in various ways. Under ordinary circumstances, however, our germ cells remain immune to the passage of time. They are as "young" (and as "old") when we are seventy years of age as they were when we were fetuses.

Other predictable events can be deduced from the basic principles of gene interaction. The first principles were established by Gregor Mendel, the nineteenth century Austrian monk whose spare-time experiments in the garden did much to launch the modern field of genetics. One of the Mendelian Laws has already been described: the preservation of genetic strains down through the generations. Another of his laws concerns *dominant* and *recessive* effects. Will the child have straight hair or curly? Will the eyes be blue or brown? These are a few of the many dimensions of genetic transmission in which dominant/recessive dynamics operate. Let's stay with eye color as the example. Brown eyes is a genetic characteristic that is dominant over blue eyes in humans. A zygote that receives one gene of each color will not wind up with one blue eye and one brown. Predictably, he or she will come into the world with brown eyes. This illustrates the *Mendelian Law of Segregation* which says genes from the two parents do not blend or compromise. Each gene keeps its individuality, although one may not express itself phenotypically. Our brown-eyed zygote, grown up, will continue to carry a gene for blue eyes. This is an example in which genotype and phenotype differ. If he or she mates with a blue-eyed partner, there is a chance (which can be calculated precisely) that some of their offspring will have blue eyes. If one of the parents has brown eyes and *no* recessive blue gene, then all offspring will have brown eyes as well. The interaction of dominant and recessive effects quickly becomes a good deal more complicated than what has been illustrated here, but this is enough for our purposes.

What about the gender of the child? This is a determination that is predictable in theory, but less so in practice. One crucial piece of information is necessary before the principle of sex determination reveals its operations to us in a particular instance. We must know something about the sex chromosome that the man is contributing to the zygote. There is no question about the woman's contribution; her sex chromosome will be of the type known as "X." But the sperm cells of the male are divided between those that contain an X, and those that contain a "Y." Physically, the Y chromosome is smaller than the X and possesses fewer genes. This means that the father actually contributes fewer genes to his son than he does to his daughter, and fewer than the mother contributes to her son. The difference is relatively small when we consider the vast number of genes that are transmitted by both parents, but it does stand as an exception to the general rule that the endowment from mother and father is equal.

When the match-up is X and X, a female child results. The X–Y combination produces a male. The odds for producing a male or a female should be equal, although population statistics usually show a slight, but clear, prevalence of male births over female.

This is an example of a principle of genetic interaction that has been established, but whose actual outcome still remains a matter of guesswork. It is possible that research advances will greatly improve our ability to detect the gender of the fetus (although why this should be considered desirable or even crucial by some people is a question that belongs in the psychosocial realm).

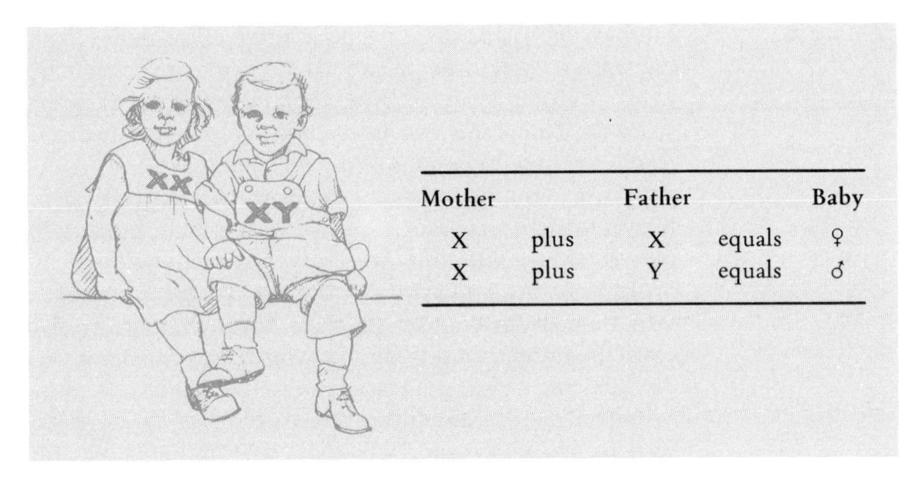

Mother		Father		Baby
X	plus	X	equals	♀
X	plus	Y	equals	♂

Figure 3-3. *Girl or boy? The mother always contributes an X chromosome. The father contributes either X or Y.*

Improved techniques for detecting gender in advance would not change the basic situation, namely, that the specific match-up between male and female germ cells that results in an offspring is so unique that it could almost be considered as a "chance" phenomenon. Of all the different combinations of genetic materials that *might* be contributed to a particular zygote, only one actually exists. The pool of possible combinations is enormous, and it affects much more than gender.

A recent review of experiments concluded that, "Genetic prediction is often complicated by the fact that many characteristics depend on complex combinations of gene pairs, rather than on a single pair, in addition, they (the geneticists) have found that genes do not simply behave dominantly or recessively, but in numerous other ways also, and that their effects may vary under different conditions" (Mussen, Conger, & Kagan, 1969, p. 43). The same authors suggested that: "The transmission of genes on different chromosomes acts like a bio-coin-tossing machine. Imagine a coin-tossing machine in which each coin represents a pair of genes—one gene on each face of the coin, and the machine is unbiased. Then imagine 500,000 coins being thrown each time. One toss of 500,00 coins represents a father's contribution. The unique combination of heads and tails from this combined toss of 1,000,000 coins will determine the genetic make-up of the child" (p. 43).

PSYCHOLOGICAL PATTERNS AND GENETIC CHANCE

Law and Chance

We see, then, that both law and chance are involved in a child's genetic inheritance. It is by "chance" that a certain combination of genetic materials is matched up to produce a given zygote. Once the combination exists, however, various principles of genetic interaction apply. If we step back for a moment from the biosphere, we can recognize that some of what the geneticist reads as chance may seem predictable or, at least, nonrandom to the social scientist. The points I wish to make here are simple ones, but ones seldom introduced into discussions of the prenatal scene.

First, we must remind ourselves that the pool of possible genetic combinations depends upon men and women who survive long enough to reach the age of reproduction. Every psychosocial process that operates to increase or decrease the probability of survival to age of reproduction thus exerts an important influence over the genetic components available for the next generation. Students of human behavior have reported an enormous range of individual and cultural phenomena that affect length of lifespan (e.g., Palmore & Jeffers, 1971; Kastenbaum & Aisenberg, 1972). Take just one example: **genocide.** This term refers to the destruction of a group of people who have a strong common history and sense of affiliation. It also implies that they possess a shared genetic configuration that is somewhat distinctive. By this time in human history we have had enough examples and near-examples of genocide to cause much alarm. A statement against genocide was a major proposal in the early days of the United Nations, and it has been debated in the United States Congress. The political, economic, and psychological considerations involved in genocide make this topic too remote to be considered in detail here. However, it should be obvious that whatever hostile actions result in the elimination of a group of people with common ancestry must also result in an altered and diminished pool of genetic materials for the human race.

Genocide is the destruction of a group with a common genetic pool

Figure 3-4. *The human genetic pool: vulnerable to catastrophe. The total pool of genetic possibilities for the human race is reduced when groups of people with special characteristics are depleted (darkly colored areas) or destroyed (gray areas) by genocide or other catastrophes.*

Secondly, it is more than obvious that mating must precede conception. Is the combination of genes in mating a matter of chance? Is it not, rather, the case that love, sex, and mating are patterned by both individual and cultural styles of life? Whatever psychosocial forces encourage mating between certain men and women, while discouraging mating between others, thereby influence the "chance" match-ups of genetic elements. Should the wealthy mate with the poor? The Oriental with the Occidental? The Negro with the Caucasian? Should "blood relatives" mate? Should couples in which both partners have demanding professional careers remain childless? These are some of the questions that our society answers, but the answers may be contradictory at any point in time and are also subject to change. A general trend toward liberalized patterns of mating might have a marked effect on the genotypes of succeeding generations. The increased ease of travel, bringing with it an opportunity to meet a greater variety of possible mates, could be one of the conditions that alters the probabilities of who will mate with whom. Would this be a trend to welcome or resist? Value judgments enter into the picture, and it is doubtful that all readers of this book would answer all the questions raised in this paragraph the same way. What we are trying to convey is the point that psychosocial influences, which themselves fall into certain patterns, have an effect upon the particular pool of genetic materials that is available for chance to work upon at the time of conception.

Finally, we might recall that some people have proposed deliberate efforts to alter the available genetic pool of the human race. This has taken various forms, including sterilization of "deviant" members of society (a program that has not always been easy to differentiate from genocide). It is no longer sufficient to speak of a single movement for **eugenics** (literally, "good" or "happy" genes). The nature of its advocates has changed with time as has the background of scientific knowledge and political pressure. Again, my point here is a simple one: decisions and actions that would select out certain people as possible parents belong to the psychosocial sphere but have major implications for the biosphere.

Eugenics advocates improving the genetic pool by intervention

Having introduced psychosocial considerations into the realm of genetics, it is only fair to take the opposite course as well. From an environmentalist viewpoint, some genetic outcomes (phenotypes) may appear so puzzling that one would be tempted to say heredity has had only a small role to play. A common example is the child who looks and behaves like neither of his or her parents. Does this prove that genetic principles have suspended their operations? No, it is more likely that the offspring is expressing characteristics that have skipped one or more generations, although they were conveyed faithfully through the genetic materials. The child may resemble a grandparent more closely than either parent, perhaps even a great-grandparent. The child's total physical and behavioral constitution will be a unique combination of characteristics that may have been both expressed and "silent" for generations. He or she will not simply be a combination of various characteristics of the most immediate ancestors, father and mother.

What might appear to be an exception to the laws of genetic transmission will on closer inspection prove to be an instance that could have been predicted with a certain level of probability (at least theoretically). More precise predictions are easier to come by when the mating pattern has been observed and controlled for generations, as is the case in animal husbandry and experimental genetics with lower forms of life.

Mutation

Mutation is a change in the genetic code between generations

Not all differences between generations can be explained in terms of differential probabilities of combination. It is known that **mutation** can take place. Changes within the DNA code itself can now be induced experimentally and may also occur spontaneously (a term that usually means we do not really know why they occur). Again, the most precise and comprehensive evidence concerns lower forms of life such as the fruit fly whose short life expectancy and modest demands have endeared him to generations of researchers.

Most mutations are disadvantageous to the individual and the species. However, mutations may also be responsible for whatever systematic change or evolution might be taking place. In humans, it is extremely difficult to identify mutations with any degree of confidence. This means that we do not know what the mutation rate is for our own species. Perhaps the rate is fairly high. We are complex and adaptive enough to survive many alterations in our external environment; the same *might* be true of our tolerance for genetic mutations. With the present growing, but still quite limited, state of knowledge in both the behavioral sciences and genetics, there is much room for controversy here. If we see differences in human behavior, does this mean that mutations are taking place, or that the environment is calling forth a different pattern of adaptive responses from us or, perhaps, something more complicated than either alternative?

Behavioral genetics studies genetic endowment, personality, and behavior

Behavioral Genetics

The discussion has now veered in the direction of behavioral genetics. Let's follow this new path just long enough to acquaint ourselves with its existence. It is one of the least-traveled paths in the study of humans developing, but one that hopefully will lead us to significant discoveries.

How much does genetic endowment contribute to behavior, temperament, and ability? This question is more difficult to approach than the inheritance of such physical features as eye color. To begin with, most of us are ready to concede that eye color is likely to be transmitted genetically. The research evidence is good. We do not argue that environmental differences have gifted one person with blue eyes, another with brown. But there is more room for doubt about the contribution of heredity when we consider complex psychosocial functions. We know that social interaction and other environmental forces do affect personality and behavior, so the analogy with eye color does not hold up. Equally important, the *outcome variable* is often more difficult to specify and measure when we are concerned with behavior. Eye color can be determined at a glance, not so for intellectual capacity, the tendency to be extroverted or introverted in personality style, or the predisposition to a certain form of mental illness. Psychological dimensions of development frequently are more difficult to identify and assess than the physical. The situation becomes even more problematical when strong personal and social attitudes enter into behavioral genetics. A relevant example is the ongoing controversy about possible relationships between race and intelligence. It is difficult to maintain perspective and objectivity on this topic.

Identical twins are two people who develop from the same zygote

Identical twins. Even with all these limitations, however, it is still possible to make headway. Studies of twins can be especially informative. We might expect identical twins to resemble each other psychologically more than fraternal twins (a pair of offspring born to the same mother, at the same time, but who develop from different zygotes). In general, results agree that identical twins do have a greater mutual resemblance. This has been found in a small study of infants as young as one year (Freedman, 1965), in children in the three to six age-range (Brown, Stafford, & Vandenberg, 1967), in older children (Scarr, 1966), and in adolescents (Vandenberg, Stafford & Brown, 1968; Gottesman, 1966; Nichols, 1965). But that is only part of the story. Environmental influences also make their mark. There are times when identical twins are separated at birth or soon afterward and reared in different homes. It has been shown that under these conditions, the twins resemble each other closely in their expressive movements—gestures, facial expressions, etc.—but do not resemble each other in their interests and attitudes as much as identical twins who have grown up together (Juel-Nielsen, 1965).

A closer look at the development of identical twins would be instructive. Although Twin A and Twin B both inherit the same genetic code, they are not necessarily absolutely equal even at the very beginning of life. As Tanner observes:

> When the single egg divides to give identical twins, it is unlikely that exactly equal amounts of cytoplasm go to each half. It is therefore unlikely that exactly the same concentration of chemical reactants will be formed in the two organisms. During subsequent development these differences could become progressively multiplied. Then, as growth continues, the two organisms are affected differently by the environment, for their positions in the uterus and their blood supplies are never quite the same. Finally, after birth, even under favorable circumstances of upbringing the two children are never identical in their total environment, for their food habits are never quite the same, their illness experiences never precisely similar. . . . (Tanner, 1970, p. 125).

It is likely, then, that prenatal influences are already shaping Twin A and Twin B somewhat differently, even though they have as much in common as any two can have in the way of genetic endowment. Recent studies also indicate that the role of genetic influences on behavior depends quite a bit on the kind of behavior that is being observed. Even within the realm of intellectual functioning alone, performance on some mental tests indicates a

These two lively creatures both began as the same egg, but they are not in any sense interchangeable. Identical twins usually have much in common, but still develop their own unique personalities. (© 1978 Photo by Thomas S. Wolfe.)

very strong contribution from inherited components, while only a moderate contribution is indicated on other mental tests (Osborn & Gregory, 1966). Furthermore, Bayley (1970) points out that the relationship between inherited components and behavior may appear different at different age-levels:

> . . . there is a clear indication that personality, or nonintellective behaviors, are also genetically determined. If this is so, then the correlations found at later ages between mental abilities and behaviors may result from genetically determined tendencies to react in characteristic ways to the environment. In other words, genetic determiners of mental abilities may be based not purely on capacities to carry on varying degrees of abstract thinking, but they may operate through inborn response tendencies which are manifested more generally as behaviors which in themselves are not classed as "intelligence." (p. 1187).

Arrested and pathological development. Sometimes clear evidence of genetic influence on behavior is easier to obtain in children who exhibit arrested or pathological development than in children who are growing up more or less normally. Mental retardation is an example. Several forms of mental retardation have been shown to be related to specific genetic problems. **Phenylketonuria (PKU),** a metabolic condition leading to retardation, results from inheriting a certain recessive genetic trait from both parents. The child will escape PKU if either parent is free from this trait, but he or she may eventually transmit this genotypic characteristic to his or her own offspring. The parents themselves usually do not suffer from the severe

PKU is a metabolic condition that can cause mental retardation if not treated early

retardation often observed with PKU because they also have a dominant genetic trait that has blocked expression of the recessive trait in their own constitution.

Much has been learned about PKU in recent years, and it has been possible to apply some of this knowledge to human welfare. The genetic mechanism is fairly well understood; early diagnosis and treatment is possible. The interested reader can now be well informed about PKU and other forms of mental retardation that are related to genetic problems (Melnyk & Koch, 1971). Perhaps some day we will know as much about the contribution of inherited factors to *normal* development!

Let's now take a single example from developmental psychopathology. *Schizophrenia* is the term given to a somewhat vaguely defined group of personality and behavior disorders. The question here is whether or not schizophrenia has a genetic base—a question that has been studied and argued for many years. One of the most careful biologists judges:

> Inheritance does no more than provide a strong predisposition toward schizophrenic breakdown. The trigger that sets the process in motion is as obscure as the nature of the process itself. It is reasonably certain that in some families, at least, a single dominant gene is necessary but it is equally clear that the primary effect of the gene does not produce schizophrenia. From the identical twin results, the secondary trigger factors cannot be genetic. There have been suggestions that somatic mutation and auto-immune processes* may act as a trigger and there is the possibility that something absorbed from the intestine could act in a way remotely like the action of barbiturate in a person with latent porphyria.† By far the most popular suggestion is that the additional factor needed is social and psychological stress but, again, there is no light thrown on either the trigger or the basis of the developed disease. (Burnet, 1971, p. 171).

This interpretation is similar to the conclusion reached by psychologists who have carefully reviewed behavioral genetics. Lindzey, Loehlin, Manosevitz, and Thiessen (1971) believe that, "While a predisposition seems to be inherited, environment (in some form) obviously plays a large role in determining whether it will be expressed as clinical schizophrenia" (p. 64).

In other words, one of the most prevalent and severe forms of personality disorder is perhaps best regarded as the outcome of interplay between genetic and environmental forces. We cannot fully explain schizophrenia by concentrating upon either nature or nurture (and we have trouble enough even when we are willing to consider all the possible factors). Again, it would be valuable to know the precise behavioral genetics operative for people who do *not* develop schizophrenic reactions, but the "breakthrough" seems more likely to emerge from research activity in the psychopathological area.

* The auto-immunity system is the way in which the body sorts out cells that "belong" from those which should be treated as undesirable aliens. Unfortunately, errors can be made in either direction.

† Porphyria is a disorder in the production of blood that can produce mild to severe abdominal symptoms. A person may have the genetic predisposition for this abnormality, but the condition itself remains latent or silent unless released by some external influence.

LIFE BEFORE BIRTH: THE PATTERN OF PRENATAL DEVELOPMENT

We have acquainted ourselves with some of the genetic and psychosocial contributions to the prenatal scene. Now it is time to focus upon the sequence of events that takes place between conception and birth. Three phases of development are usually recognized: the zygote, the embryo, and the fetus.

THE ZYGOTE

Until the moment of conception, the sperm and the ovum are two independent specks of living matter. The same may be said of the man and woman who are mating. Scientists customarily omit any reference to human love and sexuality when they are discussing processes at this microscopic level. But tradition should not deprive us of the opportunity to appreciate the admittedly imperfect analogy between sexual love and the biogenetics of fertilization. Lovers often seek to "unite" with each other, to "dissolve" in each other's arms, to "lose" themselves in the fulfillment of desire. Poets and psychoanalysts have played many variations on this theme. Sexual love certainly represents one of the few ways in which two independent adults can approach the experience of merging, of becoming one. After the sexual act, the experience becomes a memory. They are two independent adults again, although they share a continuing tide of feelings between them.

From this perspective, the rendezvous between sperm and ovum might be rated as more efficient, complete, and satisfying than the act of sexual love itself. The chance factor, of course, is whether or not the rendezvous will actually take place, and which of several million sperm will consummate the affair with the solitary ovum. The act of fertilization requires penetration of the ovum by the sperm, an obvious parallel to the sexual interaction on the part of the man and the woman. The crucial difference is that the sperm does "lose itself," the male and female elements do unite, and a cellular structure different from either of its microscopic "parents" is created. The nucleus of the sperm and the nucleus of the ovum are one.

Both the sperm and the ovum contribute their sets of twenty-three chromosomes to the new organism. These sets line up in matching pairs. Then each chromosome splits itself lengthwise. The halved chromosomes move to opposite sides of the cell. A membrane begins to form between the two halved sets—this is the process of mitosis at work. When this process is completed, two separate cells exist, each with a complement of forty-six chromosomes. The division of chromosomes and the multiplication of cells continue throughout the zygote phase. The first division of cells usually takes place within thirty-six hours, or less, after fertilization.

Growth of the zygote through mitosis is not the only significant event that occurs during this germinal phase. Another crucial process is the zygote's

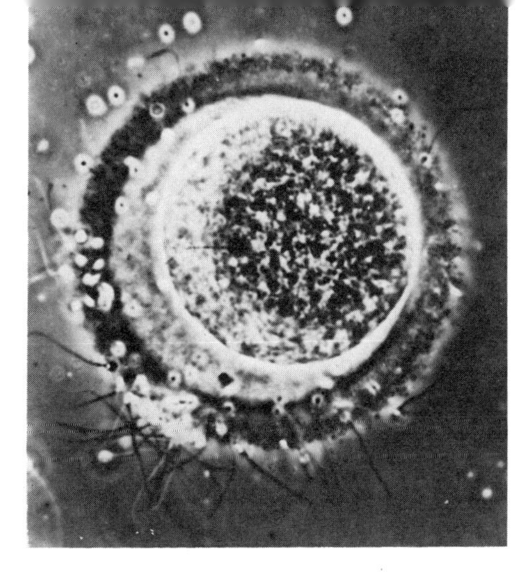

A living human ovum at the moment of fertilization. (Landrum B. Shettles, M.D.)

journey. For approximately two weeks, the fertilized and enlarging egg travels slowly inward from the oviduct to the uterus. During this first phase, then, the new organism is a traveler, not firmly attached to any fixed place within the mother's body. While enroute to its destination, the zygote develops tendrils that will be used to attach itself to the wall of the uterus. Changes also occur within the uterus in preparation for the linkage. When the zygote arrives and attaches itself to the uterine wall, it will be able to receive nourishment from the mother (having previously relied upon the yolk of the ovum). This is known as **implantation.**

The zygote attaches itself to the uterine wall by **implantation**

Successful implantation concludes the first phase of prenatal development. From now on, the new individual exists in a completely dependent relationship to its mother-and-host.

THE EMBRYO

The tempo of development picks up greatly once the zygote has become implanted. During the embryo phase, the tiny organism takes on its first resemblance, if only remote, to human form. The essential parts of the body begin to take shape. Within approximately six weeks, the embryo will have enjoyed a phase of growth that endows it with the structures and functional systems that are required for subsequent prenatal and postnatal development.

The **amniotic sac** is the gelatinous fluid that protects the embryo

The **placenta** is a temporary structure connecting mother and fetus

Two of the crucial developments are temporary, however. The embryo and its mother jointly create the **amniotic sac,** a liquid-filled space that protects the embryo and provides a relatively constant temperature. Attached to the uterine wall is the **placenta** which consists of fetal and maternal portions. Each section has its own circulation; there is no communication between the fetal and maternal circulation systems. By the end of the second month the placenta occupies a third of the total uterine surface. Capillaries (very small blood vessels) of the mother and fetus are close enough to each other in the placenta to allow for exchange of oxygen, carbon dioxide, food, water, and inorganic salts. It is in this manner that the fetus is supplied with the nourishment essential for growth, and by which it can also eliminate the waste products of metabolism. Unfortunately, toxic agents can also be transmitted to the fetus through the exchange process in the placenta. These toxic agents include nicotine, alcohol, viruses, and drugs—substances that generally injure the fetus more than the mother. The *umbilical cord* connects the placenta and the fetus. It contains two arteries and one large vein. The amniotic sac and umbilical cord are essential for the growth of the embryo, yet neither structure becomes a permanent part of the individual.

Within the embryo itself, changes occur in a predictable sequence. In much later periods of the human lifespan it will prove difficult to specify that "A" always occurs before "B," etc. Although far from simple, the development of the embryo does provide us with a clearer outline of growth than what is to occur later.

One of the first developments is the *differentiation* of the *zygote* into several types of cells. *Differentiation* is a term to remember. At the moment I am using this term to describe what happens when an individual that began with a single type of cell starts to come forth with a variety of cell types. Further along the lifespan, there will be occasion to speak of the differentiation of perceptions, thoughts, abilities, motives, and so forth. While the embry-

In just twelve days—from day thirty-three to day forty-five—the embryo loses its tail, develops arms, fingers, legs, and toes, and begins to form its facial features. (Courtesy of C. Paul Hodgkinson, M.D., Henry Ford Hospital.)

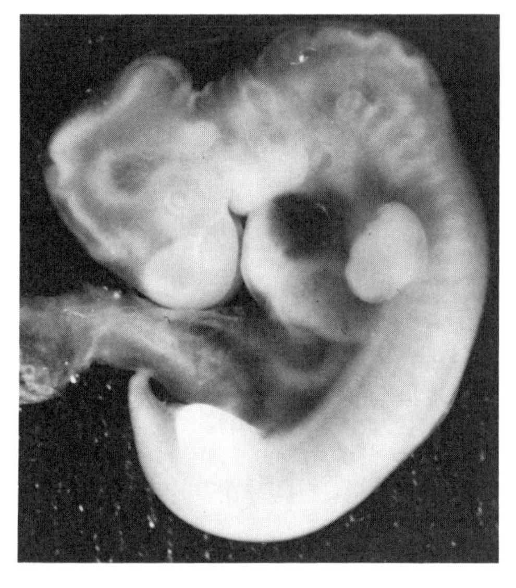

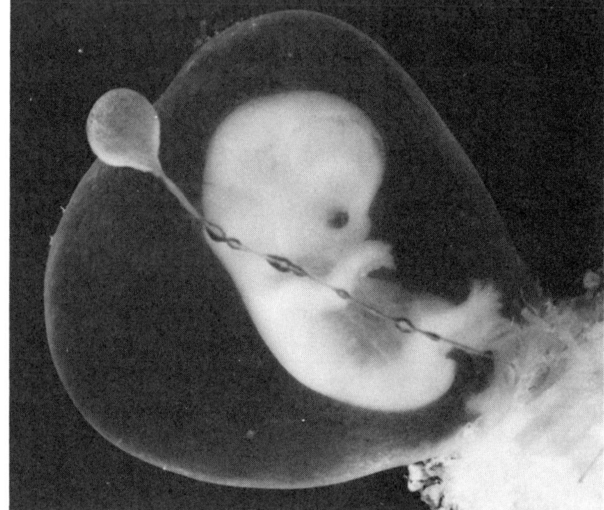

onic phase refers to a specific time and pattern of development, there are principles involved that can be applied more broadly. It is possible to be confused and misled by terms as general as *differentiation,* valuable as they can be when used properly. We ought not fall into the habit of using them as mere verbal habits. Fortunately, the process of differentiation during the embryonic phase is clear enough to move us off to a good start.

The first differentiation occurs when the inner mass of cells differentiates from the outer mass of cells in the zygote. Next, each of these groups of cells undergoes a separate process differentiation. The outer cells develop the special characteristics that are necessary to form the vital, but temporary, structures we have already mentioned, fetal sac and umbilical cord. The inner cells form themselves into three distinct layers. The outer layer, or *ectoderm,* eventually becomes skin, hair, nails, sensory receptors, and nervous system. The *mesoderm,* or middle layer, differentiates into heart, kidneys, muscles, and skeletal components. The innermost layer, the *endoderm,* gives rise to lungs, liver, digestive tract, and other internal organs.

Another important principle can be noticed during the embryo phase of development. It is the original "Operation Headstart." The embryo actually does enter its developmental career head first. The head region grows more rapidly and is etched with finer detail than the rest of the body. Other vital areas of development are also noticeable (e.g., there is a functioning heart by the third week), but the head-first pattern is dominant. The **cephalocaudal** (literally: head-to-tail) **principle,** as this pattern is sometimes known, holds true for other species besides our own. Like differentiation, it is a pattern that can be observed in many different organisms, especially during their earlier phases of development.

The **cephalocaudal principle** states development proceeds from head to tail

After about six weeks as an embryo (and about two months altogether as an organism), the "protagonist" has reached the size of about one inch. A number of organ systems are functioning, but there is not yet any substantial cooperation between the nervous system and the musculature. There is evidence to suggest that the embryo can twist itself around, and contract its arms and legs; however, these appear to be uncoordinated movements. For all of its progress, the embryo cannot yet be said to engage in *behavior* or *action* in anything that approaches the usual sense of these terms.

THE FETUS

The embryo becomes a **fetus** two months after conception

The final phase of prenatal development is also the longest—approximately seven months. It is interesting to reflect upon the fact that the *rate* of development actually slows down during this phase. Important changes continue to occur. The fetus becomes much larger, and the various organs and areas of the body re-adjust their proportions relative to each other. Nevertheless, if we were marking the tempo of prenatal development with a metronomic device, we would have to adjust it to a less hurried pace. Many people assume that development slows down after adolescence. In fact, the rate diminishes around the fourth month of prenatal life.

One of the most significant dimensions in the study of humans developing can be observed clearly at this stage, *individual differences.* C. E. Walters (1964) has found impressive differences in the activity levels of fetuses. Some move around as little as 5 percent of the time, some as much as 75 percent. One fetus is a twister and turner, while another is a kicker and thruster. There are even hiccupers and nonhiccupers among the fetal set. At least one study has found a relationship between the amount of activity observed during the later part of the fetal phase and the level of developmental progress observed nine months after birth (Walters, 1965). Obviously, we cannot start too soon if we are interested in tracing the origins of individual differences.

Regardless of its particular activity level, a fetus will develop both highly specific and more general or diffuse action-patterns. The very specialized action-patterns are known as *reflexes.* Sucking and swallowing are among the reflexes that are usually established about halfway through the total prenatal period. Diffuse or mass actions appear initially to be under the control of stimuli that are external to the fetus. Something happens, and the fetus responds by a combination of uncoordinated turning, thrashing, flexing, etc. Later, the fetus seems to "turn himself or herself on" in the absence of any obvious external stimulus. Still later, the diffuse action-pattern starts to give way to more precise movements. This could be regarded as another form of differentiation. But this differentiation might be called *behavioral differentiation.* We are not dealing here with the biochemical changes involved when one type of cell is replaced by several types with more specialized structures and functions.

The "young" fetus seems to move everything it can move when there is either external or internal stimulation. The "mature" fetus has more ability to move one part of its body in one particular way. This is a kind of differentiation. One day this fetus may be an athlete or musician who has extraordinary control over body functioning (e.g., the pianist who can call upon each of his or her ten fingers to perform in precise independence of each other). The fetus obviously has a long way to go in gaining control over his or her body, but the beginnings can be observed during the later weeks of prenatal life. Behavioral differentiation at this point is related to physical changes, true. But differentiation itself can now be described at the behavioral level; the fetus is *doing* things.

Even at this very early period of human development, it makes sense to deal with phenomena at the level that is most appropriate to them without trying to reduce one to the other. The fetus has cells, tissues, and organs; these must be described and understood in their own terms. The fetus is also starting to exhibit behavior of a rudimentary sort; this must be described and understood on its own terms as well. It is the ambition of many biological and social scientists to discover ways of understanding the relationship between various levels of developmental phenomena (in this case, between physical and behavioral development of the fetus). This is an exciting prospect. However, it is unlikely that this ambition can be achieved if we look for crude shortcuts. One of the shortcuts that might tempt us is to pay attention only to biological *or* behavioral levels of development. I am suggesting that this temptation be overcome, and a good place to begin is right here in the prenatal scene.

Perhaps the most important checkpoint in fetal development is the twenty-eighth week of pregnancy. At approximately this time, the fetus is *viable:* it could survive in the external world if given appropriate care. Every additional week the fetus spends in its intrauterine environment, however, will be advantageous to it. Improved muscle tone, periods of definite alertness, ability to cry, grasp, and move the head are among the characteristics that develop during the final prenatal weeks.

PLACENTA DETERMINISM: FROM MOTHER TO CHILD

People who are critical of the psychoanalytic approach to human development (this will be discussed in later chapters) sometimes refer to it as "diaper determinism." By this, they mean that the adult's personality is regarded as nothing more than a working-out of the dynamics established while he or she was being toilet trained! This point of view is not a really accurate repre-

Attitudes learned during a mother's own childhood may not pass through the placenta as such but nevertheless can exercise a developmental influence. (Photo by Talbot Lovering.)

sentation of psychoanalytic theory, but it is true enough that many students of human behavior do believe that early experience can have far-reaching effects.

Do these early experiences that can have far-reaching effects include the prenatal state? Expectant mothers (and fathers) have been bedeviled over the centuries by superstitions about the effect maternal experiences can have upon the fetus. There seems to be a more sophisticated set of concerns these days: Does smoking or the use of alcoholic beverages during pregnancy affect the baby? What about nutrition? Emotional stress? Questions such as these are on their way to being answered by a variety of research teams. I will summarize some of the more important findings, but the alert reader will want to keep his or her knowledge up to date as new findings continue to appear.

During the prenatal period, the mother's experiences can be shared most directly with the fetus through the umbilical cord. There is *no* structural connection between the nervous systems of mother and fetus. The answer to many questions about possible nongenetic influences on the fetus thus depends upon discovering what can and what cannot get through the placenta and what effects, if any, result. It is in this sense that we might introduce the term, *placenta determinism*. I will not be arguing that adult personality is the outcome of placental input, any more than most psychoanalytic writers believe that toilet training is the only important influence. More simply, the proposition is that life experiences (as contrasted with strictly genetic endowment) begin to affect human development before birth—and never let up afterward.

NUTRITION

There is rapidly increasing evidence that what the mother-to-be eats and drinks can have significant influence on the condition of her fetus. The influence has been demonstrated most strikingly where maternal diet has been very inadequate. The fetus cannot shop around for nourishment. If the pregnant woman herself is lacking proteins in her diet (to take one of the most critical nutritional components), the infant's mental development can be adversely affected (Davison & Dobbing, 1966). Those who live in poverty, then, may transmit normal genetic endowment to their offspring, but bring into the world a sickly or slow-developing child because adequate nourishment was lacking during prenatal development. Economics, politics, and social opportunity, thus, can be seen as dimensions of human experience that range, literally, from womb to tomb, especially when we remind ourselves that stillbirth and infant mortality are more frequent among the hungry and impoverished. Provision of more adequate nutrition after a poor start has been associated with improvement in both mental and physical status (Winick, 1972), although some of the data are still controversial (Dennenberg, 1977; Winick, 1977).

Maternal nutrition may be inadequate even when poverty is not involved. An ill-advised diet, for example, can deprive both the mother and the fetus of necessary vitamins (Scheinfeld, 1965). It is possible that much more subtle influences of maternal nutrition upon the fetus will be discovered in the future.

TOBACCO, ALCOHOL, AND DRUGS

Nicotine passes through the placenta to the fetus. It has been known for awhile that smoking during pregnancy produces temporary acceleration of fetal heart rate and activity level. But there is no direct evidence to indicate that this stimulation has lasting effects on the prenate. It has also been recognized that heavy smoking is related to premature birth (Simpson, 1957), and to smaller birth size and weight (Guttmacher, 1962). Both premature delivery and lower weight are factors associated with higher mortality soon after birth (Annis, 1978). Lags in postnatal development also are found more often in babies whose mothers smoked during pregnancy (Annis, 1978). Fortunately, there is also evidence that women who are relatively light smokers during their pregnancy or who give up smoking after the first trimester reduce the risks to their infants (Butler, Goldstein, & Ross, 1972). The research is fairly clear and consistent that cigarette smoking can be harmful to the development of the infant before birth and also impede development to some extent afterward.

Alcohol also passes through the placenta and produces temporary acceleration of fetal heart rate and activity. However, no adverse results have been found when the mother uses alcohol moderately during her pregnancy. Although smoking and drinking are often lumped together as habits, their psychophysiological effects are different. Furthermore, different types of alcoholic beverages also have different effects. There is, for example, some nutritional value in alcohol that is not found in nicotine. Concerning alcohol usage, the advice of Alan F. Guttmacher appears quite appropriate:

> There is no logical reason to prohibit the moderate use of alcohol during pregnancy to the patient who enjoys and tolerates it. Alcohol diffuses rapidly across the placenta and soon attains equilibrium in the two circulations. This knowledge may dissuade the pregnant woman from taking just one more, since she cannot know and never will know whether her fetus will enjoy the additional drink as much as she thinks she may. (1962, pp. 89–90).

It is nearly impossible to discuss in detail the effects drugs have upon the fetus. The number of drugs consumed by the American public is vast; they are used in various combinations and dosages, and new preparations continually appear. Perhaps awareness of possible adverse effects of drugs on the unborn has been heightened since the tragic results of thalidomide have become known. This drug was responsible for many babies being born without limbs or with deformed limbs. Physicians, pharmaceutical houses, and governmental regulatory agencies are aware of their responsibilities in this area. The pregnant woman herself can reduce possible risk by avoiding self-medication, and by checking for possible error in medications prescribed and dispensed to her.

PSYCHOLOGICAL STRESS

The woman who is under emotional stress during her pregnancy may change her smoking, drinking and drug-usage habits and, through these mechanisms, inadvertently affect the fetus. Can stress affect the fetus in other ways? Yes. Emotional arousal releases various chemicals into the blood stream, and these can find their way through the placenta. Some of the effects have been known for many years (Sontag & Wallace, 1935). Short term, the fetus tends to become overactive and irritable when the mother is emotionally upset. Long-term effects are more difficult to evaluate. Informed opinion, however, considers that prolonged or severe stress during pregnancy is likely to have some effect upon the child long after the prenatal phase (Ferreira, 1965). Whatever causes distress to the pregnant woman, then, has the possibility of indirectly creating problems for the child, although much remains to be learned about the precise relationships.

TIMING

Several types of deprivation and stress have been described. These may occur in combination, as well as separately as in the case of a pregnant woman with poor nutritional status who smokes heavily and is under severe emotional stress. It is also important to know *when* the stress or deprivation occurs. As we have seen, the unborn child is quite different both structurally and functionally at various phases of development. The same type of deprivation or stress is likely to have different results at different fetal ages. The same holds true for diseases that can be transmitted from mother to child (e.g., Rubella, or "German Measles," in which the first trimester of pregnancy is the most critical period).

HAZARD: ABORTION

The hazards of life are with us before birth as well as after. It would be naive to pretend that development is a process in which only the expected and desired events take place. Throughout this book I will consider examples of hazards that are worth keeping in mind at various points of the human life cycle.

These hazards differ greatly, but all involve some mixture of psychosocial and biological variables. During the prenatal period, perhaps the most salient hazard is **abortion.** This term has a clear meaning to physicians and scientists. Abortion refers to the evacuation of the prenate from the uterus before it is viable. Whether this takes place spontaneously or by induction makes no difference so far as the definition goes. *Miscarriage* is a popular term for spontaneous loss of the fetus. We will remain with the medical usage: abortion refers to both spontaneous and induced loss at any time from zygote through the sixth month of prenatal life.

Abortion evacuates the embryo or fetus from the uterus before it is viable

During the germinal phase (0–2 weeks after conception), abortion sometimes occurs because the zygote has implanted itself in an area where it cannot receive nourishment. Another of the more common causes at this time is a glandular imbalance that prevents the uterus from preparing itself to receive the zygote.

The hazard is greater during the embryonic phase (3–6 weeks after conception) and for the first few weeks of the fetal phase. The embryo can become dislodged from the uterine wall for a variety of reasons. Defective genes and other physical defects can have this result, so can severe emotional stress, malnutrition, glandular disturbances, and diseases. With so many possible sources of vulnerability, the first trimester (three months) of pregnancy appears to be the most hazardous phase of prenatal development.

During the fetal phase (7–27 weeks after conception), the prenate is less vulnerable to abortion. Deprivations and stresses that might have induced abortion at an earlier phase now either pass without harm or create nonlethal problems and defects. The hazard does remain, however, although the odds become increasingly favorable against abortion as time (and development) go on.

C. T. Javert maintains that "Spontaneous abortion is not only the most frequent complication of pregnancy, but is also the country's foremost health problem" (1957, p. 5). In one year, about 400,000 spontaneous abortions—more than 1,000 each day—occurred in the United States. "The high fetal loss due to abortion ranks it with cardiovascular conditions and cancer as the leading causes of death. . . . Spontaneous abortion affects fully half the adult population of child-bearing age, which makes it our greatest public health problem in terms of incidence" (p. 5). It should be noted, however, that many spontaneous abortions involve abnormal fetuses, so the prevention of spontaneous abortion itself is not the only problem involved.

What has been said up to this point does not take into account the number of induced abortions in this nation, estimated at upwards of one million per year (Kimmey, 1969). Those who induce abortions obviously have different attitudes than those who view prenatal mortality as a major public health problem. All the statistics and many public and private attitudes on this subject can be expected to change in response to the Supreme Court's decision to legalize abortion and efforts both to utilize and to counteract the effects of this decision.

YOUR TURN

There are many ways to apply, test, and expand your understanding of the prenatal scene. A few suggestions are:

1. Variants on Scenario 2: The Prehistory of Michael

- ***Variant A.*** Suppose Gloria and Tom are not married at the time of conception. How might this affect Michael's developmental career? Consider the possible alternatives and trace them through in light of the material that has been presented throughout this chapter and any other observations you can add.
- ***Variant B.*** Suppose Gloria and Tom's situation is essentially as described in Scenario 2, except that they are in great financial difficulty. Neither of them have money, Tom has trouble finding a job, and there are no relatives able to help them financially. How might this change in the financial situation affect Michael's prenatal development?
- ***Variant C.*** Suppose Tom's side of the family is extremely concerned about having a male heir. How might this affect the marital relationship, and Michael's development, especially if "he" turns out to be a girl?

2. Much has been learned about the prenatal phase of development from the standpoint of the prenate itself. But what phases of experience does the pregnant woman pass through? Or *is* there any particular sequence? Perhaps several mothers will be willing to share with you their experiences during pregnancy. When did they think they were pregnant? When did they know it for a fact? How did their attitudes toward the unborn child—and toward themselves—change over time? Was there, for example, an "inbetween time" as far as the woman's self-concept and body image were concerned, a time when she was not "pregnant enough" to change her activities and be recognized as expectant by others but when she was "too pregnant" to go through the day as usual? Come up with two or three further questions as you reflect upon the woman's experience of her pregnancy.

3. How does the total prenatal scene change when there are already several children in the family? Perhaps you have some friends with relatively large families, and perhaps they would be willing to share their experiences from the first pregnancy on through. What changes, if any, would you expect? What do you find is the actual case?

4. Do you plan to become a parent yourself? Under what circumstances? Try preparing an ideal scenario that describes how you hope your first experience of parenting begins. What factors are of the most importance to you? Can you foresee any particular hazards? How well acquainted are you with the whole range of attitudes and feelings you will bring to the prenatal scene? You may find a number of fresh questions coming to mind. Can you find ways of answering them as well?

From the viewpoint of the pregnant woman, an unwanted child can be regarded as a hazard. If we credit the prenate with a viewpoint, it is still impossible to know whether it would prefer death to a world of rejection, but either prospect would also seem to qualify as a hazard.

SUMMARY

The prenatal scene has two major settings: the specialized developmental chamber provided by the mother's body and the larger human environment with its expectations, resources, and hazards. In this chapter I first explored some characteristics of the larger prenatal scene. The prehistory of a child was sketched (Scenario 2), with emphasis upon the parents-to-be. Additional perspective was gained through a variant in which conception occurred at a later time than originally planned. The general point was made that attitudes, thoughts, expectations, and family resources establish an orientation toward the child even before he or she is born and, in a sense, even before conception.

This point was further illustrated with a discussion of the effects of *birth order.* Research in the area suggests that firstborn and laterborn children enter different family situations and may subsequently differ in personality, intellect, and achievement. From the early observations of *Adler,* I proceeded to the recent large-scale data analyses of *Zajonc* who has attempted to determine the *family intellectual climate* and relate it to birth order and family size. *Socioeconomic class* was also seen to be an important aspect of the prenatal scene; particular attention was given to the effects of poverty upon the well-being of the prenate.

Turning to the genetic side of the story, I laid out the basic vocabulary: *ova, sperm, chromosomes, genes, DNA, nucleotides, mitosis, zygote, fetus, embryo, genotype, phenotype, penetrance, dominant* and *recessive* traits, *eugenics, mutation.* In exploring the genetic match-ups that determine the inherited characteristics of a particular child, I noted that some outcomes are quite predictable while others are very difficult if not impossible to predict. Much of what is predictable is related to *Gregor Mendel's Laws* that: (1) genetic strains are preserved through the generations, and (2) genes from the two parents do not blend or compromise. Much of what is unpredictable is related to psychological and environmental influences as well as to the interaction of genetic complexity and "chance" factors.

The relatively new science of *behavioral genetics* was sampled by way of its contributions to the study of *twins, mental retardation,* and *schizophrenia.* Sophisticated research in this area is helping us to realize that the *interaction* between inherited characteristics and environmental influences is of prime importance and must be taken into consideration from the very beginnings of life.

I next followed the pattern of prenatal development, from the union of sperm and ovum through the stages of zygote, embryo, and viable fetus. The concept of *differentiation* was introduced to describe some of the most significant prenatal developments. Consideration was also given to the *cephalocaudal principle* and *rate of development, individual differences,* and *reflexes.*

Under the heading of *placenta determinism,* I examined some of the ways in which the mother's experiences can be shared with the prenate. Attention was given to *nutrition, tobacco, alcohol, drugs,* and *psychological stress.*

Finally, I explored one of the major hazards of this developmental phase, *abortion.* Emphasis was upon spontaneous (involuntary) abortion, known generally as *miscarriage.* The intermix of attitudes, values, and facts was noted.

Reference List

Adams, R. L., & Phillips, B. N. Motivational and achievement differences among children of various ordinal birth positions. *Child Development,* 1972, *43,* 155–164.

Allen, V. L. Theoretical issues in poverty research. *Journal of Social Issues,* 1970, *26,* 149–167.

Annis, L. F. *The child before birth.* Ithaca, N.Y.: Cornell University Press, 1978.

Ansbacher, H., & Ansbacher, R. *The individual psychology of Alfred Adler.* New York: Basic Books, 1956.

Bayley, N. Development of mental abilities. In P. H. Mussen (Ed.), *Carmichael's manual of child psychology* (Vol. 1, 3rd ed.). New York: John Wiley & Sons, Inc., 1970, pp. 1163–1210.

Belmont, L., & Marolla, F. A. Birth order, intelligence, and family size. *Science,* 1973, *182,* 1096–1101.

Birch, H. W., & Gussow, J. D. *Disadvantaged children: Health, nutrition, and school failure.* New York: Harcourt, Brace & World/Grune & Stratton, 1970.

Brown, A. M., Stafford, R. E., & Vandenberg, S. G. Twins: Behavioral differences. *Child Development,* 1967, *38,* 1055–1064.

Burnet, M. *Genes, dreams, and realities.* Aylesbury, Bucks (England): Medical & Technical Publishing Co., Ltd., 1971.

Butler, N. R., Goldstein, R. H., & Ross, E. M. Cigarette smoking in pregnancy: Its influence on birth weight and perinatal mortality. *British Medical Journal,* 1972, *2,* 127–130.

Davison, A. N., & Dobbing, J. Myelinization as a vulnerable period in brain development. *British Medical Bulletin,* 1966, *22,* 40–45.

Dennenberg, V. H. Letter in *Science,* 1977, *197,* 1134–1135.

Ferreira, A. J. Emotional factors in prenatal environment. *Journal of Nervous and Mental Diseases,* 1965, *141,* 108–118.

Freedman, D. An ethological approach to the genetical study of human behavior. In S. G. Vandenberg (Ed.), *Methods and goals in human behavior genetics.* New York: Academic Press, 1965, pp. 141–161.

Galton, F. *English men of science: Their nature and nurture.* London: Macmillan, 1874.

Gottesman, I. I. Genetic variance in adaptive personality traits. *Journal of Child Psychology & Psychiatry*, 1966, *7*, 199–208.
Guttmacher, A. F. *Pregnancy and birth*. New York: New American Library (Signet), 1962.
Hilton, I. Differences in the behavior of mothers toward first- and laterborn children. *Journal of Personality and Social Psychology*, 1967, *7*, 282–290.
Javert, C. T. *Spontaneous and habitual abortion*. New York: McGraw-Hill, 1957.
Jayant, K. Birth weight and some other factors in relation to infant survival: A study on an Indian sample. *Annals of Human Genetics*, 1964, *27*, 261–270.
Jones, H. E. Order of birth. In C. Murchison (Ed.), *A handbook of child psychology*. Worcester, Mass.: Clark University Press, 1931, pp. 204–241.
Juel-Nielsen, N. *Individual and environment*. Copenhagen: Munskgaard, 1965.
Kastenbaum, R. Fertility and the fear of death. *Journal of Social Issues*, 1974, *30*.
Kastenbaum, R., & Aisenberg, R. B. *The psychology of death*. New York: Springer Publishing Co., Inc., 1972.
Kimmey, J. *The abortion argument: What it's not about*. New York: Association for the Study of Abortion, Inc., 1969.
Kornberg, A. The synthesis of DNA. *Scientific American*, 1968, *219*, 64–78.
Lindzey, G., Loehlin, J., Manosevitz, M., & Thiessen, D. Behavioral genetics. In P. H. Mussen & M. Rosenzweig (Eds.), *Annual Review of Psychology*, 1971, *22*, 39–94.
Melnyk, J. M., & Koch, R. Genetic factors in causation. In R. Koch & J. C. Dobson (Eds.), *The mentally retarded child and his family*. New York: Brunner, Mazel, Butterworth, 1971, pp. 49–60.
Mussen, P. H., Conger, J. J., & Kagan, J. *Child development and personality*. New York: Harper & Row, Publishers, 1969.
Nichols, R. C. The resemblance of twins in personality and interests. In S. G. Vandenberg (Ed.), *Methods and goals in human behavior genetics*. New York: Academic Press, 1965, pp. 580–596.
Osborne, R. T., & Gregory, A. J. The heritability of visualization, perceptual speed and spatial orientation. *Perceptual and Motor Skills*, 1966, *23*, 373–390.
Palmore, E., & Jeffers, F. C. *Prediction of lifespan*. Lexington, Mass.: D. C. Heath & Company, 1971.
Scarr, S. Genetic factors in activity motivation. *Child Development*, 1966, *37*, 663–673.
Schacter, S. *The psychology of affiliation*. Palo Alto, Calif.: Stanford University Press, 1959.
Scheinfeld, A. *Your heredity and environment*. Philadelphia: J. B. Lippincott Company, 1965.
Schooler, C. Birth order effects: Not here, not now! *Psychological Bulletin*, 1972, *78*, 161–184.
Simpson, W. J. A preliminary report on cigarette smoking and the incidence of prematurity. *American Journal of Obstetrics and Gynecology*, 1957, *73*, 808–815.
Sontag, L. W., & Wallace, R. F. The effect of cigarette smoking during pregnancy upon the fetal heart rate. *American Journal of Obstetrics and Gynecology*, 1935, *29*, 3–8.
Stevenson, R. E. *The fetus and the newly born infant*. St. Louis, Missouri: The C. V. Mosby Company, 1973.
Tanner, J. M. Physical growth. In P. H. Mussen (Ed.), *Carmichael's manual of child psychology*. New York: John Wiley & Sons, Inc., 1970, pp. 77–156.

Terman, L. M. *Genetic studies of genius.* Palo Alto, Calif.: Stanford University Press, 1925.

Timiras, P. S. *Developmental physiology and aging.* New York: Macmillan, Inc., 1972.

Vandenberg, S. G., Stafford, R. E., & Brown, A. M. The Louisville twin study. In S. G. Vandenberg (Ed.), *Progress in human behavior genetics.* Baltimore: Johns Hopkins Press, 1968, pp. 153–204.

Walters, C. E. Reliability and comparison of four types of fetal activity and total activity. *Child Development,* 1965, *35,* 1249–1256.

Watson, J. D., & Crick, F. H. Molecular structure of nucleic acids—a structure for deoxyribose nucleic acid. *Nature,* 1953, *171,* 737–738.

Winick, M. *Malnutrition and brain development.* New York: Oxford University Press, 1972.

Winick, M. Letter in *Science,* 1977, *197,* 1135.

Zajonc, R. B. Dumber by the dozen. *Psychology Today,* January 1975, 37–43.

Zajonc, R. B. Family configuration and intelligence. *Science,* 1976, *185,* 227–236.

BIRTH AND THE FIRST THIRTY DAYS

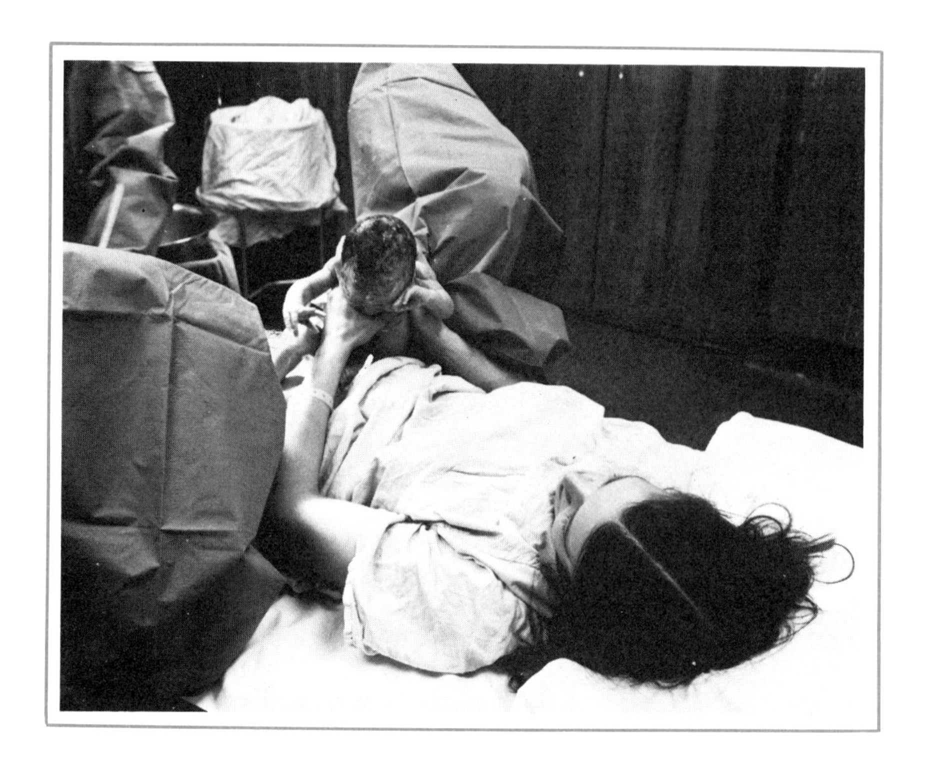

4

CHAPTER OUTLINE

I

The birth of a child has many meanings.

Communal event / Personal event and process / Rebirth / Beginning, ending, continuation

II

William James once described the newborn's world as "one great blooming, buzzing confusion."

What James really meant / Discrimination . . . comparison . . . global state

III

The neonate's repertoire of reflexes seems to be as close to unlearned behavior as anything we are likely to observe at any point in the human lifespan.

Yawning, sneezing, sucking, stretching and crying / Rooting, grasping, (palmar and plantar) / Toe flexion-fanning (Babinski) / Moro (startle) reflex / Crossed extensor, deep tendon reflexes / "Doll's eye," galant reflex, supporting and stepping reactions / Parachute response / Survival value of reflexes

IV

Reflexes may represent the unrestrained influence of a particular region of the central nervous system.

"Old brain" / Cerebral cortex ("new brain") / Umklammerungsreflex

V

Developmentalists increasingly are discovering evidence that the young infant perceives, learns, and prefers.

Seeing . . . differential response . . . visual acuity / Hearing / Learning and habituation / Classical conditioning / Unconditioned stimulus (UCS) / Conditioned stimulus (CS) / Behavioral engineering / Individual differences

VI

Postpartum depression and psychosis is a severe reaction that might exercise a strong influence on early development of the infant and the life-long pattern of mother-child interaction.

Psychosis / Developmental hazards

BIRTH HAS MANY MEANINGS

BIRTH AS A COMMUNAL EVENT

How should we think of birth? The keeper of vital statistics has one way of thinking about birth. On a certain day of the year, a male or female child is born to a certain set of parents in a particular township. Birth is one social statistic among many. It goes into the hopper along with the number of automobiles registered, the number of wedding licenses granted, and various other bits and pieces. As a social statistic, the birth of a particular child will be retrieved and used over and again: Has the balance between male and female births changed during the past century? How many of the babies who came into life in a particular area stayed there long enough to go through high school? Are there differences in population growth among various socioeconomic echelons? These are just a few of the questions for which the social statistics of birth have importance.

Still another purpose is served by the treatment of birth as a social statistic; it signifies a mutual possessiveness. There is a new person among us. He or she belongs to the community; the community belongs to him or her. Although it is not often looked upon this way on a conscious level, the certification of birth makes it clear that individual and community will be interdependent right from the start. Birth is a public as well as a private fact. As the newborn matures into an adult, he or she might develop a very private or independent kind of lifestyle. Nevertheless, right at the beginning, society has taken official notice of that individual's existence and will continue to insist upon its prerogative to regard him or her as a public as well as a private person.

BIRTH AS A PERSONAL EVENT

As individuals, most of us tend to think of birth in personal terms. Birth is an event, a happening. For the people most intimately involved, birth is a very special happening. But this does not mean that birth is the same kind of happening to all of those who are within the intimate circle. The men involved (husband, father, brother) are inclined to see birth more strictly as an event. One day they were concerned about a pregnant woman; the next day they had a mother and a child; the two situations were separated by the *event* of birth. For the mother herself, birth is more likely to be experienced as a *process*. It is not instantaneous. The moment of birth itself has been preceded by several stages of bodily preparation and experience. Furthermore, both the bodily and psychological states associated with birth continue to some extent after delivery. The woman does not immediately snap back to "normal." Birth, then, is a process and an experience that extends in time before and after the moment of delivery **(parturition)** itself. This side of birth may be appreciated by other women in the family who have experienced childbearing.

Parturition is the birth process

Some men make more of an effort to participate in the woman's experience or seem naturally more empathic to it. While it is highly doubtful that a man can approach the woman's experience of bringing forth a child, there are husbands who have learned to appreciate birth as a process. Perhaps you know a man who has accompanied his wife to parenthood classes during her first pregnancy, and who has also been present through labor and delivery. This kind of participation is not to every family's preference, and some physicians and hospitals forbid the husband's presence during delivery. But it cannot be denied that a more intimate sense of sharing in the birth process can be achieved than by the traditional route of father-to-be remaining out of the picture until he is "presented" with a "little bundle."

Today more and more fathers-to-be are becoming intimately involved in the entire process of childbearing and childrearing. (Photo by Daniel Bernstein for Newton-Wellesley Hospital.)

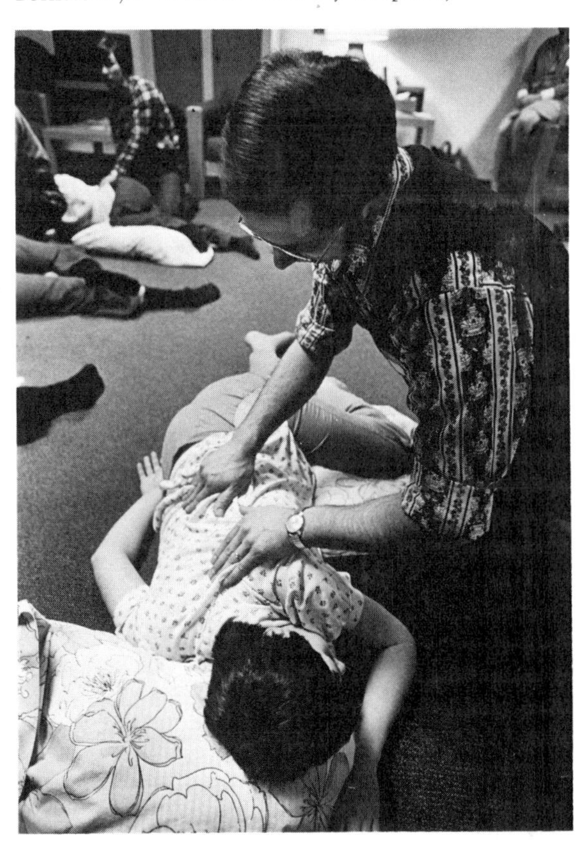

BIRTH AS AN IDEA

Birth has other meanings as well. It is not only a statistic, an event, and a process; birth is also an *idea*. In our culture (and in many others) birth is used to symbolize something that is emerging, coming forth or into being. This thing that is emerging can be a new art form ("The Southland gave birth to the blues") or a political entity ("A new republic was born in 1776"), to mention just two examples. Some of the most striking applications of this idea can be seen in religious conceptions of *rebirth*. It is seldom that religion intends a literal, physical rebirth when the term is used. Rather, the emphasis is on a spiritual awakening. Rebirth can be applied to individuals or to groups and social forms: "What this country needs is a rebirth of moral vigor"; or "After years of mediocrity, his creative spirit was reborn when he met his new love." Some people who contemplate suicide harbor the belief that this act will give them the opportunity to be reborn, to start over a life that has turned out to be disagreeable.

Usually the idea of birth is associated with a *beginning*. The birth of an infant is, of course, a beginning. The neonate takes on a new status in society and in his or her family. Psychobiologically, the neonate is on his or her own for the first time (or, at least, separated physically from the support systems that maintained and nourished previously). There is no reason to quibble about birth as the starting point for a new mode of existence.

In some ways, however, birth is also an *end*. One mode of existence (the prenatal) has ceased to operate so another can begin. Additionally, certain sequences of thought, feeling, and action are now terminated among the in-

This birthing room in a modern hospital creates an atmosphere of repose and security instead of technology and strangeness. (Photo by Charles Dixon, The Boston Globe.*)*

timate circle. The period of waiting is over. One does not have to expect; one can now adjust to the reality of the infant. People feel and behave differently when they are oriented toward a something-that-does-not-yet-quite-exist and when they are actually in the situation. Uncertainty and possibility are replaced by a particular newborn with his or her very specific and definite reality. This shift is seldom thought about. You may find it helpful at times to remember that birth represents the termination of certain psychological patterns of the family-in-waiting, as well as the beginning of new patterns of the family.

It just remains to be said that birth, in some ways, is also a *continuation.* We can be understandably impressed by the transition from prenatal to neonatal status. So much has begun that is new. Some of what existed before has come to an end. Yet all is not comings and goings. The prenate was developing in the womb; the infant continues to develop. The continuity is more evident in some ways than in others. To take an obvious example, there is no great difference between the size of an infant one day before birth and one day afterward. He or she has been growing for months, and he or she will continue to grow. The genetic components of his or her development have also been with the baby from the start and will continue with him or her throughout life. This concept of continuation can be put still another way: there is a continuity in the individual's *identity* between the prenatal and postnatal states. He or she may receive a new name at birth, but all that the prenate has been continues through to postnatal status. The principle of continuity is stronger than the transition of birth.

We are now ready to examine the psychobiological characteristics that the normal infant exhibits at and just after the time of birth.

DOES THE NEWBORN'S WORLD BLOOM, BUZZ, AND CONFUSE?

There is a tradition in developmental psychology that weighs heavily upon me at this point. I am expected first to repeat one of the most famous phrases in the history of psychology and then to forget all about it and move into more current scientific observations. The phrase is one you might already know. The neonate has been said to experience his or her new situation as a "blooming, buzzing confusion." The man who said this was William James, the first great psychologist in the United States. If we obey custom and quickly pass over this phrase, we will have learned little from this exceptional thinker. Instead, we may ourselves be left in a state of blooming, buzzing confusion. Let's examine the context in which James introduced the phrase.

James was an original thinker and writer whose brother, Henry, did a bit of writing too

Back around 1890, the fields of psychology and philosophy were more closely related than we usually find them today. As both a psychologist and a philosopher, James felt at ease in exploring problems through this double perspective. His major work in psychology was the two-volume book, *Principles of Psychology*. In the first volume, James began to focus on the questions of *discrimination* and *comparison*. He reminded us that "some men have sharper senses than others, and . . . some have acuter minds and are able to 'split hairs' and see two shades of meaning where the majority see but one" (James, 1890, p. 483). But how do *any* of us develop the knack for discriminating any object from another? And how do we become able to analyze a single object into many specific qualities? These questions may sound simple enough, but the search for adequate answers still continues today.

James pointed out that even when we are sophisticated adults we may lapse into undiscriminating states. Under the influence of chloroform or nitrous oxide, for example, we may fail to distinguish between objects that are quite familiar to us and lose track of how many objects are before us. "For one sees light and hears sound, but whether one or many lights and sounds is quite impossible to tell" (p. 487). At times, we may also have spontaneous lapses into undiscriminating states.

Perhaps there was a time in our lives *before* we could distinguish the parts of our world. James suggested that developmental accomplishments could come unravelled; we could lapse back to a more primitive state in which we had less knowledge and competence. This is a point that is worth reflection. In our discussion of prenatal development, we became accustomed to the idea that maturation involves a one-way sequence. First the zygote, then the embryo, but never the other way around. Now we see a possible exception, although on the psychological level rather than on the biological level. A person has advanced from an undiscriminating to a discriminating state, but then he or she takes a step (or a whole flight of steps) backward. And then? Why, then he or she can once more advance to his or her former position before the lapse.

James seems to have been describing a sort of dynamic oscillation that is still not well understood today. It is relevant to the newborn because we would have to establish a sort of baseline or starting point of development if we were to hope to comprehend the zigs and zags that James implied. In other words, what state are we heading back to when we slip from a higher developmental plane?

But with his distinctive turn of mind, James proceeded at once to a different implication, "Where the parts of an object have already been discerned, and each made the object of a special discriminative act, we can [only] with difficulty feel the object again in its pristine unity; and so prominent may our consciousness of its composition be, that we may hardly believe that it ever could have appeared undivided" (p. 488). In other words, when we are being our experienced, sophisticated selves we cannot easily recapture the way the world appeared to us at an earlier time. Once we saw the world through innocent eyes and took it for being an undivided whole. Now, older and wiser, we cannot see that undivided world we experienced as a neonate. There is something lost, as well as gained, through experience.

If you pick up James's book and open it to the following quote, you will see that he became so excited he launched into both italics and caps:

> . . . the undeniable fact being that any number of *impressions, from any number of sensory sources, falling simultaneously on a mind WHICH HAS NOT YET EXPERIENCED THEM SEPARATELY, will fuse into a single undivided object for that mind.* The law is that all things fuse that can fuse, and nothing separates except what must.

Variety and distinction would be wasted on a mind that has not already experienced each element by itself. The virgin mind would fuse all the lights, colors, sizes, weights, aromas, touches, etc. into a unified mental scene.

Now comes the famous part.

> The baby, assailed by eyes, ears, nose, skin, and entrails at once, feels it all as one great blooming, buzzing confusion; and to the very end of life, our location of all things in one space is due to the fact that the original extents or bignesses of all the sensations which come to our notice at once, coalesced together into one and the same space. There is no other reason than this why the hand I touch and see coincides spatially with the hand I immediately feel! (p. 448).

You should be feeling excited at this point yourself. Help yourself to some *italics* and CAPS! You can now appreciate something that has eluded several generations of readers because James's phrase was reproduced out of context. James was not really trying to persuade us that the neonate is driven to distraction by the stimulation he or she encounters. It's just the opposite! The world that blooms and buzzes is reduced in the infant's mind to a unified, *global* state. *Global* here is used in the sense of a scene that has few if any component parts, like a bubble or a marshmallow. The emphasis should be upon *"one great,"* rather than on "blooming and buzzing." The neonate has a certain way of looking upon the world. It is extraordinarily simple, from an adult's perspective. But this does not imply that the one-day-old babe is experiencing any cognitive discomfort. As adults, we can compile lists of all that exists in the world that the infant has failed to distinguish, but we would have a hard time convincing the infant that he or she is missing something.

The view of the newborn's mental state that has been offered here is the opposite of what is usually associated with the famous mini quotation from William James. I think this is James's actual position, taken within the larger context of his thought. You will notice that it has a strong resemblance to Werner's orthogenetic principle (*see* Chapter 2) which also portrays development as proceeding from a global state.

This shows the usefulness of going back to the original source

With this understanding, James's observations prepare us for an appreciation of the up-to-date findings on the characteristics of very young infants. The newborn does seem to have a set of organizing principles or strategies for meeting the world on his or her own terms. These are not necessarily perceptions, thoughts and actions in the usual sense. But the neonate is equipped with a nervous system that is already functional in many ways and which relates to the world in its own style. There is a kind of adaptive intelligence to be recognized and respected in even the youngest infant.

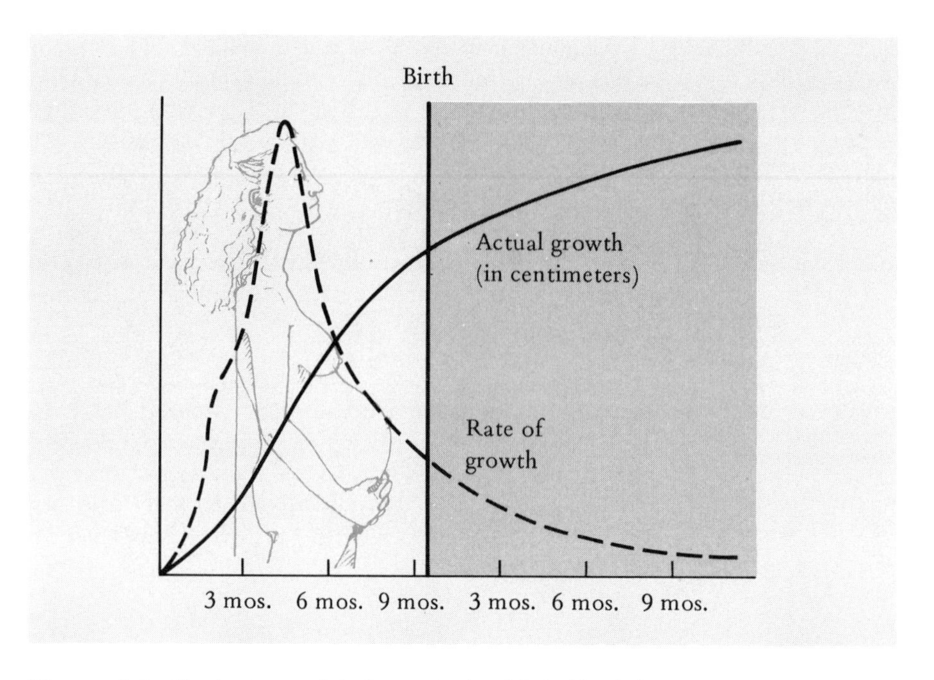

Figure 4-1. Physical growth before and after birth. *The infant continues to grow rapidly after birth (solid line), but the* rate of growth *(dotted line) declines. (Based on data presented by Thompson, 1942 and Timiras 1972.)*

UNLEARNED BEHAVIOR: THE REFLEXES

It is the observable behavior of young infants that invites our attention first. Our access to the neonate's inner state of mind is quite limited, but we can all see for ourselves what the young infant can *do*. Developmentalists have continued to extend our knowledge of the less obvious capacities of the infant through a variety of interesting, and at times ingenious, methods. We begin, however, with the obvious.

UNLEARNED BEHAVIORS

We learn quickly. Long before a child is of school age there has been a tremendous amount of interplay between his or her psychobiological systems and the environment. It becomes difficult to say, "This much of his behavior is learned and that much of his behavior comes directly from his physical endowment and level of maturity." The child's experience with his or her world declares itself in everything he or she does and the way he or she does it. Only very early in life is it possible to observe certain behaviors that truly

When something is **innate,** it does not have to be learned

appear to be *un*learned or **innate.** This is one of the considerations that make the study of reflexes interesting. The neonate's repertoire of reflexes seems to be as close to unlearned behavior as anything we are likely to observe at any point in the human life cycle. Some of these reflexes have also been observed in prenatal life. This could be regarded as strengthening the case for the neonate's status as a blank slate, free from learned behaviors. But who is to say that there might not be some type of learning or adaptation going on before birth?

Let's first look through the catalog of neonatal reflexes and then reflect further upon their possible significance.

THE CATALOG OF NEONATAL REFLEXES

1. *Yawning, sneezing, sucking, stretching, and crying.* These behavior patterns are usually fully operational in the neonate. The infant can do all these things soon after birth. Crying is, of course, especially important at this time as it constitutes the neonate's principle means of communicating a need state to the adults he or she depends upon for survival. (It has been noticed that prolonged neonatal nocturnal crying produces an increase in yawning on the part of the parents during the next day. However, infantologists have not been able to explain this strange interaction of unlearned behaviors across the generations.)

2. *Rooting.* If you gently touch the infant's cheek near the lips, the baby will turn his or her head toward your finger and open his or her mouth. This is known as the *rooting reflex* or behavior. This does not take place during sleep, but it may occur while the baby is crying. The rooting and sucking reflexes are also present long before birth; usually they appear by the third fetal month.

3. *Grasping.* Place your finger in the neonate's palm, and his or her fingers will close upon yours. Sometimes this pattern may not be found on the first or second day after birth, but usually it consistently appears thereafter. What we have just described is the *Palmer grasp.* There is also a *plantar grasp.* Place your finger or a pencil against the sole of the infant's foot right behind the toes, and the toes will flex to grasp the object.

4. *Toe flexion–fanning.* The *Babinski reflex* is one of the best known. It consists of a fanning of the other toes, while the big toe flexes. The Babinski is produced by stroking or lightly scratching the side of the sole in a heel-to-toe direction. Physicians and developmentalists are interested in the Babinski not so much for the value of the reflex itself (obviously it is not as significant to functioning as sucking and crying), but because abnormalities in this behavior pattern can possibly indicate more significant difficulties. It is best to leave the Babinski to specialists.

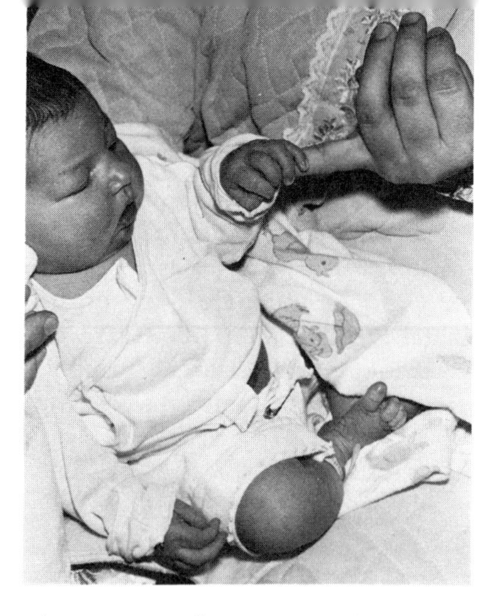

The grasping reflex is among the earliest behaviors to show itself. (Photo by Talbot Lovering. Courtesy of the Boston Hospital for Women.)

5. *Moro reflex.* This is a response pattern that is set off by loss of support (sensation of falling) or sudden, loud noises. There is an extension of the spine and lower extremities; the arms thrash out and then come together convulsively; the hands are at first fanned out, then tightly clenched. The infant may also cry. This rather spectacular pattern closely resembles the startle reflex, and for most purposes it is not crucial to distinguish between the two. Essentially, the Moro reflex is what the young baby does when startled.

The startle pattern is similar, but develops later and is part of the alarm response of adults

6. *Crossed extensor.* Perhaps this interesting reflex should have a more interesting name. If the sole of the *left* foot is stimulated while the neonate is in a certain position, then the *right* foot responds with flexion! (The same cross-over effect is found when the right foot is stimulated.) The stimulated leg itself extends and moves away, after the unstimulated leg has responded. The crossed extensor is a spinal cord reflex that can reappear in adult life after certain types of injury.

7. *Deep tendon reflexes.* There is a whole set of specific reflexes that involve the response of skeletal muscle. The well-known "knee jerk" is just one of these. The presence of deep tendon reflexes and their strength indicates how much maturity has been attained in the muscular and the nervous systems and the coordination between the two. Testing of these reflexes (and of the cross-extensor reflex) is best left to specialists.

8. *"Doll's Eye."* The neonate's head is gently turned to one side or the other while his or her body remains still. You will notice that his or her eyes remain fixed; they do not follow the directional movement of the head. The same "doll's eye" phenomenon is observed when the head is gently moved up and down. As you might guess, this lack of adjustment to shifted position must disappear before the infant can achieve good visual fixation. In fact, this is one of the first of the neonate's characteristics to change, usually within about two weeks or less after birth.

9. *Galant reflex.* This response is seen when the neonate is held securely in a prone (face-down) position, and stroked about midway down the back and slightly off to one side. He or she will arch toward the point of stimulation. This is one of the most dependable reflexes in the newborn.

10. *Supporting and stepping reactions.* Even a very young infant can "stand" if held erect with his or her feet on a solid surface. Positive supporting actions of the hips, knees, and ankles occur, which allow the infant to support his or her weight (sometimes!). A stepping reaction can often be observed next, if the infant is moved forward slightly and tilted a bit to one side.

This has been a selective description of specific behavior patterns that can usually be observed in the newborn. More detailed presentations are available (*see,* for example, Taft & Cohen, 1967; DiLeo, 1967). I have indicated specific means for producing and observing some of these patterns in order to describe them more usefully. I am not recommending that infants be toyed with to demonstrate their reflex repertoire.

Some of these very early behavior patterns remain with the individual throughout the total lifespan (e.g., sucking); others fade out quickly (e.g., doll's eye); still others fade out gradually (e.g., rooting lasts for about one year). Characteristics of these early responses provide important clues to the health and maturational status of the infant—but only when observed and interpreted by those qualified to do so. You will not want to imagine abnormalities where there are none; on the other hand, you will appreciate why a pediatrician might take a keen interest in the neonate's response repertoire.

Two other points are worth keeping in mind. Behavior patterns that are normal when they appear in the neonate can return as part of a pathological syndrome later in life. The reappearance of a Babinski reflex is a concrete sign of development moving backward on the psychobiological level.

The other point is some behavior patterns do not appear immediately after birth, but soon after. These do not fit into the framework of this chapter, which is limited to birth and the first thirty days. Developmentalists have reason to believe that reflexes emerge over a period of time in the postnatal as well as the prenatal state. The late-appearing patterns may depend on further maturation of the muscular and nervous systems than do the reflexes that appear at birth. The *parachute response* is a good example of a late-appearing pattern. Typically, this shows up some time between the seventh and ninth month of infancy, although it may come as late as one year. It is seen when an infant is held securely in the vertical plane (head up, feet down), and is then suddenly turned over to a face-down position and moved down. The infant's response involves extension of the arms and spreading of the fingers as though to protect himself or herself from falling.

We should clearly understand, then, that while the infant comes prepared with a variety of specific behavior patterns that are functional soon after birth, there are other patterns of the same general character which do not reveal themselves until later.

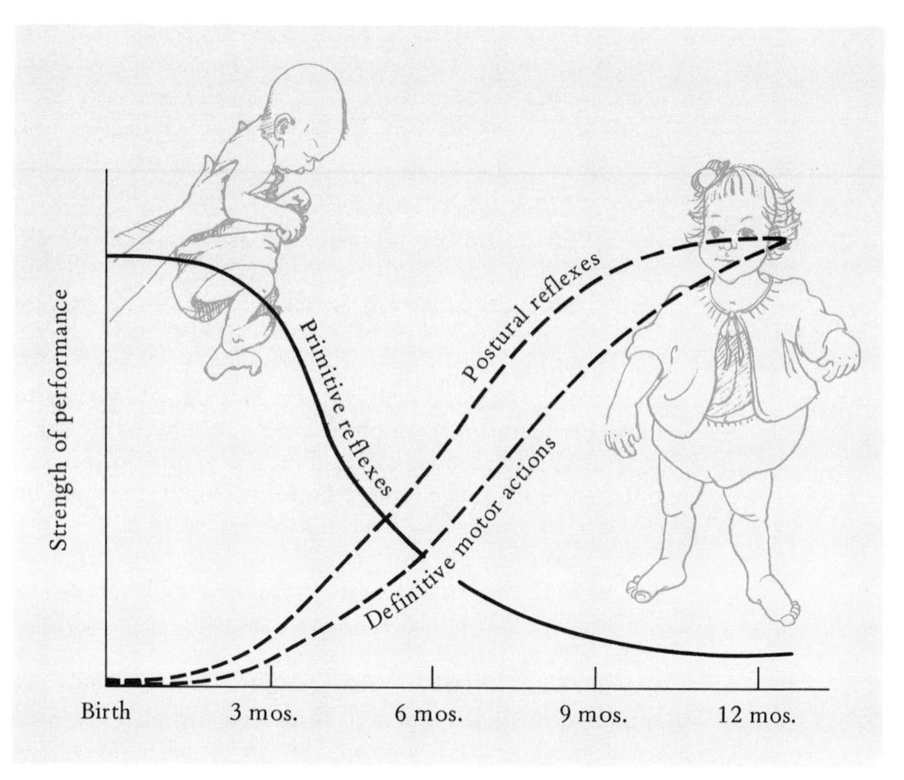

Figure 4-2. *Do reflexes become stronger or weaker through infancy? Primitive reflexes are strong at first, then they weaken. Postural reflexes, a more advanced type, strengthen with age, as does motor action—behavior that appears more complex and purposeful than the primitive reflexes. (Reproduced by permission. Capute, Accardo, Vining, Rubenstein & Harryman,* Primitive Reflex Profile. *Copyright © 1978 University Park Press, Baltimore.)*

SOME REFLECTIONS ABOUT REFLEXES

Caleocortex is the fancier name for the old brain

We might content ourselves with a cataloging of reflexes, or we might press a little further to consider the possible significance of these apparently innate response patterns. What, for example, is the biological context within which neonatal reflexes arise? Specialists today believe that the neonatal reflexes described earlier represent the unrestrained influence of a particular region of the central nervous system (Capute, Accardo, Vining, Rubenstein, & Harryman, 1978). This area is sometimes described as the *old brain.* It sounds peculiar to speak of an old brain in a brand new baby, but this term is to be taken in an evolutionary sense. The brain structures involved are relatively primitive and have appeared in creatures that exist on lower and earlier rungs of the evolutionary ladder. The old brain area is tucked deeply in and around the brain stem. Many years of research have led to the conclusion that the brain stem, cerebellum, midbrain, and basal ganglia are closely associated with the regulation of vital physiological functions. Loosely speaking, this is the kind of behavior that occurs without thinking or in a relatively involuntary and automatic manner such as breathing.

The **cerebral cortex** consists of the most highly developed cells in the CNS

The deep-lying old brain areas are covered by a mantle of more highly developed cells, the **cerebral cortex.** This mantle makes distinctively human thought possible. It takes awhile for the cerebral cortex or "new brain" to come into its own. When it does, the primitive reflexes gradually come under the control of this higher level biological structure. As the infant moves through its rapid developmental pattern, the reflexes become more inhibited by the cerebral cortex and start to take their place in more coordinated, functional, and voluntary types of action.

There are both applied implications and theoretical implications in this description. Injury to the brain after development of the cerebral cortex has been achieved can lead to a pathological reappearance of the primitive reflexes. This indicates that the cerebral cortex is no longer able to exercise inhibitory and integrative control over the lower structures. The type and extent of primitive reflexes that are released when there has been damage to the brain can provide useful information to the physician.

From a theoretical standpoint, we can see that a stimulus-response approach to human behavior fits the lower or earlier phase of development more snugly than it does the later phases. A reflex can be elicited (usually) by a specific type of stimulus. Later, however, there is much more evidence of a "mind" that is presiding over its own actions. This mind seems to have selective perceptions and preferences right from the start and, as time and development go on, it becomes increasingly unrealistic to regard the infant as a bundle of separate reflexes that are waiting to be touched off by this or that stimulus. In fact, the view of the infant as a sort of wind-up toy that puts forth reflex responses when the right button is pushed is not very adequate even for the earliest hours of neonatal life. There is room for more than one theoretical interpretation here. The main point is that close attention to these early response patterns is of importance to all those concerned with the biological status of the infant or with formulating a balanced theory of the infant as a complete psychobiosocial being.

It is also interesting to think about specific reflexes in terms of the survival functions they might serve now or the functions they served dim eons ago. The fact that crying elicits the attention of adults who can meet the neonate's needs has already been mentioned. This is one possible example of a survival function served by reflexes. There has been much speculation and some experimentation on the possible usefulness of reflex behavior for many years. One of the most intensely studied has been the Moro reflex, described earlier. E. Moro himself used the term *Umklammerungsreflex* (embracing reflex) when he first described it. He suggested that it represents a primitive tendency to cling to the mother. Argument and research on this topic was well summarized in the admirable essay on human infancy by Kessen, Haith, and Salapatek (1970). I think it would be fair to summarize their essay by saying that years of more sophisticated research have resulted in more sophisticated differences of opinion about the functional significance of the Moro reflex. It is possible to hold to Moro's original view, but alternative possibilities have also been proposed.

Research on the grasp reflex has also kept viable the possibility that it has served throughout the centuries as a survival action. In 1891 a British medical officer named Robinson claimed that newborns could support their own weight when suspended. This claim was supported by a photograph of a child clinging to a tree branch. Later research has been a good deal more sophisticated, and Robinson's observation seems to have been more than vindicated. Perhaps the most interesting study was Myrtle McGraw's observation (1940) of about one hundred children who ranged in age from soon-after-birth to seven years. She found that the very young infants were *better* able to support their own weight than the older infants. It was not until about age five that the children regained the self-support capacity they had shown in the first thirty days of postnatal life. Development started at a very high level, then it dropped off considerably, and it came up again much later. This is surely something to think about whenever we are tempted to make simple assumptions about the development of abilities. Although the grasp reflex may be a quite special case, it indicates once again the complexity of developmental patterns.

PERCEIVING, LEARNING, AND PREFERRING

Even while the infant still exhibits unlearned response patterns, he or she is already engaged in other types of interaction with the world. The process of becoming a distinctively human being is underway. Developmentalists increasingly are discovering evidence that the young infant perceives, learns, and prefers. Yes, the neonate is relatively dependent when compared with the person he or she will be in the months and years ahead, but active and interactive processes are at work from the beginning. Research in this area is fast-moving these days, with new methodologies being created to study more precisely phenomena we could only guess about a few years ago. Here we will sample a few components within the total picture of infant development.

SEEING

There are differences between an adult's eyes and those of the neonate. Nevertheless, the evidence indicates that all the necessary visual structures exist and are operational at time of birth. The neurophysiological pathways that relate the eye to the rest of the brain (for the eye is properly considered a part of the brain) also appear to be functional. The infant can receive and process visual information.

Studies have shown that the very young infant attempts to move its head in the direction of visual stimulation. Further, he or she seems to *prefer* some stimulation more than others; he or she displays a *differential response.* Eye movements of many kinds can be observed in the newborn, although these are not well controlled in comparison with the way they will function in just a few months. The infant can follow moving sights and objects with his or her eyes within a few hours after birth. It may, in fact, be easier for the infant to follow motion and to shift his or her target of visual fixation than to keep his or her eyes on the same spot (Haber & Hershenson, 1973). The infant's eye movements are jerky, and have a tendency to retreat periodically from the direction of movement, but there is little question that they do occur. The newborn is also a bit farsighted and seems to stay that way for the first four months.

How well does the neonate see? By this question we are usually making reference to *visual acuity.* Recent studies reviewed by Kessen, Haith, and Salapatek (1970) and Haber and Hershenson (1973) indicate that the newborn has impressive visual capabilities. These capabilities will continue to increase over the next few years, but it is clear that the infant starts postnatal life with all the necessary visual processing equipment in place and functional.

This knowledge allows us to inquire further into what or how the infant sees. At first the human face does not seem to "mean" anything special to the neonate, as far as his or her visual behavior is concerned. However, after about a week of postnatal experience and growth, the infant begins to show a preference for the human face and for face-like patterns of stimulation. This is one more line of evidence that suggests the existence of some guiding prin-

The pattern of a human face holds the attention of infants more than most other sights the world can offer. (Photo by Talbot Lovering.)

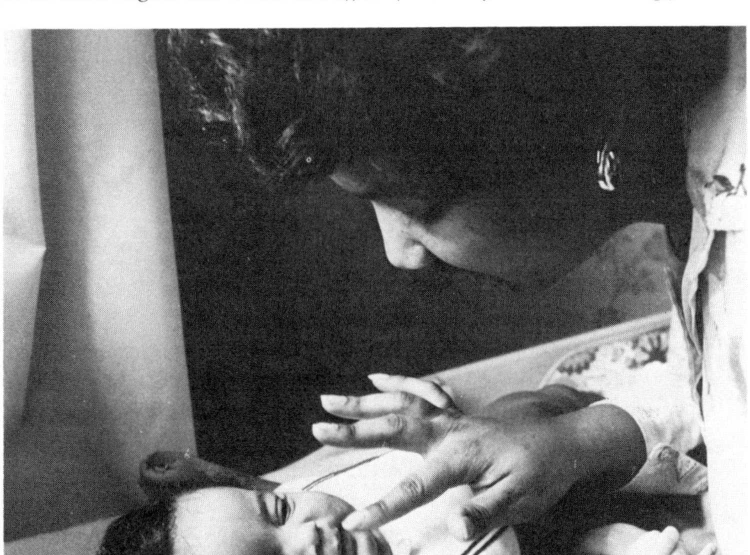

ciples in very early psychological functioning, principles that challenge the assumption that the infant is either unorganized or random in behavior. R. L. Fantz (1963) reported that an infant only ten hours old consistently looked longer at a face-like picture and other contoured pictures than at other kinds of surfaces. It is tempting to speculate that this type of preference indicates the very young infant is already warming to his or her role in the human community, attending to faces more than to other visual patterns in general.

The attention given to visual *patterns* as compared to simple or random targets also indicates a bias or preference even when faces are not involved. In other words, even the very young infant alters his or her scheme of visual scanning when he or she comes across a pattern (such as a geometrical figure). There is something for philosophers to ponder in studies such as those of Salapatek (1969) in which very young infants were given several types of visual fields to look at. These fields could be a circle or a square, for example, or a more complex shape that had either regular or irregular features. Visual patterns of this kind drew eye fixation from the infants, and eye fixation on a target is one way of measuring attention. But if an infant had a homogeneous field to look at—a target in which any one area looked like all the other areas—then he or she responded as though bored. The infant was likely to get "fussy," to fall asleep, or to scan toward the extreme edge of the field as though in hope of finding something more interesting. Do even very young infants have preferences? Do they have the capacity to be bored or interested by what comes before them? Do they perhaps even have an affinity toward contoured figures and visual organization that would make Plato nod approvingly? These are at least reasonable questions to ask as research continues to demonstrate nonrandom and organization-seeking characteristics of infant perception. The built-in bias of infants toward certain forms of visual inspection has also been shown in their tendency to move their eyes across the horizontal plane more frequently than the vertical. This horizontally oriented scanning pattern perhaps lays down the foundation that you are using at this moment as you read across this page.

Plato held that all humans have an inner conception of perfect form

In a way it is surprising that methodology has been developed to study the eye fixations and displacements of babies who are only a few days into this world. But it is even more remarkable that useful findings have emerged in so new an area of study. It is already fairly clear that the infant scans his or her visual environment on the basis of certain strategies or preferences and favors visual displays that offer a certain degree of stimulation or "interest." The infant's eye goes to details, but it also searches out larger forms. Even within the first few weeks of postnatal life there is an increase in the size of "visual hunks" that the infant can take in. There are very early perceptual dispositions, then, and they develop and expand rapidly during their first weeks of exposure to the extra-uterine world.

Some studies have brought visual patterns to the inspection of infants who are themselves in a stationary position. But it is recognized that babies also adjust their own positions, perhaps "to get a better look" at what is happening, although there are other possible explanations for postural shift as well. Tronick and Clanton (1971) placed babies as young as three weeks (along with older babies ranging up to fifteen weeks) in a specially designed lightweight harness that did not restrict their motion. They were able to record both eye and body movements through an ingenious apparatus and determine where the baby was looking at any point in time. Brightly colored cubes were placed in various positions within the potential visual field. The researchers were impressed by the coordination or smoothness of the head-body adjustments these very young babies made as they looked at the cubes. Very early in life, infants not only have all the visual equipment needed, but also the capability to actively adjust their bodies to line up with a visual target. The youngest babies explored less of the environment and took "littler looks," yet they approached the visual opportunity in much the same way as the older and more skillful infants in the study. It was concluded that even at three weeks of age, infants could explore and extract information from the environment. Here is one more—and a rather important—indication that our minds are in purposeful action from the very beginning!

When adults look at something interesting, they are not limited to eye movements and supportive postural shifts. They behave more organismically; they are involved or engrossed in what they are seeing. The same could well be true for infants as young as fifteen days. In a rather elaborate study, Stechler, Bradford, and Levy (1966) measured some aspects of physiological reactivity and body motility (moving around) when infants either did or did not have a visual target to look at. When there was a pattern to be seen in their visual field, the babies were more likely to be quiet and unfidgety. This finding by itself might have suggested an inhibitory state, but a measure of electrical activity taking place on the surface of the infants' skin indicated a state of arousal or excitation. It was as though the infants were so engrossed in what they were seeing that their attention was concentrated in their eyes and minds rather than diffused throughout their bodies. The researchers spoke of this as a "vigilant-like state." This could be an early model for vigilant or engrossed behavior in older children and adults, and it certainly makes sense as part of a person's strategy for survival ("What's going on here? I'd better look carefully!").

Quiet does not mean turned off. Even in infancy, close observation repays the effort

Although there is much to learn about visual perception in early infancy and its subsequent course of development, perhaps it is not too soon to form the impression that seeing is a complex and adaptive component in our total relationship to the environment from our earliest days. Newborns cannot traverse the world with their feet or take it into their hands, but they can cast active and curious eyes upon the world so that they come to later actions with at least a spectator's experience.

HEARING

More research has been directed to the early visual capabilities of the neonate than to any of the other sensory modalities. This selectivity may be related to our society's general reliance on and preference for the visual ("Seeing is believing!"). For a more balanced perspective, let's now "look" at hearing.

The physical structures that are necessary for hearing (or *audition*) are usually well developed before birth. The prenate's sensitivity to sound is muffled, however, by the presence of a gelatinous liquid in the middle ear. This gelatinous liquid or tissue is reabsorbed during the first week or so after birth, and hearing then seems to improve markedly (Timiras, 1972). The external ear also unplugs itself during the transitional period between prenatal and postnatal existence.

This does not mean that the prenate is completely unresponsive to sound. It has been known for sometime that exposure to music performed at a high volume or to sudden, loud noises can change or increase fetal activity (Forbes & Forbes, 1927). The mother-to-be who is in her eighth month might be making an impression upon her prenate when she listens to Beethoven or The Rolling Stones, although we have no evidence about the possible differential effects. There is some question as to whether the babe's excitement is based upon actual auditory experience or upon a vibratory and tactual sensitivity. More recent studies have confirmed the earlier findings that the prenate responds to at least some auditory stimulation in the weeks before birth (Johansson, Wedenberg & Westin, 1964).

Intense sounds capture the neonate's attention, as indicated by changes in heart rate and respiration, as well as body movements (Steinschneider, 1968). Within the first thirty days of postnatal life, the infant also shows responsivity to rhythmic stimulation. The rhythm of the mother's heartbeat has sometimes been observed to have a quieting effect upon the young infant, but it now seems probable that a variety of other auditory rhythms also have the same comforting influence (Brackbill, Adams, Crowell & Gray, 1966).

Is there a type of pattern recognition in hearing that is roughly parallel with visual pattern recognition? Perhaps the most obvious and most important place to look would be the perception of speech or speech-like sounds. The human voice, like the human face, comes to take on particular meaning throughout development. But how could speech perception be studied in very new babies? The methodological challenge is formidable. And yet, speech perception has now been studied, and studied rather successfully, in infants as young as thirty days. One of the more significant studies of infant speech perception was done by Peter D. Eimas and his colleagues in 1975. These researchers found a way to translate what the infant heard into an observable response that was clearly related to the auditory input. They did this by utilizing a type of behavior that comes naturally to the infant: sucking. Babies of various ages (starting at thirty days) were given non-nutrient nipples to suck upon while being held comfortably. Preliminary studies were made to determine the sucking tempo when nothing else was going on. A procedure was then devised by which the infant could influence the auditory input by his or her rate of sucking, literally "sucking in sound," if you like. Speech-like sound patterns would be heard when the infant increased his or

How to make "private" experience "public" is a frequent challenge for researchers her sucking rate to a certain level. You will notice that what was to happen was in the baby's own hands or, rather, in his or her mouth. Apart from the fact that the study was concerned specifically with the perception of auditory events, it also represented an experiment in providing the instrumentality for control on behalf of a being who is usually highly dependent.

What else might babies do if able to put their perceptions into action?

This particular study and a number of variations on it have demonstrated that even the young infant can discriminate speech from non-speech sound patterns. There is an early-manifesting ability to process speech patterns *as* speech patterns. It can be said that the month-old baby can perform some simple analyses of the sounds he or she hears. Eimas noted that this involves the ability to break down a series of continuous sounds into smaller elements. Without overemphasizing the similarities, this might be a process that is more or less parallel to the infant's ability at this same time to pick out details in a visual field.

We have all the more reason, then, to respect the infant as an alert, active being whose limited ability to act upon the world does not entirely stand in the way of experiencing and learning. The interest in speech-patterned sound and the ability to give special attention to it probably plays a significant role in social learning and general cognitive development, although almost everything about the details of these processes remains to be learned.

We also have more reason to appreciate that human development does not start from a zero baseline. If we seriously underestimate the young infant's mental activity, biological readiness, and learning potential, then the gap between infancy and adulthood is made to seem larger and more mysterious than necessary. Continued research into the earliest phases of human development may well reshape our ideas of development across the entire lifespan.

LEARNING AND HABITUATION

The neonate changes rapidly. It is obvious that physical maturation is taking place. But are some of the changes related to the way he or she is using experience? Is the neonate modifying his or her behavior based upon what he or she sees, hears, feels, smells, and tastes? Two psychological concepts are important here: learning and habituation. An organism can show a behavioral change after experience by coming up with a new pattern of response. In many psychological experiments, for example, rats have developed the knack for finding the desired food or water by scurrying to the right location at the right time. They "know" that the good stuff will always be behind the middle door, let's say, and they will not ordinarily waste their time with the other doors. When an organism shows a new response (or a new patterning of previous responses) after a certain experience, we say that *learning* has occurred. Sometimes, however, experience seems to teach the organism to reduce or

Habituation is becoming less responsive to a stimulus as it becomes familiar

omit a response. We then speak of **habituation.** The stimulus appears, and the organism pays attention to it. But nothing much happens. The stimulus is repeated, and repeated and repeated. Eventually the organism—man or beast—behaves as though the stimulus is no longer worth its attention.

Habituation is particularly important in studies of infant behavior. If the neonate no longer appears interested in a certain stimulus, then it seems appropriate to credit him or her with certain psychological processes such as discrimination, memory, and the capacity to be bored. Both learning and habituation represent ways of adapting to the world. There is evidence that both occur during the first thirty days of postnatal life. Habituation is also of special interest in this early developmental phase because the psychologist cannot ask the infant what he or she is seeing or hearing, much less what is on the infant's mind, or why he or she is doing this rather than that. By studying habituation, developmentalists can obtain information about the infant's abilities that perhaps could not be obtained in any other way.

There have been many studies of habituation in newborns, and the general conclusion is well established: it does happen! Keep presenting the same sounds, sights, or even odors and touches, and the infant will keep paying less and less attention to them. But how do we know when the newborn has be-

Figure 4-3. *The gourmet in the cradle: taste discrimination comes early. Discrimination between sweet tastes on the one hand and bitter, acid, or salty tastes on the other can be observed shortly after birth. Sweet substances elicit sucking and swallowing responses in young infants. Bitter, acid, and salty substances elicit grimacing and contracting responses.*

come habituated? He or she cannot tell us, "This is terribly boring." More subtle, yet reliable forms of observation are required. Many different measures of the habituation response have been devised. One of these, changes in heart rate, is especially interesting. Heart rate changes when we become excited. When a new sound or sight is presented to a neonate, his or her heart rate quickens (along with other indications of arousal). As the stimulus becomes increasingly familiar, the cardiac response smooths out. If another new stimulus is introduced among the now-familiar stimuli, the cardiac response will show a fresh peak. By reading his or her heart, then, we can demonstrate that experience is not lost upon the newborn. He or she is capable of becoming familiar with the world and of being surprised and attracted afresh. The cardiac response to novel stimulation seems to be an example of an unlearned behavior, somewhat akin to blinking our eyes at a puff of air. Because this unlearned behavior can be controlled and set aside when the same stimulus is repeated, it is possible to describe the phenomenon as habituation. In the broad sense, this is a display of intelligence on the part of the very young infant. It is as though he or she "knows" what is new and worth paying special attention to and what is familiar enough to ignore.

An ***uncondi-tioned stimulus*** is an event followed by unlearned behavior (puff of air—eye blink)

A **conditioned stimulus** is an event whose power to elicit behavior requires learning

One of the simplest forms of learning is the type known as *classical conditioning.* It has been the subject of a great many experiments since the historic work of Ivan Pavlov (1928; 1957). As you probably know, conditioning involves two kinds of stimuli. One is known as the ***un*conditioned stimulus** (UCS). This refers to an event that elicits a reflex or unlearned behavior pattern. A nipple placed in the mouth of a neonate is a UCS for the sucking response.

The other kind of stimulus is the **conditioned stimulus** (CS). At first, a CS has no power to elicit the response. A baby will not begin sucking when he or she hears a musical tone, for example. Why should the baby? There is no direct or built-in connection between the tone and a sucking response. L. P. Lipsitt and his colleagues conducted a series of experiments to determine if very young infants could learn a relationship between CS and UCS. In one of these studies (Lipsitt & Kaye, 1964), they paired presentation of a nipple with the sound of a tone for some infants and presented only the tone to another control group of infants. Sure enough, even three- and four-day-old infants showed the type of behavioral change that can be described as classical conditioning. Those who had experienced the tone-and-nipple pairing learned to begin sucking when the tone was presented without the nipple.

In general, studies of classical conditioning indicate that it can be established in the neonate when the experimental situation is carefully attuned to the infant's level of functioning. Conditioning is stronger and more impressive in older infants, but it is reasonably clear that the process can also be observed in very young infants. The experimenters have learned to respect individual differences among the neonates. In the preceding chapter we mentioned the large variation among individuals in activity level *before* birth. Differences exist after birth as well, and these affect the infant's response to conditioning attempts.

There is more to learning, even in early infancy, than forming associations between CS and UCS. This was demonstrated, for example, in a study by Engen, Lipsitt, and Kaye (1963). The focus of this study was on the newborn's ability to distinguish among several types of olfactory (smell) stimuli. The researchers found that newborns were able to make such distinctions. Perhaps more interesting, however, was the observation that the infant's response was at first: ". . . diffuse and seemingly disorganized, similar to a mild startle, and that in the course of the experiment the response became a smooth and efficient means for escaping from a noxious odor stimulus. In the early trials the infants apparently responded with their entire bodies, while in later trials a simple retraction or a turn of the head from the locus of the odor was executed."

This suggests that infants as young as thirty-two hours postnatal can differentiate their own responses and differentiate incoming stimuli. It is as though the infant were thinking, "Here comes that awful smell again, but I can get away from it by turning my head." This would seem to involve a rudimentary process of discrimination, perhaps of comparison and memory, and the ability to exercise selective control over body action. It can be misleading to read adult concepts into the mind of the infant. The words I have given the newborn should, of course, not be taken literally. Nevertheless, it is difficult to escape the impression that some form of intelligence is being exercised in the sequence of behavior described above.

The study of learning in very young infants can eventually contribute much to our understanding of humans developing. It is now a vigorous area of research with a variety of methods employed, some of them quite fascinating. It would be premature to go into more detail at this time, as the most active researchers are still trying to sort out and evaluate the accumulating observations. But it is not too soon to mention some of the opportunities and problems that could be just around the corner.

One of the ways we can go wrong is to consider learning and habituation as processes that are independent of the state of the organism (whether the organism studied is rat or human, or within humans, infant or adult). This is a temptation that has lured some otherwise very competent researchers. Classical conditioning, for example, is a type of research strategy. It can be (and has been) used with many types of animals, and with humans from the prenatal state all the way to old age. An expert or advocate of classical conditioning can be justifiably impressed by the findings obtained by using more or less the same technique with so many different kinds of organisms. But this enthusiasm can go so far that some people may conclude we are dealing with an **empty organism.** Who the organism is (e.g., rat, ape, or human) and what level of development it has attained may seem less important than the technique or concept that has been applied to it. A better approach is to consider the specific physical and behavioral characteristics of the organism, both the species it represents and the individuality of the particular representative.

Empty organism describes the view that the organism is less important than how it is studied

Another way we can either go wrong or open up new opportunities is in the area of so-called behavioral engineering. Social values can influence the way that knowledge about early development is utilized. Plato outlined a plan for influencing children from the earliest days of infancy so they would fit in better with his favorite vision of society (see *The Republic*, and also Chapter 7 of this book). In our own century Aldous Huxley warned of the dangers inherent in such systematic influence of individual development by a totalitarian regime in his brilliant novel, *Brave New World*. The possibility of applying behavioral engineering to humans from infancy onward is serious enough to deserve being taken seriously. As we have seen, there is evidence that we can be influenced by our experiences soon after birth. Whether or not a systematic structuring of human experience is applied by society, then, might significantly affect the future development of individuals and entire generations.

It is heartening to see that many experts in the area of early development are aware of the morality or values problems and risks—as well as the opportunities—posed by the emphasis on influencing a person while he or she is still, literally, in the cradle. One of the most knowledgeable students of infant learning, for example, has had this to say:

> . . . it would be unfortunate if the investigators of infantile behavior in general, and infant learning in particular remove their investigations from organismic and developmental contexts. It is to be hoped that investigators will not ignore the information now becoming available from studies of behavioral genetics, and that they will pay close attention to the rich and detailed descriptions of infant *states*. Furthermore, research programs dealing with infant behavior must also take into account neurophysiological and biochemical data, and the relationships between rearing environments and biological status. . . .
>
> The study of infancy must be concerned with individual differences and their longitudinal implications. Associated with the question of individual differences are social and moral, as well as theoretical, issues surrounding the emergence of an effective and technically proficient art of behavioral engineering. . . . If a major purpose of the new technology becomes the elimination of individual differences, either because they are troublesome from the vantage point of experimental convenience, or because they are at variance with the social value system of the behavior manipulators and shapers, then a warped and potentially obscurant picture of human potential may arise. Furthermore, the longitudinal relevance of early individual differences will be masked . . . many organisms may be brought to make responses that they would not otherwise make. . . . How much effort (at what cost to the individual as well as to the trainer) is involved in eliciting or shaping a particular response pattern is a question that may not be avoided. . . . For example, with respect to crying behavior, does it make sense, in terms of present knowledge, to eliminate this response in an infant who is rated as an active cryer? Is it not possible that the very activities associated with crying play some important role in stimulating growth and development? (Gollin, 1967, p. 260).

TV commercials that bombard children are also subject to this kind of criticism

Eugene S. Gollin has also pointed out that a decision to make determined use of an infant's waking time for systematic learning purposes might have undesirable consequences. In truth, we do not know precisely what would be the consequences. The more we learn about the newborn's capacity to learn, hopefully the more responsibility we will acquire to use this information wisely.

SCENARIO 3: MICHAEL-THE-NEW

We have been considering some of the characteristics and capacities of the newborn. Unlearned behavior patterns, habituation, and the other factors that have been mentioned are important for our understanding of development. Yet, they do not have much resemblance to the newborn as seen by his or her parents. How do parents tend to view their new baby, especially when it is their first offspring? What questions enter their minds? What conclusions do they reach? Let's take a parent's eye view of the neonate, and for this purpose recall Gloria and Tom from Scenario 2. Michael has arrived and turns out to be nobody else but Michael—a normal, healthy, full-term baby.

Traditionally, Tom's first glimpse of his son takes place through the window of the nursery on the maternity ward. A nurse holds Michael for his inspection: a tiny, red-faced figure wrapped in a blue receiving blanket. Among the other thoughts that crowd into his mind, Tom regards Michael as big, hardy, firm, and alert. If the infant had been born a female and presented to him in a pink receiving blanket, chances are Tom would have regarded her as small, delicate, soft, and fine-featured, even though no objective physical differences actually existed apart from gender! A team of Boston psychologists has found that fathers (and, to a lesser extent, mothers) begin to sex-type their infants psychologically from the first day of life, although there is no appreciable difference in size, alertness, etc. between male and female neonates (Provenzano, Rubin & Luria, 1973).

Move the scene now to the mother's bedside. The nurse has brought Michael to her for the first time. After a moment or two, the nurse departs, leaving mother and child together. Perhaps she also draws the curtains to afford them more privacy. Naturally enough, Gloria is curious about what she has been carrying around for nine months and has now brought forth. She looks at his face, his head, his skin. Michael's complexion at this moment is dark pink. As Gloria touches him gently she notices a creamy substance in the folds of his skin. It does not concern her because she knows that this is a secretion of the sebaceous glands that protected Michael's outer surfaces during the prenatal state.

Taking courage, Gloria disrobes Michael further. She notices slight, downy, fine hairs over parts of his body, especially the back and shoulders. This is also common with newborns, and will disappear within a few weeks. Gloria has also learned that she might expect some flaking and peeling of the

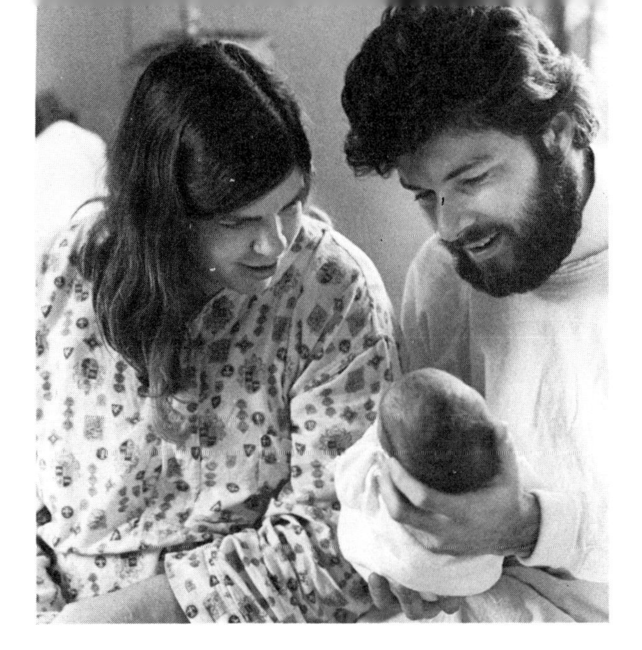

Even in a scientific age, it is not unusual for a parent to look upon this tiniest human as a miracle. (Photo by Bradford F. Herzog.)

outer skin surfaces in the first few weeks, especially on Michael's nose, knees, and elbows. Again, nothing to worry about. If Michael's parents were of Oriental, Negro, or Mediterranean background, she might also have noticed slate-colored spots on his lower back and buttocks. These would fade within the preschool years.

Truth to tell, Michael's face looks a bit swollen to Gloria. This is common and no cause for concern. She is a little surprised, though, to notice that Michael's chin is receding—where did he get that from? It is not a family trait! His nose is flat and his cheeks are pudgy. They will stay that way as long as he depends upon sucking for his nourishment. These facial contours make it easier for Michael to take his meals. There will be time later for him to develop an heroic jutting chin or "Roman" nose. Gloria may be so overcome by this first opportunity to be with her child that tears begin to drop upon her cheeks. Whether or not Michael is equally overcome, he will not reciprocate in kind. It will be several months before he can shed tears (from his indeterminate slate blue-grey eyes).

As Gloria holds Michael, she is tempted to touch the top of his head. Is it really soft, as she has been told? Indeed, the newborn's cranium (skull) is not sealed up tight. There are openings where the skull bones will unite. This allows for flexibility and growth. It will be a year or perhaps eighteen months before these **fontanels** have closed. Gloria settles for just kissing him gently on top of his head.

Fontanels are "growing places" in the neonate's cranium that close and harden

Something rather unattractive catches her eye next. There is a shrunken and discolored object on his abdomen: the umbilical stump. Once vital to Michael as his lifeline, what is left of the umbilical apparatus is now unimpressive and kind of ugly. It will turn black in a few days, and then slough off a few days later. Gloria will be advised by her physician or nurse that the umbilican stump can be a locus of infection, so she will keep this in mind.

What catches Gloria's eye next is Michael's sexual organs. He really *is* a boy after all! There are large individual differences in the proportion of genitalia to total body size among newborns of either sex, but Gloria will not be surprised if Michael's genitals appear somewhat oversized. Proportion of both the head and the genitalia to the rest of the body will take some time to reach their adult balance.

For a moment, Gloria's heart sinks when she gazes upon her son's lower extremities. Tom is already buying baseball gloves and football cleats for his future all-star. But Michael's short, bowlegged physique seems to limit his suitability to bronco busting (on very small, stout broncos). Then she remembers that this infantile feature will also pass away, to be replaced by the normal shape and musculature of an active boy, when he starts walking.

Just before the nurse arrives to return Michael to the nursery, Gloria takes a good look at his fingernails (having previously assured herself that he does have fingers, in the appropriate number and configuration). Yes! He does have fully developed fingernails on all of his short, plump fingers (ditto for the toes).

Michael is really here! He is no longer Michael-the-Possible, but a real, new person down to his fingernails. Gloria talks and coos to him a little more, and then it is time for both to nap.

Unfortunately, the scenario does not always run this smoothly. There are a number of hazards encountered by some mothers and their newborns. We will now consider one for illustrative purposes: postpartum psychosis.

HAZARD: POSTPARTUM PSYCHOSIS

News from the maternity ward: "Mother and baby are doing very well." Everybody relaxes. The baby is normal; the mother is in good health. The physical health of the mother is obviously of much importance both for herself and for the care of the child. However, "doing very well" also assumes a favorable state of mind. Is the mother pleased with herself? Pleased with her baby? Confident in her ability to care for the newborn? Free of mental stress and concern? To go one step further, is she "thinking straight," is she mentally clear, alert, and coherent?

In the first hours or days after birth, the answer to these mental-and-emotional-status questions will not always be yes. It is not unusual for the woman to have passing moods and fears early in the **postpartum period.** One does not give birth every day. The experience can stir up many thoughts and feelings that are not typically a part of the woman's life. Furthermore, the new mother may still be vulnerable to bodily states that tilt her thoughts and feelings.

The **postpartum period** is the time soon after birth

Depression and other negative psychological states sometimes make an ordeal out of the postpartum period. (Photo by Talbot Lovering.)

The majority of women seem to rebound quickly from the "postpartum lows," and some never seem to have even the temporary difficulties that have been mentioned. But there are exceptions. Enough women experience mental and emotional disturbances during the postpartum period to merit our acknowledgement of the problem. The disturbances range from mild to severe on a scale of psychiatric disorder. I will focus upon the more severe reactions, as these might be expected to exercise a more formidable influence upon the very early development of the infant and, perhaps, upon the lifelong pattern of the mother-child interaction as well.

SCOPE OF THE PROBLEM

Precisely how many women experience significant psychological disturbance in the postpartum period is not known. It has been estimated that at least one birth in every twenty is followed by an obvious manifestation of maternal mental illness (Kline, 1965). Perhaps one birth in four hundred is followed by a psychotic* reaction. This latter statistic is probably an underestimation, being based entirely upon women who are admitted to mental hospitals with

* Psychosis is a mental condition in which the person is out of touch with reality. Actually, psychotic individuals differ markedly in their reality-awareness and symptoms. *Postpartum psychosis* is a diagnostic category in many European nations, but is used only descriptively in the United States.

postpartum psychosis; some women receive treatment at general hospitals or do not receive treatment at all. Yet, enough cases are admitted to mental hospitals to constitute approximately 10 percent of the female patients in the typical institution. Most of these women (75–80 percent) show their disturbance within the first week or two after delivery (Madden, Luhan, Tuteur & Bimmerle, 1958; Baker, 1967).

There is some evidence that more disturbances occur during and after the first pregnancy (e.g., Thomas & Gordon, 1959; Vislie, 1956). The sex of the baby has *not* been shown to have any important relationship to the mother's reaction (e.g., Jansson, 1963). Maternal age turns out to be a complicated question. Some women develop postpartum psychoses in their early twenties; by and large, these are people who are under more psychological stress than they can handle from all sources. Other women develop the reaction in their mid-thirties; these women seem to be fatigued, overworked people who already have a houseful of children (Flohil, 1967).

CHARACTERISTICS OF POSTPARTUM PSYCHOSIS

It cannot be said that all women with severe disturbances after delivery think, feel, or behave in the same way. Yet some characteristics have been observed often enough to be worth singling out for attention:

- *Sense of inadequacy.* The mother feels she just cannot do what is expected of her. Her self-image is that of a bungler, a failure, a person who cannot rise to the demands of life.
- *Sense of despair.* Feeling blue, hopeless, and drained of energy, the mother makes little or no effort to cope with her situation. She is withdrawn, hard to reach.
- *Suicidal/homicidal.* Thoughts of "ending it all" may occur during this time. There is a real danger of suicidal attempt. The same person may also have an impulse to do away with her infant, an impulse that can frighten her enough to increase her motivation for suicide.
- *Shame and guilt.* The woman may be convinced she has done something terrible and unforgivable in having a baby, or she may hold herself responsible for real or imagined abnormalities in the child.
- *Denial and other bizarre thoughts.* She may deny either the pregnancy or the birth, even deny the fact that she is in the hospital. She may believe strange things are taking place in her body, etc.

Symptoms such as these make it very difficult for the mother to take care of herself properly, let alone begin to form a positive relationship with her newborn.

DEVELOPMENTAL HAZARDS

The newborn may sometimes feel the effects of the mother's postpartum difficulties long after his or her first thirty days of life. The effects are not necessarily limited to the newborn itself.

If the mother does commit suicide, then the infant has not only lost a parent (and the husband a wife), but he or she grows up with the stigma of suicide over his or her head. Stigma is also possible if the mother is given the identity of a *mental patient*. Unfortunate and unfair as it may be, the child can become tainted in society's eyes by the mother's sufferings. And if the newborn is a female, her own subsequent adjustment to the role of motherhood might be affected by this critical incident in family history.

In some instances the mother's disability is temporary and soon recedes into the past. In other instances, however, the episode may continue to haunt the family. Furthermore, the postpartum psychosis itself may vanish, but leave behind a woman who lacks confidence in her own worth and competence. The relationship between wife and husband (or between wife and other relatives) can also be strained and may or may not recover.

What is the direct effect of the mother's psychological disturbance upon her infant? In many instances another person takes over the mothering role during the neonate's first few weeks of life. Whether or not this constitutes a hazard depends, of course, upon the quality of care that the substitute provides. But there is reason for concern when the infant (and other young children) depend for their care upon a disturbed mother. The hazards range from physical neglect and assault to more subtle forms of rejection, all of which can leave their marks on the developing child.

WHAT CAN BE DONE?

Perhaps the best indicator for identifying a woman who is especially vulnerable to major disturbances after delivery is previous life experience. The following facts and events tend to occur more frequently than the average in the personal and family histories of women who suffer a postpartum psychosis (Gordon, 1959):

1. Family history of emotional disorder, divorce, separation, or early death of a parent
2. Personal history of previous emotional disorder
3. Physical complaints during pregnancy
4. Recent changes in place of residence
5. Husbands unavailable during the weeks after delivery
6. Less outside help available in the first few weeks at home with the baby

It would be useful for physicians, nurses, and others who counsel pregnant women to be aware that this type of personal history tends to be associated with postpartum disturbance. In fact, whether or not the woman develops a full-blown psychosis, all of these indices suggest that here is a person who needs a little extra understanding and support.

One might recommend psychotherapy for the pregnant woman who fits the profile sketched above. But one might also provide help by such measures as:

1. Teaching her details of home management and child care that could build confidence and make her, in fact, a more competent mother.
2. Assisting in other reality arrangements, for example, helping to plan for help with the baby at home if the husband is likely to be unavailable.
3. Bringing the husband more actively into the picture, through counseling or instruction. What are his expectations and fears, strengths and weaknesses? How can he be helped to become the kind of father who will make it easier for his wife to overcome her postpartum vulnerabilities?
4. Finding a way to take some of the extra load off the parents that may exist in certain families (e.g., a neighborhood plan to look after other small children in the family for awhile).

Men can have postpartum reactions too (e.g., a kind of dethronement)

It has not been my intention here to frighten prospective mothers and fathers with horror stories about postpartum psychosis. But we should recognize that the psychological hazards every infant faces include the mental state and competence of the parents. Although the hazards become intense and operational in a minority of situations, some of the factors described here may be present and troublesome to a lesser degree for other mothers and their families. Our opportunity to avoid or minimize these problems depends upon our willingness to face up to their reality, a willingness that does not have to include fatalism or the inclination to reject those who fall victim to this form of suffering.

SUMMARY

Birth is in one sense a social fact and statistic, indicating that the new human is firmly related to his or her community. The *event* of birth is distinguished from the *process*. Birth is also an *idea* that symbolizes emergence. Although coming into the postnatal environment represents a new *beginning*, it also involves the *termination* of some factors, and the *continuation* of others. A core of *identity* passes from the prenatal to the postnatal state.

What can the newborn infant experience and do? *William James* suggested many years ago that the neonate has his or her own unified way of looking at the world. Some kind of organized nervous system/mental activity is with us from the start of postnatal existence. In more recent years much research has been conducted into the newborn's *unlearned behaviors.* These *reflexes* are: *yawning, sneezing, sucking, stretching, crying, rooting, grasping, toe flexion-fanning, Moro (startle), crossed extensor, deep tendon, "doll's eye," galant, and supporting and stepping* reactions, as well as the later appearing *parachute response.* These behavior patterns normal for the neonate can return as part of a pathological syndrome in later life. Reflections on reflexes were considered. These centered around the possible *survival value* of the early-appearing responses and their relationship to the balance between "old brain" and "new brain" dominance.

An active view of the infant does more justice than one that is limited to a stimulus-response package of reflexes. This was seen through an exploration of perceiving, preferring, and learning in the first weeks of postnatal life. The *visual* apparatus is functional at the time of birth, and *acuity* is often well developed. After about one week, the infant begins to show a preference for the human face as a visual stimulus. The visual scanning behavior of young infants shows evidence of coordinated movements and definite preferences and "search strategies," for example, an inclination to scan horizontally rather than vertically and to fixate more on patterned configurations than on homogeneous fields. *Audition* (hearing) occurs to a limited degree before birth. The neonate responds to intense sounds and is apt to be quieted when exposed to the mother's heartbeat or other rhythmic stimulation. Ingenious studies have shown that even month-old infants can distinguish speech-like sounds from other sounds and have some ability to analyze what they hear into categories.

Much knowledge of the neonate's abilities has been obtained by studies of *habituation* and *learning.* The very young infant pays less and less attention to stimulation that is repeated without variation, as though he or she knows the difference between what is new and worth paying attention to and what can be ignored. *Classical conditioning,* a simple form of learning, has been demonstrated in babies less than one week old, but requires favorable circumstances (e.g., an attentive, stress-free child). Neonates also show the ability to differentiate their own responses to a limited degree (i.e., exercise selective control over their behavior). The prospects of *behavioral engineering* to produce certain types of behavior in neonates was raised. Dangers in this approach were cited, including the potential threat to *individual differences.* In every dimension of neonatal research, individual differences, as well as group trends, must be acknowledged and respected.

Scenario 3 portrayed the arrival of Michael on the postnatal scene. Tom's differential reaction to the gender of his first child was noted, and we explored Michael's physical appearance as seen through Gloria's eyes.

Postpartum psychosis was examined as one hazard that could develop during the period immediately following birth. The stress of childbearing can

YOUR TURN

1. There are several variants on Scenario 3 that you might find worth your effort to develop.

Variant A: Premature birth. What would Michael look like and be like (in terms of reflexes and other capacities) if he were born at 8 months? 7 months? 6 months? How do you suppose Gloria and Tom would have responded? What kind of help might they need at this point? You will have to do some reading and observing on your own to develop this scenario. Start off with one additional fact: the timing of birth itself is not as important as the level of development that has been achieved. Although gestation time and fetal development are very closely related, they are not identical. Low birth-weight is often a better measure of immaturity and possible difficulties than prematurity alone.

Let me also share one of my own biases; perhaps you will agree with it, perhaps not. Physicians and nurses often speak of the early or low-weight neonate as a *preemie.* This is a shortened form of *premature.* I suggest that this is dehumanizing. This impression is strengthened by my observation that referring to a newborn as a *preemie* is often accompanied by other verbal and nonverbal behaviors that indicate a desire on the part of the speaker to be emotionally aloof from the infant. Try speaking of a *preemie* yourself. Isn't it easy to persuade yourself that you are talking about something other than a human being? I have said, this is a bias that you do not have to share, but it is another element for your consideration in developing Variant A, and it recalls some of the issues touched on in Chapter 1.

Variant B. The situation is as described in Scenario 2 (and Scenario 1), but Michael turns out to be plural: twins. How might this affect the family constellation? Are there ways in which the life experiences the twins have might be different in the first thirty days from those of a single child? Might the fact that a single child was expected make a difference?

Variant C. The mother is unmarried and has decided to place her baby for adoption. In some hospitals the infant is never shown to the mother. In other hospitals (an increasing trend), the unmarried mother is given her choice. What might be the consequences for the woman and child of her seeing or not seeing the infant she has decided to relinquish? Consider the possibilities not only for the first thirty days, but for the long run.

Again, there may be other variants that come to your mind when you think through Scenario 3 and the related material presented in this chapter.

2. Are you preparing for a career in nursing, physiology, medicine, or a related field? You might then find it rewarding to do a more thorough study of the neonate's unlearned behaviors. When does each reflex come into operation? What are the anatomical and physiological bases of each reflex? When does the reflex normally disappear? How can we account for its disappearance at this particular time? Under what conditions does a reflex fail to develop, or fail to disappear? You probably will not be able to answer all these questions satisfactorily, but you should learn some useful material and also deepen your appreciation for the complexity of humans developing.

3. Recently a woman described the birth of her daughter under the care of a French physician by the name of Frederick Le Boyer. Caterine Milinaire (1974) reported that her physician has a new and perhaps revolutionary approach to the event of birth. He takes special care to provide an atmosphere of comfort and security to the neonate during the very first minutes of life outside the womb. One of his techniques, for example, is to immerse the newborn up to his or her neck in bath water slightly warmer than body temperature. Le Boyer states:

> What happens then is wonderful: the child truly relaxes. You can see he feels at ease, often a real smile appears on his face. . . . The baby in the warm water literally begins to play and to discover inner and outer space, moving arms and legs. You look at him and you really get the feeling of someone discovering the world. Often I turn my eyes away so as not to impose my gaze upon the child, his searching is so great that it almost brings tears. Some babies become calm straightaway, stop crying and play at stretching their arms and kicking their legs; others take longer. . . . We then take him out, dry him and wrap him very loosely in cotton cloth. Then another important moment comes: we let him experience stillness. It is quite an unknown feeling for the baby since he has lived in a storm (of amniotic fluid motion) for nine months. . . . No more screams and tears but a smile. When a child is born this way everything becomes voluptuous instead of painful. (Milinaire, 1974, p. 208).

Read Milinaire's book and reflect on questions such as these: Is it possible that for years the standard procedure of bringing a neonate into his or her new environment has imposed unnecessary stress and alarm? If so, why have so many people, including members of the health professions, taken this condition for normal? What are the possible short-term and long-term effects on the infant's development when introduced to the world in such an apparently delightful manner, as contrasted with the usual approach? What do Dr. Le Boyer's innovations suggest about the relationship between our own expectations and values and what we observe and do to each other, starting from the moment of birth? How might we evaluate carefully the technique reported by Dr. Le Boyer? Should it be used more widely and, if not, why not?

be accentuated by other situational factors and produce a dangerous vulnerability in both mother and child. The discussion of maternal disturbance after birth included suggestions for identifying possible problems in advance and attempting to alleviate them.

Reference List

Baker, A. A. *Psychiatric disorders in obstetrics.* Oxford, England: Blackwell Scientific Publishers, 1967.

Brackbill, Y., Adams, G., Crowell, D. H., & Gray, M. L. Arousal level in neonates and preschool children under continuous auditory stimulation. *Journal of Experimental Child Psychology,* 1966, *4,* 178–188.

Capute, A. J., Accardo, P. J., Vining, E. P. G., Rubenstein, J. F., & Harryman, S. *Primitive reflex profile.* Baltimore: University Park Press, 1978.

DiLeo, J. H. Developmental evaluation of very young infants. In J. Hellmuth (Ed.), *The normal infant.* New York: Brunner/Mazel, 1967, pp. 121–142.

Eimas, P. D. Speech perception in early infancy. In L. B. Cohen & P. Salapetak (Eds.), *Infant perception from sensation to cognition. II: Perception of space, speech, and sound.* New York: Academic Press, 1975, pp. 193–232.

Engen, T., Lipsitt, L. P., & Kaye, H. Olfactory responses and adaptation in the human neonate. *Journal of Comparative & Physiological Psychology,* 1963, *56,* 73–77.

Fantz, R. L. Pattern vision in newborn infants. *Science,* 1963, *140,* 296–297.

Forbes, H. S., & Forbes, H. B. Fetal sense reaction: Hearing. *Journal of Comparative & Physiological Psychology,* 1927, *7,* 353–355.

Flohil, J. M. A preliminary study of the interrelationships of certain clinical data in postpartum psychoses. *Psychiatrica, Neurologia, Neurochirurgia,* 1967, *70,* 191–196.

Gollin, E. S. Research trends in infant learning. In J. Hellmuth (Ed.), *The normal infant.* New York: Brunner/Mazel, 1967, pp. 241–266.

Gordon, R. E. Sociodynamics and psychotherapy. *AMA Archives of Neurology & Psychiatry,* 1959, *81,* 486–503.

Gozali, J., & Demorest, A. S. *A review of the literature on postpartal mental disorders.* Milwaukee, Wisc.: Milwaukee Jewish Vocational Service, 1970.

Haber, R. N., & Hershenson, M. *The psychology of visual perception.* New York: Holt, Rinehart and Winston, 1973.

James, W. *The principles of psychology.* New York: Holt, 1890.

Jansson, B. Psychic insufficiencies associated with childbearing. *Acta Psychiatrica Scandanivica,* 1963, *39,* suppl. 172, 1–168.

Johansson, B., Wedenberg, E., & Westin, B. Measurement of tone response by the human fetus. *Acta Otolaryng.,* 1964, *57,* 188–192.

Kessen, W., Haith, M. W., & Salapatek, E. H. Infancy. In P. H. Mussen (Ed.), *Carmichael's manual of child psychology.* New York: John Wiley & Sons, Inc., 1970, pp. 287–446.

Kline, C. L. Emotional illness associated with childbirth. *American Journal of Obstetrics & Gynecology,* 1965, *69,* 748–757.

Lipsitt, L. P., & Kaye, H. Conditioning sucking in the human newborn. *Psychonomic Science,* 1964, *1,* 29–30.

Madden, J. J., Luhan, J. A., Tuteur, W., & Bimmerle, J. F. Characteristics of postpartum mental illness. *American Journal of Psychiatry,* 1958, *115,* 18–24.
McGraw, M. Suspension grasp behavior of the human infant. *American Journal of the Disabled Child,* 1940, *60,* 799–811.
Milinaire, Caterine, *Birth.* New York: Harmony Books, 1974.
Pavlov, I. P. *Lectures on conditioned reflexes.* New York: International Publications, 1928.
Pavlov, I. P. *Experimental psychology and other essays.* New York: Philosophical Library, 1957.
Provenzano, F., Rubin, J., & Luria, Z. Research in progress. Reported by Ellen Goodman in her column in *The Boston Globe,* Feb. 20, 1973.
Salapatek, P. Visual scanning of geometric figures by the human newborn. *Journal of Comparative and Physiological Psychology,* 1969, *66,* 247–258.
Stechler, G., Bradford, S., & Levy, H. Attention in the newborn: Effect on motility and skin potential. *Science,* 1966, *151,* 1246–1248.
Steinschneider, A. Sound intensity and respiratory responses in the neonate. *Psychosomatic Medicine,* 1968, *30,* 534–541.
Steinschneider, S., Lipton, E., & Richmond, J. Auditory sensitivity in the infant: Effect of intensity on cardiac and motor responsivity. *Child Development,* 1966, *37,* 233–252.
Taft, L. T., & Cohen, H. J. Neonatal and infant reflexology. In J. Hellmuth (Ed.), *The normal infant.* New York: Brunner/Mazel, 1967, pp. 79–120.
Thomas, C. L., & Gordon, J. E. Psychosis after childbirth: Ecological aspects of a single-impact stress. *American Journal of Medical Science,* 1959, *238,* 363–388.
Thompson, D. A. *On growth and form.* Cambridge, England: Cambridge University Press, 1942.
Timiras, P. S. *Developmental physiology and aging.* New York: Macmillan Inc., 1972.
Tronick, E., & Clayton, C. Infant looking patterns. *Vision Research,* 1971, *11,* 1479–1486.
Vislie, H. Puerperal mental disorders. *Acta Psychiatica et Neuologica Scandinavica,* 1956, *31,* suppl. 111, 3–42.

THE FIRST TWO CANDLES

Growing Toward People

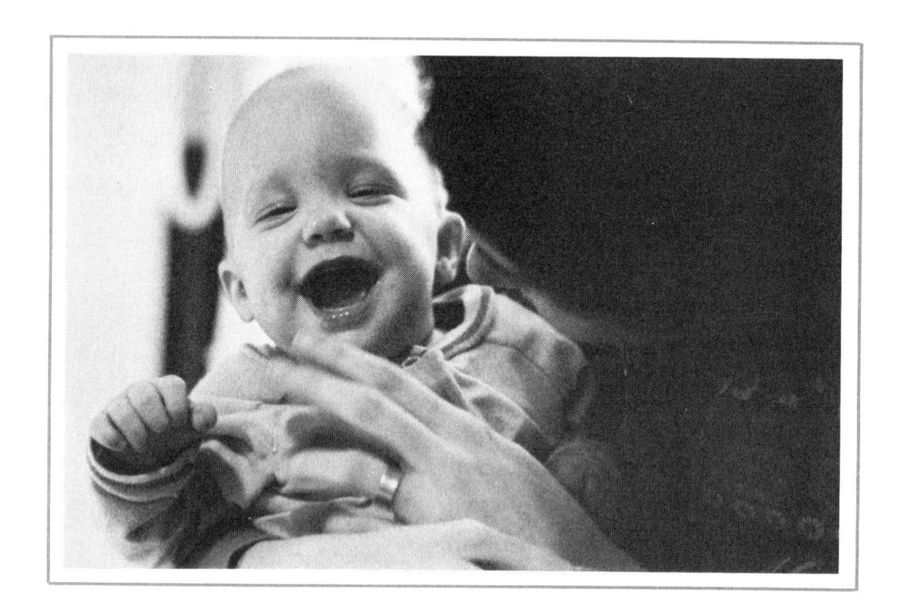

5

CHAPTER OUTLINE

I

Growing toward people—and in the process becoming a person oneself—is a fundamental pathway for development in the first two years of life.

Developmental time / Dependency (objective and subjective) / Attachment (Bowlby) / Goal-directed partnership / Exploratory behavior / Visual-motor orientation / Clinging

II

Development occurs between people as well as within people.

Parenting / Affective climate / Feeding / Let-down reflex / Response to crying / Maternal overprotection / Fathering / Separation anxiety / Average expectable environment / Stimulation / Parental assault

How long is a minute? An hour? A year?

The stopwatch, the clock, and the calendar lack imagination. According to the stopwatch, the minute that elapses while the dentist readies his drill for your molar is equivalent to the minute that you stretch, yawn, and turn over before getting out of bed in the morning. The clock fails to distinguish between the hour that is taken up with an impossible final exam in developmental psychology and the hour spent in loving intimacy. If we are to believe the calendar, the "distance" between age forty-six and age forty-seven is the same as that between ages one and two.

But we know better. Periods of time may feel quite different to us although their objective durations are identical. The concept of *developmental time* (*see* Chapter 2) is relevant here. It is not just the case that more *seems* to happen during some periods of time than others; in human development, more *does* happen at certain periods. This is worth keeping in mind, despite the fact that major individual differences exist in developmental time. For a particular person, the twelve-month span between the forty-sixth and forty-seventh birthdays may be consumed in a profound renewal and redirection of life-style. For another person of identical age, the "same" period might be merely another twelve months of standard operating procedure.

The first two years of life throb with developmental time. So much occurs during these months that I have decided to devote two chapters to the same period and, even so, I must be highly selective in what I include. Let's begin with a focus on the relationship between child and mother. This choice is intended to illustrate the point that development takes place *between* as well as *within*. It would be unreal to describe the young child's behavior and growth as though it were contained entirely within itself, like toothpaste waiting to be squeezed from the tube at the right time. Furthermore, some of the most fundamental concepts in developmental research involve relationships; terms such as *dependency* and *attachment* will soon become familiar to you, if they are not already.

Growing toward people—and, in the process, becoming a person oneself—is a fundamental pathway of development in the first two years of life. Important in itself, this topic also provides a useful context for viewing the physical and mental growth that is described in Chapter 6.

THE INFANT AS WARY ADVENTURER: EXPLORATORY AND ATTACHMENT BEHAVIOR

GETTING AROUND MOTHER: EARLY EXPLORATIONS OF THE ENVIRONMENT

We may think of the older child as "getting around mother" in the sense of achieving his or her own goals in the face of mother's opposition. For the very young child, however, "getting around mother" can be taken more literally. Mother provides a home base, both physically and emotionally. An example of this is found in the research of Harriet Rheingold and her colleagues at University of North Carolina (Rheingold & Eckerman, 1970).

The components of this research scenario were: infants ten months of age, their mothers, two adjacent rooms, and a few toys (plus, of course, the investigators themselves). Mother and baby would enter the "starting room" and sit on the floor. The door leading to the second and larger room was left open. The study attempted to answer these questions: Would the infant leave his or her mother to enter the "open field"? How much difference would it make to some babies to place a toy in the second room?

One of the advantages of the laboratory approach is the opportunity to establish a relatively controlled or constant environment, where only those variations occur that the experimenter has planned. Another advantage is the opportunity to make careful, detailed observations. In this study there were two observers who looked through a one-way glass (to see without being seen) and independently recorded a variety of facts: how long it took the infant to enter the second room, how long he or she stayed there, what he or she touched, etc. A tape recorder was also used for later determination of "distress" or "nondistress" sounds that the young traveler might emit.

As normal ten-monthers, all twenty-five babies in the study had developed ways of getting around. The group included many proficient creepers, a few belly-crawlers, and a few early toddlers (again, individual differences!). Each mother-child pair started the session facing each other on the floor, the parent smiling and speaking softly to her child. In half the situations, the second room was empty; in the other half, it contained a pull toy.

What happened? *All* the infants left their mothers and crept, crawled or toddled into the open field. There was no fussing or crying. Toy or no toy, the babies stayed just about as long in the second room. But they proved they were not too young to surprise the researchers: the toy-finders did not linger in the open field; they "decided" to bring their prizes back to mother in the starting room. As might be expected, those who did not return with toys spent more time in direct contact with their mothers than those who had their new prizes to play with. Furthermore, the infants did not settle for a single trip; they would go out again, alternating between the two environments.

Do babies really "decide"? We don't know

This study (and others like it) have demonstrated that infants will engage in **exploratory behavior** with apparent spontaneity and pleasure as long as there is a home base for a safe return. For some babies, eye contact is sufficient. They are happy and adventuring if they can return to see mother, while others seek the comfort of physical contact. A baby placed *alone* in the same environment will not behave in the way that has been described here. Without the reassuring presence of mother, the child will show obvious signs of distress and little or no exploratory behavior (Rheingold, 1969).

Exploratory behavior is action motivated by wanting to know more about the environment

We should not be deceived by the apparent simplicity of what was observed in this study. The process of gradually detaching oneself from home

Mother is often the point of security around which the earliest environmental explorations occur. (Photo by James R. Holland, Stock, Boston.)

base may begin in small ways. But begin it does—and very early as we have seen—and it continues until the once-infant has become an independent adult. The ten-monther creeps from one room to the next. Later, in the words of Rheingold and Eckerman:

> . . . he walks out of the house, plays in the yard all morning, goes to school, goes still farther away to high school, then to college and to work. He crosses the country, and now he may even go to the moon. Eventually he sets up his own home and produces infants who, in turn, repeat the process.
>
> The infant's separating himself from his mother is of biological importance. It is of consequence for the preservation of both the individual and the species—of the individual, since it confers the advantage of greater familiarity with the environment and thus increases the likelihood of adaptation . . . of the species, since it allows the mother to care for the next offspring and leads eventually to the formation of breeding pairs.
>
> The infant's separating himself from his mother is also of psychological importance for it enormously increases his opportunities to interact with the environment and thus to learn its nature. For, while he is in physical contact with his mother, his universe is confined to her person and the environment near her. There are limits to what the most attentive mother can bring to him. Even when he is carried about, his contacts with the universe are necessarily circumscribed. When, however, he leaves her side by himself, many new kinds of learning can occur. (1970, p. 78).

The child's earliest locomotions away from the parents can be regarded, then, as links in a chain of development that is crucial both for the individual and for the species. But this is only one side of the situation. The mother obviously represents security and home base in Rheingold's studies, as she does in every day life. To deepen our appreciation of what takes place between mother and infant, it is important to consider the ties that bind them, as well as those processes that lead to separation and independence.

ATTACHED OR DEPENDENT?

Objective dependency is a need for others' protection based on reality

Subjective dependency is a need for others' protection whether or not it is based on reality

Every human begins in a state of **objective dependency.** He or she truly requires care from others to survive. This state is not the same as **subjective dependency,** in which a person *feels* that he or she needs the support of others. In our society it is often a put down to describe a person as dependent. The implication is that he or she is unwilling or unable to stand on his or her own feet psychologically. This bias should not be applied to the infant. The infant is realistically dependent upon the care-taking behavior of those who are more experienced and competent in adapting to the world.

As the child grows up he or she becomes more self-reliant. Eventually he or she becomes an adult who can meet the dependency needs of a new generation. Dependency, then, is a characteristic that is most powerful at the start of life and progressively diminishes thereafter. (Illness and advanced age may, of course, return the adult to a position of objective dependency on either a short- or long-term basis).

Attachment is a relationship in which closeness and interaction are sought on a continuing basis

Attachment is quite another story. It is a bond that unites a particular offspring with a particular adult. In the broader sense, attachment refers to any relationship in which the participants show a pattern of recognition, preference, and loyalty over a period of time. A very young infant is not likely to show consistent preferences among various adults. He or she is highly dependent, but unattached. You could "play mother" to him or her and be acceptable. Long before the second birthday, however, the child has fixed upon mother as a very special person and distinguished between intimates and strangers.

Dependency and attachment move in opposite directions. Dependency lessens and attachment behavior increases as the infant matures and learns. These opposite currents of development may not always be obvious to those who are caring for a young child on a daily basis. The dependency needs remain high, while growing signs of attachment may be small and subtle. Nevertheless, every passing month witnesses the child moving away from a state of basic dependency and toward a more particularized and complex set of relationships with humankind.

Mary Ainsworth, one of the leading reasearchers in the area of attachment behavior, observed infant-mother relationships in Baganda, Africa. Although she had reason to believe Bagandan babies were somewhat ahead of most infants in U.S. society in their development of attachment behavior, she also judged that "the same principles of development apply to infants regardless of specific racial or cultural influences" (1964). What is most relevant here are the examples of attachment behavior she observed; these are summarized in Figure 5–1. The earliest age at which a given behavior was noticed is listed in the middle column. The typical age for the same behavior appears in the far right column.

The first three of the thirteen behaviors involve the ability of the infant to discriminate mother from other people. A baby who cries when held by

Behavior	Earliest Observation	Commonly Observed
Differential crying	8 weeks	12 weeks
Differential smiling*	9 weeks	32 weeks
Differential vocalization	20 weeks	?
Visual–motor orientation*	18 weeks	?
Crying when mother leaves	15 weeks	?
Following	17 weeks	25 weeks
"Scrambling" over mother*	10 weeks	30 weeks
Burying face in mother's lap	22 weeks	30 weeks
Exploration from mother as secure base	28 weeks	33 weeks
Clinging	25 weeks	40 weeks
Lifting arms in greeting*	17 weeks	22 weeks
Clapping hands in greeting*	28 weeks	40 weeks
Approach through locomotion*	26 weeks	30 weeks

*This behavior may occur even earlier than revealed by the original study.

Figure 5–1. *The pattern of attachment behavior shown by the infant toward mother. (Reprinted from M. D. Ainsworth. Patterns of attachment shown by the infant in interaction with his mother.* Merrill-Palmer Quarterly, *1964, 10, Table 1, p. 52.)*

somebody else, but who stops crying when taken by his or her mother, is showing differential behavior; he or she is more attached to one person than to people in general. As Figure 5–1 indicates, differential crying, smiling, and vocalization are among the first indices of attachment behavior to appear. The behaviors that follow seem closely related to the infant's concern for his or her mother's whereabouts. The nature of these behaviors is fairly obvious from Figure 5–1, but it might be helpful to describe a few.

Ainsworth describes *visual-motor orientation* toward the mother as follows:

> The baby, when apart from his mother but able to see her, keeps his eyes more or less continuously oriented towards her. He may look away for a few moments, but he repeatedly glances toward her. When held by someone else, he can be sensed to be maintaining a motor orientation toward his mother, for he is neither ready to interact with the adult holding him, nor to relax in her arms. (p. 54).

Clinging is a pattern that develops much later.

> The clinging pattern which is so conspicuous in infant monkeys was not observed in these infants until twenty-five weeks at the earliest. The most strik-

> ing instances of clinging in the first year of life were clearly associated with fright. The only clear-cut fear-arousing stimulus . . . was the stranger. . . . If the mother tries to hand him to the stranger . . . the baby screams and clings desperately, resisting all efforts to disengage him. This panicky clinging in response to strangers was not observed in any child younger than forty weeks of age. A less intense kind of clinging was seen in somewhat younger children. With one six-month-old child, for example, the cause seemed to be separation anxiety, for he wanted to be with his mother the whole time, and sometimes clung to her, but in an intermittent way and not so desperately and tightly as did the infants who were frightened by a stranger. . . . (p. 55).

Perhaps a mother feels annoyed or embarrassed at times when her offspring displays some of these behaviors. "Be nice to Aunt Nellie!" But the infant knows enough to be aware that Aunt Nellie is not his or her mother, and the burden is upon the "stranger" to gain the baby's trust if she can. It is worth observing that the earliest forms of attachment behavior do seem to bring along some inconveniences. The infant may seem more temperamental in demanding Mother and no substitutes. We should not be surprised that developmental achievements (such as perceptual and emotional discrimination) exact their price from the world. In fact, the student of human development might find it a useful exercise to determine what new problems are created for all parties concerned *every* time the maturing person displays a new pattern of thought or behavior.

The attachments formed by the very young child provide the foundation for the most profound relationships he or she will establish as an adult. This statement does not mean that there is a simple and direct connection between differential crying, for example, and choice of spouse two decades later. But it does mean that the complex web of human relationships known to adults has early beginnings. And, just as frustration and distress may accompany the pleasures and comforts of adult relationships, so we might expect the attaching process to pose its challenges and problems right from the start.

I will be concentrating on attachment somewhat more than on dependency in the following section. But whatever the shifting balance between dependency and attachment happens to be, there is no question about the continuing need of the infant and young child for responsible care from adults. Keep the distinction between the two in mind, however. There are important individual differences in the progression from dependency to attachment and in the environmental factors involved.

SCENARIO 4: MICHAEL ATTACHES

The process of attachment can be illustrated by calling on Michael again. We will observe his development of relationships with his mother and with other intimates over the first two years of life. Much of the information I am drawing upon here derives from the work of John Bowlby and his colleagues at the

Tavistock Institute of Human Relations in London (Bowlby, 1958, 1969; Heinicke & Westheimer, 1965). It is Bowlby, a perceptive and far-ranging psychiatrist, who introduced the concept of attachment into the study of child development. This condensation and interpretation of Bowlby's work does not substitute, of course, for your own exploration of the original sources.

Expressive behavior reveals the individual's state of mind

Instrumental behavior directly affects the environment or others

We saw in the previous chapter that right at birth or soon after, Michael could use his eyes, ears, and other sense organs. He could also engage in **expressive behavior** (such as crying) and **instrumental behavior** (such as sucking). Before long, he also started to smile and babble. Because he was fortunate enough to be in an affectionate family setting, Michael often had company when he gazed, bawled, smiled, and babbled. In a very broad sense, his behavior was social. But he had not yet aligned himself with particular individuals in a particular way—he was dependent, but not attached.

Looking at Michael's development through Bowlby's eyes, we recognize four phases in the attachment process (although the phases may overlap each other).

1. Even during the first two or three months of postnatal life, Michael is human enough to take notice when he is in the presence of another person. He recognizes his own species (more or less), but not specific individuals. When another human is nearby, Michael attempts to track him or her with his eyes; he grasps and reaches, smiles and babbles. According to Bowlby, this is phase one: *"orientation and signals without discrimination of figure."* This might be said another way: "Hello—I know you're there, and I'm mighty glad to see you (whoever you are)!"

2. Over the next three months or so, Michael gradually becomes more "turned on" by a particular person. In this case, it is his mother. It is *not* just because she is the one who usually feeds him and meets his physical care needs. This is only part of the attachment story. The other part of it is that mother has been receptive to Michael's social overtures. It is natural for him to favor the company of somebody who notices and reinforces his actions. The infant continues to be friendly and responsive toward other people as well, but his mother's voice (and, a little later, her face) especially delights and comforts him. Father also receives special notice, but the beginning of a clear attachment to mother is the dominant development. Bowlby describes this second phase as *"orientation and signals directed towards one (or more) discriminated figure(s)."* Michael's response to his mother at this time is, "It *is* you, isn't it—I am so pleased!"

3. Michael is now about half way toward his first birthday. An interesting shift is starting to take place in the way he relates to others. He is beginning to care about the difference between one person and another. It is now possible for somebody to become a "stranger." Until this time, Michael favored mother, but accepted most other people as though they were all one of the family. These days, however, he is becoming more cautious in sizing up

people. If you are new to him, then he is likely to give you a long, unsmiling stare. Sooner or later, he will probably have episodes of alarm and withdrawal when approached by somebody other than his parents.

At the same time, Michael is showing his attachment to mother in more and more obvious ways. He follows after mother when he can, greets her joyously when she returns, and starts to use her as home base for explorations. Since Michael also sees a lot of his father, he has accepted him as a "subsidiary attachment figure." In other words, "You're OK, too, dad (but where's mother?)." This third phase of attachment is designated by Bowlby as *"maintenance of proximity to a discriminated figure by means of locomotion as well as signals."* This definition emphasizes the growing attachment to mother, but the parallel discrimination against strangers should also be kept in mind.

4. The next phase in Michael's relationship to humanity is more difficult to describe and pinpoint in time. It seems to begin as he nears his second birthday. During the preceding months, his attachment to both his mother and father has become more obvious. He has also become adept at influencing *their* behavior in various ways—for example, the smile that brings a smile in return. But a major discovery is just beginning to dawn on him. He is starting to comprehend the fact that other people have lives independent from his own. They also have goals and plans. This means that he must have his own "strategy" if he is to influence their behavior in his favor.

Michael now tries seriously to figure out his mother, so as to predict and influence her behavior. If Michael were a scientist, we might say he is attempting to develop a *working model* of mother, a mental representation of the way his mother thinks and functions. Interestingly, Bowlby does not hesitate to use this term in describing the developmental situation of the young child:

> Inevitably, the earliest attempts that a child makes to change his partner's behavior are primitive. Examples might be pulling and pushing, and such simple requests or commands as "Come here" or "Go away." As he grows older, however, and it dawns on him that his mother may have her own set-goals and, moreover, that his mother's set-goals may perhaps be changed, his own behaviors become more sophisticated. Even so, the plans he makes may be sadly misconceived, owing to the inadequate working model he yet has of his mother. An example is a little boy, just short of two years, who, having been deprived of a knife by his mother, attempted to get her to return it by offering her his teddy bear. (Bowlby, 1969, p. 352).

In a **goal-corrected partnership,** one person alters his or her behavior to better accord with another's

Bowlby describes this fourth phase as *"formation of a* ***goal-corrected partnership."*** In other words, Michael is attempting to suit his behavior to the behavior of his mother, to establish a relationship that at least approaches a partnership. The more he can learn about mother's ways, the better he can adjust his actions to them.

This will be a long and complex process. Michael will be working on this challenge for several years, well into his school days. He will have to develop considerable mental skills to read the clues properly, and to create his plans with adequate consideration of what is in his mother's mind. As Bowlby notes: "A main reason for this is that, in order to grasp what another's goals and plans are, it is usually necessary to see things through the other's eyes. And this is what children are especially bad at doing. Evidence suggests not only that competence in so doing develops extremely slowly but that it is not even reasonably well developed until after a child has passed his seventh birthday" (p. 352). Michael's parents could tell us a good deal more about his attachment behavior, whether or not they use this term. They might report, as researchers have, that the little one enjoys the company of many different people when he is in good spirits, but when hungry, tired, or distressed, it is mother that he really wants. They have also observed that Michael's strong attachment to mother does not seem to stand in the way of gradually making friends with others. This is in keeping with the observations of child developmentalists who have found that an infant who has an intense attachment relationship to one person is more likely to broaden his or her realm of attachments in succeeding months than the infant who has only a weak attachment to his or her mother or mother-figure. We have here, then, the suggestion that outgoing sociability as a personality characteristic throughout life might be facilitated by a strong attachment relationship in early childhood.

Michael himself will both show and tell us that he is attached to some objects as well as people. His mother may be hard put to borrow his blanket to wash it because Michael will be calling and looking for "bankie" when he feels in need of comfort. As a very young baby, Michael seemed to care only about his bottle and his thumb. But now that he is heading toward his second birthday, he has gathered a group of favored objects. It is not easy to substitute any old blanket or cuddly toy for the particular one that he has taken to heart. Even at this early age, we see the beginning of attachments to things and places, pets and people. Michael has preferences; he can be pleased or frustrated by objects and people in the world. In short, he has rapidly become one of us.

PARENTING: THE EFFECTS OF PARENTAL MOOD AND BEHAVIOR

So far I have concentrated on the infant. But the people he or she is depending on and attaching to are also essential to the scenario. Development, as I have said, occurs between people as well as within people. How the parents interpret their baby's behavior, how much time they spend with him or her, what type of affection and discipline they provide—factors such as these recommend themselves to our consideration.

THE AFFECTIVE CLIMATE

Affective climate is the emotional tone of a certain group or situation

Would you choose to live in a soaking hot and humid region of the earth, or one that is frigid and blustery? There is little doubt that we are influenced by the prevailing climate and would prefer to live in "our kind" of climate, whatever that happens to be. The same appears to hold true for what Rene Spitz (1947) has characterized as the **affective climate.** The emotional tone of a particular family, neighborhood, office, factory, restaurant, classroom, etc. may be very much to our liking or it may constitute a source of distress. The food may be equally good (or equally bad) at two different restaurants, for example, but one place has a warm and friendly atmosphere; the other is edgy and uptight. A moment's reflection tells us that the infant has little if any choice about the affective climate that surrounds him or her on entering this world. Further, the infant is more the captive of his or her environment than we are and, most probably, a great deal more impressionable. He or she has known no other climate of human relations than the one being experienced at the moment. To understand the development of a particular infant, then, it is helpful to consider the prevailing affective climate. This does not mean, by the way, that the baby is only a passive recipient, basking

The affective climate an infant grows up in can nourish or impede personality growth. (© 1978 Photo by Thomas S. Wolfe.)

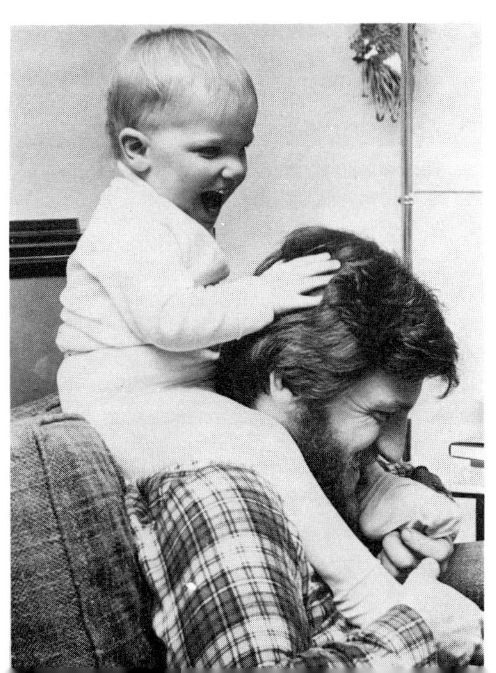

in the warmth of a serene and loving home or cringing in a hostile environment. Baby's own appearance and behavior will influence the affective climate to some extent. In some family constellations, in fact, baby may become the dominant shaper of the prevailing climate.

Spitz sees the affective climate as the most significant and sensitive framework of infant development:

> Affective signals generated by maternal moods seem to become a form of communication with the infant. These exchanges between mother and child go on uninterruptedly, without the mother necessarily being aware of them. This mode of communication between mother and child exerts a constant pressure which shapes the infantile psyche. I do not say that this pressure produces anything in the nature of unpleasure for the infant. I speak of "pressure" only because the words to convey these extraordinarily subtle and intangible exchanges have never been coined. (Spitz, 1965, p. 138).

Words such as *uninterruptedly* and *constant* are important in the appreciation of Spitz's views. He does not go along with the notion that personality

Figure 5-2. *Three is what kind of company? How infants interact with older siblings and parents. Eighteen-month-old infants are more likely to smile at, approach, vocalize to, and make offering gestures to their parents than to their preschool siblings. (Based on data presented by Lamb, 1978.)*

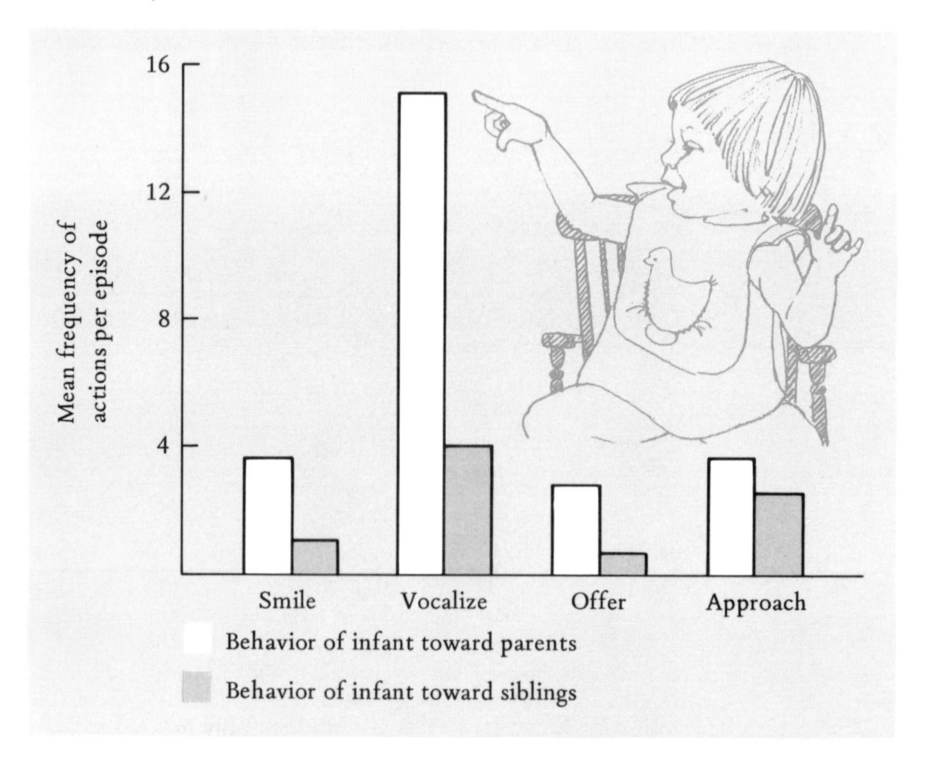

The trauma theory has been a media favorite because it *seems* simple and dramatic

is shaped for better or worse by specific traumatic events. One bad experience in childhood does not explain an individual's hang-ups twenty or thirty years later. Instead, personality is formed through the baby's pattern of constant *exchange* and *communication* with the environment.

Mother is the most powerful force in the environment, as she generates so many of the stimuli and responds to so many of the infant's attempts to communicate. It would be acceptable to define the affective climate for the infant as the basic mode of communication between mother and baby. The infant knows and shares the world through the feelings that pass back and forth between mother and himself or herself. Feelings of some kind are *always* present, now more intense, now more relaxed, varying from time to time, but usually within the general climate that has been established between them.

This means there is good reason to study mother and her moods as well as baby. If Spitz is correct, "The existence of the mother, her mere presence, acts as a stimulus for the responses of the infant; her least action—be it ever so insignificant—even when it is not related to the infant, acts as a stimulus" (Spitz, 1965, p. 123). The mother will ordinarily go through a variety of moods as she interacts with her baby. In return, baby sends many types of signals out to mother: drowsiness, playfulness, rage, etc. The close student of infant development should then learn to appreciate subtle interchanges between the partners in this twosome. (The term *dyad* is often used to denote the relationship between two individuals and will be encountered occasionally on these pages.)

THE FEEDING SITUATION

We can learn something about mother's mood and baby's response by focusing on the feeding situation. Food is provided to the infant several times a day, every day of the week. The infant will not develop adequately, will not even survive, unless his or her nutritional needs are met. In other words, the feeding situation is a frequent and significant behavior situation. For many years there has been controversy over the relative merits of the breast versus the bottle. Either can meet the baby's nutritional needs, but which is more satisfying to him or her emotionally? Does the infant need to nurse from the mother's breast in order to develop a sense of basic security? Or is the bottle just as good, perhaps better, and in some situations more convenient to use? The breast-versus-bottle controversy remains a dependable topic for popular magazines and newspaper advice columns. But research has provided a clear answer. As Rappaport puts it:

> Various studies of the problem done over the past thirty years all converge on the conclusion that breast or bottle feeding *per se* is less important than the manner or style of the feeding process. . . . The relative merits of breasts and bottles as apparatus are not psychologically important. The infant is certainly

> liable to be influenced in many ways by the quality of his feeding experience, but this will depend on more complex things than the mere fact of feeding through breast or bottle. (1972, p. 128).

The **let-down reflex** is a change in nipple position during nursing due to mother's relaxed state

A mother who is usually relaxed and comfortable while nursing her baby exhibits a **let-down reflex.** This refers to a change in nipple position that has the effect of allowing the baby to draw milk more easily. The infant nursing in the arms of a relaxed mother will appease his or her hunger needs without becoming especially tense and urgent. The infant's satisfied gurgles testify to the success of both mother and child. Next time they will feel confident, the mother will again be relaxed, and her reflex will be in good working order. By contrast, the mother who tenses up as nursing time approaches may not provide the helpful nipple reflex. This makes nursing a tougher job for her infant. The infant's own increasing tension affects the mother; she is likely to become even more uncomfortable, perhaps frustrated, disappointed, or angry. And the more upset the mother becomes, the harder it is for her to provide milk, and so goes the vicious cycle (Newton, 1958). It should also be added that the baby's own ease or difficulty in drawing nourishment from breast or bottle might be partially responsible for the mother's level of tension in the first place. A firstborn infant who is a good nurser might quickly reassure the new mother, while an infant with problems in feeding can have the opposite effect.

Thinking about these variations in the affective climate of feeding, we can appreciate that the physical object employed (bottle or breast) does not tell us what is taking place psychologically. The baby with the relaxed mother and the baby with the tense mother have different experiences although receiving nourishment in the "same" way. The mother who is tense during the feeding situation may also be tense in her other interactions with baby. If this is the case, then her child starts life with a pattern of interpersonal cues that bespeak rigidity and anxiety. Around these cues he or she begins to organize his or her own personality. This baby's attachment and dependency behavior might well differ from that of the child whose mother is relaxed in most of their encounters, and differ again from that of the child whose mother conveys relaxation in some situations but tension in others.

THE BABY IS CRYING

Although I have been using the feeding situation to illustrate how mother and baby affect each other's moods and responses, almost any kind of mother-child interaction would yield examples. Take another illustration: babies cry and mothers respond. Some babies cry longer and more often than others. Might this difference have something to do with the maternal response and, if so, in what way? Silvia Bell and Mary Ainsworth (1972) explored this question. They noted that the human newborn does not have the physical ability of offspring in some other species who can reach their mothers directly; the newborn cannot follow like the chick or cling like the monkey. But he or she can—and does—send vocal signals. Crying is one of the most conspicuous, and for the first months of life, probably the most important of these signals. Interestingly, crying is effective largely because it is disagreeable to adults. "Unlike smiling, which gratifies a caretaker, crying arouses displeasure or alarm and elicits interventions aimed at terminating it and discouraging its recurrence. Herein lies the power of crying to promote proximity more effectively than other early signaling behavior" (Bell & Ainsworth, p. 1172). The baby's first reaching out for human contact, according to this view, is by way of "disagreeable" behavior.

Just as in the breast/bottle controversy, there has been popular as well as scientific disagreement about the proper adult reaction to crying. How should a parent respond to the crying infant? Should mother come promptly and try to make the child comfortable, or would quick and consistent response spoil him or her? Let's acquaint ourselves with the way in which Bell and Ainsworth measured maternal response:

> The following measures of maternal behavior were obtained: The number of crying episodes that a mother ignores; the duration of maternal unresponsiveness, as measured by the length of time that a baby cried without obtaining a response from her; the types of interventions produced by the mother in terminating crying. Maternal interventions were classified as follows: picks up, holds; vocalizes, interacts; feeds; approaches, touches; offers pacifier or toy; removes noxious stimulus; enters room; and other. A crying episode was considered to have been terminated successfully when the baby remained quiet for more than 2 minutes thereafter. Whereas the maternal unresponsiveness measure reflects the promptness with which the mother intervened, the measure of maternal effectiveness reflects the mother's perceptiveness and willingness to give the type of intervention that will serve to terminate the cry. Maternal effectiveness was measured by the mean number of interventions a mother undertook before the cry was properly terminated.*

Subdividing a concept into many observable dimensions is useful in research

What did these researchers find in studying maternal response to crying? Consider first the mother-infant interaction during the first three months of life. Some mothers came quickly when their young babies cried and did so consistently. Others were deliberately unresponsive, in the belief that to respond would make the baby too demanding and dependent. Consider now the crying behavior of these infants as they approached their first birthdays: Those with promptly responsive mothers in their "youth" cried less often and for shorter periods of time than the infants who had been left to cry unattended or forced to wait longer for the maternal response. The more responsive the mother to her infant's signals of discomfort, the fewer such signals he or she exhibited at a later time.

When the infants were very young, they were more likely to cry when mother was away, not in physical contact with them, or out of visual range.

*From "Infant Crying and Maternal Responsiveness" by S. Bell and M. D. Ainsworth, *Child Development,* 1972, 43, 1175.

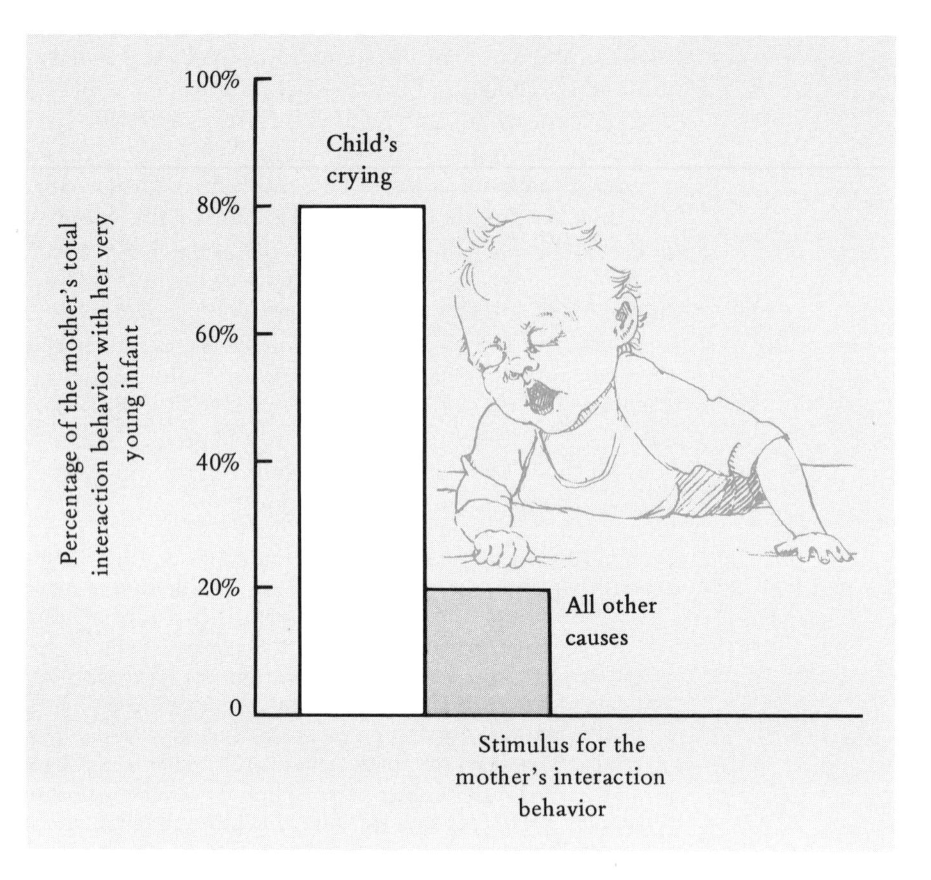

Figure 5-3. *Crying: a very effective behavior. Films of mothers and their newborn babies show that 80 percent of the mother's interaction behavior with her infant is in response to the baby's crying. (Based on data presented by Korner, 1974.)*

Goal correction is changing one's own behavior to achieve a goal as new information is discovered

By their ninth month or so, they were more likely to cry when near mother. This was interpreted by Bell and Ainsworth to mean that the child had begun to make **goal corrections** as if in order to influence mother's behavior more effectively, as Bowlby suggested. In other words, the older baby was more likely to behave (in this case, cry) when the person who this behavior might influence was close by. In this sense, the baby's behavior was taking the situation of the other person into account. The babies with responsive mothers were quicker to develop other social signals to replace crying. The researchers also had the strong impression that: "Mothers who promptly heed an infant's cries are sensitively responsive to other signals as well. Mother and infant form an interactional dyad: the more responsive she is, the less likely he is to cry and the more likely he is to develop more varied modes of communication" (p. 1188).

This study therefore came down strongly on the side of encouraging prompt and sensitive response to the young child's communications.

> Infants whose mothers have given them relatively much tender and affectionate holding in the earliest months of life are content with surprisingly little physical contact by the end of the first year; although they enjoy being held, when put down they are happy to move off into independent exploratory play. In contrast, those held for relatively brief periods during the early months tend to be ambivalent about contact by the end of the first year; they do not respond positively when held, but yet protest when put down and do not turn readily to independent activity. (p. 1189).

I have touched upon feeding and crying as though they were two different situations entirely. Yet we all know that hunger is one of the conditions that prompts crying, especially in the very young. Let's take one illustration of mother-infant dynamics concerning both feeding and crying, even if it seems to contradict some conclusions already presented.

MATERNAL OVERPROTECTION

Maternal overprotection is expression of a mother's own anxiety in excessive control of the child

In 1943, David Levy described a pattern of mothering to which he gave the title, **maternal overprotection.** It is the type of behavior that others have characterized as "smother love." Following up on this problem in 1965, Spitz observed several varieties of overprotection. The example I will use here is the mother who displays "primary anxious overpermissiveness" (p. 213ff.).

As Spitz's term suggests, the mother in this case indulges her infant too much, not necessarily because her love is so deep and boundless, but because she must do so in order to overcome a state of discomfort within herself. Spitz believes that *exaggerated* solicitude for baby actually leads to ill health if the infant happens to be physically vulnerable. The infant screams and screams. Feeding calms him or her only for a little while. Before long he or she is screaming again and showing symptoms of pain in the digestive tract. Changing over from bottle to breast, or vice versa, makes no difference, nor does any change in the formula. Astonishingly, however, this disturbance often disappears as mysteriously as it started (usually around the third month of postnatal life).

What has been going on? According to Spitz, the overanxious mother really cannot tell the difference between when her baby is crying because of hunger pangs and when he or she is crying for any other reason. She feels so guilty and anxious that she does not pause to look for alternative explanations. "Baby is screaming her head off: she must be hungry, I'd better feed her!" The mother's concern results in her "pushing food" into her infant at a time in development when excess food produces a great deal of intestinal activity. The baby's discomfort from overactive innards leads him or her to scream and become agitated. This results in more food service!

Does this actually contradict what we learned from the Bell-Ainsworth study? Not really. The sensitive, responsive mother looked for a variety of causes behind her infant's cries. Mostly she held the infant in her arms, comforting by physical proximity, gentle words, and caresses. She was free enough from her own anxieties to become a good reader of the infant's state. The overanxious mother was not spoiling her child with too much love; instead she was limiting her own response repertoire to an almost automatic type of behavior: "He cries, I feed."

MOTHERING, FATHERING, AND PARENTING

I have sampled some of the interaction that takes place between mother and infant. Most of the data on parental behavior with infants and very young children is limited to the mother. We do not know for certain how much of adult interaction should be thought of as *mothering,* as compared with *fathering,* or with *parenting* in general. Apart from breast feeding, is there anything a mother does with her infant that a father might not also do? Are the differences mostly quantitative—is the mother around the baby more often and thus more in contact with him or her? Or are there certain qualities that distinguish fatherly interactions from motherly interactions? These are questions that remain for future research to answer. One of the corollaries of the feminine liberation movement seems to be a fresh appreciation of men as care givers as well as wage earners, just as society is also working toward a view of women as much more than just homebodies.

One important line of research has already indicated that the father does play a significant role in the development of his child. Absence of the father during a boy's first few years of life seems to be associated with a more dependent and less aggressive child (Biller, 1971). The effect is strongest when the father is absent during the preschool years. Some developmentalists regard this finding as supporting a *modeling theory* of personality formation. The young boy usually "models" himself in part after his mother, but also in part after his father. Without Dad on the scene, the boy must develop his early core of attitudes and behavior patterns solely around the model provided by his mother. What remains to be learned are the precise ways in which sons learn to be more or less like their fathers. But it is already fairly clear that they grow up differently when father is absent. (Boys sometimes outgrow early dependent and docile characteristics as they vie with each other to demonstrate their masculinity in peer interactions, but traces of the original father-deprivation state may remain in the personality.)

Girls are not boys, however, and the effects of father deprivation might be expected to be somewhat different for girls. Hetherington (1973) found that fatherless girls showed most of the effects of their fatherlessness in adolescence, rather than during the early childhood period. The girls she studied

How important is father's attention for the young child's development? Recent studies suggest both mother's and father's are important. (Photo by Talbot Lovering.)

did not differ much from other girls in basic feminine identification. What they did manifest was: ". . . mainly an inability to interact appropriately with males. . . . The lack of opportunity for constructive interaction with a loving, attentive father resulted in apprehension and inadequate skills in relating to males. Their tension in relating to males was supported by their own reports of insecurity around males and by their self-conscious gestures when they talked to a male interviewer" (p. 52). Hetherington cautioned, however, that the girls she studied were living with mothers who had not remarried, which is likely to be a contributing factor.

The daughters and sons Hetherington studied were similar, however, in that the effects of father-absence were strongest when this situation prevailed during the earliest years of life. Fathering, as well as mothering, seems to be especially critical early in life.

Even when the father technically is available to the infant and young child, this is no guarantee that he actually will be spending much time in close contact. Rebelsky and Hanks (1971) taperecorded all interactions with babies over twenty-four-hour sampling periods through the first three months of postnatal life. (A miniature microphone was used in an effort to keep the study as unobtrusive as possible and so as not to interfere with what mother and father would ordinarily do.) The average father in this study had only three vocal interactions each day with his baby, for a total average time of thirty-eight seconds. Furthermore, the fathers spent less time speaking or cooing to their babies as time went on, while the mothers vocalized more often week by week. Lynn (1974) suggested that fathers may not realize how important their interactions with very young infants may be to later development. If findings such as these are supported by later studies, then it would seem that differential patterns of parenting by mother and father start very early in our society and may have more influence on total development than has usually been appreciated.

The question of father absence or attenuated contact also serves to remind us that *all* separations from parents are of potential significance in early childhood. This lesson was learned well by the British during World War II when London was under heavy air attack. Some very young children were moved out of the city for their greater safety. Although treated with kindness and consideration by their caregivers, many of the youngsters became deeply disturbed. Dorothy Burlingham and Anna Freud (1944) recognized that separation from their parents and familiar environmental objects was at the heart of the disturbances, and directed remedial effects that were at least partially successful.

Separation experiences in adult life may stir up old anxieties from earlier separations

During the past three decades, appreciation of *separation anxiety* has increased. It is now recognized that separation can be more or less *traumatic* (upsetting, harmful) to the young child, depending upon the circumstances and the total support system available. The problem has taken on new dimensions today as an increasing number of children are placed with babysitters and at day care centers. The interested reader will find much to think about in John Bowlby's *Separation* (1973), the sequel to his earlier *Attachment* (1969). Separations obviously are part of life and cannot be avoided at all times. Both the parent and the developmentalist are apt to be more successful, however, if they remain sensitive to the possibility that the young child they are concerned with might be caught up in separation dynamics of one type or another.

STIMULATION: HOW MUCH AND WHAT KIND?

Parents are major components of the infant's environment and are responsible, directly or indirectly, for much of the stimulation that is available. When developmentalists are concentrating on some particular aspect of behavior and growth they tend to assume that the setting is within certain normal or expectable limits. There is stimulation available, but not too much or too little. There is some reasonable organization and orderliness in the infant's sociophysical world, some variety, some consistency. This can be called the assumption of an **average expectable environment.** Awkward though the phrase may be, this term serves as a useful reminder that we are

Average expectable environment is the situation found when no emergency factors are operating

making a number of assumptions about the environment when we speak of individual developmental phenomena. At times these assumptions are examined and their limits tested.

Harry Harlow's justly celebrated experiments with primates led to important revelations about those features of the environment that normally provide support to the infant. These studies are too well known to require detailed presentation here, but their main points deserve our attention. A few monkeys were raised in complete social isolation (not, incidentally, as part of a deliberate research design). When they were introduced into a monkey colony, these isolates just could not function as part of the group. They could not defend themselves, respond to social or sexual advances, or, in general, be a monkey among monkeys. One might describe their situation as deviant, arrested, or psychotic; their normal development had not taken place because they had not experienced the normal resources of an average expectable environment (Harlow & Harlow, 1962). The lack of appropriate stimulation at the appropriate time had led to a miserable, dysfunctional state.

Even more striking were the results of Harlow's famous "cloth mother" experiment (Harlow & Harlow, 1962). Young monkeys were provided with wire-mothers who provided milk or with terry-cloth mothers who did not. This experimental arrangement in effect separated out two components of the "average expectable mother": the nutrient and the feels-comfortable-to-be-next-to aspects. The infant monkeys made their choice clear; comfort of

Harlow's research showed that young monkeys need a mother's comfort even if the only one available is made of cloth and spare parts. (H. F. Harlow, University of Wisconsin Primate Lab.)

contact was a greater attraction. It seemed to call forth more attachment behavior than the bare-wire provision of nutrition. This experiment was of particular importance because the subsequent lives of the poorly mothered monkeys were followed closely. Harlow arranged for a kind of group psychotherapy for the surrogated monkeys by introducing them to the monkey island in a zoo. There they did learn many of the typical monkey-social behaviors and became integrated into the colony. However, they seemed quite unprepared for, or uninterested in, sexual activity. Smiley, described as the "most experienced, most patient, and most kindly breeding male," eventually found his way to the heart of some of the females, who then did become mothers.

But did they really become mothers? They bore infants, but they did not seem to have a clue about how to care for them, nor did they have any special interest in doing so. Deprived of the stimulation and learning experience of being mothered themselves, these lost souls of the primate world could not mother their own infants. Here, then, was an example of the probable influence of early deprivation on the behavior of the individual as an adult.

Monkeys are not people, it is true. There have been numerous observations made on the human level as well, however. These suggest profound long- and short-term effects resulting from understimulation early in life. Rene Spitz (1945, 1946) observed young infants who were raised in two institutions that differed appreciably in the kind of stimulation available. Those in the place he called *Nursery* were held, touched, spoken to, and cared for either by their own mothers or by other women who took this as a full-time responsibility. They also had toys to play with and ample opportunities to see the world about them from their cribs. Those in *Foundling Home* had few toys available to them and could not see much because of sheets hung around them. Instead of regular "loving times" with mothering people, these infants usually interacted with other humans only at feeding time; they had busy nurses, not mothers. It must have been painful for Spitz to observe the changes in the Foundling Home infants as time went on. The other babies developed rather normally. But those in the deprived situation (few interactions, few objects, little to see or do) became seriously retarded physically and socially. At first they had been as lively and ready for interaction as their more fortunate peers, or as babies anywhere. There even came a period near the end of their first year of life when the Foundling Home infants were unusually sociable, reaching out with perhaps a desperate eagerness for human contact. When this was denied them, they began to show a profound decline in most areas of functioning. In fact, many became seriously ill and some died despite excellent precautions taken by the Foundling Home staff.

Conditions improved a great deal at Foundling Home not long thereafter, and the infants who physically survived their early period of general and maternal deprivation were transferred to an enriched and certainly more normal environment. But it appeared to have been too late. The usual course of development had been crippled because of the early deprivations, although how much could be attributed to general lack of stimulation and how much to lack of adequate mothering in particular could not be clearly determined.

Other studies (e.g., Provence & Lipton, 1962; Goldfarb, 1949) added further observations on both the immediate and long-range effects of early deprivation on child development. There has been controversy and criticism in this area. Methodological flaws in the classic studies have been pointed out (e.g., *see* Pinneau, 1955), and there has been a lack of complete agreement on the precise nature of the deprivation that made the difference. Yet with all the limitations that could be noted and all the questions that remain to be answered, the significance of early experiences, particularly of understimulation, is taken very seriously by contemporary developmentalists.

We turn now to parental behaviors that are extreme—but, unfortunately, not extremely rare.

HAZARD: PARENTAL ASSAULT

Parents love and protect their young children. Parents also attack, injure and kill their young children. Both statements are factual. I cannot avoid an examination of child abuse, unpleasant as this topic may be. The practical significance of this problem is obvious: children are bruised, maimed and slaughtered. But assault upon the young poses a challenge to our understanding as well. How are such actions possible? Are current developmental theories adequate for predicting, explaining and preventing child abuse?

SCOPE OF THE PROBLEM

Child abuse has been defined as "Nonaccidental physical attack or physical injury, including minimal as well as fatal injury, inflicted upon children by persons caring for them" (Gil, 1969). The severity of injury ranges from minor, self-healing wounds to fractured bones, permanent scarring, loss of sight or hearing, brain trauma, and death. The victims may be crippled or otherwise disfigured and impaired for the rest of their lives, and their life expectancies reduced. Physical assault is also likely to affect mental and emotional life. It would be naive to expect a battered child to progress through his or her developmental career as though nothing exceptional had happened along the way.

How many children are victims of parental assault? Accurate and comprehensive statistics have proven difficult to obtain on this sensitive question. It is clear, however, that the problem is frequent enough to constitute a public health menace. Pediatrician C. Henry Kempe and his colleagues found so many cases among their own patients that they introduced the diagnosis, **"battered child syndrome"** (Kempe, Silverman, Steele, Droegenmuel-

Battered child syndrome is the general term for physically assaulted children

Child abuse takes both physical and emotional forms, with long-range as well as immediate effects on development. (Courtesy of the Massachusetts Society for the Prevention of Cruelty to Children.)

ler & Silver, 1962). In a subsequent nationwide survey, 33 deaths and 85 permanent brain injuries were discovered among more than 300 cases observed by 71 hospitals—in a single year. Another one-year survey (by The American Humane Society) uncovered 179 fatalities among 662 cases. When the State of California established a central registry for reporting incidents of child abuse, an average of almost 100 cases per month resulted (Gil, 1969). Another study indicated that in New York City alone, there was an average of one child fatality per week resulting from physical assault (Wyden, 1963). At least one expert maintains that parental abuse and neglect is the most common cause of death in childhood (Fontana, 1964). Many cases, especially those with nonfatal outcomes, either do not come to the attention of the health authorities or are not reported as abuse. There is little question that child abuse is among the leading hazards to the young.

HISTORICAL BACKGROUND

Although it is only in recent years that child abuse has become widely recognized as a developmental hazard in the United States, world history is replete with examples of mistreatment of the very young. Radbill noted that assault upon children has been "justified for many centuries by the belief that severe physical punishment was necessary either to maintain discipline, to transmit educational ideas, to please certain gods, or to expel evil spirits" (Radbill, 1969, p. 3). His examples included the "Whip Man" in Sumerian schools of five thousand years ago, the ritualistic beating of children (in the memory of the innocents massacred by Herod) to make them better Christians, flogging as "treatment" for childhood epilepsy, and the liberal use of the birch rod by English and American teachers. Furthermore, outright murder of the very young has been practiced in many societies. Infanticide has often targeted "surplus" females in societies whose marginal survival status impelled them to engage in this extreme form of population control. However, infanticide has also been prevalent in many of the culture centers of the world, and not only in remote times (Leasor, 1963).

At least one distinguished scholar has proposed that the Book of Job is an attempt to combat the impulse toward infanticide (Bakan, 1968). Certainly, a careful reading of the Bible suggests that sacrifice of the young was an important theme whose meanings go deep into parent-child relationships.

In many times and places the child has been regarded as the property of the parent. He or she was, thus, to be disposed of as the parent saw fit. Legislation to protect the basic rights of the child has not always been with us. Even today, such measures are not always enforced with consistency, and new advocacy efforts are being made to confirm and extend the essential rights of the child. It is clear that conflicting attitudes toward the young child have been part of our cultural mix for a long, long time. This does not excuse the parent who attacks his or her child. But according to our cultural heritage, child abuse is one of the "solutions" or "alternatives" that an adult can choose to exercise when he or she is in a problem situation. If this practice were not already in our culture's repertoire, it is doubtful that so many incidents would be taking place. That leaves us with the mighty question: why do some adults (but not others) select this particular solution? We must consider both the characteristics of the victim and the characteristics of the abusing parent.

CHARACTERISTICS OF THE VICTIM

The victim is most often a very young child. Newspaper surveys reveal that 45 percent of child abuse fatalities occur to children aged two years or younger. More than two thirds of all child abuse deaths befall children aged five years or younger. It is between the ages of one and two that the child seems to be most vulnerable to a fatal physical assault. Most nonfatal incidents also occur in the preschool range. There is a tendency for boys to be abused more often than girls, and for all incidents to be reported at a higher rate in major metropolitan areas (Gil, 1969).

In our culture the two year old is celebrated as a demanding, stubborn, temperamental, independence-seeking little person. As we have seen, it is around the two year mark that severe attacks on the child are most frequent. We do not know for a fact that most victims of parental abuse are in the typical "terrible twos." But there might be a particularly deadly kind of interaction between the developmental needs of the two year old and certain kinds of parent.

The physical vulnerability of the very young may also be a factor

Often, however, it is only one child in the family who is "picked on." It is not possible to specify a single type of victim. At least five different patterns have been noted (Kastenbaum & Aisenberg, 1976):

1. The oldest child may be abused because his or her birth generally causes a greater dislocation in the husband-wife relationship and the family routine than does the birth of subsequent children.
2. The abused child may be the one who most clearly resembles the parent who administers the mistreatment. This resemblance can be physical, temperamental, or both.
3. By contrast, the abused child at times may be the one who seems most "different" from everybody else in the family.
4. A child who has the "wrong" characteristics may be punished for this "failure" (e.g., the desired boy proves to be a girl, or vice-versa).
5. Apart from these other considerations, some young children seem to fit the victim's role more easily than others. Just as a particular child may come in for more than his or her share of teasing or bullying by age-mates, so a particular child may have some characteristics that make him or her especially vulnerable to abuse by a parent (e.g., a child who is "too trusting," "too clumsy," etc.).

CHARACTERISTICS OF THE ABUSER

Most child abuse is at the hands of the parents. The father is somewhat more likely to mistreat his son, the mother to mistreat her daughter. While the father is more likely to be reported as a child-beater, the mother is more likely to be the child-slayer (Gil, 1969; Simons, Downs, Hurster, & Archer, 1966).

Is there a typical child-abusing parent? The available information does not permit me to draw such a portrait with assurance, but some common features have been observed. Steele and Pollock (1969) found that the physically abusive parent tends to doubt his or her own worth as a person. Furthermore, the child beater is apt to need somebody else to depend upon and to be uncertain of his or her own adequacy as an adult. Perhaps, then, child assault becomes a greater prospect when an immature adult, subject to doubt and depression, is faced with the demands of another human being. "How can I be a real parent to the baby, when I still need somebody to baby me?" It should be added that the physically abusive parent, on the average, is neither more nor less intelligent than other adults, nor do all such parents have the same kind of basic personality.

Clinical studies suggest that the "murdering mother" is often either psychotic or unusually impulsive and immature (Myers, 1967). We must remind ourselves, however, that immaturity and psychosis usually do not result in murder by physical assault. Neither of these conditions is sufficient explanation for abusive behavior directed against children. Some clues regarding the relationship between psychosis or immaturity and child slaying have been given by Bromberg (1965) and Kastenbaum and Aisenberg (1976).

A more probing view of child abuse would have to take into account at least two additional factors: (1) the precise nature of the abuse, and (2) the parent's own developmental history.

1. Attacks upon a small child sometimes occur as highly unusual episodes. In other families, however, a particular child is subject to abuse, even to systematic torture, over a prolonged period of time.
2. Many (but not all) parents who abuse their children once suffered similar assaults themselves. The passing of child-beating behavior from generation to generation is one of the facts that needs to be understood more adequately. Abuse might be prevented if attempts were made to help the victim before he or she steps into the role of parent.

PUNISHMENT OR ASSISTANCE?

Our sympathy goes out to the child who is the victim of physical abuse. On the other hand, we are likely to be furious at the assaulting parent. These reactions are understandable. Nevertheless, the assaulted often becomes the assaulter: when this happens do we lose *all* concern and hope for the victim-unto-assaulter? It is possible for the community to provide intervention at critical points, even when the abuse pattern has already started:

> I was determined that I would not raise my children the way my mother raised me. But when my second child was born, I started hating my two-year-old daughter. Everything she did upset me. I constantly yelled at her, beat her, even choked her. . . .
>
> After the little (murdered) girl was found in the garbage bag I really got scared. When I read about Parents Anonymous I started going to the meetings. . . . The next week I really went after my daughter. For a minute I almost thought I killed her. I was so overcome with guilt and depression I could hardly move. Somehow I got up enough energy to call the PA chairman.
>
> She talked to me and calmed me down. She told me that sometimes it helps to sit beside your sleeping child and make yourself look at her innocent helplessness.
>
> After I hung up I sat beside my daughter and touched her lightly. She woke up and started to cry.
>
> I took her in my arms and told her I didn't know why I hurt her but I was going to try to stop.
>
> For a week after that I really loved my daughter. And she was such a perfect loving child. Then she started being really bad. I learned this is expected—she is testing me.
>
> With the help of PA and a private counselor, I am not letting her push me into hurting her again. But it's too much of a strain on my new-found love. I can't say I love her now, but I don't hate her either. I believe we will get through this testing period and find love again.
>
> My husband is no longer afraid of me. I have found that he can love me; he is helping me now. I hope that anybody else who is in my shoes or knows anybody like me reads this and calls PA for help. ("Learner," 1972).

YOUR TURN

Here are a few of the ways in which you might increase your knowledge of the infant's development toward people and toward his or her personhood and perhaps make some discoveries that will be of interest to others as well.

Parent-Child Interaction

Direct observation. Perhaps you have already seen many babies and young children interacting with their parents. You may be a parent yourself. Try to take a different perspective, however. Become an objective observer whose responsibility is to see and hear what is taking place as fully and accurately as possible. Skilled observation—and reporting—is critical for both clinical and research purposes. Whether our primary intention is to help or understand, neither purpose will be served unless supported by dependable and sensitive observations. What I am talking about, then, is an exercise in observing parent-child interaction. Observational techniques can become rather complex and may at times require use of specialized equipment. But it is possible to make useful observations with nothing more than a parent, an infant, and some idea of what you are doing. Here are some points to consider as you plan your observation.

1. What will be my understanding with the people I am observing?
2. What will I be looking for, and why?
3. How will I organize and record the observations?

With regard to the first point, put yourself in the place of the people you are observing. If you were the observed rather than the observer, what conditions would be necessary for you to feel comfortable and respected? Once you put yourself in the other person's place, you are off to a good start in establishing the observational framework. The precise contact you establish will depend upon many factors that are specific to you and your situation. With a close friend or relative, you may have to do nothing more than indicate your wish to observe and record an "ordinary" period of time when parent and infant are together. No matter who you observe, however, it is both courteous and proper to request consent and convey a reasonable idea of what you will be doing and why. (This general rule of behavior applies as well to the research efforts of highly trained developmentalists.) Once your purpose is known and approved, the parent you will be observing may have helpful suggestions about the time, place, and situation.

As an observer, you do not have to wear a white laboratory frock and a blank expression. On the other hand, you will compromise yourself as an observer if you turn into a babysitter, therapist, conversationalist, etc. instead of concentrating upon the task at hand. Perhaps the parent will want to read your observations when they are completed. Why not? But if you prefer not to share your observations with the parent for some reason, this might also be part of your mutual understanding.

The purpose of your observations? Usually I would emphasize careful planning of observations around a central problem or guiding hypothesis. You might ask, for example, do baby girls "attach" sooner than baby boys? Or you might attempt to repeat somebody else's study to see if you obtain the same results, for example, Bell and Ainsworth's inquiry into crying behavior and maternal response. But I suggest a different approach for your first few observations. The purpose for these would be to acquaint yourself in general with how parent-child interaction looks and sounds when you yourself are not a direct participant. It would also be to acquaint yourself with the problems and pleasures of making and recording observations. After one or two observational exercises you might well come up with some clues or ideas that you wish to pursue, and which might also lead you to change your observational or recording techniques. For the beginning, however, just concentrate on accurate observations.

Observational Unit	Baby's Behavior	Mother's Response	Time Elapsed
1	Nurse brought baby who was awake and crying.	Mother cooed, turned to left side.	
2	Baby placed by mother's side.	Mother gave him left breast.	
3	Baby grasped it immediately and began sucking, slow and strong.	Mother cooed, told baby to look at her, giggled, cooed, said "You're so cute."	
4	Baby released nipple.	Mother laughed, put it back in his mouth.	
5	Baby took it and sucked again. Rested every ten to twelve sucks.	Mother jiggled breast sometimes to urge him to suck. Watched baby, made kissing sounds with mouth, cooed.	
6	Baby rested a long time.	Mother said, "Now don't stop. You know you can eat more." Mother had right hand on breast.	
7	Baby sucked again. Closed eyes. Sucking periods grew shorter, rests longer.	No response.	8 minutes
8	Baby released nipple.	Mother put nipple back in his mouth. Coaxed baby hard to start him sucking.	15 minutes
9	Baby sucked about ten times when coaxed.	Mother jiggled breast and patted his cheek to urge him.	
10	Baby released nipple.	Mother brushed nipple across his mouth.	20 minutes
11	Baby smacked lips but didn't take nipple.	Mother tried to insert nipple again.	
12	Baby refused nipple.	Mother said, "I guess you've had enough." Covered breast.	
13	Baby smiled in sleep.	Mother said, "Hey, cookie. What are you laughing at? What's so funny?"	
14	Baby grunted.	Mother said, "Yes, indeed." She moved back a bit, looked at baby and smiled and cooed.	
15	Nurse took baby.	Mother said, "He was so good and ate a lot. Bye, bye, cookie."	30 minutes

Figure 5-4. *A sample mother-infant interaction. (From Levy, David M.,* Behavioral Analysis, *1958. Courtesy of Charles C Thomas, Publisher, Springfield, Illinois.)*

How to do it? You will want to be in a position where you can see and hear adequately without being in the way. A ruled pad or sheets of paper on a clipboard and a writing implement (and a spare) are all you absolutely need for equipment, although a timepiece of some sort would be very helpful. Let's take an example now of a straightforward format for recording parent-child interactions. This excerpt is from a useful book on *Behavioral Analysis* by David Levy (1958). The observer stood at the foot of the bed and recorded the interactions between mother and newborn infant, either during a feeding or a nonfeeding situation. These observations took place in the maternity units of several hospitals.

The simple tabular arrangement shown in Figure 5-4 allows the observer to record the interactions in convenient numbered units, place the infant-mother behaviors in parallel, and keep track of the elapsed time. Some variation of this format may be more useful to you, or you may try it first exactly as shown in Figure 5-4.

Your own observations, of course, can be made at home or in some other convenient situation, not necessarily in the maternity ward. You can even try out this observation format if you have the opportunity to see a film of parent-infant interaction in class or elsewhere. And there is no necessity to limit yourself to the nursing or feeding situation, or to the very young infant. Just find a situation in which parent and young child have the opportunity to interact with each other over a reasonable period of time with a minimum of interference and distraction. Your report should include a brief description of the environmental setting, for example: "Mrs. T's living room. Baby is in playpen . . ."

You might find it helpful to make at least one set of observations along with another classmate or friend and compare your records and experiences later. In any event, comparison of records and of the observational experiences is a useful interchange in class itself.

When you feel confident in your observational skills, you might like to formulate your own questions or research problem and make further observations of a more specific nature.

Indirect observation. Although there is no substitute for direct observation, it is also useful to obtain information from other sources. Perhaps the simplest alternative is to interview one or both parents. The ethics and courtesy involved in direct observation should be applied here as well. Many parents will be quite receptive when they understand your purposes. What questions should you ask? It could be that some questions are already in your mind, coming either from this chapter or from your previous thoughts and experiences. Or you might prefer to use an established questionnaire. One possible model is the Guided Interview Schedule developed by the University of Nottingham (England) Child Health Research Unit (Newson & Newson, 1963). Question categories include orientation (family size, parents' occupation, etc.), birth, feeding, sleeping habits, behavior problems and toilet training, and father's participation in child rearing. The latter category, for example, includes such questions as:

> How much does his father have to do with him? Does he (mark each category: often: O; sometimes: S; or never: N):
> Feed him? _____ Change him? _____ Play with him? _____ Bathe him? _____ Get him to sleep? _____ Attend to him in the night? _____ Take him out without you? _____
> (Newson & Newson, 1963, p. 247).

You might browse through the child development literature until you find the kind of items you would prefer to use. Or, you might develop a list of questions by yourself, with several classmates, or as a class project. The questions should be clear and direct, sensibly organized, free of value-bias, and nonembarrassing to either you or the respondent.

Experience in developing and trying out both direct and indirect observational methods may prove helpful to you later when there is a specific problem you wish to look into for yourself. At the very least, it should give you some appreciation for the art and science of people watching.

Learning from Animal Research

Fascinating studies on dependency and attachment behavior in a variety of animal species have been made. Some of these observations took place in the natural environment of the animals, others in controlled experimental situations. You probably would find some of this material interesting in itself. But it is also worth thinking about the possible relationship between growing up as a fish, bird, puppy, or chimpanzee, and growing up as a human. Experts in comparative psychology and ethology (two of the relevant research fields) caution us to bear in mind that we should not automatically jump conclusions from one species to another—even from one bird to another, let alone, from bird to baby. Nevertheless, perspective on the early behavior of humans can be gained by studying our feathered, furry, and finned kin.

Among the sources that are worth your attention are:

- Kellogg & Kellogg's *The Ape and the Child* (1933/1967), detailing the growth of Gua, an infant chimpanzee and Donald, the Kellogg's own son, within the same home environment
- Harlow's series of experiments on the development of affectional patterns in monkeys, (*see* Harlow, 1961; Harlow & Harlow, 1962)
- The useful survey of various studies in Bowlby's *Attachment* (1969).

There are many other books and articles on this topic; there are bound to be at least a few that will snare your attention if given half a chance.

Why not (1) select two animal species that are clearly different from each other, (2) read up on their patterns of early development (and do some first hand observation yourself if possible, or see a film), and (3) compare and contrast their patterns of dependency, attachment, and independence-seeking with each other and (4) with the human infant and his or her mother.

Do you have the sort of mind that enjoys the fantastical? If so, then borrow from what you have learned about animal infancy to redesign human development. What would happen if, for example, the human newborn had some of the physical maturation characteristics of the newborn calf, colt, or fawn? Some mental experimentation, based upon actual facts, might lead to a fresh appreciation of how and why humans develop as they do.

Other Hazards

In this chapter I considered one hazard to the well-being of the infant and young child: parental assault. What other hazards exist? Can the infant be overstimulated? Understimulated? Are there certain types of problems that an infant might be born with, and which parents might find difficult to identify? What kind of environmental hazards might be of special concern?

These are some of the lines of inquiry you might care to follow. Is there a type of hazard that could be illustrated with photographs as well as words, or with a sound recording? Find the method of observation and reporting that best fits the problem you have come up with, whether in the realm of hazard, or any of the other exercises suggested here.

SUMMARY

Development occurs between people as well as within people. For the very young child, two interpersonal processes are of particular importance. All infants begin life in a state of *dependency* on more experienced and competent members of their species. Although the dependency relationship continues, with modifications, for a long time, there is also the gradual establishment of *attachment*. The child develops particular interpersonal bonds with particular people, usually starting with the mother. Dependency-attachment dynamics provide an important foundation for the child's emerging style of relating to people throughout the lifespan.

Specific examples of dependency-attachment dynamics include the infant's tendencies to use mother as home base for explorative behavior. This is seen by some developmentalists as highly significant for the child's subsequent adaptation to life. A four-phase description of early attachment behavior suggests that the infant begins with undifferentiated recognition of other humans and soon achieves what Bowlby characterizes as a *goal-directed partnership*. With Ainsworth's observations we see a specific pattern of attachment behavior that might be typical of young children around the world.

The *affective climate* in which the young child develops refers largely to the exchange of moods and feelings between infant and parents. In the feeding situation, for example, the mother's shifting moods of anxiety and tension seem to be communicated to the infant. The mother's moods are more influential than the choice of breast or bottle as feeding device in determining infant mood. Different patterns of maternal response to the infant's crying behavior also affect the infant. Research indicates that the more responsive the mother is to her infant's signals of distress, the fewer such signals he or she exhibits later and the more likely he or she is to spend time in independent activities. The importance of *fathering* as well as mothering is noted, along with some evidence that fathers do not seem to appreciate how influential their interactions with their young children can be. The effects *separation* and *separation anxiety* find examples drawn from British World War II experiences, and research into the effects of father-absence upon the development of boys and girls.

Studies at both the subhuman and human level indicate that the *kind and quality of stimulation* in infancy can have a profound influence on the subsequent course of development. Examples are taken from the classic studies of Harlow and Spitz.

Parental assault—the "battered child syndrome"—is a developmental hazard for children. It has a long history in our culture. Both the victim and the abuser are shown typically to have certain characteristics.

Reference List

Ainsworth, M. D. Patterns of attachment shown by the infant in interaction with his mother. *Merrill-Palmer Quarterly*, 1964, *10*, 51–58.

Bakan, D. *Disease, pain, and sacrifice*. Chicago: University of Chicago Press, 1968.

Bell, S., & Ainsworth, M. D. Infant crying and maternal responsiveness. *Child Development*, 1972, *43*, 1171–1190.

Biller, H. B. *Father, child, and sex role*. Lexington, Mass.: Lexington Books, 1971.

Bowlby, J. The nature of the child's tie to his mother. *International Journal of Psychoanalysis*, 1958, *39*, 350–373.

Bowlby, J. *Attachment*. New York: Basic Books, 1969.

Bowlby, J. *Separation*. New York: Basic Books, 1973.

Bromberg, W. *Crime and the mind*. New York: Macmillan, Inc. 1965.

Burlingham, D., & Freud, A. *Infants without families*. London: Allen & Unwin, 1944.

Fontana, V. J. *The maltreated child*. Springfield, Ill.: Charles C Thomas, Publisher, 1964.

Gil, D. G. Incidence of child abuse and demographic characteristics of persons involved. In R. E. Helfer & C. H. Kempe (Eds.), *The battered child*. Chicago: University of Chicago Press, 1969, pp. 19–42.

Goldfarb, W. Rorschach test differences between family-reared, institution-reared, and schizophrenic children. *American Journal of Orthopsychiatry*, 1949, *19*, 625–633.

Harlow, H. F. The development of affectional patterns in infant monkeys. In B. M. Foss (Ed.), *Determinants of infant behavior* (Vol. 1). New York: John Wiley & Sons, Inc., 1961.

Harlow, H. F., & Harlow, M. K. Social deprivation in monkeys. *Scientific American*, 1962, *207*, 136–140.

Heinicke, C. M., & Westheimer, I. *Brief separations*. New York: International Universities Press, 1965.

Hetherington, E. M. Girls without fathers. *Psychology Today*, 1973, *6*,(9), 46–52.

Kastenbaum, R., & Aisenberg, R. B. *The psychology of death*. New York: Springer Publishing Co., Inc., 1976.

Kellogg, W. N., & Kellogg, L. A. *The ape and the child* (facsimile edition). New York: Hafner Publishing Co., 1967. (Originally published, 1933.)

Kempe, C. H., Silverman, F. N., Steele, B. F., Droegenmueller, W., & Silver, H. K. The battered-child syndrome. *Journal of the American Medical Association*, 1962, *181*, 17–24.

Korner, A. S., Kramer, H. C., Hassner, M. E., & Thoman, E. B. Characteristics of crying and noncrying activity of full-term neonates. *Child Development*, 1974, *45*, 953–958.

Lamb, M. E. Interactions between eighteen-month-olds and their preschool-aged siblings. *Child Development*, 1978, *49*, 51–59.

"Learner." Letter in *Detroit News*, June 6, 1972.

Leasor, T. *The plague and the fire*. New York: Harcourt, Brace, 1963.

Levy, D. M. *Maternal overprotection*. New York: Columbia University Press, 1943.

Levy, D. M. *Behavioral analysis*. Springfield, Ill.: Charles C Thomas, Publisher, 1958.

Lynn, D. *The father: His role in child development*. Brooks-Cole, 1974.

Myers, S. The child slayer. *Archives of General Psychiatry*, 1967, *17*, 211–213.

Newson, J., & Newson, E. *Infant care in an urban community.* London: George Allen & Unwin, 1963.

Newton, N. The influence of the let-down reflex in breast feeding on the mother-child relationship. *Marriage & Family Living,* 1958, *20,* 18–20.

Pinneau, S. The infantile disorders of hospitalism and anaclitic depression. *Psychological Bulletin,* 1955, *52,* 429–452.

Provence, S., & Lipton, R. C. *Infants in institutions.* New York: International Universities Press, 1962.

Radbill, S. X. A history of child abuse and infanticide. In R. E. Helfer & C. H. Kempe (Eds.), *The battered child.* Chicago: University of Chicago Press, 1969, pp. 3–18.

Rappaport, L. *Personality development.* Glenview, Ill.: Scott, Foresman & Co., 1972.

Rebelsky, F., & Hanks, C. Fathers' verbal interaction with infants in the first three months of life. *Child Development,* 1971, *42,* 63–68.

Rheingold, H. L. The effect of a strange environment on the behavior of infants. In B. M. Foss (Ed.), *Determinants of infant behavior* (Vol. 4). New York: Barnes & Noble Books, 1969.

Rheingold, H. L., & Eckerman, C. O. The infant separates himself from his mother. *Science,* 1970, *168,* 78–83.

Simons, B., Downs, E. F., Hurster, M. M., & Archer, M. *Child abuse, a perspective on legislation in five middle-Atlantic states, and a survey of reported cases in New York City.* New York: Columbia University School of Public Health and Administration, 1966.

Spitz, R. A. Hospitalism: An inquiry into the genesis of psychiatric conditions in early childhood. In *Psychoanalytic study of the child* (Vol. 1). New York: International Universities Press, 1945, pp. 53–74.

Spitz, R. A. Hospitalism: A follow-up report. In *Psychoanalytic study of the child* (Vol. 2). New York: International Universities Press, 1946, pp. 113–117.

Steele, B. F., & Pollack, C. B. A psychiatric study of persons who abuse infants and small children. In R. E. Helfer & C. H. Kempe (Eds), *The battered child.* Chicago: University of Chicago Press, 1969, pp. 103–148.

Wyden, P. *The hired killers.* New York: William Morrow & Co., Inc., 1963.

THE FIRST TWO CANDLES

Moving and Experiencing

6

CHAPTER OUTLINE

I
In learning to stand and walk, the youngster demonstrates his or her competence to self and others.

Verticality / Danse reflex / Hitching-creeping-crawling sequence / Large muscle coordination / Small muscle coordination / Bowel and bladder functioning / Myelination

II
Mental development in life is supported by physical changes in the nervous system.

Neural development / Brain weight / Sense organs and sensibilities / Sleep-waking cycle

III
Language is a significant part of becoming and being a person.

Baby talk / Vocalization / Phonemes / Babbling / Single-word statement (holophrastic speech) / Contextual interpretation versus linguistic structure / Expressibility / Telegraphic speech / Pivot words / Open words / Language Development Sequence (Brown) / Imitation / Stimulus-response / Cognitive-nativistic view

IV
The assessment of mental development in very early life poses many methodological and conceptual challenges.

Continuity / Quantitative or qualitative / General intelligence / Sensorimotor intelligence (Piaget) / Presymbolic / Egocentric beings / Decentering

The story is told of a king whose curiosity exceeded his sensibility. He wondered if infants would learn to speak the language of the realm if they were deprived of human companionship. The king commanded that two infants be isolated on an island to grow up by themselves. All too predictably, the babies died, proving nothing about language development, but perhaps something about the weird turns of mind that can be associated with unlimited power. More responsible people have also wondered about the relationship between human interaction and the development of language, thought, and other abilities. Many observations have been made, both in naturalistic and controlled situations. It is now clear that in growing toward people, the infant is also discovering and exercising new abilities of his or her own. Personal and interpersonal development occur simultaneously. In this chapter, I will emphasize the emergence of new mental and physical characteristics of the infant and toddler. But we should not forget that development of particular abilities takes place within the context of the social as well as the physical environment.

ON THE MOVE: THE CHILD BECOMES MOBILE

There is very little the newborn can "do" in the usual sense of the word. He or she cannot move about under his or her own power, nor can he or she perform coordination actions in that tiny sector of the world he or she inhabits. Let's follow the typical course of the newborn's development in locomotion. It is one of the most commonplace and taken-for-granted developmental sequences, yet it does not lack in both practical and theoretical interest.

THE PERILS OF VERTICALITY

What goes up must come down. This remains a dependable rule, unless we are dealing with some of the exotic phenomena that occupy the attention of theoretical physicists. The obverse of this proposition is: What does not go up, will not come down. If the infant were to play it safe, he or she would confine his or her position and locomotion to the horizontal plane. Hitching along on his or her bottom, creeping on the tummy, or crawling about on all fours are means of moving along while remaining securely supported on the ground. The creeper, for example, is taking no liberties with the law of gravity. He or she does not have to worry about falling without support through space.

Standing up is one of the first risk-taking behaviors in which a human engages. With uncertain feet way down there, and head way up here, the infant is taking chances with the perils of verticality. When he or she tries to step as well as stand in the vertical position, the audacity of the infant's action is increased—retaining position in one dangerous dimension (vertical) while simultaneously attempting to maneuver through another dimension (horizontal).

Is it silly to take the achievement of verticality so seriously? Consider these points if you will:

1. Fear of falling is generally recognized as a basic—perhaps *the* basic—fear of infants and young children. It is probably the combination of fear and sense of security that provokes delight in babies who are playfully tossed in the air and caught by the parents.

2. Concern about falling has its adaptive value. The adult who senses himself or herself about to slip on an icy sidewalk, for example, may instantly convert alarm into a balancing reaction that prevents injury. A person of any age can be injured by uncontrolled movement from the vertical to the horizontal plane.

3. *Homo sapiens* are not perfectly designed for moving about on only two limbs. I have heard, for example, learned discussion about the difficulties in working up a good sneeze through a respiratory system that might function more smoothly if nose and mouth were pointed earthward. A case can also be made for the unfavorable influence of gravity upon the erect human over a period of time: the lower parts of our bodies are more vulnerable to the sagging pull of the gravitational field, and all our parts are subject to this directional effect whenever we are in the upright position. The most relevant point is this: standing up and walking on two limbs probably has its price and requires a system of psychophysiological supports to become well established in the first place and then maintain itself over the years. It is not as easy as it looks.

Humpty Dumpty and "the cradle will fall" began as political rhymes—another connection between infant and adult perils of verticality

4. Tribute is paid to the perils of verticality in the everyday language and thoughts of adults. The sinner *falls* from grace, as the once mighty do from positions of power. When things go wrong, we feel *down*. As prospects improve, we say that things are *looking up*. The family that is bettering its financial position is *up*wardly mobile. A person who attempts to free himself or herself from the shackles of a disadvantageous social position is still described in some quarters as *uppity*, a thinly veiled warning that somebody might attempt to *put him or her down*. Heaven and bliss, of course, are sky-high, while hell and despair are in the lower depths. We achieve progress and success by moving onward and *up*ward. The financial status of a failing corporation is represented by a line that plummets downward on the graph. These are not isolated examples. A thorough exploration would reveal a network of language and thought in which high upness is regarded as a favorable but dangerous status. These usages can be found in many cultures. This is not surprising; infants everywhere begin on the horizontal and must mature and take risks before they can achieve the triumph of verticality.

Standing and walking, then, are significant achievements for the young child. These skills are acquired at some risk to pride and hide, and they also serve as a common source for human symbols of victory and success. Having emphasized the perils and costs of verticality, let's remind ourselves of the positive side. What would become of an infant who became such a virtuoso creeper or crawler that he or she "decided" not to bother with walking? At the least, he or she would be at a serious disadvantage occupationally; most lines of work and, of course, most human activities require the free use of hands. Becoming a biped liberates the hands from being just another pair of feet. You can probably think of many other ways in which human society as we know it would be difficult or impossible if, as a species, we decided to remain on all fours.

The sheer joy of conquering gravity and one's own fears should be not overlooked. In learning to stand and walk, the youngster demonstrates competence to himself or herself and others. He or she overcomes obstacles, and can enjoy (more keenly than older companions) the delights of stepping about on his or her own two feet. It is likely that this success experience helps to fortify the child as he or she moves on to other triumphs in the sociophysical environment.

FROM REFLEXING TO WALKING

We have already seen that the newborn has a set of response sequences known as *reflexes*, some of which were available to him or her even before birth, and some of which emerge soon afterward (*see* Chapter 4). One of these reflexes bears resemblance to the act of walking. Hold a one-month-old baby upright with his or her feet on a hard surface. If the infant lifts first one foot and then the other in a series of stepping movements, he or she may be said to have evidenced the *danse reflex*. The early appearance of stepping behavior can mislead a rookie parent into believing that his or her offspring is an incredible prodigy. True walking behavior is a long way off, however, and the danse reflex will have vanished long before the prodigy fulfills his or her promise.

The usual sequence of infant behaviors that lead to walking is outlined in Figure 6–1, taking a month-by-month approach. Individual differences exist in the timing of each stage. Some infants seem to devote themselves wholeheartedly to a particular behavior that is intermediary between reflex and walking and linger there for awhile. Similarly, some of the stages mentioned in Figure 6–1 may be passed through very rapidly by a given baby. There is no reason for parents to be disturbed if their infant's timing and style of prewalking behavior differs somewhat from the typical sequence I have described. Individuality asserts itself from the very start of life, and developmental guides are not intended to trouble or intimidate parents.

Age	Typical Behavior
1 month	*Danse reflex* (lifts feet alternately when held upright with both feet on a hard surface).
2 months	Raises head and chest slightly when lying tummy-down.
3 months	Sits, with back rounded and knees flexed when supported. *Danse reflex* is gone.
4 months	Sits, propped up, holding head steady. When held upright, will sustain some of own weight.
5 months	Sits with only slight support. Holds back straight when raised to sitting position.
6 months	Pulls self to sitting position. Some babies *hitch* (i.e., move backward in a sitting position by pushing with hands).
7 months	Sits briefly, leaning forward on own hands. Control of trunk is more advanced. Bounces actively when held in a standing position. (Grandmothers often disapprove, "That will make him bowlegged!" It won't.)
8 months	Sits alone steadily.
9 months	*Creeping* begins (i.e., pulls body along by moving elbows). Tummy and legs sort of drag along. *Crawling* begins a little later: trunk is raised above the floor and propels self forward with hands and knees. Can pull self to feet with parent's help.
10 months	Now can pull self to feet with crib rail or other secure objects for leverage. Makes stepping movements when parent holds his or her hands. Becomes a proficient *cruiser* (i.e., walks sideways while holding onto furniture or playpen rails).
11 months	Finally a two-footed person. Stands erect steadily with the help of parent's hand.
12 months	The yearling stands alone for a short time and, with assistance, walks. When let go, sooner or later he or she sits down. Thick diapers prevent injury.
15 months	Proficient at walking alone by now, balancing on a wide-based stance. Continues to crawl or creep when it suits own purposes, such as going up or down stairs.
18 months	Runs as well as walks. Still uses wide stance, but not as wide as before. Seldom falls. Now can walk sideways and backwards.
2 years	Walks and runs more steadily. May use new locomotive abilities to run away from Mom or Dad as well as toward them. Tries to jump, often falls, because cannot control forward momentum. Climbs up and down steps by placing both feet on a step at the same time while holding the wall or a railing.

Figure 6-1. From reflexing to walking. *(Based on information presented by Marlow & Sellew, 1961.)*

A few comments might be added to the material described in Figure 6–1. The hitching-creeping-crawling sequence is not followed by all infants. Some infants move readily to the more advanced types of locomotion without hitching, for example. Early attempts at creeping may be noticed around the fourth month or so, although this maneuver is usually not effective until approximately the ninth month.

By the time the infant is ready to stand alone, he or she will probably measure about twenty-nine or thirty inches in height and weigh approximately three times what he or she weighed at birth (currently in the United States between twenty-four and thirty pounds). If the yearling is typical of his or her age echelon, he or she will be sporting six teeth at this time—the better for punctuating a proud smile while standing tall against the law of gravity. The first step is often taken within a few weeks of standing upright without support. With much coaxing, Mom and Dad convince the child that the time has come to march under his or her own power from one parent to the other. The congratulations for that first tottering step and the later rewards of vertical mobility (e.g., getting to toys or the cat's tail more quickly) keep the child walking, then running, climbing, hopping, and skipping into the school years. Even for the adept two-year-old walker-runner, such virtuoso variations as hopping and skipping remain in the future.

OTHER MUSCLES, OTHER ACTIONS

Maturation of small muscles. Standing without support, walking, and running are actions that require the development and coordination of some of the body's larger muscle groups. The smaller muscles are also important to normal functioning. It takes awhile, for example, before the infant has enough muscular coordination to pick up small bits of food and put them in his or her mouth. Physical growth of the muscles is necessary, but so is practice and learning. The enthusiastic self-feeder will often decorate his or her face and bib with near and not so near misses. An immature or anxious parent may become tense when the baby "fails" in a particular effort to coordinate muscle action. It is unfortunate when a recurring situation, such as feeding, takes on a high tension atmosphere because baby is an imperfect, "messy" eater. The understanding parent not only expects the offspring to make his or her share of "errors" in establishing new actions, but enjoys following baby's progress, misadventures, and eventual triumphs.

Selected aspects of the infant's progress in using small muscles are presented in Figure 6–2. The descriptions are not intended to be comprehensive, and individual differences are again important.

You will notice that as the infant increases his or her competence in the use of small muscles, he or she also becomes a more interesting playmate for older children, and a fledgling do-it-yourselfer in feeding, dressing, and getting toys, etc. Physical development thus facilitates social interaction and

Age Plateau	Typical Actions
1 month	Grasps your finger or an object placed in his or her hand, but drops it quickly. No reaching behavior.
3 months	Reaches for objects, but usually misses them. Holds onto rattles and other small toys. Brings hand to mouth. (Ouch! if a hard or sharp object is in that hand. Might also swallow small objects.) Plays with own fingers and hands and seems to enjoy staring at them.
6 months	Bang! Uses rattle, spoon, or other object to rap out sounds, apparently the louder, the more enjoyable. Uses thumb in apposition to other fingers and so can grasp more effectively. Reaches for objects that are beyond his or her grasp. Will accept the object you hand to him or her.
9 months	May begin to show clear preference for one hand over the other. Skillful in holding own bottle and connecting it to his or her mouth. Transfers toys from one hand to the other.
12 months	Takes off own socks, puts his or her arm through a sleeve. "Writes" with crayons. Drinks from a cup, eats from a spoon, but may prefer to eat with fingers, picking up small bits of food and transporting them to his or her mouth. Can let go of toys as well as pick them up.
15 months	Can throw as well as drop objects. Enjoys pitching toys and picking them up again. If older brother or sister will play catcher, can play a rudimentary sort of game. Pokes fingers in holes, opens boxes, places one block on top of another. Uses spoon and cup held in own grasp, but with many a spill.
18 months	Scribbles enthusiastically, using both straight and circular strokes. Attempts to build block towers, makes good use of pull toys. More skillful in drinking from a cup. Still has some trouble with the spoon (such as turning it upside down at just the wrong time).
2 years	Can turn knobs so closed doors are no longer impossible obstacles. Scribblings show more control as he or she tries to imitate the strokes demonstrated by older brother or sister. Makes block towers successfully. Has mastered the spoon!

Figure 6-2. *The development of small muscle coordination in the infant. (Based on information presented by Marlow & Sellew, 1961.)*

makes it possible for the young child to broaden his or her range of psychological experiences. We see that social and physical development are not to be separated too rigidly. And we can also appreciate that infants who have unusual difficulty in establishing bodily coordination may also have problems in other spheres.

Bowel and bladder functioning. Even a selective discussion of muscular coordination and control in the first two years of life would not be complete without some attention to bowel and bladder functioning. Once again we must focus on the psychological and social aspects as well as the physical side. Elimination of waste products is a biological need of every organism. Perhaps, then, we should have the same attitudes toward bladder and bowel functioning that we do toward other biological processes necessary for survival, such as breathing and taking nourishment. In practice, however, Americans (and people in most other parts of the world) have developed special attitudinal overlays toward biological functions. Not only are there differences among cultures, but there are differences even within the same nation. The middle-class value structure in the United States, for example, has been notable for its emphasis on being clean and tidy. Body odors are supposed to be covered up or disguised. This is one of the value biases that can influence human development. An infant's experiences with urination and defecation are influenced by the attitudes and practices of the culture and of the particular family into which he or she is born. These experiences, in turn, affect that person well beyond the age at which adequate control over elimination needs was achieved. Development of the infant's total personality is influenced to some extent by what he or she learns and experiences during the long months that elapse before reaching the culturally required standard of tidiness and control.

There are important rewards for the youngster who masters bladder and bowel control quickly and easily. He or she probably pleases his or her parents by so doing, gains satisfaction for having achieved control, and finds himself or herself more at liberty to participate in a broad range of activities (such as being more socially acceptable to older children). He or she feels more grown up and has, indeed, mastered one of the many rungs on the ladder of success in our culture.

Potential difficulties also exist in the toilet training situation. The young child who is not mastering bladder and bowel control rapidly enough to meet parental expectations may become tense and confused. Tension rises between parent and child as well as within the child. Toilet training may come to overshadow everything else that is going on when, in actuality, it represents only one of the many components of total development. There may result a battle of wills that is unpleasant for all concerned, escalating beyond the limits of two-year-old "negativism." Parental support and approval are very important to the young child. The alert and relaxed parent will not permit a situation to develop in which the child's self-concept and access to affection is dominated by successes and failures in bathrooming.

Do we have "dirty words" because the oral stage moves into the anal with its tensions?

Time out for a friendly tickle. Body senses are especially acute in early childhood. (Bobbi Carrey/THE PICTURE CUBE.)

I am hesitant to offer a month-by-month schedule of bowel and bladder control. Not only are individual differences quite important but, frankly, I am concerned that any such developmental guideline might be used by some parents more strictly and literally than intended. Let's be content with the following approximations:

A nerve is **myelinated** when a sheath of neural tissue has formed to enclose and insulate it

1. Around the end of the first year, many children have the neurophysiological readiness to attempt elimination control. The most important element here is the sheathing of the lower end of the spinal cord with a covering of neural tissue that is indispensable for sensitivity and control in that area of the body. Attempts to initiate control before the spinal cord is completely **myelinated** are doomed to frustration and failure.
2. Some children are not ready for toilet training at this time, even though the neurophysiological underpinnings may be established. Much depends upon the pattern of the child's bowel functioning and the mother's success in adapting to it. While successful toilet training can begin at the end of the first year for some children, it is far from a life-or-death matter.
3. Bowel control is often established about halfway through the second year. It is a little easier to establish bowel control than bladder control, and mothers-in-law usually recommend that toilet training begin with the bowels (they're right).
4. Bladder control during the daytime is often established around the end of the second year. It may still be another six months, year, or even more before the child has dependable control over bladder functioning at night.

When toilet training is accomplished smoothly, it is an achievement both for parent and for the child. The mother has been patient, relaxed, sensitive to her child's total needs, and reasonably consistent. The child has performed what has been asked, augmenting mastery of gravity and the upright position with mastery over the inner workings of his or her body.

EXPERIENCING AND EXPRESSING

What are you experiencing at this moment? If you do not choose to tell me, then I must attempt to "read" your experiential state indirectly, perhaps by the expression on your face, the way you have positioned your body, your rate of breathing, the presence or absence of fidgetings, etc. In daily life we use both verbal and nonverbal clues to fathom what other people are experiencing. Some people are fairly easy to read, for example, the youngster hopping up and down, hand on crotch, who volunteers, "I don't *need* to go potty! I'm just dancing!" Others are difficult to penetrate ("I've worked with him for years, but I still don't know what he's thinking, when he is bored, or when he is angry").

There are special challenges in attempting to discover what the infant and young child are experiencing. We do not have language clues to go on for the first several months, and it is quite awhile before the toddler can put his or her thoughts, feelings, and needs into words. Fortunately there are many nonverbal hints available, such as smiling, crying, wriggling, yawning, etc. And enough observations of various kinds have been made to construct a rough guideline to mental development in the early months of postnatal life. Mental or intellectual development is no more uniform from child to child than is physical development, so we must continue to respect individual differences. Several important theories concerned partly, or wholly, with mental development have been advanced. In this section, however, I will describe and explore basic phenomena rather than launch into theoretics.

THE STARTING POINT: NEURAL DEVELOPMENT

Mental functioning requires the same basic services that must exist for other body functions to flourish. There must be systems for nutrition and for elimination of waste products, for example. In this sense, mental development has already come a long way between conception and birth. The network of physical systems that can support mental activity becomes increasingly elaborated in the months preceding birth (*see* Chapter 3) and this surge of growth continues for several years. The body is preparing itself rapidly for the support of mental processes.

The growing brain. An obvious example of how far mental development progresses after birth can be found in an obvious place, the head and brain. The newborn's brain weighs only about one-fourth of what the adult's weighs (brain weight is one of the standard measures of neural development). Neural tissue continues to mature so rapidly after birth that by the age of nine months, the weight at birth has doubled. By the child's second birthday, there will have been another substantial increase in brain weight; the child's brain will now weigh up to three-fourths of the adult's heft (Eichorn, 1963). This increase, however, already reflects a slight slowing down of growth surge, and it will taper off even more as time goes on.

Some of our knowledge concerning mental development comes from the study of physical structures and processes, brain weight being just one example. These studies indicate that much is going on at the neural level during the first few years of life. Parallel observations on the behavioral level indicate that much is also going on in the youngster's experience of life. Biologically oriented and psychologically oriented scientists need each other's observations. The significance of physical changes (such as brain weight increase or myelination) can best be evaluated by observing what new behaviors infants show at various stages of neural development. For example, biological findings can help us avoid such behavioral errors as insisting upon toilet training before the neural equipment has matured. I will be combining information from both biological and behavioral sources in this section.

Sense organs and sensibilities. The young infant, then, does not "start" mental functioning on a particular date. He or she has been functioning in at least a premental since way before birth, registering sensations and snapping off reflex actions. The infant's mind is already a going concern. Evidence suggests that his or her sense organs are serving him or her well very early in postnatal life. By the third month, vision, hearing, taste, and smell are all in good working order. The body senses seem to be quite acute. Temperature, touch, pain, and pressure stimulation are experienced with much sensitivity. This means that baby is probably more "tickle-able" now than when grown up, but it also means that he or she can be made very uncomfortable by variations in temperature, touch, and pressure that might pass an adult almost unnoticed.

Almost all our body surface is sensitive to stimulation (although not equally so). Further, those body parts we usually think of in terms of other major functions also have their sensibilities to contribute. The infant's lips and mouth are vital in securing nourishment, yet they also provide a variety of sensations. The same body parts also become important in expressive communications. Experiential states such as joy, astonishment, disappointment, and rage can be read, in part, by observing the child's oral gestures. The mouth and its surrounding structures have three different, although related, roles to play: (1) nourishment-intake, (2) a rich source of sensation or input, and (3) a means of expressing and communicating experiential states.

Kissing, a logical invention, has its origins in nourishment and contact pleasure

Sleeping and waking. Another clue to the infant's early progress in mental development can be obtained from the sleep-waking cycle. At first, the newborn spends much of his or her time asleep. It is said that neonates sleep about twenty hours per day on the average. However, the most intensive study of sleep and wakefulness during the first three days of postnatal life indicated that sixteen hours of sleep for each twenty-four-hour cycle is the more typical pattern (Parmalee, Schulz & Disbrow, 1961). Individual differences are great. In the study cited, for example, one of the seventy-five infants slumbered for only ten and one half hours while another snoozed twenty-three hours per day. While individual differences continue to express themselves, there are also general trends in sleep and wakefulness as the infant develops. Kleitman and his colleagues (Kleitman & Engleman, 1953) made a study from the third week of postnatal life to the twenty-sixth of nineteen babies who were reared under family home conditions. On the average, these babies increased their wakefulness by about one hour during this period of time. They also began to shift their sleep patterns away from many short naps during the daylight hours toward a long period of relatively unbroken sleep during the night. During the second half of the first year, the infant typically reduces even further the amount of time asleep during the day, while sleeping about the same amount of time during the night as before. The two year old generally takes an afternoon nap to supplement his or her long night's sleep, for a total wakefulness time of about eleven hours in the twenty-four-hour cycle.

The relationship between mental functioning and the sleep-wakefulness pattern should not be oversimplified. It would be erroneous, for example, to maintain that the mind (either of infant or adult) is turned off during sleep. Research advances make it clear that sleep is a complex phenomenon that repays intensive investigation (for a useful introduction, *see* Luce & Segal, 1966). Yet it is obvious that the type of mental functioning we associate with the waking state is available only in rather limited amounts to the newborn. Wakefulness gradually emerges from the original pattern of extended slumber. The relationship between biological and psychological aspects of development is quite intimate here. Kleitman's authoritative summary of research on sleep duration informs us:

> The gradual decrease in the total hours of sleep per 24 hours can be expressed as an increased ability to remain awake, parallel with anatomical maturation and physiological development. The relative wakefulness capacity of the individual is increased fourfold, as the young infant has to "pay" with two hours of sleep for each hour of wakefulness, whereas in the adult corresponding payment amounts to only half an hour. The absolute capacity for remaining awake increases from a few minutes at birth to 15 to 17 hours in the adult—the "natural" maximum in a sleep-wakefulness rhythm of 24 hours. (Kleitman, 1963, p. 121).

THE CHILD SPEAKS: THE EARLY DEVELOPMENT OF LANGUAGE

The infant has experiences and modes of expression before he or she has words at his or her command. But language is so significant a part of becoming and being a person that the progress of its development merits special attention.

"BABY TALK"

Speaking to an infant as "one of the guys" elicits a different response than babytalking at him or her

Mother, father, uncle, aunt, older brother, older sister, all sorts of people, direct remarks to the infant. The term *baby talk,* in fact, usually refers to that characteristic vernacular and tone of voice with which we often address the young. Here, however, I am more interested in the kind of baby talk that comes from babies themselves.

Infant speech is significant for several reasons. When the use of language first appears, it provides observers with a new access to the child's mental processes. They have a better opportunity to be "let in" on the child's way of thinking. Access to the child's mind through speech is important for parents who want to relate their behavior and expectations to their child's ever-changing level of functioning. It is also important for those who are attempting to study the nature of mental development scientifically.

Another side of baby talk is the expressive function it serves for the young child. Now he or she can begin to communicate needs and interests more fully and precisely through the verbal medium, as well as through tears, smiles, and shouts. Infant speech, then, is both an index (although only a partial one) of mental development and a means of self-expression and communication. From the interpersonal standpoint, speech behavior helps to strengthen and extend the infant's bonds with humanity. He or she starts to participate in the rich network of verbal communication by which we relate to each other. Talking is a further way of demonstrating that he or she is "one of us," not just a small presence in the house that requires a lot of attention.

A duck can make a good listener. Even early vocalizations tend to be beamed at an available listener. (© 1978 Photo by Thomas S. Wolfe.)

Most families are pleased—even delighted and fascinated—when words start to come forth from baby. It might be said that infant speech reinforces parental and sibling behavior. It gives them pleasure, makes them proud. I have observed preschool children, for example, who were enthralled with the words offered by a yearling. Through speech, the infant became of greater interest to his "elders" who now chose to spend more time with him, trying to teach him new words. Infant speech can thus influence the listeners and shape their behavior. A recent experiment has found that most adults recognize when a young child does not comprehend what they have been saying and accordingly speak in shorter and simpler sentences (Bohannon & Marquis, 1977). The young child's relatively simple verbal expressions call forth simple language from adults.

This mutual influence process operates even before the infant speaks his or her first words. The vocal behavior of three-month-old babies has been increased by social-expressive cues from adults, cues introduced as minimally as one cue lasting one second of time per day (Rheingold, Gewirtz & Ross, 1959). A big smile from an adult, accompanied by a little tsk-tsk-king, and a gentle touching of the baby's skin elicited more vocalization (the forerunner of language) than an adult just standing there without any particular expression on his or her face.

We might also speculate about the effect of infant speech upon the infant himself or herself. Observation suggests that the baby is often delighted by his or her own verbal triumphs. He or she seems to experience direct enjoyment from being able to form sounds and utter words, as well as secondary enjoyment from the attention given to this achievement. It would be good to know more about the consequences of this self-reinforcing use of sound and words for cognitive and personality development.

FROM VOCALIZING TO TALKING: A DESCRIPTIVE APPROACH

Vocalizing is making sounds that are not words or wordlike structures

There are several interesting theories of early language development, all of which are supported by some evidence. But we will look first (with as little theoretical bias as possible) at the transition from **vocalizing** to babbling and talking. The descriptive approach cannot be 100 percent pure because the observational method and type of data analysis usually has some relationship to the researcher's priorities and expectations.

Infants are makers of sound right from the beginning. The parents of a two month old can often distinguish several different kinds of cry, as well as a variety of softer and more contented sounds. Careful listening usually reveals that vowel sounds dominate the vocal repertoire (Ervin-Tripp, 1966). The "aaahs" and "ooohs" are played with and shaped—a phrase drawn out here, a pause, a repetition, a change in pitch. One might think of the jazz musician who has discovered an intriguing phrase and is enjoying the pursuit of its possibilities.

Consonant sounds begin to appear more often around the end of the second month. Additional variety emerges in the form of assorted gurgles, clicks, and other sounds that are difficult to convey by the written word. It is interesting to recognize that the infant tends to introduce an assortment of sounds that have no "official" place in the language of his or her society. Each language system emphasizes certain sounds more than others. The infant has no way of knowing whether those fun-to-make clicks and clucks will be part of his or her standard speech pattern in years to come. I have said that it is difficult to separate description entirely from theory. Here is an example in point. Young infants around the world seem to exhibit much the same wide variety of sound-makings. But by the end of the first year, some sounds have already started to drop out—the sounds that have little place in their parents' particular language. At the same time there is a general increase in the number of specific **phonemes** (speech-like sounds) that the adult's ear can discriminate. By his or her first birthday, the infant is already favoring sounds that appear more often in the language of the realm. This pattern of selective phonemic utterance appears to favor theories that emphasize *imitation, social learning* and the significance of *culture-specific linguistic systems.* And yet, the fact that the infant's *early* vocal repertoire is fairly similar in all societies studied also implies something about an innate or built-in tendency to produce certain phonemes somewhat independently of cultural influences. We will see that the nature-versus-nurture issue comes up in many other developmental contexts besides language growth (*see* Chapter 7).

Phonemes are sounds that resemble speech

Usually the six month old is an accomplished *babbler.* There is a more elaborate pattern of vocalizations at this time. The rhythms and sound combinations more closely resemble adult speech. Consonant-vowel combinations are likely to be repeated over and again (e.g., "da-da," "ga-ga"). It can be tempting to interpret some of these utterances as words, but most investigators of early language development do not share this view.

The first recognizable word usually appears somewhere between the ninth and twelfth month. Once that first word is out, others typically follow in short order. It is characteristic that single words make up the unit of utterance at this time. "Kitty!" is an entire unit of exclamation. One word at a time may not sound like much, but it is a remarkable improvement over zero. "May I have a second helping, please?" is a phrase that is much beyond the yearling's capacities. However, a forcefully declaimed syllable—"More!"—can produce the same desired effect.

The single-word statement is also more powerful and useful than one might think because the words can be made to serve several purposes. "Kitty," for example, could mean either that the infant has just spied the cat, or that he or she *wants* to see the cat. Often we have to see the young child in the immediate situation to determine what the statement means. The *contextual interpretation* of early speech behavior has perhaps been underemphasized in current theory and research that analyzes *linguistic structure* in great detail. We have the choice, in other words, of paying very close attention either to to the way in which the young child puts words together and

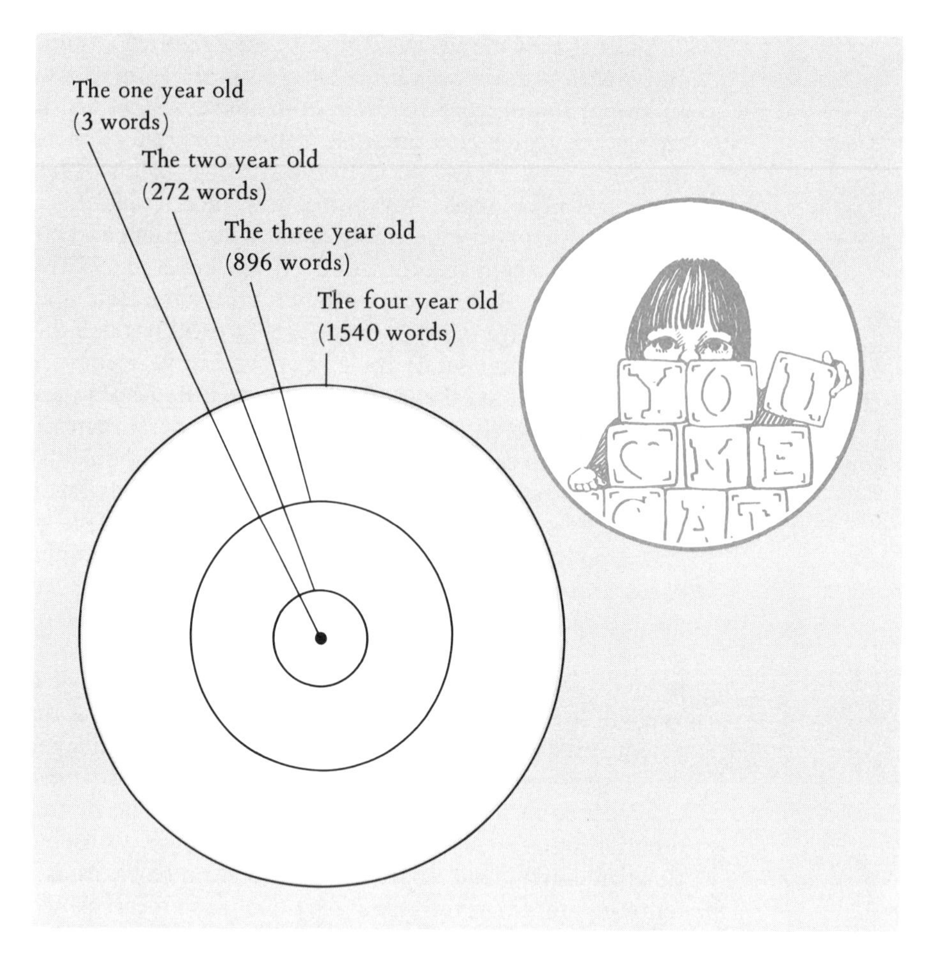

Figure 6–3. *words, Words, WORDS: vocabulary growth in early childhood. Vocabulary growth occurs at a rapid pace during the preschool years. The average number of words in the vocabulary of a child at ages one, two, three, and four appears in parentheses below each age. (Based on data presented by Ausubel & Sullivan, 1970.)*

the rules that are implied by this process, or to how the language flashes out into communication within a particular, concrete situation. Those who emphasize linguistic structure generally have to wait until the child has moved beyond the one-word-at-a-time phase in order to practice their analytic craft.

Even when the child is in command of a larger vocabulary and does know how to put words together, it remains helpful to understand the situational context of his or her statements. The rules of language that we as adults use in our own speech liberate us from concrete situations and make communication more accurate and standardized. We use different words to express different facts and emotions. The yearling has fewer words and few, if any, of the rules that govern the way in which words are put together. We have to decode his or her messages as best we can.

Holophrastic speech is made up of one-word utterances that may mean more than the literal meaning

The young child's characteristic of using a single word at a time is sometimes known in the technical literature as **holophrastic speech.** One word represents a message that we ordinairly would convey in a structured combination of word units. It is useful to distinguish between the *fact* that yearling's use one-word utterances and the *theoretical implications* of the term, holophrastic speech. Probably, we should not use this term at all unless we share the views of some researchers (e.g., McCarthy, 1954) that children are consciously trying to load a whole set of meanings on a single word-utterance. This credits the year-old child with a degree of comprehension that is far ahead of his or her communicative skill. The child knows and intends more than he or she can say, but does not have the communicative expertise to say it. Theorists who use the term holophrastic speech usually do not mean that the child understands about sentence structure or has precise representations of the world and how it operates. Rather, the young child is viewed as working with perceptions, feelings and motivational promptings that are not clearly distinguished from each other, but which lead him or her to attempt verbal expression.

Observers often have noticed that *expressivity* seems to be more at the core of some holophrastic utterances than an effort to communicate information as such. The word lets the world know that the child feels terrific or terrible, or it may convey the thrill of a fresh discovery. In very early childhood, then, an utterance may primarily express feeling-oriented states of mind.

Telegraphic speech is made up of two- or three-word utterances using simple grammar

The young child does not move directly from single word utterances to complete sentences. There is an intermediate stage of short word combinations. This is sometimes known as **telegraphic speech** (Brown & Fraser, 1963). The child sends out key words, but omits many of the parts of speech with which we decorate our conversation. "Where Kitty?" is question clearly enough asked, although the grammarian would have preferred, "Where is the Kitty?" As you will notice, adding even one word to the single-word unit makes a big difference. "Kitty," by itself, does not tell us precisely what the child is trying to communicate. "Where Kitty?" or "*There* Kitty!" make more effective communications.

Lennenberg (1967) reported that word combinations usually begin between the eighteenth and twenty-fourth months. As might be expected, two-word combinations come before larger units. However, the ability to form sentences usually does not show up until well beyond the second birthday. When the young child is at the phase of using words but is not yet able to form sentences, he or she is often very adept at communicating the essence of what he or she has to say. Nouns, verbs, and adjectives are favored, while prepositions, conjunctions, and articles of speech are seldom heard. Nouns and verbs are especially popular in the vocabulary of young children, perhaps because these are generally the high-information words. As the child imitates the speech usage of his or her parents, he or she preserves the most significant words and usually keeps them in the same sequence he or she heard them in (Brown & Bellugi, 1964). In the English language, for example, the child is likely to say "You bring!" rather than "Bring you!" Although the number of words in a single utterance remains small, with many parts of speech omitted, the essential structure of direct statement in the parents' language is generally brought forth by the child.

When the young child reaches the point of making two-word utterances, it is possible to undertake fairly detailed analyses of the "rules" that seem to govern his or her speech. Here again direct observation and theoretical interpretation work together. Theoretical inclinations may lead, and have led researchers, to different types of analyses from essentially the same data.

Braine (1963) offered a pioneering analysis of the child's first pattern of combining words. A distinction was made between two types of terms collected from the speech of children who were about eighteen months old. **Pivot words** have action implications, express possession, or are qualifying (e.g., "big," "here"). **Open terms** are often nouns and refer to people and objects. The child works with a relatively small number of pivot terms and adds to this set slowly. He or she has a larger set of open terms and adds to them more rapidly, but usually the child does not employ any particular open term as frequently as he or she employs particular pivots.

Pivot words express action or possession, or describe nouns

Open words refer to people and objects

In the typical two-word statement of the eighteen month old, there is a fixed relationship between open and pivot terms. A particular pivot term will always come either before or after an open term. This indicates that there is a lawfulness already present in the child's use of language. Furthermore, the child will not combine two pivot words. He or she will use a pair of open terms together, or a pivot word and an open word. "See," for example, is likely to be a pivot word for the young child. He or she will say, "See Mommy," or "See Daddy," or "See Kitty." He or she will not say, "See big" or "See my," which are paired pivot words.

Findings of this type whet the curiosity of those with a keen interest in psycholinguistic development and arouse some interesting alternative interpretations as well (e.g., Bloom, 1971, who takes a more complex view of the situation). But for our purposes here perhaps the most striking impression is that even the very young child is forming word structures according to some kind of regularity, some kind of organizing principle.

Brown's analysis is also a good example of a psychologist's mind at work

An important series of studies by Roger Brown (1973) and his colleagues led to the formulation of a five-stage model of language development in young children. These investigators were more attuned to *developmental time* than to chronological age. Children differ somewhat with regard to the chronological age at which certain behaviors come forth and the rate at which development continues to proceed. However, according to the developmental perspective, the existence of a developmental process can be demonstrated by the fact that the *sequence* is the same whether the child happens to be an earlier maturer or a later maturer.

The first phase of language development, as analyzed by Brown, reveals the child's ability to establish relationships among different clusters of meanings. Who is doing what and where is the type of information the young child is attempting to sort out and convey in early holophrastic and telegraphic speech. In studying three young children in great detail, Brown and his colleagues came to the conclusion that the children had a greater linguistic competence than might first seem to meet the eye or ear. The researchers therefore set themselves the task of, in effect, writing a grammar that expressed the implicit rules the children were using. Even at the first phase of development, the grammar is somewhat more complex than can easily be summarized here (*see* Brown, 1973; McNeill, 1970, for more detail). By the second stage, utterances are longer and the grammar governing those utterances a good deal more subtle and complex. It is here that the child begins to use inflections in ways that add to the communication value of his or her speech, to use more parts of speech, and to make certain distinctions more readily (e.g., singular and plural). The child at stage three (still a prekindergartener) is even more resourceful and versatile with language. He or she can ask questions as well as make statements. Furthermore, the child learns the *negative* case. "No Kitty!" is a statement the child can easily volunteer, while he or she can understand such sentences as "It's not cold" (Klima & Bellugi, 1966). The remaining two stages would take us beyond the two year old. They feature an increasing ability to produce and understand complex sentence structures.

SOME THEORETICAL APPROACHES TO EARLY LANGUAGE DEVELOPMENT

Some major features of early language development have been described. But how can we explain the infant's progression from howl and coo and cry to articulated speech?

There are three approaches to this question, the first two of which are attractive for their simplicity or theoretical economy. They do not seem to make a lot of assumptions and hover fairly close to common-sense experience at a number of points. One possibility is that infants *imitate* what they hear. Children learn to speak because they have been spoken to. It is clear enough that what is said in the presence of the child and how it is said does have its effects; a few research examples have already been cited. It would not be realistic to deny that exposure to adult speech has an influence. However, this approach falls much short of complete explanation. On a philosophical level, we might wonder how people ever learned to speak in the first place, let alone cultivate that complex system of symbols known as language. A parallel question arises for each and every baby born. How could he or she imitate (or learn by other such simple means) if he or she did not already have certain capacities in or dispositions toward the use of language? The baby sees birds fly, but he or she does not take wing himself or herself no matter how exciting flight seems. The complexity of our language systems and much research (e.g., McNeill, 1970) makes imitation an incomplete explanation at best.

The *stimulus-response* or behavioral habit approach is another contender. There have been attempts to explain the development of language—and of all thought and mental activities—in terms of a few stimulus-response or associationistic principles since the eighteenth century. Although the theoretical principles are fairly simple, they lend themselves to broader and more intricate applications than the more commonsensical imitation approach. The stimulus-response approach has an additional merit; at the hands of resourceful researchers it can lead to useful methodological innovations. Whether or not one is satisfied that speech develops through one or another form of conditioning (e.g., Staats, 1966), the methods and findings of such research help to sharpen everybody's ideas. Furthermore, people who find cognitive and nativistic types of theory too convoluted and complex might take comfort in regarding language as just one more behavior pattern that is shaped by reinforcement, strengthened by repetition, and otherwise conforms to the customary rules of stimulus-response psychology.

The *cognitive-nativistic* approach is a third and much more complex alternative. It seems more nearly equal to the task of explaining language development in young children and its subsequent flowering in adulthood. Theories of this type hold that an ability to process experience in terms of language structure is part of our fundamental human equipment. It's the way we are made (although research into the biological aspects of early language development is itself in the early and inconclusive stages). Cognitive-nativistic approaches are less likely to seek **reductionistic explanations.** Brown, for example, respects the complexity of young children's word processing and attempts to represent this in an appropriate form. Noam Chomsky's influential work on the "deep structure" of language (Chomsky, 1965) has the effect of making imitation and stimulus-response analyses appear quite superficial.

Reductionistic explanations try to account for complex events in simple ways

The last word has yet to be spoken about the first word. Complex theories that hold we bring language development structures with us into life are more convincingly rich in detail and level of analysis. Yet, the effect of the environment in general and of specific interpersonal transactions can hardly be denied, and definitive research of an interactive type is a goal that is still before us.

MENTAL DEVELOPMENT IN THE VERY YOUNG CHILD

The very young child's development of language is an impressive and exciting process. But there is more to mental functioning at this time of life than can be revealed by the use of words alone. Much of the child's mental activity is expressed nonverbally through behavior. It appears to be only partially under verbal control. This generates questions both of theory and method.

Although the type of mental operations that characterize adult thought remain far in this boy's future, there is little doubt mental activity is as natural to him as physical activity. (Photo by Talbot Lovering.)

One important theoretical question concerns the *continuity* of mental development from early childhood to later periods of life. Is the difference between the two year old and the adult mostly a matter of *quantitative* growth in mental functioning? Or do profound *qualitative* transformations take place? The first proposition would picture the very young child as a miniature adult who has a lot to learn and many intellectual, as well as physical, skills to master. Historically, many people have taken the "tiny adult" view of the young child and engaged in parenting and disciplinary practices attuned to that view. The second proposition is perhaps more truly developmental (which does not necessarily make it more true). The structure and function of the entire psychobiological organism is seen as unfolding in time. The transformation is not quite so obvious as caterpillar-into-butterfly, but there are developmental laws to describe and explain the differences from infant-into-adult.

A practical side of this question is the attempt to predict from mental activity in infancy or very early childhood how "bright" the individual will be later in life. This type of question is of interest to many people in our achievement-oriented society.

For both theoretical and applied reasons it would be useful to know the starting point or baseline of mental functioning. What can the mind of the very young child do, and what appears to be beyond its capacities? Unless a baseline can be established, it becomes very difficult to test competing theories of development or make practical predictions.

The methodological question centers around our ability to assess mental functioning so early in life. *What* should we be assessing? *How* should the assessment be made? *Which* aspects of mental functioning are of greatest importance? We turn now to a brief consideration of the problem of assessing mental functioning in the very young child. For more detail you will find the historically oriented survey by Brooks and Weinraub (1976) valuable.

MEASURES OF MENTAL FUNCTIONING IN INFANTS AND VERY YOUNG CHILDREN

Each procedure for assessing mental functioning in the very young has had its own history, purposes, and assumptions. In the early days of mental assessment, it seemed natural to assume that the individual starts life with a certain amount or level of intellectual potential. If we developed the appropriate assessment techniques then we could classify the very young with respect to their level of intelligence. Years later, especially bright babies should be especially bright adults and so on. The most-used and best-known measures of mental assessment in the very young have usually shared these assumptions of a *general level* of intelligence that is *continuous* from infancy into adulthood.

Even with these relatively simple assumptions, the science of constructing, and the art of administering, assessment procedures proves to be challenging. As Honzik notes: "The perfect assessment . . . takes place when the squalling, sucking, chewing, ever-moving baby is relatively quiescent, attentive, wide-eyed with interest, and above all, responsive to the tester and his toys. This idyllic situation is only achieved with effort on the part of all concerned" (Honzik, 1976, p. 60). The tests or scales must be designed to do justice to the large differences in behavior shown by children even within such a small time range as infancy-to-two-years-old. In measuring adult intelligence, the same task presented to a thirty year old and to a thirty-two year old might indeed be the *same* task. But the rapidly changing configuration of development in the very young makes it difficult to build stability into any assessment procedures while also being responsive to developmental changes.

Test situations are not always idyllic with older children and adults either!

In the United States, perhaps the best-known measures of early mental functioning have been the procedures developed by Gesell and Amatruda (1962), Cattell (1940, 1960), and Bayley (1933, 1969). Another procedure inspired by Gesell's work (Griffiths, 1954) has also seen service, as well as a French version that has added some features of its own (Brunet & Lezine, 1951). These have been used for many clinical and research purposes. The Gesell Developmental Schedules have provided inspiration and specific items for most of the subsequent approaches. "DQ's" (developmental quotients) are determined on the basis of the infant's or young child's performance in the areas of adaptive behavior—motor, language, and personal-social. The later techniques have added refinements of scale construction and analysis that were not known to Gesell when he started his work in the 1930s.

(g) is a symbol for general intelligence

The research findings of Gesell and others have called the assumptions of general intelligence (**g**) and simple continuity from infancy to adulthood into serious question. Assessments of children two years of age or younger do not show a dependable relationship to mental test performance in the same children at a later age. This was discovered by the original scale developers themselves (e.g., Bayley, 1933), and it has been well documented in recent years (e.g., McCall, 1972). Not only is there a lack of strong correlation and prediction across time, but one aspect of mental functioning does not necessarily relate well to another aspect that is assessed at the same time.

An **emergent** characteristic is a new characteristic that appears during development

After considerable research and thought, developmentalists tend to believe that the difficulties are not entirely with the assessment procedures themselves. The procedures yield information that can be useful for a number of purposes, and they can be administered and scored with adequate reliability. The problem, rather, is that mental functioning in the very young just doesn't lend itself to description on a single, general level. More than one type of ability or process is in operation. Furthermore, the evidence indicates that so much of mental functioning is **emergent** (not present in "tiny adult" form) in very early life, that there is no way to go from the first manifestations of intelligence to more mature forms with any high degree of comprehensiveness. There are still unanswered questions about how much of subsequent mental development is quantitative (improving along early-existing dimensions) and how much is qualitative (new levels or dimensions of functioning coming into play). But the previous simple conception of infants being more or less bright along a single dimension and preserving this level throughout their lives is difficult to maintain in the face of accumulated research.

The fact that predictions made on the basis of early mental assessment are not very dependable should serve as a caution for those who have programmatic social plans in mind. For example, attempts to *intervene* in the early development of children are difficult to evaluate because, among other reasons, the structure of mental functioning at this time of life is not as unitary and simple as often assumed. A program of environmental enrichment might be judged either a success or a failure on the basis of assessment scores that actually have little predictive value. Some developmentalists are concerned about the *sociopolitical misuse* of mental assessment of the very young (e.g., Lewis, 1976). During the past few years there has been a gradual heightening of awareness about misinterpretations and misapplications of IQ-type tests in adulthood. These considerations also apply to the testing of very young children as well. Lewis charges that: "IQ scores have come to replace the class systems or feudal systems that previously had the function of stratifying society and distributing the goods and wealth of that society. . . . The 20th century technological society's stratification device has become the intelligence test" (1976, p. 16). Along with a number of other experts in early mental development, Lewis cautions that we should not rush into classifying infants and very young children into IQ-like pigeonholes that may limit their options and opportunities later in life.

Another type of assessment procedure has been developed in recent years. It is based upon a somewhat different view of the nature of human intelligence and its lifespan development. In particular, these newer procedures show the strong influence of Jean Piaget. The contributions of Piaget will be considered at various points throughout this book. I begin now with a summary of his work with the very young.

SENSORIMOTOR INTELLIGENCE

Jean Piaget—a philosopher of knowledge (epistemologist) as well as a psychologist—has made prodigious contributions to developmental psychology. The breadth and detail of his work cannot be captured here. A recent *selection* from his writings comes to almost nine hundred pages (Gruber & Vonèche, 1977), and it might easily have been doubled or tripled in length.

Piaget (1954) did not give psychometric tests to the very young. Instead, he observed their behavior carefully and provided them with small challenges (such as finding a hidden toy). He was not interested in expressing an infant's mental functioning in terms of quantitative scores. Piaget sought insight into the processes of mental life and the rules of developmental change. His observations have suggested that the earliest period of mental development in the child can be understood as the **sensorimotor stage,** which itself can be divided into six fairly distinct substages.

The **sensorimotor stage** is Piaget's earliest period of mental development

The concept of developmental stages, whether found in the writings of Piaget or in those of other scientists, almost always involves a specific sequence. Stage 1 *is* Stage 1 because it is the first to appear in the organism's life history. In a rigorous stage theory, the sequence will be the same from individual to individual, although other details may differ, including how long it requires particular individuals to advance from one stage to the next. If Piaget is correct, then, all of us began at the sensorimotor period or sensorimotor stage and moved through this stage in the same sequence through six substages.

I will first describe the sensorimotor substages and then attend to Piaget's views on the relationship between these phenomena and the child's total pattern of thought.

Substages of Sensorimotor Development in Early Childhood

1. *Reflexes.* Approximately from birth to six weeks. The newborn begins by adjusting to his or her environment by using innate reflexes, many of which have already been described in this book.

2. *Habits.* Approximately from six weeks to four or five months. The first habits appear. Thumb sucking exemplifies the simple habits that are established and maintained. During this substage the infant seems to perform simple actions for their own sake rather than to accomplish anything through them. These early behaviors are called *habits* because the infant tends to repeat them over and over again.

Prehension is seizing or holding onto, as in understanding

3. *Coordination between vision and* ***prehension.*** Approximately from four to nine months. The infant now seems to have more "purpose" in his or her actions. In the first of these more intention-filled substages, the infant shows an increased ability to *coordinate* what he or she sees with what he or she does. Furthermore, he or she begins to take an interest in the *effect* of his or her actions. Bodily actions that produce an observable result are repeated (perhaps for pleasure). The infant grasps for objects in a manner that indicates he or she really wants them, as distinguished from an "empty" gesture that is repeated over and again.

4. *Coordination between means and goals.* Approximately from nine to twelve months. Intentional behavior becomes easier to observe at this substage. The infant can, for example, search for a lost or hidden object although his or her search logic remains very limited. How well the child at this substage can solve such problems as the hidden toy is less significant than the fact that he or she is actually making a problem-solving effort. Old habits are pressed into the service of newly-recognized problems and challenges. By this point the infant seems to have some notion of *space* as the place where things are and *time* as before and after. It is a limited beginning of appreciating such challenging concepts as time and space, but even Einstein started this way (or did he?).

To see something, then to move toward it, then to do something with it. The ability to coordinate perception and action comes after the substages of reflexes and habits. (Courtesy of Fisher-Price Toys.)

5. *Discovery of new means.* Approximately twelve to eighteen months. The motto of this substage might be called, "try and try again—but with a new plan!" The infant is now able to try more than one approach to the same problem. This is obviously a marked improvement in flexibility of adaptation to the environment. If the infant can't reach what he or she wants with hands, he or she might try feet, or a stick held in his or her hands, or crying a lot. This ability to explore alternative ways to reach a goal will be important throughout the total lifespan. Cultivation of this ability remains the dominant aspect of sensorimotor intelligence from the time it appears until approximately the eighteenth month. The child's actions at this point may strike us as clever and innovative. And indeed often that is just what they are, even though generations of children have preceded in making the same or similar discoveries.

6. *Insight.* Approximately eighteen to twenty-four months. In the sixth and final sensorimotor substage, the youngster is now able to perform some actions mentally, in his or her head, as well as with his or her hands and feet. A kind of **interiorization** is taking place. In a sense, the child can behave inside his or her own head as well as out in the external world. Some of the experimentation is carried out in the absence of any observable trial and error. Piaget describes, for example, a little girl who discovers for herself that she can push a doll carriage from either end. This realization seems to occur as though by "sudden understanding" or insight, rather than by accidental or labored movements of the carriage. These moments of insight on the child's part can be rewarding to the observer, especially, if it is one's own child who has suddenly stopped, lit up, and then demonstrated his or her own solution to a problem. (Piaget, by the way, gained some of his first developmental insights by such observations of his own children.) The new accomplishments at this substage seem to require some ability to represent objects and actions mentally. The child who is approaching his or her second birthday can form and manipulate images of the world in his or her head. This enables him or her to think before acting.

Interiorization is representing external objects and processes within one's own mind

Parents and other child watchers know that two year olds do not *always* seem to think before they act (a characteristic their elders sometimes share as well). But the point is that at this age the young child is already on his or her way as a "thinker."

Why the Sensorimotor Stage Is Important

Much of the significance of this first stage of mental development is obvious from what has already been described. But several other points also merit our attention.

What is the relationship between the sensorimotor form of intelligence and language? We have seen that some milestones of mental development occur at an age when language is lacking or rudimentary. In this sense, the sensorimotor stage might be regarded as preverbal: it comes before language does.

Presymbolic is a level of mental functioning at which objects, events, etc. cannot be inwardly represented

Piaget has a different way of looking at this, however. He regards the sensorimotor stage as **presymbolic.** Preverbal and presymbolic do not mean the same thing. Piaget regards language as just one way, although a socially dominant way, of conveying thought. More fundamental than language is the ability to think symbolically, to *represent* objects, events, and relationships in the mind. It is only toward the end of the sensorimotor stage that the young child starts to cultivate the capacity for interiorization and mental representation. At this point, language development can move ahead more rapidly, but this is only one of the more obvious signs that the child has started to process his or her experience symbolically.

Let's reflect a moment longer on the culminating substage of sensorimotor or presymbolic intelligence. The child now possesses a mental structure with improved connections between perception and response. Because he or she is more experienced and coordinated, the child can engage in fairly smooth and coordinated actions. But there is another striking advance that also deserves our appreciation. *The child's mental structure itself has started to structure the world.* We do not know precisely how the young child represents the world in his or her mind's eye. Perhaps we will never know the answer to this question. Yet, at this point, he or she has started to use the experienced world as raw material for his or her own innovations, intentions, and interpretations. We could, if we wished, emphasize the differences between the thought experiments of the two year old and those of the mature adult. The differences are, indeed, appreciable. More sensitively, however, we might instead honor the early mental adventuring of the child. Every mind that has extended the knowledge and achievements of our species probably began with the same kind of modest experiments in structuring the world through its own rapidly changing mental structure.

It is also during the sensorimotor stage that the infant begins to form an awareness of what is known as *the permanent object.* Very early in this stage, the infant behaves as though out of sight is truly out of mind. Later, as the hidden toy experiments reveal, the child seems aware that the toy, and other people and objects, continue to exist even when not in view. From such small beginnings, the child eventually develops a broader and more substantial appreciation of the fact that there is a world which exists independently of his or her own perceptions and needs. Full appreciation of the existence and properties of external reality will take a long time to develop. Conceptions of time and space require progressive differentiation from the infant's initial sense that all the action is where he or she is.

Egocentrism is viewing the world only in relationship to one's self

This brings us to another basic characteristic of the very young child's mental life. Most observers agree that we begin as **egocentric** beings. The world seems to revolve around us, somewhat as planets revolve around the sun in the heliocentric system. The infant can see the world only as he or she can see it—not as others can. He or she cannot put himself or herself in the place of mother, father, or Kitty. Events in the world exist only to the extent that they affect him or her. The infant is cold, or wet, or hungry—and he or

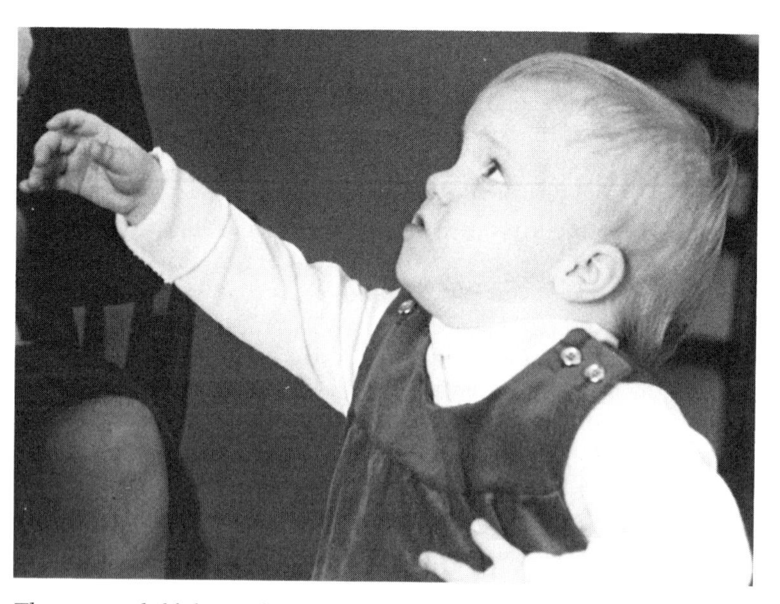

The young child, limited to an egocentric view of the world, seeks immediate release of tension no matter what might be happening from the standpoint of others. (© 1978 Photo by Thomas S. Wolfe.)

In times of stress adults may become situational persons too

she lets everyone else know about it. But the idea of coldness, wetness, or hunger, either in its general form or as experienced by others, is not in the infant's repertoire. This tends to make the infant a situational kind of person. He or she is at the mercy of inner and outer forces that operate upon him or her at the moment. Adults also experience inner and outer stimuli, but they have a mental structure and a self that provide insulation against the spurs of the moment. The infant is embedded or trapped in the immediate situation. His or her experience of the world is not only *personalized,* but also *situationalized.* He or she does not yet possess that strong continuity of selfhood that enables him or her to move securely from situation through situation, or to distinguish between what is happening to oneself and what is happening to others.

Egocentrism, in this sense, has little to do with popular phrases such as *egotistical* or *going on an ego trip.* These expressions assume the existence of a self-image that is being inflated or elevated beyond its most appropriate level. The very young child does not yet have the type of ego implied by these expressions. Egocentrism in the infant and young child should be understood as a kind of close-up relationship to immediate reality. The infant is sensitive to momentary influences. He or she perceives, responds, and copes to the best of his or her ability. Gradually this coping behavior (evidences of advancing sensorimotor intelligence) becomes smoother and more versatile. But it will be a long while before the infant has—or, rather, *is*—a self that is master of its own experience.

Decentering is shifting from an egocentric to a more relational orientation

In Piaget's view, sensorimotor intelligence begins from a state of high egocentrism and then evolves into a more sophisticated awareness of self-world relatedness. The shift from egocentrism is characterized by Piaget as a process of **decentering.** It is through this process that "the child eventually comes to regard himself as an object among others in a universe that is made up of permanent objects (that is, structured in a spatio-temporal manner) and in which there is at work a causality that is both localized in space and objectified in things" (Piaget & Inhelder, 1969, p. 13). In other words, the child has entered into *the construction of reality.* Each human must do this for himself or herself, for each begins with feeling states and reflexes and must somehow develop a mental framework that acknowledges both the ways of the world and the ways of the self. The world comes to appear as a more or less stable, dependable place when one's own mind has become a stable, dependable, adaptational system.

The Piaget approach to early mental development has inspired new assessment procedures. While these are indebted to Piaget's work, they also bring in observational methods and techniques for systematic data gathering that are not found in Piaget's work as such. Scales of Sensorimotor Development have been prepared by Corman and Escalona (1969). Mehrabian and Williams (1971) developed the Cognitive Development Scale. Another promising procedure is the Infant Psychological Development Scales (Uzgiris & Hunt, 1975). All of these Piaget-based scales show that sensorimotor functioning improves with chronological age, which is no more than what we would expect. More interesting findings about the conditions of early mental development are likely to be forthcoming as these scales attract widespread application. There is an emerging body of research, for example, on the association between various kinds of home environments and the rate of sensorimotor development as indexed by these scales. Piaget's theory of mental development has itself entered a new phase, then, one that encourages insight and new coordinated actions among researchers themselves.

SUMMARY

The infant begins with very little ability to cope with his or her environment physically or mentally. Yet his or her mind and body are active from the start and develop rapidly together. Within the short span of two years, the infant begins not only to walk, run, and manipulate objects, but also to carry out mental manipulations and to participate in the exchange of words. Attention was given in this chapter to the risk and achievement involved in transforming one's self from a four-limbed to an erect organism (the "Perils of Verticality"). The typical sequence of behavior development *from reflexing to walking* was described, as was the parallel development of *small muscle control and coordination. Bowel and bladder functioning* was taken as one example in which biological, psychological, and social dimensions of devel-

YOUR TURN

1. The perils of verticality have been suggested. Some basic material has been offered on the developmental sequencing that we pass through before we become adept at standing and moving around on two limbs. I also explored some of the less obvious implications of attaining verticality. You could do the same for other aspects of early development. Take, for example, the *size differential* between the world of the infant and young child and that of the adult. What are the implications, social, psychological, and biological, of starting life undersized? How might this affect perception? Ability to manipulate objects? Fears? Goals? The growth of self-esteem and confidence? Does the early difference in stature between child and adult produce remnants or derivatives that influence people in later life? (When we say, "This is a person I can look up to," for example, are we continuing a pattern of height worship that was acquired very early in life?) Observe and think about the obvious fact that every human begins life as such a miniaturized specimen. Assemble facts, including, perhaps, the comparative size at birth and adulthood between humans and other species. But also give some rein to your fantasy. How would the course of individual development and of society differ if newborns were as large as their parents? Or if they started out large and grew "down" instead of "up"? These are examples you can easily improve upon. The general idea is to seriously and systematically consider one aspect of early development that is so "obvious" we seldom think about it at all. The size differential is one such aspect—what are some others? Photographs, charts, and other illustrations might add to your presentation. If it is the size differential problem you take on, then you probably would enjoy reading Alfred Adler's views on the development of inferiority feelings and the striving for power (Ansbacher & Ansbacher, 1956).

2. Separate attention was given in this chapter to the development of standing and walking, small muscle coordination, and thought and language. Obviously, however, these areas of development exist within the same individual. If you have access to children in the age range discussed in this chapter, why not examine the relationship between these three areas of functioning, or between any two of the areas? This does not have to be a controlled scientific or clinical study. You might find it more practical to combine methods, such as asking parents to help you chart some of their child's past development, while making fresh observations on the child's current status yourself. It would be especially interesting if you could compare these relationships in several children. Does one pathway of development seem to be more accelerated or dominant for all the children, or is one youngster surging ahead in small muscle coordination, while another is excelling in the acquisition of language? Organize and present your observations as clearly as possible. Offer whatever conclusions seem justified on the basis of your observations, but also share questions and problems that have arisen and have not been resolved. This latter consideration is an important dimension of the scientific process: recognizing and admitting what we do not yet know. Remember also that if a number of your classmates independently carry out this exercise, you will wind up with an appreciably larger sample than any of you would have been likely to obtain by yourself and, thus, a better basis for discussion and generalization.

3. Why do adults use "baby talk?" Does this really perform a service for the infant or is it for the benefit of the grown-ups themselves? Think about the phenomenon of baby talk. Then collect some examples if you can, perhaps with the help of a portable tape recorder. Speak also to the baby-talkers themselves and ask them some intelligent questions on the subject. The specific questions you ask will depend upon the specific ideas and hunches that come to you when you first explore this topic. In addition to the basic materials you gather (samples of baby talk and discussion with the baby-talkers), you should note basic facts in the situation (e.g., age and sex of the child, the behavior setting in which the conversation took place, and whatever else seems important in interpreting the communications). Furthermore, you might experiment by speaking to the youngster in a more adult manner and observing the effects. (For a better-controlled experiment, you might prevail upon the baby-talker to speak in a more adult manner after you have heard and taped the original monologue or dialogue. You yourself might then try both baby talk and adult speech with the child.) As you create a presentation from this material, try to come up with a clear definition of what baby talk *is*. If you do manage to make tape recordings and have permission to play them for others, this could be an instructive and perhaps entertaining experience.

4. The first of the two chapters on this early age period (Chapter 5) concentrated on patterns of interpersonal development: how the infant or young child relates to his or her parents. This chapter has concentrated on physical and mental development. What relationships can you see between the material presented in this pair of chapters? How does the "stranger reaction" (baby over forty weeks of age clinging desperately to mother in response to strangers), for example, relate to the level of physical and mental development characteristic of of that age? Glance through Chapter 5 again, with the material of Chapter 6 still fresh in your mind. Observe, think a bit and see what you can come up with, especially in the way of questions that are worth further study.

opment could be seen as closely interrelated. Parental behavior and expectations are more helpful to the young child when guided by knowledge of the bio-developmental timetable as well as respect for the individual characteristics of the youngster.

The powerful surge of mental development early in life is supported by physical changes in the nervous system, which include a doubling of brain weight between birth and the nine-month plateau. The newborn spends much of his or her day asleep. But there is a steady increase in *wakefulness* throughout the early months and, therefore, more opportunity for the kind of mental activity that is associated with the waking state.

The development of language during the first two years of life is a remarkable and complex process. From early vocalizations, the baby progresses to babbling at about six months, and to his or her first word a few months later. He or she is already selectively keeping or discarding speech sounds depending upon the frequency with which they appear in the parents' language. Utterances are at first *holophrastic*, one-word units that often convey some of the meanings that older children and adults would express in sentences. Listeners must be aware of the context in which the child speaks in order to decode these messages. *Telegraphic speech* comes as a transition between holophrastic speech and the construction of complete sentences. Short word-combinations (usually two words and mostly nouns and verbs) come before sentences. Even at this early stage of language development, it is evident that the child is following certain rules of word combination and grammar. Braine's analysis of *pivot* and *open* terms and Brown's charting of five stages of early language patterning both indicate that simple stimulus-response or imitation theories of language acquisition probably will not prove adequate.

The parent-child interaction is important to encouraging and shaping language development, as shown by the successful *reinforcement of vocalization* in children as young as three months of age. The infant and his or her parents seem to reinforce each other and to shape each other's speech. With the increased use of language, the young child has an improved means of expressing needs and a new source of pleasure.

The *assessment of mental development* in very early life poses many methodological and conceptual challenges. It has been found that scales for assessing intelligence in infancy through the entire below-two-years-old period do not predict subsequent level of mental functioning very well. The problem is not entirely in the assessing procedures, but rather in the assumption that intelligence is unitary and proceeds in a direct line throughout life. Newer assessment scales are now being developed and applied based upon the work of Jean Piaget.

Piaget's analysis of the *sensorimotor stage* and its six substages was given close attention. During this *presymbolic* phase of mental development, the young child shows increasing *purposefulness* in behavior, greater *coordination*, and a new *flexibility* in adapting to the environment. Toward the end of this stage a very important development shows itself, the ability to *interior-*

ize, to conduct manipulation inside the mind as well as out in the world. The emerging ability to *represent* objects, events, and relationships is crucial for the development of language.

By the time the child has reached the two year mark, he or she has some awareness of *the permanent object,* has partially *decentered* from the deep *egocentrism* with which everybody begins life, and is well off on the long road to a mature *construction of reality.*

Reference List

Ansbacher, H., & Ansbacher, R. (Eds.), *The individual psychology of Alfred Adler.* New York: Basic Books, 1956.

Ausubel, D. P., & Sullivan, E. V. *Theories and problems of child development* (2nd edition). New York: Grune & Stratton, Inc., 1970.

Bayley, N. Mental growth during the first three years: A developmental study of 61 children by repeated tests. *Genetic Psychology Monographs,* 1933, *14,* 1–92.

Bayley, N. *Bayley scale of infant development: Birth to two years.* New York: Psychological Corporation, 1969.

Bloom, L. *Language and development: Form and function in emerging grammars.* Cambridge, Mass.: Massachusetts Institute of Technology Press, 1971.

Bohannon, J. N., & Marquis, A. L. Children's control of adult speech. *Child Development,* 1977, *48,* 1001–1008.

Braine, M. D. S. On learning the grammatical order of words. *Psychological Review,* 1963, *70,* 323–348.

Brooks, J., & Weinraub, M. A history of infant intelligence testing. In M. Lewis (Ed.), *Origins of intelligence.* New York: John Wiley & Sons, Inc., 1976.

Brown, R. *A first language: The early stages.* Cambridge, Mass.: Harvard University Press, 1973.

Brown, R., & Bellugi, U. Three processes in the child's acquisition of syntax. *Harvard Educational Review,* 1964, *34,* 133–151.

Brown, R., & Fraser, D. The acquisition of syntax. In C. F. Cofer & B. S. Musgrave (Eds.), *Verbal behavior and learning: Problems and processes.* New York: McGraw-Hill, 1963.

Brunet, O., & Lézine, P. U. F. *Le développement psychologique de la premiere enfance.* Issy-les-Moulineaux: Editions Scientifiques et Psychotechniques, 1951.

Bryant, P. *Perception and understanding in young children.* New York: Basic Books, 1974.

Cattell, P. *The measurement of intelligence of infants and young children.* New York: Psychological Corporation, 1960.

Cattell, P., Halverson, H. M., Ilg, F. L., Thompson, H., Castner, B. M., Ames, L. B., & Amatrudo, C. S. *The first five years of life.* New York: Harper, 1940.

Chomsky, N. *Aspects of the theory of syntax.* Cambridge, Mass.: Massachusetts Institute of Technology Press, 1965.

Corman, H. H., & Escalona, S. K. Stages of sensorimotor development: A replication study. *Merrill-Palmer Quarterly,* 1969, *15,* 351–360.

Eichorn, D. H. Biological correlates of behavior. In H. W. Stevenson (Ed.), *Child psychology, part I, 63rd yearbook National Society for Study of Education.* Chicago: University of Chicago Press, 1963, pp. 4–61.

Ervin-Tripp, S. M. Psycholinguistics. In *Annual Review of Psychology,* 1966, *17,* 435–474.

Gesell, A., & Amatruda, C. S. *Developmental diagnosis: Normal and applied child development, clinical methods, and practical applications.* New York: Harper, 1962.

Gesell, A., & Thompson, H. *Infant behavior: Its genesis and growth.* New York: McGraw-Hill, 1934.

Griffiths, R. *The abilities of babies.* New York: McGraw-Hill, 1954.

Gruber, H. E., & Vonèche, J. J. (Eds.). *The essential Piaget: An interpretive reference and guide.* New York: Basic Books, 1977.

Honzik, M. P. Value and limitations of infant tests: An overview. In M. Lewis (Ed.), *Origins of intelligence.* New York: John Wiley & Sons, 1976, pp. 59–96.

Kleitman, N., & Engelman, T. G. Sleep characteristics of infants. *Journal of Applied Physiology,* 1953, *6,* 269–282.

Kleitman, N. *Sleep and wakefulness.* Chicago: University of Chicago Press, 1963.

Klima, E. S., & Bellugi, U. Syntactic regularities in the speech of children. In J. Lyons & R. Wales (Eds.), *Psycholinguistic papers.* Edinburgh: Edinburgh University Press, 1966, pp. 183–207.

Lennenberg, E. H. *Biological foundations of language.* New York: John Wiley & Sons, Inc., 1967.

Lewis, M. (Ed.), *Origins of intelligence.* New York: John Wiley & Sons, Inc., 1976.

Luce, G. G., & Segal, J. *Sleep.* New York: Lancer Books, 1966.

Marlow, D. M., & Sellew, G. *Textbook of pediatric nursing.* Philadelphia: W. B. Saunders Co., 1961.

McCall, R. B. M., Hogarty, P. S., & Hurlburt, T. Transitions in infant sensorimotor development and the prediction of childhood IQ. *American Psychologist,* 1972, *27,* 728–735.

McCarthy, D. Language development in children. In L. Carmichael (Ed.), *Manual of child psychology* (2nd ed.). New York: John Wiley & Sons, Inc., 1954.

McNeill, D. *The acquisition of language; The study of developmental psycholinguistics.* New York: Harper & Row, 1970.

Mehrabian, A., & Williams, M. Piagetian measures of cognitive development up to age two. *Journal of Psycholinguistic Research,* 1971, *1,* 113–124.

Parmelee, A. H., Jr., Schulz, H. R., & Disbrow, M. A. Sleep patterns of the newborn. *Journal of Pediatrics,* 1961, *58,* 241–250.

Piaget, J. *The construction of reality in the child.* New York: Basic Books, 1954.

Piaget, J., & Inhelder, B. *The psychology of the child.* New York: Basic Books, 1969.

Rheingold, H. L., Gewirtz, J. L., & Ross, H. W. Social conditioning of vocalizations in the infant. *Journal of Comparative and Physiological Psychology,* 1959, *52,* 68–73.

Staats, A. W. Integrated-functional learning theory and language development. In D. I. Slobin (Ed.), *The ontogenesis of grammar: Facts and theories.* New York: Academic Press, 1966.

Uzgiris, I. C., & Hunt, J. McV. *Assessment in infancy.* Urbana, Ill.: University of Illinois Press, 1975.

INSIDE OUT OR OUTSIDE IN?

A Choice of Perspectives on Humans Developing

7

CHAPTER OUTLINE

I

Development may be seen as the unfolding of characteristics that are entrusted to the individual by nature. Or development may have more to do with the experiences an individual encounters, with the total pattern of nutrition, stimulation, social opportunity, and pressure.

Nature versus nurture / *Instinctual behavior (Darwin)* / *Reflex actions versus behaviors (McDougall)* / *Purpose* / *Psychological dispositions* / *Instinctual drives versus society (Freud)* / *Oral stage* / *Anal stage* / *Psychosexual development* / *Pleasure Principle* / *Ego instincts* / *Reality Principle* / *Imprinting (Lorenz)* / *Critical period* / *Fatalism* / *Theological orientation* / *Mentalistic orientation* / *Royal lie (Plato)* / *Walden Two* (Skinner) / Sensation, reflex experience (Locke) / Social compact (Rousseau) / Equal opportunity / Inside-out or outside-in

II

The child is not simply the instrument by which inner biological forces express themselves. Nor is he or she an infinitely shapeable apparatus at the command of environmental stimuli and pressures.

Selfhood / *Identity* / *Interactionist's fallacy* / *Active psychobiological field* / *Self-control* or *guidance function*

THE PERSPECTIVES IN OVERVIEW

The two year old has blown out his or her birthday candles (perhaps with some help) and devoured a slice of the cake (with no help at all). What's next?

Nature versus nurture concerns the importance of inborn factors versus life experience in development

How we answer this question depends upon our general view of human development. Perhaps you have already come across what is known as the **nature versus nurture** controversy. It has been going on a long time, taking different forms as our knowledge changes. Some people have seen development as the unfolding of characteristics that are entrusted to the individual by nature (the *nature* side of the controversy). The acorn will become an oak tree and nothing else; at maturity it will display the major features of its species. The same general expectations would hold true for human development. But perhaps what an infant becomes has much more to do with the experiences he or she encounters, with the total pattern of nutrition, stimulation, social opportunity, and pressure. (This is the *nurture* side of the controversy.) Between the nature and nurture extremes there are also many intermediate and interactional views. The question of whether it is most appropriate to regard human development as a process that moves mostly from the inside out (nature) or from the outside in (nurture) is of practical and theoretical interest. The kind of decisions we make about child rearing and education, for example, often reflect our assumptions about the respective influence of inner and outer forces. Here is an example. In our society we often distinguish between "preschoolers" and "school children." This distinction is functional. It tells us where a child spends much of his or her time during the weekday and also something about the experiences, opportunities, and demands that come his or her way. However, the differences between the preschooler and the school child are not entirely biological or chronological. The differences are created in part by our society's decisions to: (1) introduce compulsory education, (2) make that compulsory education of a certain type, and (3) introduce children to formal education at a certain age level. In this sense, we—the outside, the human environment—draw the young child into a set of patterned experiences. The developmental sequence is established by society. It is there, waiting for him or her. The child's subsequent development can then be described, at least in part, in terms of grade levels and scholastic achievement. Clearly, this is a much different orientation from one that emphasizes built-in sequences of physical and psychological events that manifest themselves according to their own private blueprints.

While the two year old naps, then, let's awaken ourselves to the continuing interplay between inside-out and outside-in approaches to human development. This background will enable us to form our own opinions more soundly when we continue to observe the individual's traversal of his or her lifespan.

The two year old gives us an especially good opportunity to pause and think this question over. He or she is young enough for the processes of physical development to seem very important. This encourages the inside-out view of growth. But he or she is already old enough to bear obvious marks of learning and experience. Let's begin by exploring the case for an inside-out interpretation.

FROM THE INSIDE: DEVELOPMENT IS INSTINCTUAL

DARWIN: THE INSTINCTUAL BASIS OF BEHAVIOR

The inside-out tradition has a long heritage. I will limit myself here to its more recent history, reaching back only the length of a single century. The concept of *instinct* was familiar to people well before the time of Charles Darwin. But it was Darwin's observations and thoughts about the instinctual basis of behavior that proved so extraordinarily stimulating, controversial, and influential. He believed that instinctive behavior is relatively simple to comprehend in general, although close and patient observation is necessary to establish the facts for a particular species:

Both Darwin and Freud were exceptional observers and reflectors-on-observation

> Everyone understands what is meant, when it is said that instinct impels the cuckoo to migrate and to lay her eggs in other birds' nests. An action, which we ourselves should require experience to enable us to perform, when performed by an animal, especially by a very young one, without any experience, and when performed by many individuals in the same way, without their knowing for what purpose it is performed, is usually said to be instinct. (Darwin, 1859/1966, p. 409).

Patiently building up observations on species after species of the organisms that share the earth with us, Darwin found much evidence suggestive of the instinctual or inside-out development of behavior:

> If we were to see one kind of wolf, when young and without any training, as soon as it scented its prey, stand motionless like a statue, and then slowly crawl forward with a peculiar gait; and another kind of wolf rushing round, instead of at, a herd of deer, and driving them to a distant point, we should assuredly call these actions instinctive. (p. 412).

Darwin made an impressive case for the proposition that much of the behavior exhibited by a particular species has an instinctual basis. In other

Are laughing and smiling derivatives of a tooth-baring instinct to look threatening in a dangerous situation? Or are they culturally learned behaviors? Inner and outer explanations have been offered for many types of behavior. (Photo by Talbot Lovering.)

words, the behavior appears to be *un*learned, at least in its core features. He did not deny that the individual animal could learn from experience. Life, in fact, poses a constant challenge for adaptation and survival. But creatures both great and small *begin* with certain knacks or strategies that give them at least a fighting chance to adapt and to survive.

Controversy about Darwin's findings and views continues to swirl around us today. Some of the phrases associated with his contributions have significant implications not only in psychobiology, but also in the realms of politics, economics, and philosophy. The "survival of the fittest" notion, for example, has sometimes been used to rationalize the exploitation of the less fit. Why try to help or even feel sorry for people who deserve to fail because obviously they are not successful competitors or predators? Darwin did not sanction such irresponsible and misleading, not to mention inhumane, uses of his work. But after important ideas are introduced into the mainstream of thought, they are likely to be used and misused in many ways.

What is most important for us here is the impetus Darwin provided to an inside-out approach for understanding human development. One does not succeed in training a bird to swim under water nor a turtle to soar into the sky. There must be appreciation for those patterns of functioning that are characteristic of a particular species. This means, for example, that coming into the world as a human infant implies distinctive limitations and potentialities. Certain basic features of development closely resemble what takes place in neighboring species. The earlier the phase of development, and the more neighboring the species, then the closer the resemblance (e.g., mammals resemble humans more than invertebrates do, and primates resemble humans more than other mammals). But other physical and behavioral properties of the developing human are specific to our own species. These properties establish some limits to our functioning. However, they also open many possibilities for adaptive behavior that other species do not possess. In fact, it is the range and flexibility of human adaptive behavior that provides a crucial link between the inside-outness of the Darwinian approach and an interactionist view.

Darwin would not have us ignore the environment. It is only in the interaction between organism and environment that the concept of adaptation and its evolutionary dynamics can be understood. But he would have us identify and respect the instinctual equipment with which living creatures, including ourselves, begin life.

MCDOUGALL: BEHAVIOR IS INTEGRATED AND PURPOSEFUL

At one time in psychology, the concept of instinct assumed great importance. Theories emphasizing the instinctual origins of human behavior were commonplace. It seemed for awhile as though every other psychologist had his or her own list of instincts, and as if almost every question could be answered by "instinct." Some advocates of the instinctual approach offered premature conclusions in their enthusiasm and were insufficiently self-critical of their findings and interpretations. Others eventually supplied the criticism and, along with the technical retorts, a withering hail of ridicule. Even the strengths of the instinct approach were discredited or disregarded as a rising tide of *behaviorism* (*see* the following) began to dominate in the United States.

Both McDougall and John Watson (Skinner's spiritual ancestor) thought they were behaviorists although their views were far apart

This historical background is relevant because many contemporary psychologists are acquainted more with the lingering waves of criticism and ridicule than with the particulars of the instinct approach itself. There is a tendency to smile condescendingly and turn off the subject. This does an injustice to the best of the instinct-oriented psychologists, and it deprives us of what we might still learn from them.

William McDougall was one of the most resourceful and thorough of the instinct theorists. His richly detailed approach and the far-ranging scope of his thought cannot be adequately represented in a summary. We can, however, gain some idea of what he judged to be essential in understanding development, both animal and human.

Purpose is a central idea in McDougall's approach. The actions of man or beast are not under the control of the stimulus. We act to achieve goals; we are not merely reactive. This view is much different from the type of inside-out position that argues a person is somehow made up of mechanical reflexes that combine to form a total being (a position that had influential advocates in McDougall's day). The distinction between a purposive and a mechanistic view of human nature and development can be seen when *reflex actions* are compared with *behaviors*.

Yes, agreed McDougall (a physician by training), we certainly do have mechanical reflexes such as the knee-tendon reflex. But we should not confuse these simple and limited response tendencies with what we are mostly interested in: humans *behaving* in the service of their goals. The knee-jerk response, for example, ends almost as quickly as it begins. This is not so for behavior; we are likely to keep on going after being aroused or alerted by a stimulus. The knee jerks only when it is stimulated; behavior can be spontaneous. Purposive movements are "indefinitely variable," while reflex actions are "stereotyped or fixed, the movements evoked by the same stimulus falling on the same sensory nerve are, approximately, the same on all occasions. Perhaps most significantly, reflex actions do not present that appearance of seeking a goal which is common to all behavior, and of which the essential feature is the persistence of movements with variation until, and only until, that goal is attained" (McDougall, 1923, p. 53). Behavior implies purpose, and purpose means that some kind of mental life must be attributed to the organism (although not necessarily a high level of conscious experience).

Psychophysical dispositions are instinctive tendencies to seek need satisfaction by selective action

Where does "instinct" come in? McDougall attempted to explain complex human experiences and actions on the basis of **"psychophysical dispositions"** that are instinctual in origin. These dispositions do not concern anything as simple as the knee-jerk reflex; the instinctual core of human functioning is more subtle and pervasive. As humans, our behavior is not pulled about by the environment like a piece of taffy, nor do we find ourselves equally attracted to all aspects of the environment. We have certain dispositions that involve both our physical and psychological structures. The dispositions guide our development, but not in a strictly automatic or stereotyped way. A display of the "combative instinct" in one person, for example, may come in the form of physical assault, while for another person a sarcastic reply will serve the same purpose. Both of these people began life as infants whose combative instinct or disposition was displayed through kicking and crying, but they subsequently followed somewhat divergent courses of individual development.

At any particular time, our behavior may involve one or more of the instincts McDougall considered to be characteristic of our species. These include, for example, the instincts of gregariousness (seeking the company of others), curiosity, mating, escape, and the parental or protective disposition. But there is something in common in our behavior, no matter which instinctual disposition is dominant at the moment: we become excited or aroused, having a store of energy at our command that is seeking some form of release through action. Some instinctual dispositons are stronger than others, according to McDougall, but they all work essentially in the same way. When an instinct is aroused, a certain amount of energy is liberated or made available for our use. We experience this surge of energy (in many, if not all instances), and we then behave with the goal of achieving the purpose associated with this disposition.

If McDougall were alive and writing his instinct theory today, he might describe instinctual arousal as a state of being "turned on." This, in fact, is very close to what he did say.

> When an instinctive impulse is liberated or evoked, the organism becomes absorbed in the endeavor toward the goal of the instinct; its reaction to the exciting object is a total reaction; its energies are concentrated on the task in hand and the functioning of the various organs is subordinated to and harmonized with the dominant system of activity. This absorption of the organism in any particular task or mode of activity is what we call in ourselves "attention"; and that general excitement whose indications we observed in the animals, when their instincts are strongly excited, is what we call in ourselves "emotion." (pp. 109–110).

McDougall continued with a more formal definition of an *instinct* as "an innate disposition which determines the organism to perceive (to pay attention to) any object of a certain class, and to experience in its presence a certain emotional excitement and an impulse to action which find expression in a specific mode of behavior in relation to that object."

The instinct, in other words, has three closely related components: (1) energy flow that influences our perception, (2) an altered state of inner experience (becoming more "emotional" in some way), and (3) a readiness for action of a particular kind. It is *not* that the eye has one instinct, the hands another, and the "mind" still another. *An instinctually-based state of arousal involves our total being.* McDougall's approach, then, made it appear essential for psychologists to interpret *human behavior as an integrated totality.* In this sense, many of us today who are concerned chiefly with the development and functioning of the *person* owe a debt to McDougall and his psychology-with-a-purpose. To fulfill specific research aims, it is often necessary to focus on one or two elements of the total, but it would never do to lose

Curiosity is an instinctual disposition according to McDougall, but this does not mean it is a simple, reflex-like response. (Photo by Talbot Lovering.)

sight of the whole human being who is attending, experiencing, and acting within his or her environment.

The idea that behavior can be instinctual and also continue to develop is important

At least one other aspect of McDougall's approach should be mentioned here. He did not regard instincts as coming to us as "finished products" fresh from the factory. Our psychophysical dispositions grow along with us. They are truly capable of development, in contrast, for example, with simple reflex mechanisms such as the knee-tendon jerks which keep on doing the only thing they "know" how to do. In outlining the "life-history of an instinct," McDougall states that each instinct develops gradually. The instinct is expressed in partial and incomplete form before it is perfected. The period of development is not the same for all species. Even within the insect kingdom, some organisms have "but little youth, that period of development in which skill and knowledge are acquired while the instincts are still slowly maturing." Others, like the wasp, "enjoy a period of free wandering before they begin their principal life-task, that of laying the eggs under conditions suitable for their development" (p. 112). The youthful wasp is guided by the instinct of self-nutrition. At a later time, the wasp's behavior changes radically—because *she* has changed or matured, not because of anything the environment has done in the meantime. This is a clear example of inside-out dynamics.

Humans spend more time being young and, thus, more time in maturing than other organisms. A thorough study of instinctual dispositions would have to be developmental, carrying us well beyond infancy and early childhood (although McDougall himself did not pursue a systematic lifespan developmental psychology). McDougall's instinct theory can be criticized on many counts. The field of social psychology that he did so much to establish has found other ways to approach and study, as well as to account for, behaviors that he considered to be instinctive. But it cannot be denied that McDougall provided a broad framework for observing and interpreting the behavior of humans and other organisms, one that respected physical constitution, species differences, individual differences within a species, and at least some of the complexities associated with the developmental process.

FREUD: INSTINCTUAL DRIVES AT WAR WITH SOCIETY

"It is probably true to say in general terms that the exercise of all its instincts in normal degree is essential to the fullest vigor and health of any animal."

This statement is one that Freud might have made, had he chosen to express himself so mildly. The founder of psychoanalysis repeatedly called attention to ways in which basic human urges were subjected to controls and pressures from society. Yet the quotation is actually a left-over from McDougall. I have introduced this passage to emphasize some of the points of consensus among inside-out theorists who otherwise differ in significant ways.

Along with McDougall and some others, Freud was concerned with life at many levels—lower animals as well as humans, political and historical dynamics as well as individual dynamics. He also had the strong conviction that psychological understanding must be developed within the context of our physical and biological properties. And, like McDougall, Freud had a far-ranging and subtle mind that requires appreciation in depth. Even his view of "instinctual drives" (which is only one aspect of his work) continued to change during the course of his long and active life.

Instinctual energies are the tensions generated by organ systems which seek release

Freud emphasized the *origins* of our **instinctual energies.** Every organ system in the body generates tensions that demand to be released or discharged. Essentially, these organ tensions *are* the instincts. We feel a sense of relief or pleasure when the tension is discharged. As organisms, we therefore attempt to reduce the tensions generated by our body systems (e.g., by taking nutrition when we experience hunger pangs). At this level, Freud's instinct theory is fairly simple and straightforward. The complexities of human functioning, however, require further specifications in his theoretical model. We will consider a few of these.

The conflict between instinctual drives and society. There appears to be an inherent conflict of interests between the instinctual drives with which we begin life and the nature of the society in which we participate. Our biologically rooted tensions and appetites demand immediate relief. This is true of the infant and young child, as we can all see. But Freud observed that this infantile characteristic of urgency and impatience continues to manifest itself throughout adult life as well. We develop more sophisticated psychological structures within ourselves (many of which are known collectively as *ego functions,* the abilities we use for coping with daily realities). Yet, our bodies crave immediate gratification for their instinctual drives much the way they did back in the first weeks of life when we consisted mostly of churning states of need.

This condition sets up perpetual conflicts on two fronts: between ourselves and society, and between the "grown-up" and the "baby" in each of us. It is neither possible, nor socially acceptable, to meet the demands of our instinctual drives whenever and wherever they arise. The adult part of us knows this. We make certain compromises between our need for instinctual gratifications and what it takes to keep us in society's graces.

The relationship between instinctual tension and gratification. Where the organic tensions come from and what is required to satisfy them are two somewhat separate questions. It is possible that the same "object" can satisfy more than one type of tension, and that the same tension can be satisfied by more than one type of object. The two year old, for example, may seek his or her mother when in general need of comforting, as well as for help in meeting a specific organ tension. Mother is one "object" whose presence soothes several of the youngster's organic tensions. This was implied at an even earlier age level, as when we discussed the "stranger reaction" and the baby's sense of freedom in exploring space when the mother is clearly available. Similarly, more than one type of food (several "objects") may satisfy hunger pangs (one organic tension). In later phases of development, the object of sexual strivings may not only include various members of the opposite sex, but also members of the same sex, animals or inanimate objects, and even abstract causes, systems, or crusades.

This open relationship between origin of instinctual tension and the object that provides gratification leads to many problems and possibilities. Much of the developmental process is concerned with discovering, selecting, or creating opportunities to satisfy instinctual tensions within the complex world of physical and psychological realities. One does not automatically proceed from a felt state of inner tension to a specific behavior pattern that will produce relief and pleasure.

Instinctual drives and psychosexual development. The psychoanalytic theory of instinctual drives requires a comprehensive and elaborate account of human development. Taking into account the complexities already touched upon, how do we get from the infant ruled by immediate and urgent needs to the adult who functions adequately in society? Part of the answer is that we do *not* always negotiate the developmental process in good shape. People may reach the adult years without having "worked through" earlier needs and challenges. Suffering, pathology, and incompetencies in adult life are often traced by psychoanalytic observers to problems in early development. But we must distinguish between the many problems that befall individuals and a normal or ideal progression. This can be a difficult distinction. It is easy to be distracted by the various quirks and miseries one finds described on so many pages of the psychoanalytic literature. "Normal" development is not always easy to discover, but Freud does try to set it into perspective.

To begin with, it is "normal" to experience organ tensions. This is just part of being a living organism. It is also normal to seek the release of these tensions without delay; the healthy infant does so. No matter which particular organ system gives rise to the tension that is being experienced at the moment, there are pleasure-pain dynamics involved that can be described as *sexual* in a very broad sense of the term. There is a sense of pleasure, gratification, and well-being when the body has become (at least temporarily) contented.

As development proceeds, certain areas of the body become the focus of both tension and gratification. The **oral stage** is first. It is normal for the baby to relate to the world chiefly through his or her mouth—taking in nutrition and crying for attention. Later, around age two, the child is said to enter the *anal stage* in which control of bowel movements becomes a significant issue. The child can be "good," for example, and conform to the parents' expectations, or he or she can be "stubborn." Both of these stages (and those which follow) involve psychological, as well as physiological, processes. Because ten-

The **oral stage** is the first of Freud's psychosexual stages. It centers around the mouth

sions and gratifications are thought to center around particular bodily zones, which change in significance from stage to stage, the sequence is known as **psychosexual development.** One must solve the physiological and psychological challenges of the oral stage in order to move successfully to the anal stage, and so on. The adult who can be accurately described as an "oral character" or an "anal character" is a person who has not fully mastered these early developmental challenges. But the infant who appears quite "oral," and the two year old who is caught up in "anal dynamics" are entirely normal for their stages of development.

Psychosexual development is Freud's maturational pattern in which tensions focus first on one and then on another body area

Classifying instincts. Freud opposed the notion of describing and naming a great variety of instincts. He tried instead to classify human instincts into a few broad categories and to trace their developmental careers. He proposed that we begin life with instincts that crave *immediate* gratification. As a matter of fact, although bodily tension may be generated by a variety of organ systems, there is fundamentally only one instinct, and it imperiously seeks its own release or expression (Freud, 1935).*

The instincts do not remain this simple for very long. The baby must come to terms with external, as well as internal, reality. Instinctual tensions cannot be gratified unless the individual somehow connects with the world. He or she must, for example, let the world know he or she is hungry. The original caldron of instincts then gradually shapes itself into two forms.

Some inner tensions remain hooked on the **pleasure principle.** Throughout the individual's entire lifespan, these tensions persist in seeking direct and immediate gratification. But other tensions are drawn to the service of the developing **ego.** In Freudian terminology, *ego* refers to those functions of the individual by which he or she copes with environmental challenges and attempts to identify and master problems. The emerging *ego instincts* begin to take into account the *reality principle.* One thinks and behaves as one must in order to survive in the world. It is not adaptive to just go along with the impulses felt from within.

The **pleasure principle** is the impulse toward direct and immediate gratification

Ego is made up of an individual's coping and adjusting functions

Development of the ego instincts is a long and gradual process, certainly far from complete in the two year old. Nevertheless, the child at this age has already come some distance from his or her original state of bondage to the raw pleasure principle. One of the most significant indications that the child is developing ego instincts can be seen in any behavior that involves a *delay of gratification.* The ability to wait a little, to tolerate delay between impulse and gratification, is a critical step in psychological development.

Freud later had second thoughts about his instinct theory. His revised theory centered around what might be regarded as a "life instinct," ***Eros,*** and a "death instinct," ***Thanatos*** (Freud, 1920). He also added more detail and made other revisions in the basic instinct theory that has been described

Eros is a broadly conceived life or love instinct

Thanatos is a broad instinct directed toward death or cessation

* Freud's early thought on instincts predates the (Freud, 1920) reference below. The 1935 reference is to a book that is a good statement of Freud's overall instinct theory as well as a useful introduction to his total theory.

here. The main point, however, is that Freud's powerful influence on psychology and other fields clearly involved an emphasis upon inner forces which undergo complex development, especially in the early years of life.

MORE RECENT CONTRIBUTIONS

Inside-out dynamics continue to be investigated today, although some of the issues, concepts, and methods have changed. Konrad Z. Lorenz has been honored with a share of the Nobel Prize for his work in this area. His naturalistic experiments have had considerable charm, as well as scientific influence:

> I decided to try experimenting on the [Mallards'] call note which is happily well within the powers of the human voice to imitate. I took seven young Mallards and while they were drying under the electric heater I quacked to them my imitation of the mother Mallard's call. As soon as they were able to walk, the ducklings followed me quite as closely and with quite the same reactional intensity that they would have displayed toward their real mother. I regard it as a confirmation of my preconceived opinion about the relevance of the call note, that I could not cease from quacking for any considerable period without promptly eliciting the "lost peeping" note in the ducklings, the response given by all young anatides on having lost their mother. (Lorenz, 1961, p. 59).

Imprinting is the establishment of a behavior at a specific developmental time if the necessary stimuli are present

Lorenz demonstrated that he could be accepted as "mother" and elicit the following-response not only of Mallard ducklings but of other fowls as well. The serious point behind these "Mother Goose" games has to do with the concept of **imprinting.** Lorenz and others have found that there are certain limited periods of time early in the development of some creatures when they are ready to respond to particular stimuli. In the specific case of the following-response, the duckling accepts any of a variety of possible "mothers" as long as "mother" fits into a general pattern—about the right size, moves the right way, makes the right sounds. Once the duckling has hit upon a particular "mother," this is the one that it will stick with, be it the biological parent, a toy, or a professor (Lorenz, 1952).

A **significant other** is a person whose presence and caring are of special importance

The young animal's response to a mother-object that is formed in this way is extremely firm and intense. As the term *imprinting* suggests, it is as though the response has become engraved permanently. Studies of imprinting are important because they emphasize a built-in or instinctive tendency for a very young creature to select its **significant other** (borrowing a term from personality theory) from a certain range of stimuli available in the environment. Theorists continue to disagree as to whether this should be interpreted as *un*learned behavior or as a very special kind of learning. In either event, one can neither ignore the highly specific range of preferences that show themselves within a highly specific period of time, nor the firm and intense nature of the response-bond once it is made.

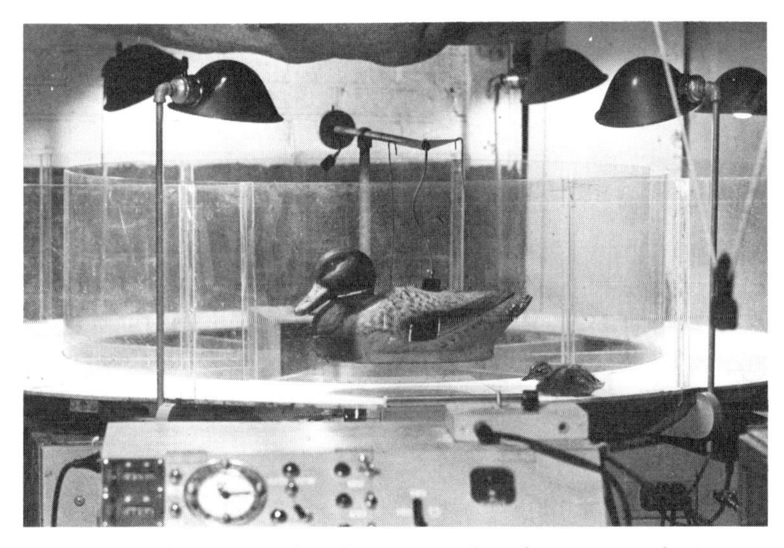

While Konrad Z. Lorenz played Mama Duck to demonstrate the imprinting process, later research has turned to more traditional methods to study imprinted behavior. (Photo by Eckhard H. Hess.)

Critical period is a time when a person is ready to be strongly influenced by certain stimuli

The term **critical period** has been given to developmental phenomena that occur within specific and limited periods of time. The organism appears to be ready for a certain event or experience. The ducklings, for example, were ripe for "Momma" Lorenz at a particular time in their maturation; at a later time he might not have won their hearts. There can be critical periods for vulnerability as well as growth. An injury or stimulus at one phase in the development of an embryo, for example, has a different effect that it does at an earlier or later time (Scott, 1962).

Researchers try to be certain that they are dealing with a true critical period by determining how regularly a particular event occurs and how closely linked it is with a fixed time in the organism's developmental history. A particular bird, for example, may be ready to show pecking behavior, but it needs to have food available at a certain time in order to learn what pecking is good for (Padilla, 1930). If the food is available earlier, the bird will not peck because it is not ready.

It is very difficult to translate data gained from studies of critical periods in lower animals to complex behavioral situations in humans. Nevertheless, the concept of critical periods is sometimes introduced into theories of human development. The researcher usually looks for indications of a new physical readiness to make use of certain experiences. Evidence concerning maturation of the brain, and the central nervous system in general, provides a useful starting point for discovering possible critical periods. Perhaps in the future, for example, specific maturational changes in the central nervous system (C.N.S.) will be related to the emergence of specific forward leaps in language and thought. The possibility that attachment-seeking behavior in young children might be understood according to critical period terms has also been explored in a preliminary way (Caldwell, 1962). A useful introduction to this literature with some interesting speculations is provided by Bronson (1965).

The man with whom Lorenz shared his Nobel Prize has also made major contributions to the modern study of instinct-like behavior. Nikolaas Tinbergen (1951) performed some of his best-known studies with the male stickleback fish. Tinbergen did not beguile the fish to swim after him a la Lorenz. Instead, he created fake stickleback. He learned that the male sticklebacks' fighting and courting advances were cued into a very specific stimulus on the underbelly. What the rest of the fake fish looked like did not matter very much; the male would literally see red and attack a fish (or fake) with a red underbelly. A fish (or fake) with an "un-red" and somewhat smaller underbelly would, by contrast, receive amorous advances from the male stickleback. Although the fighting and courting behaviors were complex, they were "released" by simple visual stimuli.

Releasers is a term often used for stimuli that are followed by instinctual behaviors

Tinbergen's studies have called attention to very *specific* factors in the interaction between an organism and its environment. In the stickleback studies, for example, he differentiated among five types of fighting behaviors. The attacking fish might show one type or another, depending upon cues that could be experimentally studied. Systematic studies such as those of Tinbergen have led to fresh and ever more precise observations of behavior patterns that characterize certain species. The term *instinct* is not used in a loose or philosophical sense by Tinbergen. He identified numerous varieties and levels of behavior and then related them to specific stimuli and interactions. It is a more precise and experimental approach than we usually associate with either McDougall or Freud.

FATALISTIC AND MORALISTIC VIEWS

I have been emphasizing the biogenetic aspects of the inside-out approach. The mental or behavioral dimensions of development have seemed to come across as the consequences or products of innate physical structure. There are other ways of looking at this, however.

Both in ancient times and in our own, there have been people devoted to the *fate* theme. The future course of one's life is predetermined. It is written in the cards, in the stars, or wherever. Life has, in effect, been laid out in advance. The individual has to move through life in a certain way and has no choice about what is yet to happen. He will have a long, serene life, or she will die violently. He was meant to have an ugly temper, or she just can't help being so charming. Personality does not really develop, nor do environmental inputs exert decisive influence. The major characteristics of the individual's life have been predetermined by powerful forces in the universe.

Fatalism is often associated with a sense of helplessness

Theological and *moral* views of the world have sometimes led to fatalistic outlooks, but not invariably. According to some theological orientations, certain peoples have been chosen as favorites of God, and others excluded. In its more extreme forms, this view sees people as fixed permanently in a particular status. In its more flexible forms, the outcast has the possibility of earning redemption through faith and good works. Another theological orientation attributes intrinsic evil or weakness to the human species in general. Even before we grow old enough to commit our individual sins of thought and action, we receive, by our very nature, the moral taint that is common to all humankind. An individual might be more or less successful in overcoming personal frailties, but the central fact about human nature is that it is morally tainted and prone to temptation.

Other theological orientations take a brighter view of human nature. Every person is created in the image of the Creator. There is a spark of the divine in everyone, therefore no person can be entirely worthless. Sometimes we feel a person's fate hangs in the outcome of a contest between the innately divine and the innately evil characteristics within that person's human nature. The fate theme, taken out of its religious context, is not necessarily associated with being a "good" or an "evil" person. The individual might instead be a "lucky" or "unlucky" person.

Both the fatalistic and the theological-moralistic approaches locate developmental control some place else than in the individual's own powers. There are, however, a variety of other orientations that see guidance and controls as basic properties of the person. We enter this world with minds that already know their way. Sometimes known as the **nativistic tradition,** this view has been advanced by many influential thinkers. *Immanuel Kant* in his *Critique of Pure Reason* (1781/1953) and other writings is one of the most important philosophers in this tradition. He probed the possible relationship between mind and nature just at that point in our cultural history when the emerging sciences were starting to raise exciting and difficult questions. Kant emphasized the active structure of the human mind. Experience itself is not enough! There must be a mind with the power to register and work upon experience. Kant's arguments have had great influence over the years in many areas of thought, including what has since emerged as developmental psychology. In general, his approach emphasized that we have mental as well as physical structures "inside" us during our developmental journeyings. For humans to develop, he might have said, we must begin with a human!

The **nativistic tradition** is a position that emphasizes the importance of built-in characteristics

FROM THE OUTSIDE: WE DEVELOP THROUGH EXPERIENCE

If we really want to know what the two year old will be like—in fact, if we really want to know *who* he or she will be—we must concentrate on his or her world of experience, past, present, and future. This is the outside-in message and it, too, has taken many forms through the years.

THE PREPARED SOCIETY: FROM PLATO TO SKINNER

Human development *can* and *should* be controlled by society. From birth onward, the individual should be shaped into one of several possible life-styles that society requires for its own continued survival. Does this plan require what today might include a computerized conditioning of behavior and personality? Is the individual (and his or her parents) to have no choice in the matter? Must the state function as a factory for tooling its citizens into fixed role-functions? And wouldn't all of this control be the height of totalitarianism?

Plato would have answered these questions, Yes, yes, yes, and yes! His radical proposal can sound heartless when reduced to its fundamentals. Yet *The Republic* displays spirited minds at work and play, with the authoritarian scenario unfolding itself in a genial and witty manner. A leisurely reading of this dialogue is a treat that you should not deny yourself, whether or not you are in sympathy with the ideas as summarized here.

Society and the individual are intimately related in Plato's view. The state is itself something of an organism—a vital, interconnected system of systems, all functioning for the health and survival of the total configuration. The individual should be able to recognize his or her own being "writ in large" in the form of the state. Notice that this is a much more intimate, almost biological image of the individual-society relationship than what most people envision today in a nation such as ours. The neighborhood is not just a place we leave and go back to each day; the government is not "They" and "Them." Rather, the individual and the state in which he or she lives (*state* in the sense of political entity) are both organisms that depend on each other if either is to prosper.

In a playful mood, Plato proposed a "royal lie" to reinforce this sense of togetherness and to advance his proposal for mass shaping of individual development. "I will speak, although I really know not how to look you in the face, or in what words to utter the audacious fiction, which I propose to communicate . . . first to the rulers, then to the soldiers, and lastly to the people. They are to be told that their youth was a dream, and the education and training which they received from us, an appearance only. . . ." In other words, Plato had in mind something like a posthypnotic suggestion: train people the way we want them to be, and then have them forget all about the training process itself. Should they have been curious about their past, Plato would have had his fellow citizens believe that "they were being formed and fed in the womb of the earth . . . when they were completed, the earth, their mother, sent them up; and so, their country being their mother and also their nurse, they are bound to advise for her good, and to defend her against attacks, and her citizens they are to regard as children of the earth, and their own brothers" (Plato, *circa* 385 B.C./1953, p. 340).

This "royal lie" or "audacious fiction" would help prepare the citizens to recognize individual differences among themselves, without resenting the different life-styles linked to these differences. "You are brothers, yet God has framed you differently. Some of you have the power of command, and in the composition of these he has mingled gold, wherefore also they have the greatest honor; others he has made of silver, to be auxiliaries; others again who are to be husbandmen and craftsmen he has composed of brass and iron . . ." (p. 340). Plato proposed a developmentally graded series of screening procedures to identify the gold, the silver, and the brass-iron children in each generation. The state's total resources of childrearing, education, employment, and propaganda were to be used in the effort to shape each child according to the kind of metal he or she displayed and the kind of functionaries required by society. Usually gold would be born to the gold, silver to the silver, and so on. "But as all are of the same original stock, a golden parent will sometimes have a silver son, or a silver parent a golden son . . ." (p. 340). Rank in society would depend upon the mental and physical qualities with which each child entered the world. Individuals would be carefully assigned to one or another of the levels of rearing and education, the final product being a ruler, a soldier, an artisan, a farmer, etc. The individual would know precisely who or what he or she was going to be from very early in life.

The now-flourishing infant/child testing establishment was advocated by Plato

The general image, in retrospect, is of a society attempting to keep itself stable over time by a powerful social planning process with plenty of muscle behind it. Characteristics of the person as a psychobiological organism are not neglected. This is clear in the distinction among levels of quality, using the metal analogy. But once the individual has been "typed," society has the right, the responsibility, and the means for bringing this person along a predetermined developmental path. The prepared society endorsed by Plato places great responsibility on the philosophers and rulers who have to decide: What should be taught? In what way? To whom? How should babies be raised, and by whom? Whether or not we agree with the specific answers offered by Plato or with his general proposition, it is evident that this "outside-in" approach makes early childhood a highly significant (because it is a highly influenced) time of life. The inside-out view of individual development often (although not invariably) implies a limit to modification. If we are ruled by instinct, fate, or any other in-dwelling and fixed pattern, then all we can do is play out our individual scripts. Plato's vision asserts an intimate association between the individual and the type of society in which he or she lives. Wise people might devise a society that both demands and creates individuals who will enjoy moral and effective lives. There is enough flexibility in human nature for the social environment to have a decisive effect. A person can live in harmony with his or her own nature and with society only when he or she has from the beginning been educated, trained, or shaped to participate in a rational and prepared environment.

It is interesting to know that the prepared society, for all its change potential, was favored by Plato as a conservative measure, to preserve his own city-state from threats to its stability.

Many variations on the prepared society have appeared in literature and philosophy. Two of the most influential visions in our own time are those of

Aldous Huxley in his *Brave New World* (1950) and George Orwell in his *1984* (1949). Both of these novels project future societies in which all-powerful regimes tightly control the lives of individual citizens. Privacy, choice, freedom, and inner depth are qualities that The State systematically seeks to avoid or eliminate. Apart from their value as creative works of literature, these books put us on the alert. Perhaps the "brave new world" that modern technology has to offer will prove instead to be a new and surpassingly cruel form of enslavement. These books have given rise to much dialogue and are still very much worth our attention today.

It is not fair, however, to identify the prepared society with an oppressive regime. Plato might admit that these nightmare visions have roots in *The Republic.* Yet we cannot imagine that Plato, himself, would have enjoyed such dehumanizing and humorless states nor accepted them as fulfillments of his social philosophy. A more challenging and appropriate vision of outside-in shaping of the individual can be found in another influential book of our times, *Walden Two.* The author is B. F. Skinner (1948), a psychologist whose research and thinking have had impact in many areas of society. *Walden Two* is also cast in the form of a novel. It presents Skinner's conception of a desirable form of human community, one that is created on the basis of behavioral technology. Skinner attempts to show how the application of behavioral principles can shape both the individual and his or her community into a way of life much superior to what we currently experience. Skinner's authoritative knowledge of behavioral technology gives his vision of a prepared society much of its special character. Along with *Brave New World* and *1984,* it is a book that a thoughtful person will want to read.

But consider how Skinner answers the question, "What is man?" in a more recent book, *Beyond Freedom and Dignity* (1971). He believes it is vital that we change our thinking about this question:

> Science has probably never demanded a more sweeping change in a traditional way of thinking about a subject, nor has there ever been a more important subject. In the traditional picture a person perceives the world around him, selects features to be perceived, discriminates among them, judges them good or bad, changes them to make them better (or, if he is careless, worse), and may be held responsible for his action and justly rewarded or punished for its consequences. In the scientific picture a person is a member of a species shaped by evolutionary contingencies of survival, *displaying behavioral processes which bring him under the control of the environment in which he lives, and largely under the control of a social environment* [italics added] which he and millions of others like him have constructed and maintained during the evolution of a culture. The direction of his controlling relation is reversed: *a person does not act upon the world, the world acts upon him* [italics added]. (p. 211).

According to Skinner's view, then, our behavior is shaped from infancy onward by the social environment constructed by those who have gone before us. We are under the stimulus control of somebody else's past. If we fail to recognize this fact, then the behavior-shaping process remains unregulated by systematic rational control. The newborn's development is at the mercy of cultural biases and inconsistencies. Skinner does not deny the physical and genetic side of human behavior. But he does emphasize cultural practices as the critical determinants. The behavioristic methods he proposes are intended to replace the inadequate and frequently counter-productive techniques that are commonly in use. If we do as he urges, then he feels we would have a more enlightened and more scientifically based pattern of interaction.

Skinner is acutely aware that some people consider his views offensive and dangerous. In examining these objections, Skinner comes to the conclusion that traditional ideas and sentiments about the "self-directed person" stand in the way of progress. Skinner is, thus, not only a preeminent contributor to behavioristic technology, but also an articulate opponent of the inside-out approach.

THE UNMADE MIND: SOME REVOLUTIONARY IDEAS

Plasticity is the tendency to be influenced or shaped by experience

One can emphasize the power of society and the environment. But one might dwell instead upon the **plasticity** and the potentialities of the human mind. In a way, of course, these are two sides of the same approach. But we see the outside-in tradition a little differently when we begin with the individual.

> The understanding, like the eye, whilst it makes us see and perceive all other things, takes no notice of itself; and it requires arts and pains to set it at a distance and make it its own object. But whatever be the difficulties that lie in the way of this inquiry; whatever it be that keeps us so much in the dark to ourselves; sure I am that all the light we can let in upon our minds, all the acquaintance we can make with our own understandings, will not only be very pleasant, but bring us great advantage, in directing our thoughts in the search of other things. (Locke, 1690/1953).

This statement launches one of the most important philosophical-psychological inquiries into the human mind, *An Essay Concerning Human Understanding* (Locke, 1690/1953). *John Locke's* far-ranging thoughts encompassed many topics. But he believed our understanding of the world should properly begin with an effort to comprehend the workings of our own minds. His lucid, closely reasoned exploration did much to advance the *empirical* outlook. We know what we know and become what we become as a result of experiences. Locke insists that we do not enter this world with a headful of ideas, or even with a set of basic principles. The infant's mind is better likened to a pure, blank tablet *(tabula rasa)* upon which nothing has been written (yet).

Much of Locke's argument attacks the innate or nativistic position that we are born with mental properties which express themselves as time goes by.

Will society try to hold these children in the past, or enable them to grow into distinctive and liberated people? (Photo by Patricia Hollander Gross, Stock, Boston.)

But if this nativistic approach is mistaken, how *do* we acquire our mental furnishings? Locke tells us that all ideas come from either of two sources, *sensation* and *reflex*. Observing external, readily perceptible objects gives rise to sensations. But we also observe *"the internal Operations of our Minds, perceived and reflected on by ourselves"* (p. 121). (Note this use of the *mental operations* terminology more than two centuries before Piaget). *Experience* is the ultimate source in either case. It is experience that yields us either sensory impressions of the outer world or the opportunity to reflect upon our mind itself by comparing, believing, doubting, and so forth. Reflection is a sort of internal sense. Together with the sensations transmitted through eyes, ears, and other modalities, reflection helps to make up our entire source of potential ideas.

Locke's conception of the human mind came like a fresh breeze to many people of his own and succeeding generations. Others were disturbed. This new psychology had strong implications for the social order. If *nothing* was written on the individual's mind at the start, then, perhaps *anything* could be! This was either an exhilarating or a frightening prospect, depending on one's viewpoint. For example: if all humans begin life equally unformed, should they not have *equal opportunity* to develop themselves through experience? This idea set much of the establishment a-tremble. The writings of John Locke, starting as a philosophical and psychological inquiry, became a powerful force in the reshaping of thought and society. The minds and lives that have been influenced by this outside-in doctrine include our own. The founding fathers of the United States drew inspiration from Locke in their efforts to establish a constitutional government that respected the rights and potentialities of the individual as well as the state.

Locke influenced both stimulus-response psychology and the U.S. Constitution

In Locke, we have again seen how a connection is made between conceptions of individual development and visions of a desirable society. In contrast with some of the other examples given, however, Locke's approach encouraged a more open type of society. Not many years later, another important voice pressed for social reform on the basis of still another interpretation of what is fundamental in human nature.

Jean Jacques Rousseau was himself both troubled by society and regarded, in turn, as something of a troublemaker. He was greatly displeased by the political and cultural scene that he observed all about him in eighteenth century Europe. "Modern man" seemed to be the product of an artificial and distorted way of life. People flattered themselves by speaking of their high "civilization." Not Rousseau. He was enraged by social inequalities, and the perpetuation and abuse of wealth and power. Rousseau believed that society could not be improved until people discovered what had gone wrong with the "natural man."

Rousseau warned that it would be very difficult to discover whatever is—or was—natural to humankind. "Every advance made by the human species removes it still farther from its primitive state," including even "our very study of man." This concern led him to raise two methodological questions that were far in advance of their time: *"What experiments would have to be made, to discover the natural man? And how are those experiments to be made in a state of society?"* He concluded that the truth-seeker would be better off "throwing aside . . . all those scientific books, which teach us only to see men as they have made themselves." Instead, we should contemplate "the first and most simple operations of the human soul" (Rousseau, 1754/1953, p. 330).

Rousseau then undertook to describe the development of human thought, language, motivation, and family and communal interaction from early historical times to his own day. This included a scathing attack on the custom of declaring natural resources to be private property.

> The first man who, having enclosed a piece of ground, bethought himself of saying, *This is mine,* and found people simple enough to believe him, was the real founder of civil society. From how many crimes, wars and murders, from how many horrors and misfortunes, might not any one have saved mankind, by pulling up the stakes, or filling up the ditch, and crying to his fellows, "Beware of listening to this imposter; you are undone if you once forget that the fruits of the earth belong to us all, and the earth itself to nobody." (1754/1953, p. 348).

Rousseau asserted that all people had natural rights to life, liberty, and the pursuit of happiness. If this proposition is familiar today, we have Rousseau to thank for it. In Rousseau's own time, the idea was radical. He saw these basic rights as being infringed by the existing social structure. To correct the abuse of property, power, and privilege, it was necessary to develop a more representative form of government, one in which the State derived its strength from the will or consent of all members. By going back to what is most basic in human nature and unspoiled by the distortions of "civ-

Social contract is the arrangement by which a person lends consent to the group, yet remains individual

ilization," Rousseau felt we should create a new **social compact** or **contract.** *"Each of us puts his person and all his power in common under the supreme direction of the general will, and, in our corporate capacity, we receive each member as an indivisible part of the whole"* (1754/1953, p. 392).

Like John Locke before him, Rousseau exerted a strong influence on social and political movements that still affect our lives today. Especially relevant to us here is his view of human development. His concept of development was philosophical, moralistic, and political, as well as psychological. He attacked the dominant tradition which held that human nature is sinful and not to be trusted. Instead, according to Rousseau, it is the child who can be trusted. Humans begin life with the potential to love, share, and reason. The villain? Society! Corrupted by his or her experiences with society, the child becomes still another grasping, scheming, unhappy adult wretch.

Interestingly, the social history of childhood has only been examined and described in detail in the past few years. Aries (1962) and, even more so, deMause and his colleagues (1974) brought to light much material that supports Rousseau's negative view of what society has often done to the young child.

Today Rousseau's view of the natural person is criticized, but the power of his words is hard to surpass

The solution to society's corruption? We must do away with the restrictions that interfere with human development. The child should not be forced to move along a straight and narrow pathway, buffeted at all sides by constraints, threats, and punishments. Instead, he or she should be allowed to explore freely an environmental setting that is as natural and as open as possible. In this way, the child will select what is of genuine interest to him or her, and learn to become a happy and worthy adult through the exercise of his or her own capacities (Rousseau, 1762/1911).

Free and exuberant—the natural person is this way until stifled and warped by civilization. This was part of Rousseau's critique of society. (Photo by David Kelley.)

Rousseau, then, is an outside-in advocate, but in quite a different way from Plato. Rousseau felt that the trouble with human development is *too* much patterned experience—and of the wrong kind—has been forced upon the young. We do not need a prepared society that "knows better" what is good for the child. What we need instead is an "outside" through which the child can venture and adventure, capturing his or her own experiences and forming his or her own personality. This view does, in fact, make some assumptions about what is *inside* human nature. But the practical emphasis is upon ways in which an encouraging, minimally restrictive environment can enable the child to make his or her own mind as he or she goes along.

Many other variations on the outside-in theme have been sounded since Rousseau. You might enjoy searching out the extremist views of Ellsworth Huntington (1920) who insisted that both individual and social character are shaped by basic environmental forces. Geography and climate make us what we are and keep us from being anything different. Few scientists accept Huntington's thesis as presented. Nevertheless, it has alerted some observers to environmental influences that were under-appreciated until just a few years ago, and are perhaps still not fully recognized. Today there is a growing body of knowledge in environmental psychology (Barker, 1968; Proshansky, Ittelson & Rivlin, 1970). And related fields such as architecture, public health, and sociology are also giving serious attention to outside-in dynamics.

VISIONS OF CATASTROPHE

Those who foresee environmental deterioration and catastrophe may also be included among the contemporary corps of outside-in advocates. Human development—even survival itself—is seen chiefly in terms of what conditions prevail in the social and physical environment. Ehrlich and Ehrlich (1970), for example, ask us to think of "the Colombian mothers forced by hunger to practice infanticide; the Biafran children in the last stages of starvation; the Indian women who, during the recent Bihar famine, spent days sitting in the sun picking up grains of wheat one by one from railroad beds; and the several hundred thousand residents of Calcutta who live in the streets . . ." (p. 3). Meanwhile, in developed countries (DCs): "Overpopulation . . . is lowering the quality of life dramatically . . . as their struggle to maintain affluence and grow more food leads to environmental deterioration. In most DCs the air grows more foul and the water more undrinkable each year. Rates of drug usage, crime, and civil disorder rise and individual liberties are progressively curtailed as governments attempt to maintain order and public health" (p. 3).

Continuing in this doomsday vein, Ehrlich and Ehrlich assert:

> The global polluting and exploiting activities of the DCs are even more serious than their internal problems. Spaceship Earth is now filled to capacity or beyond and is running out of food. . . . The food-producing mechanism is being sabotaged. The devices that maintain the atmosphere are being turned off. The

> temperature-control system is being altered at random. Thermonuclear bombs, poison gases, and supergerms have been manufactured and stockpiled by people in the few first-class compartments for possible use against other first-class passengers in their competitive struggles for dwindling resources—or perhaps even against the expectant but weaker masses of humanity in steerage. But, unaware that there is no one at the controls of their ship, many of the passengers ignore the chaos or view it with cheerful optimism, convinced that everything will turn out all right. (p. 4).

Obviously, the outside-in view takes many forms. It recommends either prepared or open societies and offers visions of ideal human development or catastrophic nightmares. The specific facts and arguments vary in their persuasiveness, as do the facts and arguments that can be mustered by those who regard human development essentially as an inside-out process. It is time now to see what can be done about *integrating* both of these approaches into a view that does justice to the inner and the outer, but also to the child.

BECOMING A PERSON THROUGH INTERACTION

The two year old himself or herself will help lead us to a new level of understanding. As the child grows older, it becomes less and less adequate to describe what is happening as either "inside" or "outside." The child is not simply the instrument by which inner biological forces express themselves; nor is he or she an infinitely shapeable apparatus at the command of environmental stimuli and pressures. The child constitutes a major sphere of influence. Concepts such as *selfhood* and *identity* begin to demand our attention. Evidences of emerging selfhood can be seen before the two-year mark, of course. The "stranger reaction" (*see* Chapter 5), for example, suggests an early awareness of who I am and where I belong. Much of what follows in the course of human development (and in this book) centers around the continued emergence of the *person*.

INTERACTION AND PERSONALITY

Most observers of human development today agree that personality is formed through *interaction*. Theorists differ among themselves about the importance assigned to various inner and outer dimensions and many other things. The central point, however, is that what we recognize as the human personality is considered to arise from patterns of interaction rather than from either the relatively automatic unfolding of inner potentials or the equally automatic impress of external designs and influences. Furthermore, it is usually agreed that the interaction process begins with the first appearance of the person-to-be and continues throughout the total lifespan. This much being said, however, we still must sharpen our basic ideas about the relationship between personality and the interaction process.

The bare fact that two or more processes are interacting with each other does not guarantee anything about the emergence of personality. Begin with Process A, based upon the physical properties of the human zygote, the fertilized egg. Add Process B, representing environmental influences, first those inside the mother's body and, later, those in the world outside the womb. Processes A and B interact with each other . . . and interact . . . and interact. Why should this interaction produce a *person?* There must already be a person on the scene. Once we see this, it becomes much easier to comprehend how this person *continues* to develop. But if we tend to think instead of two processes or sets of influences that are working on each other directly, then there is nothing in between that is available to be shaped or developed.

The **interactionist's fallacy** is the error of concentrating on external and internal systems to the exclusion of self

Two kinds of error are possible here. The first type is implied in what has just been said. Let's call it the **interactionist's fallacy.** This fallacy (or logical error) consists of assuming that we can account for development simply by describing the influences of two independent systems on each other. It talks about an inside system (innate characteristics such as genetic code) and an outside system (environmental conditions such as nutritional supply) but there has been no clear frame of reference established. Inside *who?* Outside *who?* The relative contributions of innate and environmental factors in development can only be discussed intelligently once the person is there, in one form or another. Development implies the existence of a person or organism that is available to be developed, not just two or more sets of processes that are working upon each other. With a person on the scene, we can then distinguish between inner and outer influences. A naive or exaggerated interactionist approach fails to consider the organism itself as an organism. It relies overmuch on the interaction process without coming to terms with the problem of who or what is at the center or crossroads of the interplay.

The other kind of error is simply to *assume* that there is a kind of spirit, soul, or psyche there right from the start. It is one thing to hold this position from a religious or general philosophical orientation, but it is something else to insist upon the existence of an invisible little personage who cannot be observed and studied, but who nevertheless makes things happen. Actions can be "explained" by referring to the whims and needs of a source of control that somehow inhabits the body. This view can lend itself to careless thinking about human behavior and experience. We can say whatever we like about this unobservable internal personage without fear that our assumptions can be tested out in the open. An elaborate mythology can develop around the assumed inner person. Critics of psychoanalysis, for example, sometimes have argued that **id** and **superego** are prime specimens of psychomythology. Picture an actual child whose hand is about to plunge, so it would seem, into the cookie jar. This is a behavioral tableau we all can observe. What is "really" happening, though, exists in the hidden, invisible psyche. "Yes, go ahead!" the id is urging; "No, better not!" the superego rules in re-

Id is a person's most primitive, least developed set of strivings for pleasure

Superego is a person's judge or conscience that tries to inhibit id functions

ply. There is no limit to the number and type of little beings with which we might populate the body, all safely beyond the test of direct observation. Years ago, when much less was known about human physiology, theorists would sometimes locate soul or psyche in a particular organ whose biological function was at the time not well understood. As knowledge advanced, the in-dwelling little spirit would be evicted from one place to another until it has today nearly run out of obscure crannies to call its own.

There is an interesting variation of this error. We might begin with interaction between biogenetic and environmental processes. Only observable or potentially observable phenomena are being included. But then we suddenly introduce person-ality at a later point. "Where did personality come from?" one is tempted to inquire. "And why at this point?" This strategy merely delays the interactionistic fallacy to a later point in development. It might be compared with promptly bestowing the full demands and rights of adult citizenship at age eighteen after previously treating the individual as lacking in moral responsibility and competence. Similarly, it is difficult to believe that interacting environmental and biological processes suddenly produce a person.

There is still another approach that has won a share of advocates, especially since the rise of industry and technology. We might stick grimly with the observation that life begins with various processes that influence each other but with no person as such on the scene. The essentially impersonal processes continue to interact indefinitely—no person is ever really created nor is the person needed. This is an automaton or machine model of the human being, differing from other complex machines only in details. Human development is seen as a sequence of interaction between processes that have their own fixed rules or programs. We can call the product a *person* or anything else, but this is just a fashion and a conceit. From this standpoint, human behavior can be regarded as the outcome of mechanistic principles. It is unnecessary and misleading to introduce the concept of person.

I suggest that none of these alternatives is sufficiently faithful to the facts of human development, nor does any provide a useful framework for study. Students of philosophy will appreciate some of the difficulties that are involved in approaching the question of personality development. We are touching on a whole set of riddles that are nested within each other. These include the *mind-body* problem and the questions probed by *epistemology.* A secure foundation for the study of human development requires the ability to avoid absurdities about the relationship between mind and body in general. It also requires sophistication in determining *how we can truly know anything,* not just the particular facts of human development. Somehow we must avoid the interactionist's fallacy and yet not introduce undefined concepts that raise more questions than they answer. There must be a way to appreciate the contributions of the person in his or her own development without flying in the face of both fact and logic.

The mind-body problem questions: If the mind and body are fundamentally different, then by what rules do they interact? Or, are *mind* and *body* clumsy terms misrepresenting a single reality?

A FRAME OF REFERENCE: THE PERSON AS AN ACTIVE PSYCHOBIOLOGICAL FIELD

I suggest now a frame of reference that might be useful in looking at the long sweep of human development from conception through the most advanced years. This view is deliberately broad and general. Most of the specifications that are necessary in a detailed theory of human development would only get in our way at the present time.

The **psycho-biological field** is the organism seen as an active, integrated reality

Organized activity is there at the start. The human being is from the start of life an **active psychobiological field.** From the inception of life, there is already a *system* of processes that function with some measure of integration and directionality. In other words, it is not a question of determining when the *in*-active or the *re*-active becomes active; activity is there from the start (or it is not a live organism that we are talking about). Furthermore, it is not a question of when *organized* activity begins. In some basic form, the activity *is* organized and systematic.

It takes only one more step to think of organized activity as a *psychobiological* field. This term is not intended to intimidate anybody, but a simpler expression would not do the job. The *psychobiological* part of this term is meant to convey the idea that the organism is best considered as an entity in which the so-called *psychological, behavioral,* or *mental,* and the so-called *physical* dimensions are inseparable. Through our scientific tradition, we have become accustomed to distinguishing between the *psycho* and the *bio.* Useful as this distinction can be in some circumstances, it also lends itself to an artificial separation through language of a reality that probably is more unified than either of the two component terms implies.

The *field* part of this term is meant to call several other properties of the organism to mind. The human being (or other organism) has certain boundaries and limits, shapes and contours. It is here, not there; it has this shape or form, not that one. It never was a completely shapeless "blob" of protoplasm, or a "psycho-blob" of functional impulses going any which way. Furthermore, as a psychobiological field, the organism functions as a *whole.* There are component properties within the organism. But at any and all points in its developmental career, there is an overall direction, selectivity, and set of operating principles. It can be misleading to make too close a comparison between electromagnetic or gravitational fields with the concept of psychobiological field. Yet some of the most basic principles are potentially applicable to all. A gravitational field exercises its influence at all possible points within the field, although not equally so. Similarly, the organism exercises its influences upon everything "inside" or "outside" that it can encompass ("mentally" or "physically"). In a sense, the organism can be *defined* in terms of the total scope of its influence. It exists as an influence in the world, along with many other influences or fields. The activities of a psychobiological field, or any other kind of field, are not random. Patterns can be observed, and laws or principles of functioning can be established. But this is not identical to insisting that everything can be explained in terms of specific cause-and-effect sequences. "This caused that" is a powerful and familiar way of thinking about the world. However, it is not the only way; it is not even the only way to think about the world from a scientific frame of reference. In attempting to understand field phenomena, it is often more useful to think in terms of multiple lines of causation, patterns, waves, or dynamics, instead of specific one-to-one cause-effect sequences.

Theoretical physicists have found field conceptions more useful than cause-effect connections

As a psychobiological field, then, the human being starts off active and organized, maintaining a kind of dynamic equilibrium. With continued growth, the field becomes more complex. It develops more and more component subsystems and functions. Recall the pattern of cells subdividing during which process the total organism becomes more specialized and differentiated within itself. As development continues, it also becomes more distinct as a *particular* individual, in comparison with other developing organisms. This general pattern of differentiation and complexification occurs for the organism as a total psychobiological field; it is not limited to cell growth. More specialized kinds of behavior show themselves and, *even at the two-year-old mark, thoughts and feelings are more differentiated than they were at an earlier time.*

Self-control emerges. The more that is "going on" in and around the developing psychobiological field, the more an organizing structure is necessary to keep everything together. Chaos would be the alternative. This means that a *self-control or guidance function* comes increasingly to the fore. It is useful to think of this emergence of a self-guidance function dynamically. In other words, it is not that one day there was only a lot of activity churning about in the organism, and the next day a system of controls came into being. Rather, the simpler principles of organization with which the organism began grow along with the total field. The subsystems differentiate from each other and become integrated into a more "advanced" type of system.

Continued maturation is marked by even more differentiation and integration. There is "more to" the psychobiological field, and it is organized in a more complex manner. (Remember Werner's Orthogenetic Principle?) In general, the more advanced the level of organizing structure, the more flexible and adaptable is the organism. Yes, at a very early point in development, one could analyze the "raw components" (inside-out) of the organism and make some useful descriptions and predictions. Or one could analyze the organism's environment outside-in and do the same. But now one makes increasingly more significant errors if the organizing structure, or self-guidance functions of the organism itself, is neglected. These self-guidance functions were always there, if in rudimentary form. *As time and development proceed, both the exaggerated inside-out and outside-in interpretations become less and less appropriate.* One has to begin looking at the organism in terms of its own way of working upon the world.

The demands of ballet go beyond ordinary body control to requiring exceptional mental discipline. Self-control also means a more flexible and accomplished self. (Photo by David Kelley.)

THE ORGANISM AS ITS OWN FRAME OF REFERENCE

Where has this exploration taken us? We are now at the point of being able to reconcile the traditional inner and outer views with a frame of reference that comes closer to the known facts of development and which leads to a keener appreciation of the organism as the center of its own activity. At first, the relatively low level of organization within the psychobiological field makes it possible to analyze the developmental events fairly well by concentrating on either inner or outer dimensions, or both. Actually, it might be more accurate to say that *all* dimensions are "outside" at this point. The organism's own frame of reference has not strengthened to the point where it can interpose itself between internal processes and external stimuli.

Development of the organism's integrating functions establishes a frame of reference. The two year old, for example, is integrated and sophisticated enough to recognize "This is mine, this is not" or "You are one person, I am another." Many differentiations and boundary-settings remain to be made, but we can see the beginnings of the individual's own frame of reference. Some of the hard-to-get-along-with aspects of the two year old probably relate to his or her own encounter with resistances in the world. He or she is forced to distinguish between internal need states and tensions and what is actually out there, facilitating or blocking the way. At the same time the child is trying to work out the differentiation and coordination of his or her physical subsystems (e.g., using both hands independently and together), he or she has parallel challenges on the psychological and interpersonal planes.

What has become of the biogenetic components that contributed to the first substance and activities of the individual? And what has become of the social expectations and the vast array of environmental stimuli that had so much effect upon the biological building blocks? Both realms remain available—but to the emerging self. Inner and outer no longer rub up against each other with only a rudimentary organizing structure inbetween. Instead, inner and outer are both registered and worked upon by the active psychobiological field we call a person. The self hones and flourishes through interactions with what were formerly inner or outer dimensions. More and more, the self makes itself at home in the world and makes its presence felt as a field that influences the world in return.

A self that can reflect upon experience. I have not yet mentioned the most striking and distinctive qualities of the emerging human self. I am referring, in general, to the ability to reflect upon experience. The self gradually comes to know what it is doing, instead of just doing it. Experience can be anticipated and experience can be examined in retrospect. Stimuli from inside or outside can be represented within the self. Once represented, the materials of the world become resources that the growing self can manipulate in an ever more flexible and effective manner. In becoming self-conscious, the psychobiological field is also becoming what we recognize in daily life as a human personality. Much of human development after earliest childhood centers on the maturation of selfhood and identity. It is not enough to pay attention to specific changes in dimensions we call *physical* or *psychological*. It is also crucial to focus on the new capacities of the individual to establish and elaborate its own frame of reference. This is one of the reasons why the work of people such as Jean Piaget is so significant. We must learn from whoever can teach us precisely how the individual moves toward mastery from his or her origins as an organizing tendency within a relatively primitive psychobiological field.

The human as a self-conscious field living in a universe that functions as a field of fields?

In the chapters to come I will be giving heightened attention to the active and integrative functions of the organism, the developing self that is intimately related to inner and outer forces, but which is not reducible to either. Recognizing the self as an authentic and vital participant in the developmental process—as a true protagonist in the scenario—will help to bridge the gap between our daily world of experience and the facts and theories offered through the scientific study of human growth.

SUMMARY

Two major traditions have been expressed over the centuries: humans develop mostly from the inside-out, or humans develop mostly from the outside-in. Some of the most important views from both traditions were sampled in this chapter. This was an appropriate time to pause for a little historical

YOUR TURN

1. Observe the behavior of an infant or very young child in his or her natural or familiar setting. Between fifteen and thirty minutes observation time should be enough. Record what you see and hear in any accurate way you prefer. Now, do the same for a child who is a bit older, around age four or five.

Treat your observations of both children as *raw data* (factual bits without interpretations added). The information you have recorded is an attempt to describe what took place in the actual sequence with a minimum of inference from the observer. Now, *interpret!* Decide that you will be an inside-out advocate. You want to demonstrate that early development springs chiefly from forces within the individual. Try to interpret the behavior of both children from this perspective.

Next, shift your thinking to an outer-in framework. It is the environment that has called forth the children's behavior in one way or another. Work over the same raw data with this opposing orientation in mind.

Finally, examine both sets of interpretations you have made. Which one was easier to substantiate or seemed more appropriate for the very young child? For the somewhat older child? Did you convince yourself from this little experiment that one approach is more useful than the other? If not, what would it take to persuade you? Perhaps, from this attempt to take both sides, you now have a more balanced view of the behavior of very young children. As in most other projects I have been suggesting, this one will prove particularly useful if you will commit yourself to setting down your thoughts and interpretations on paper, the way scientists and authors must do, and if you will also share your experiences with others.

2. This beginning of this chapter introduced a number of basic sources pertaining to human nature and its early development, along with several that emphasize subhuman species. Select one of these authors and read the material for yourself. In addition to the people mentioned in the chapter, you might also investigate the theoretical contribution of Kurt Lewin (1935) who first made psychology aware of the "field" way of conceptualizing the human personality in its environment. Summarize an author's views in your own words, evaluate his or her position in the light of your own words, and then evaluate that position in the light of your own perspective and knowledge. One useful device in such an evaluation would be to see how far you could apply the views of Darwin, McDougall, Locke, etc. to problems that interest you but which they did not deal with explicitly.

3. Some of the most important outside-in thinkers of the past gave us visions of how human development could be fostered by improving environmental quality (especially social institutions). By contrast to Locke and Rousseau, however, many contemporary writers emphasize the *degradation* of life as a result of environmental changes (Ehrlich and Ehrlich, for example, but there are many others as well). Precisely why has the mood shifted? Are the optimistic ideas of the past worn out? Disproven? Overwhelmed by the accumulation of error? Is the current generation of doom speakers an over-reaction? How much of the environmental catastrophe outlook is based upon external fact, and how much upon insecurity, frustration, and pessimism?

What favorable and unfavorable influences are at work in our society, so far as their effect upon the very young child is concerned? Are we taking advantage of the potentialities that an infant brings with himself or herself into postnatal life, or are we exposing the youngest humans to a sequence of distorting and debilitating experiences? Think about these large questions until you are ready to share your thoughts with others.

and philosophical perspective. The two year old is still close to his or her biogenetic origins, yet has been around long enough to have been exposed to an appreciable amount of experience.

Our survey from the inside-out tradition began with *Darwin,* whose observations and thoughts about the *instinctual basis of behavior* have influenced society at many levels. Creatures both great and small begin their lives with specific unlearned behaviors or adaptational strategies that give them at least a fighting chance to survive. One must appreciate patterns of functioning that are characteristic of a particular species. The environment is not ignored, for only in the interaction between organism and environment can the concept of adaptation and its evolutionary dynamics be understood. But Darwin would have us identify and respect the instinctual equipment with which living creatures, including ourselves, begin life.

In psychology, *McDougall* presented a theoretical orientation in which the concept of instinct is closely linked with the organism's *purpose.* Complex human experiences and actions have subtle instinctual components at their core. In McDougall's approach, the individual is seen as an active force who has developed and blended his or her innate dispositions into a unique constellation. These dispositions continue to grow along with the individual throughout life.

Freud emphasized the origins of our instinctual energies. Organ tensions that crave complete and immediate release are generated throughout our bodies. But physical and social reality is such that one cannot always satisfy these needs directly or immediately. During the course of development, *ego functions* appear to moderate between inner demands and those of society. This conflict, between the desire for immediate satisfaction and the desire to be acceptable to others, continues throughout adult life and can be at the basis of severe frustrations and maladaptive patterns. Much of the developmental process is concerned with discovering, selecting, or creating opportunities to relieve instinctual tensions within the complex world of physical and social realities. Freud's theory of *psychosexual stages* is one major aspect of his instinctual-developmental theory.

In more recent years, the inside-out approach has taken a turn toward precise experimental investigations. The work of *Lorenz* and *Tinbergen* was cited as illustrative of the newer contributions. Lorenz's study of *imprinting,* and Tinbergen's demonstration of highly specific factors in the organism's mode of interacting with the environment, emphasized the innate characteristics position while adding interesting new lines of evidence. Finally, we reminded ourselves that a variety of nonbiological arguments have also been advanced for the inside-out tradition, including *fatalistic, theological,* and *mentalistic* orientations.

In *The Republic, Plato* proved himself one of the great champions of the outside-in tradition. He proposed a kind of "prepared society" in which all of a culture's resources would be drawn upon to shape the developmental career of its children. Society was seen as having the right, the responsibility, and the means for bringing the infant along a predetermined developmental path. Apart from his specific theories and proposals, Plato generally asserted an intimate association between the individual and the type of society in which he or she lives. In our own time, *Skinner's* behavior modification approach (as expressed, for example, in *Walden Two),* provides an interesting up-dated version of the prepared society. We are under the stimulus control of our environment anyhow, Skinner argues; it is to our benefit to make this stimulus control work to our better advantage.

Other outside-inners emphasized the potentialities of every fresh human mind that comes into the world. *Locke* argued that our ideas come, directly or indirectly, from experience. This means that our minds are essentially "unmade" until exposed to whatever the world has to offer them. This view also implies the desirability of *equal opportunity* for each mind to develop itself—a doctrine that made waves on both sides of the Atlantic. Another whose ideas endorsed a more open type of society was *Rousseau,* an outspoken critic of so-called civilization. He put forth the radical notion that all people have the natural rights to life, liberty, and the pursuit of happiness. Society owes its existence to its constituent individuals, not the other way around. Our finest hope is in the child, who is closer to the "natural state" than is the adult who has been warped by poor education and an evil society. The child should be allowed to explore freely an environmental setting that is as natural and as open as possible.

One of the strong trends in recent years has been the vision of environmental deterioration and catastrophe. The outer world is in trouble and so, too, is the emerging individual. The warnings of *Ehrlich and Ehrlich,* centering around world overpopulation, were taken as illustrations.

After considering both the inner and the outer traditions, I explored the possibility of an *interactionistic* approach. Most developmentalists today agree that personality is formed through constant interaction, rather than by either inner forces or outer forces taken separately. There are some problems to be resolved in working out an adequate interactionistic position, however. How can selfhood and identity emerge from the bare encounter between inner and outer systems, neither of which themselves have personality? After a little discussion, I proposed a frame of reference in which the human being is regarded as an active *psychobiological field* from the very beginnings of its existence. Basic properties of a psychobiological field were mentioned, including its functioning as a total configuration with the capacity to change and develop over time. The individual starts active and organized and becomes more complex as development continues. Differentiation within the field is accompanied by a more advanced type of integration or self-guiding function.

According to this view, the distinction between inner and outer determinants of behavior is more useful during earlier stages of development. At first, one can be fairly accurate in describing development in terms of separate realms of inner and outer factors. Gradually, however, the organism's own organizing structure asserts itself. The individual develops his or her own frame of reference within which both the inner and the outer become resources to be interpreted and utilized for his or her own emerging purposes. Selfhood and identity now become important concepts. And, as we move along the lifespan with the two year old, we find a striking enrichment of the self's resources for representing and reflecting upon the world (including the self reflecting upon the self). The human self, then, can be seen as a dynamic creation from the relatively simple organizing structures that first held the organism together in its early history. This overview is presented in something of an abstract manner, but relevant details will be abundant in the chapters to come.

Reference List

Aries, P. *Centuries of childhood.* New York: Alfred A. Knopf, Inc., 1962.

Barker, R. G. *Ecological psychology.* Stanford, Calif.: Stanford University Press, 1968.

Bronson, G. The hierarchical organization of the central nervous system: Implications for learning processes and critical periods in early development. *Behavioral Science,* 1965, *10,* 7–25.

Caldwell, B. M. The usefulness of the critical period hypothesis in the study of filiative behavior. *Merrill-Palmer Quarterly,* 1962, *8,* 229–242.

Darwin, C. *On the origin of the species by means of natural selection, or the preservation of favoured races in the struggle for life.* In R. J. Herrnstein & E. G. Boring (Eds.), *A sourcebook in the history of psychology.* Cambridge, Mass.: Harvard University Press, 1966, pp. 408–413. (Originally published, 1859.)

deMause, L. (Ed.), *The history of childhood.* New York: The Psychohistory Press, 1974.

Ehrlich, P. R., & Ehrlich, A. H. *Population, resources, environment.* San Francisco, Calif.: W. H. Freeman and Company, 1970.

Freud, S. *Beyond the pleasure principle.* New York: W. W. Norton & Company, Inc., 1920.

Freud, S. *A general introduction to psychoanalysis.* London: Liveright Publishing Corporation, 1935.

Huntington, E. *World-power and evolution.* New Haven, Conn.: Yale University Press, 1920.

Huxley, A. *Brave new world.* New York: Harper, 1950.

Kant, I. The critique of pure reason. In Encyclopaedia Britannica, Inc., *Great books of the western world* (Vol. 34). Chicago: Author, 1953, pp. 1–252. (Originally published, 1781.)

Lewin, K. *Principles of topological psychology.* New York: McGraw-Hill, 1935.

Locke, J. An essay concerning human understanding. In Encyclopaedia Britannica, Inc., *Great books of the western world* (Vol. 35). Chicago: Author, 1953, pp. 85–395. (Originally published, 1690.)

Lorenz, K. Z. *King Solomon's ring.* London: Methuen, 1952.

Lorenz, K. Z. Imprinting. In R. C. Birney & R. C. Teevan (Eds.), *Instinct.* Princeton, N.J.: D. Van Nostrand Co., Inc., 1961, pp. 52–64.

McDougall, W. *Outline of psychology.* New York: Charles Scribner's Sons, 1923.

Padilla, S. G. Further studies in the delayed pecking of chicks. *Journal of Comparative Psychology*, 1930, *20*, 413–433.

Orwell, G. *1984*. New York: Harcourt Brace Jovanovich, Inc., 1949.

Plato. The republic. In Encyclopaedia Britannica, Inc., *Great books of the western world* (Vol. 7), Chicago: Author, 1953, pp. 225–441. (Originally published, *circa* 385 B.C.)

Proshansky, H. M., Ittelson, W. H., & Rivlin, L. G. (Eds.). *Environmental psychology*. New York: Holt, Rinehart and Winston, 1970.

Rousseau, J. J. *Emile, on education*. New York: Dutton, 1911. (Originally published, 1762.)

Rousseau, J. J. A dissertation on the origin and foundations of the inequality of mankind. In Encyclopaedia Britannica, Inc., *Great books of the western world* (Vol. 38). Chicago: Author, 1953, pp. 323–366. (Originally published, 1754.)

Rousseau, J. J. The social contract. In Encyclopaedia Britannica, Inc., *Great books of the western world* (Vol. 38). Chicago: Author, 1953, pp. 387–439. (Originally published, 1762.)

Scott, J. P. Critical periods in behavioral development. *Science*, 1962, *138*, 949–958.

Skinner, B. F. *Walden two*. New York: Macmillan, Inc., 1948.

Skinner, B. F. *Beyond freedom and dignity*. New York: Alfred A. Knopf, 1971.

Tinbergen, N. *The study of instinct*. London: Oxford University Press, 1951.

THE PRESCHOOL SCENE

8

CHAPTER OUTLINE

I
Development continues to be fast and fascinating on the preschool scene.
Use of hands / Locomotion / Increasing understanding of communication / Nightmares / Adventuring

II
Mental development is extensive during the preschool years.
Verbal control over behavior / Semantic (meaning) properties of words (Luria) / Stimulus-Response learning (Kendler & Kendler) / Memory processes / Episodic memory / Semantic memory / Childhood / Childhood amnesia / Perspective-taking ability / Representational (Preoperational) stage (Piaget) / Symbolic functions

III
Play in one form, or another, takes up much of the preschooler's time.
Play-is-practice view / Letting-off-steam view / Psychoanalytic approach / Play-work / Sibling rivalry / Explorative play / Pretend games / Parallel play / Cooperative play

IV
The preschooler's important relationship with parents is a continuing challenge to balance opportunities for assertion of competency and independence with the need for guidance and protection.
Independence versus protection / Pleasures of parenting / Identification / Oedipal situation (The family romance) / Castration anxiety

V
Alternatives to the mother-at-home, father-at-work pattern of parent availability lead to the need for and the use of day care centers.
Factors involved / Problems / Effects of day care

VI
Bereavement (loss of a parent) can accentuate some hazards in early childhood.
Scope of the problem / Effects of bereavement / Helping behavior

Child watchers gain many rewards when they set their sights on the preschool scene. Development continues to be fast and fascinating. I will begin by attempting to capture something of the child's own experience of the world at this time by way of the scenario that follows immediately. Next I will examine several specific areas of development that are salient at this time within the general realms of thinking, playing, and socialization. I will then consider two very different types of separation experience: day care and bereavement.

SCENARIO 5: OUT-OF-THE-BODY EXPERIENCES

Suppose you made the following discoveries about yourself: you can walk through solid walls, change the physical world to suit your fancy, read other people's minds, see into the future, and travel your mind beyond your body.

Suppose further that these discoveries are all being made at about the same time, almost tumbling over each other as they emerge. Yet all is not smooth going. Your expanding self-universe is creating problems of a new magnitude both for yourself and others. You are beset by perils the likes of which you never dreamed. You have the added problem that your new abilities are far from perfected. Frustration, misunderstanding, even the prospect of total disaster seem to hover about your efforts to exercise and exploit your new-found talents.

You are, in short, both a joy and a menace to yourself. This does not make you an easy person to live with—although certainly a fascinating person for those who can recognize and appreciate the transformations you are experiencing.

This science-fantasy scenario is not the hodgepodged invention of an overworked novelist. It is simply one way of looking at common, taken-for-granted experiences through which hundreds of millions have passed enroute to becoming men and women. The early childhood years are remarkably eventful. And, remarkably, as adults we usually remember little of what made these years so special to us (Schactel, 1959). Experiences of such novelty and power seldom intrude into our adult lives—and when they do, they astonish us.

Perhaps an act of imagination is needed to appreciate how fresh and transfiguring are the experiences of the young child. Even if we get the basic facts straight about child development, we will be missing something very important if we look on the process as strictly routine and ordinary. True, it is common for infants to move into and through early childhood. But this does not detract in any way from what a particular boy or girl is experiencing and encountering. It is powerful and novel for *this* child. It is the first time he or she has been in this kind of situation or in partial command of such fantastic new abilities.

TRANSFORMATIONS IN THE CHILD'S WORLD

Consider what is happening in the world of the young child, the world you were once discovering:

Finally secure on your own two feet, you can now survey the environment from a more elevated perspective. Your hands are free to function independently. You see places where you want to be. But there are barriers that wall you out (or in). A solid physical barrier separates you from the inside of the closet, from the outside of the house, from mother or father in the next room, from the who knows what you might find some place else. Now you can penetrate these barriers, at least some of the time. With your free hands, you learn how to turn the door knob. This is a simple and routine action for your elders, but a significant achievement that opens many possibilities for you. Indeed, now *you can walk through solid walls.* One such discovery after another continues to expand your opportunities to travel through space; to close distances between yourself and others; to move through, above, under, or around physical barriers; and even to establish barriers of your own to keep others out.

Child-proofing the environment is important at this time

Once you had to cope with the world as you found it. In fact, you had to make a series of discoveries in order to recognize that there actually were places and they had things in them. Previously you had not seen much difference between yourself and the outer world, let alone between different places and things. It made you feel good to learn about these "thing-places"; it gave you a sense of security and dependability that you could build on. Now, however, you are moving to a new level of discovery. These thing-places can be changed (at least on occasion). The world is a more congenial place if you take your blanket with you. And so you do. The pots and pans belong behind a certain cabinet door or on the range. But it is more entertaining to pound them together in the dining room. The stool also belongs in the kitchen, but you need it in the living room so you can reach way up to water the plants (or almost). Thing-places do not have to remain entirely as they are; *you can change the physical world to suit your fancy.*

People have not always behaved as they should. Sometimes they did precisely what they were supposed to do (for you), but sometimes they did not. This annoying pattern continues. However, now you are starting to pick up strangely helpful vibrations. Big sister enters the room after you have lavishly decorated your face and her mirror with lipstick and makeup. There is something about the way she looks at you that gets a message across even before a word is said. Those vibrations are working. Or you have the feeling it is time to switch to Plan B when mother and father are not paying proper attention to you, although they have not actually said, "Go away!" Although only a beginner at this exotic mode of communication, *you are discovering how to read other people's minds.*

The feeling that others can read my mind may also begin at this time

When you do something, something else happens (sometimes). A case is when you bump the table and the flower pot breaks and water drips down the walls. What other people do also has consequences. You notice that catching a ball brings you an approving smile, while tearing pages out of a book does not. Slowly you develop the knack of seeing where a particular action will lead. And you also know what to expect when a familiar sequence of events unfolds in the household, for example, the going-to-bed directives or negotiations. With a developing sense of what is to come, you can make simple plans and devise simple strategies of your own. These will eventually seem primitive when compared with what you will come up with as you continue to mature. But you have at least started to exercise the gift of prophecy: *you can see into the future.*

All of these new talents involve in some way the ability to *travel your mind beyond your body.* Example: You are a young child standing here, but what you want is in the closet, behind that closed door. Mentally, then, you reach out to encompass the thing behind the closed door. Chances are that your body posture will give away your intention. You have not yet mastered the civilized art of pretending to be disinterested, or simply sitting quietly while thinking of other times and places. Your body language is fairly straightforward and often can be read by adults with ease. But the point is that your mind is reaching out to touch and encompass people, objects, and situations that are not part of your physical body. There is a relationship of seeking, knowing, or intending—in any case, a psychological relationship—that connects you, standing here in one particular place with one particular body, with other thing-places. Scientifically minded adults in our society probably would insist that your mind doesn't go any place without taking its nervous system along (or vice versa). This proposition may be true, but irrelevant. The psychological reality here is your thoughts and emotions have begun to operate so as to *intend.*

Just as you have more physical freedom to explore and act upon your environment than you did as a *very* young child, so you are now discovering a new sphere of mental freedom. Even when you are asleep, your mind travels (or does it open its doors to let outside images in?). Awake or asleep, you are a mental traveler.

PERILS AND FRUSTRATIONS

The perils and frustrations of your new abilities must be spoken of also. Now that you have dreams, you can have bad dreams. Nightmares may, in fact, be the most common sort of dream that is remembered and reported by preschool children. This is one of the conclusions reached by John E. Mack (1970) on the basis of his own research and that of other investigators. The central importance of the *oral* area (mouth, lips teeth, etc.) in early childhood can show up in a most frightening form in nightmares. "Being eaten or eaten up, and equivalent forms of devouring, such as being ground up inside large machines, occur with particular frequency in the nightmares of children under five" (p. 26). Mack adds the provocative interpretation: "Often such themes of the nursery years may reappear in the nightmares of older children and even of adults. Anxieties derived from the child's early mode of expressing affection by devouring, incorporating, and thus retaining the other person 'inside,' or venting aggression by biting or eating someone, readily find their way into nightmares" (p. 26).

Pavor nocturnus or night terror is less common than the usual nightmare and more disturbing

You may find yourself very much alone in the midst of nightmares or even *daymares.* Your thoughts have traveled to a menacing scene or have been invaded by unwelcome visitors. But the adults in your life do not seem to fully recognize your peril. Mack tells us, for example, about a three-and-one-half-year-old girl whose night terrors and fears of sleeping alone brought her eventually into the office of a psychotherapist. When she felt comfortable with the therapist, Rachel described "grabby monsters" that come into her room after she is asleep. The monsters try to take her away. Rachel screams for her mother—who responds by spanking her! Yet, the spanking does help; it drives the monsters away for awhile. But Rachel fears the growling, biting monsters during the day as well. (The therapist believes she sometimes confuses them with characters she has seen on television monster movies.) As darkness falls, Rachel becomes increasingly anxious because this is when the monsters turn out in full force, and when she is most alone and vulnerable.

In this instance, mother (and father) were having their own problems, and these were reflected in Rachel's nightmares. But the fact that there were parental tensions and that Rachel herself was seen by a psychotherapist does not make her experience all that exceptional; it just makes her experience available for study. As preschool children, most of us experienced states of fright both sleeping and waking. Anticipating that monsters might come again is part of Rachel's emerging sense of futurity; each of us may have had our own version of fearful expectations. At this time in her life, Rachel has not yet established a clear distinction between those nocturnal events we call *dreams* and those daylight events we call *realities.* Indeed, there are significant differences from one adult culture to another in how this distinction is made and how seriously it is taken (Von Grunebaum & Caillois, 1966). Her mother finally "had to do something" about Rachel's behavior because it was disturbing others. The mother did not seem to appreciate that Rachel herself was undergoing experiences of terror that perhaps should have been the focus of concern.

The hours of daylight and companionship can bring their own woes. There is so much you can *almost* do! Now that you can open some doors, it is infuriating to come up against a door that resists your efforts. Now that you can catch a ball (under favorable circumstances), it is upsetting to be excluded from the ballgames that older children want to play by themselves. The physical world you are trying to re-arrange may collapse on top of you, either figuratively, or all too literally. And, as we have already seen, the ability to glimpse a little way into the future can fill you with alarm or reduce you to tears. You have, then, the heady feeling of entering new worlds of actions and experiences, but with many a misadventure along the way.

CHILD AND ADULT ADVENTURING

The young child's emergence as a mental traveler perhaps tells us something about "far-out" phenomena in adult life. As grown-ups, we tend to take much for granted. What was fresh and exhilarating for the young child is routine for us. We have long forgotten, for example, what a thrill it was to turn the door knob and pass from one room to another under our own power. Yet we all have had the experience of going beyond what we once could do and moving into a realm of strange and marvelous powers. This may have given us the taste for mental adventuring. But as adults we need other forms of experience to achieve the same effect that we knew in childhood.

Reaching into places, seeing over, rising above—emerging abilities such as these in early childhood probably play a role in adults' curiosity and adventuresomeness. (Photo by Bobbi Carrey.)

Consider just one of the many possible examples that can be drawn from history. For centuries people were fascinated by maps and map making. What might the earth contain apart from the familiar people, animals, and terrain that one could see everyday in his or her own locale? Map makers projected seas, islands, cities, nations, and whole continents on the surface of the earth even when nobody had actually laid eyes on them (Ramsay, 1972). Actual observations and careful approximations appeared side by side with exuberant specifications of fabulous lands, peoples, and animals. The mixture of fact, possible fact, and sheer fancy concerning the geography of our planet was an early and popular form of what today we are inclined to call *science fantasy.* It served many purposes, but among these was the adult equivalent of the child's adventuring beyond the limited physical and mental world of infancy.

The ability to travel freely through space and time has, of course, been a stock in trade of science fantasy for many years. Both in serious adult novels and in popular comic strips and television programs, we encounter people who possess powers that go beyond normal adult capacities, perhaps in the same way that the child surpasses the infant. Parapsychologists (e.g., Rhine, 1967) study what they call **precognition** (the knowledge of events before they occur) as well as other mental phenomena that are extraordinary, if authentic. Artists such as Dali have "bent" time and space to create mental landscapes that do not have their physical counterparts in the world or, like Chagall, have liberated men, women, and children to float in the air.

Precognition is the ability to know an event before it occurs

I am suggesting, then, that the developmental experiences of the preschool child remain with us to whet our appetite for marvelous powers. And who can say how much of the motivation for exploration of outer space and highly sophisticated forms of communication owe a debt to the young child's wish to complete the adventure he or she has begun?

It is time to take a closer look at the development of mental functioning in the preschool years. I will begin by examining some of the power of words, then I will examine memory processes, perspective-taking ability, and Piaget's overall view of this period of life.

MENTAL DEVELOPMENT IN THE PRESCHOOL CHILD

THE POWER OF WORDS

Words—even a few of them—greatly enhance the young child's ability to influence and interact with other people. The mutual influence process begins well before words, of course, with coo's and cries and all manners of vocalization. **Receptive language** (the ability to understand something of what is being said to one's self) often appears to be ahead of the child's ability to express his own thoughts and feelings in words.

Receptive language is the ability to understand words

But in addition to their important role in communication, words also wield another kind of power. They become critical links in thought-and-action sequences. To some extent our actions are controlled or mediated by verbal symbols. *Control* and *mediate* have different nuances, and there are particular situations in which one of these terms appears more appropriate than the other. The main point, though, is that what happens between stimulus and response depends more and more upon internal representations. A tickle stimulus may elicit a reflex response in an infant without seeming to involve much internal mediation. However, as development proceeds it becomes increasingly useful to credit the child with some kind of internal process that mediates or controls action. Words are not the only possible internal representations. The young child may also be using a more perceptually oriented form of mediation, a memory-picture or image. But words do become increasingly dominant. These include both the words that others say to the child and the words that the child says or thinks to himself or herself.

VERBAL CONTROL OVER MOTOR BEHAVIOR

In the opening scenario, we glimpsed some of the preschool child's emerging liberation from the close-in environment. The child becomes increasingly free from the strong controls exerted by the environment and by his or her own physical needs and predispositions. This freedom is only relative, of course. There is more opportunity to explore, less necessity to stay put; more ability to act, less obligation to respond through simple reflexes.

Luria (1961) and Vygotsky (1962) conducted a stimulating series of studies that emphasize the increasing significance of words in the regulation of action as the child matures. They found that the individual develops the ability to regulate his or her own behavior through the control of verbal symbols that are attached to stimuli and responses. This means that the individual child's own level of competence in manipulating verbal symbols (as well as his or her motivation and style) become critical determinants of overt behavior. We have to know the child's mind better and better in order to predict behavior.

According to Luria, there is an orderly sequence to the child's development of self-regulation through verbal symbols. The very young child is capable of being stimulated by the words that are directed to him or her. What mother or father says can stimulate a child's actions, but it will not usually terminate a behavior pattern. Speech is exciting. The infant responds either by starting to do something, or by intensifying what he or she is already doing. This shows a certain power of words over action. Yet it does not demonstrate any fine tuning of behavior as a consequence of speech, nor does the young child have the tuning knob in his or her own control.

The second and third stages in this sequence are located in the preschool years. In the second stage (from about age two to age four or four and one half), the child becomes able to guide some of his or her own actions, to excite himself or herself by his or her own words. The control is still weak and limited. Other forces, such as stimulation and distraction from the environment, may gain dominance over his or her own words. Furthermore, his or her ability to *excite or incite* himself or herself into overt action remains well ahead of the ability to call a *halt* to his or her own actions by verbal symbols. Luria proposes a neurological explanation for this differential. Speech—whether other people's or one's own—leads to a generalized excitation of the central nervous system. The brain areas that release motor behavior are stimulated. The child may engage in actions even though they are not appropriate to the situation.

In one study, Luria asked children in this age range to squeeze a balloon when a blue light went on, but not when a yellow light went on. Most of the children squeezed to both lights. He then asked them to *say* "Squeeze!" to the blue light and "Don't squeeze!" to the yellow light. The result? They made even more "wrong" squeezes than they had before. Studies of this type suggest support for the theory that speech generally excites behavior in young children but does not give it much directional control.

Somewhere in the four-and-one-half- to five-and-one-half-year range, Luria found that most children can place this kind of behavior under their own verbal regulation. The instructions now obtain the effect that might otherwise have been expected. In his terms, the **semantic properties of speech** come to dominate the purely excitatory qualities of words. *What* is said becomes more important than the fact that something is said.

Semantic properties of speech are the meanings of words not the sounds that make them up

In its general outline, Luria's view is consistent with Piaget's (1973) that sees the developing child as becoming more influenced by his or her own internal structures (or schemata) and less by external structures. Vygotsky's work suggests that at least part of this process is facilitated by the child's in-

Figure 8-1. *Verbal control of actions in childhood. Words excite and intensify actions before they take on the power to control and inhibit. (Based on data presented by Luria, 1961.)*

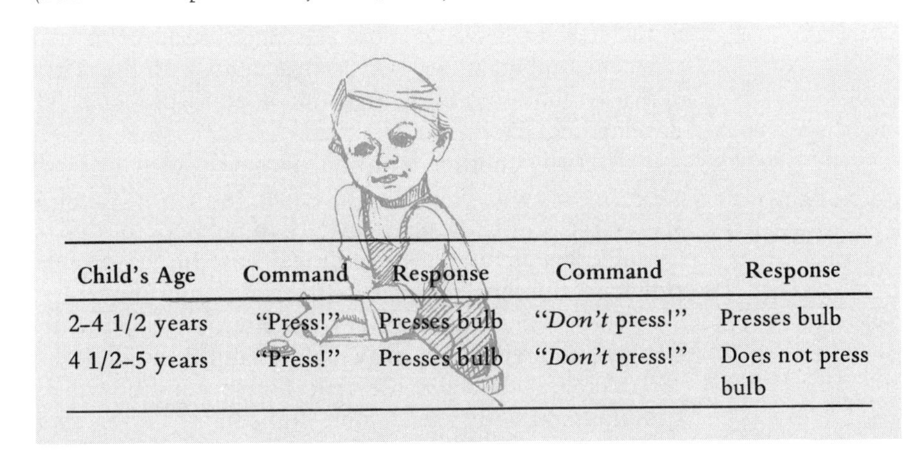

Child's Age	Command	Response	Command	Response
2–4 1/2 years	"Press!"	Presses bulb	"*Don't* press!"	Presses bulb
4 1/2–5 years	"Press!"	Presses bulb	"*Don't* press!"	Does not press bulb

ternalization of verbal instructions from the parents and other adults. What others have said to the child, the child can now say to or for himself or herself, and enjoy much the same effect.

Katz (1970) has confirmed findings of this type, and has also added information on individual differences. He found that three or four year olds who scored higher on an impulsivity scale than their cohorts were less likely to inhibit their actions by use of verbal regulators in the experimental situation. This reminds us again that personality or life-style differences remain important even when general developmental trends can be seen. Katz repeated his research with old people as respondents. He found that the ability to regulate one's actions with verbal commands remains intact well into advanced age, although it did weaken for the very oldest (Katz, 1974).

Reflecting on his 1970 study, Katz explained one of its implications for child rearing:

> In many cases, when a child is told not to do something by his parents, but does it anyway, he is punished because his behavior is construed to be the result of negativism, disobedience, or hostility. An alternative interpretation . . . is that it is not the result of willful disobedience, but of the developmental immaturity of neurological and psychological mechanisms responsible for self-control. (Katz, 1974, p. 143).

AN ALTERNATE APPROACH: STIMULUS-RESPONSE LEARNING

The kind of theory and research that has been considered so far in this chapter (like the work of Roger Brown mentioned in an earlier chapter) views words as taking on more power because the child himself or herself is changing in important ways. Neurologically and psychologically, the older child is really different from the younger child. Although they differ in emphases and in many details, Piaget, Luria, Werner, and a number of other theoreticians share the overall view that the organism itself becomes organized at a higher level, becomes more capable of resourceful adaptation to the environment, and makes its own arrangements with life. They feel that the use of verbal symbols to gain the freedom of self-regulation should be understood in developmental terms.

There is another tradition, however, which attempts to find simpler ways to account for observable changes in words and actions. Stimulus-response types of learning theory try to predict and explain behavior without speculations about the neurological state of the organism and without introducing the concept of qualitatively different stages.

Those who are attracted by this alternative approach will find some of its best research samples in the work of Kendler and Kendler (1962; and Kendler, 1963, 1967). They propose that simple conditioned associations between stimulus and response become linked into larger chains. The **chained**

Chained stimulus-response bonds are a series of psychological connections built from simple stimulus-response associations

stimulus-response bonds are acquired from repeated experiences in the world. They are stored as memory traces that are called forth either by outer stimuli or by other covert (internalized) stimuli.

This is a step beyond the most basic itch-twitch theory. It is not sufficient to account for human behavior on the basis of an overt stimulus that elicits an overt response. Something does have to happen inbetween. The something inbetween is not mental processing in the usual sense, let alone the intentional, purposeful action of an organismic personality. Instead, the hypothetical inner scene consists of acquired associations that set each other off until an observable behavior occurs. It is essentially a mechanical sequence of events.

Older children have more verbal mediators to link external stimuli with possible responses. The exposure to stimulation that living longer brings increases the repertoire of mediators. One does not have to speak of the child shifting to higher levels of functioning and developing entirely new capacities.

The interplay between cognitively oriented, organismic theories and the stimulus-response approach stimulates both camps to vigorous and innovative research.

DEVELOPMENT OF MEMORY IN THE YOUNG CHILD

It may seem curious to even be curious about memory in the preschool child. With such a short past, what does he or she really have to remember? But the processes involved in memory are closely related to mental functioning in general. In fact, James (1890), Bartlett (1932), and other pioneers have insisted that what we call *memory* cannot be separated from the total picture of cognitive activity. Remembering is a dimension of most, if not all, of our cognitive activity. It is not too early, then, to examine memory in the preschool child. Studies of memory have the potential of **indexing** the general level of mental functioning and the pattern of its continuing development. As we will see, it remains useful to respect the advice that we should not isolate memory from the overall pattern of the individual's thought and social interactions.

Indexing is a marking or pointing toward other phenomena

Episodic memory is the recalling or recognition of specific events and facts

Semantic memory is the system by which memories are stored and retrieved

Episodic and semantic memory. A useful distinction has been made between two general types of memory (Tulving, 1972). **Episodic memory** is just what the term suggests, the storage and retrieval of specific events that once happened at a certain time in a certain place. **Semantic memory** refers to the individual's *system* for storing and utilizing information. The word, *Kitty*, for example, and the rules or grammar of language usage are aspects of the semantic memory system. There is no reason to assume that both aspects of memory develop in the same way or at the same tempo. As the

To **encode** is to register a fact or experience into memory storage

child gains more experience there will be more episodes to **encode** (register into storage). Furthermore, there will be an increased ability to *use* these stored episodes. Semantic memory will change in complex ways. Piaget (1968) and others believe that the character of semantic memory changes as the individual moves from one stage or substage of mental development to another. If we look on memory as a *process* rather than as a fixed, automatic kind of mental reflex, the significance of memory in the total developmental pattern stands out clearly. Whatever influences the way in which an individual's semantic memory system develops will also influence mental development—and personality—throughout life. We use semantic memory to *reconstruct* experience, not simply to pluck a particular episode from the past. This means that at various times in our lives we may reconstruct our own pasts in light of our current capacities, needs, and pressures.

Some characteristics of memory in preschool children. Both episodic and semantic types of memory can be found in preschool children. As we might expect, the older the child, the more efficient and accurate the memories that can be produced. Experimenters usually prefer to create new memories for their participants under controlled conditions. This is one of the many methodological devices that enable students of early memory and cognition to focus clearly on a specific question and avoid contamination by uncontrolled variables. However, it also has the effect of emphasizing memory performances under specialized circumstances rather than the more spontaneous play of memory in the child's familiar life setting.

These children are all ears in listening to the story. But precisely what will they remember of it five minutes from now? Or five days or five years from now? (Photo by Talbot Lovering.)

From experimental studies, it has been learned that young children are relatively deficient in strategies for retrieving episodic memories (Brown, 1975; Reese, 1976). It does not help a young child's performance to alert him or her to the fact that memory will be tested. Older children and adults seem to have a variety of strategies for making their memories readily available to them when they are cued into the need for doing so. They can use their familiarity with clock and calendar time, for example, to pinpoint a particular event in the past. Rehearsing and grouping strategies (*see* Chapter 9) are also more common among older children and adults. The preschool child does not seem to have these strategies.

Research that is closer to the real child in real world has made it clear that young boys and girls can indeed show accurate episodic memories. But the situation must mean something to them; it must have personal relevance. This has been one of the major themes of research and theory pursued by a number of Russian developmental psychologists such as Yendovitskaya (1971). The active involvement of the child in some situation of interest to him or her is much more likely to result in accurate memories later than any set instruction or request to learn and recall. The first deliberate attempts to remember in young children seem to occur amidst their own meaningful activities (Smirnov & Zinchenko, 1969).

The semantic memory system in early childhood is limited by a number of factors, including either a lack of concern or a lack of grasp for formal time relationships. Older children and adults can organize experience and thus retrieve it more efficiently because they are able to retain the original sequence of events. The young child often will not retell a story in a manner that seems logical to the adult listener. The child emphasizes those facets of the story that caught his or her fancy for some reason. The temporal structure either does not seem as important, or it has not been mastered at the young child's existing level of cognitive functioning.

With advancing maturation and experience, the child will have a greater resourcefulness in organizing his or her memory. What the adult world seems to expect in the way of accuracy and sequence will be better understood and the child will be more able to meet these expectations. Memory, like other features of mental functioning at this time, tends to bear the mark of egocentricity. This does not necessarily mean that the child will excel in self-generated memories, however. Preschool children have difficulty in recalling items that they themselves have prepared to learn and recall (Tenney, 1973).

Memory functioning in preschool children is enhanced when the material is conveyed by an external organizing structure. Songs, poems, and stories with readily comprehensible structures transmit information that the child can not only receive but hold onto and retrieve. Preschool children retain a grip on such meaningful material even when the experimenters place other material between the learning and recall conditions (Brown & Murphy, 1975). It has been argued that external organizing structures such as songs have served as vital memory aides for preliterate societies as well as young children of our own time (Colby & Cole, 1973). Certainly, the barrage of singing commercials and repeated buy-words in the advertising media indicate that the memories of both young and old continue to be exposed to external organizers whether or not we seek this influence. Perhaps it is time for vigorous research on the ability of young minds to *resist* memory-shaping influences from the outside.

Why do adults have difficulty recalling their childhoods? The richness of current research into early memory development has been lightly sampled here. I have touched on some of the questions that occupy experimentalists. There is a very different question, however, that, for all its obviousness, has seldom been asked: Why do we, as adults, have such a difficult time recalling our early childhood experiences? Freud (1938) identified this question clearly and proposed an answer in terms of socially encouraged repression of infantile sexuality. He explained that around three or four years of age the child starts to respond to parental and societal pressure for organizing his or her thoughts and feelings along conventional lines. Many of the earliest impulses and experiences are forced out of awareness; there is no place for them in the mind of the "good" boy or girl. Whatever one makes of this theory, it does address a question that is still rather neglected, and it does suggest that some kind of unremembering process might be set into motion. It is interesting to think that at the same time traditional researchers are finding evidence for growth of memory in the young child, there might be a coexisting process headed in the opposite direction. This raises many theoretical questions and, perhaps, requires an even greater respect for the child's intellectual task.

Childhood amnesia is the adult's difficulty in recalling early childhood experiences

A stimulating contribution to the problem of **childhood amnesia** was made by Ernest G. Schactel (1959). Like Freud, he was struck by the fact that it is normal (in the sense of common) for adults to come up nearly empty when seeking early memories. His core explanation is:

> *The categories (or schemata) of adult memory are not suitable receptacles for early childhood experiences and therefore not fit to preserve these experiences and enable their recall. The functional capacity of the conscious, adult memory is usually limited to those types of experience which the adult is consciously aware of and is capable of having.* (p. 284).

Western civilization has no use for the experiential world of the child, according to Schactel. It is not just a question of sexually oriented experiences; openness to pleasure and joy conflicts with our society's need for work-a-day adults who will be satisfied by doing their jobs and child-rearing functions well. With few exceptions, over the years, the child's actual fresh joys and experiences are transformed into barren cliches. "Experience increasingly assumes the form of the cliche under which it will be recalled because this cliche is what conventionally is remembered by others" (p. 288).

Locke's "blank tablet" may become self-erasing because of socialization pressures

Schactel's provocative view is worth reading in its entirety. Of particular developmental interest may be the implication that memory does not so much develop after childhood, but shrivel. This is an extreme statement, and Schactel does not make it as such. But he does make a case for the possibility that our minds become so tight and rigid as a consequence of socialization pressure that we become less capable both of present pleasure and pleasure remembered. Schactel's clinical approach is not that of the experimental researcher, although he respected and utilized the research available at the time he discussed childhood amnesia. Because his orientation is somewhat different from most active researchers it has tended to suffer neglect. It merits consideration, however, not just for its specific hypotheses, but for the general note of caution it sounds about development. It is easy to assume that *all* change in mental and emotional development from early childhood on represents a developmental progression, an improvement of some sort. Perhaps, though, some of our important human qualities and potentialities are constricted and arrested relatively early in life. Sophisticated research that would combine experimental and clinical methodology might yet shed significant light on this kind of question.

PERSPECTIVE TAKING: SEEING THE WORLD AS OTHERS DO

Infants do not passively receive whatever stimuli the environment happens to send their way. The youngest humans already have an active orientation toward the world, scanning the environment with certain types of strategy, expressing more interest in some stimuli than others. Throughout that early period of development that Piaget calls the sensorimotor stage, there is rapidly increasing evidence that the young child interacts systematically with the environment. The child has started constructing reality.

But *whose* reality is under construction? Most indications are that the young child constructs the world *egocentrically*. It is achievement enough to grasp the world in at least some of its features with such a new mind. Being able to see things as others do is an achievement of additional magnitude. Piaget is not alone in regarding the ability to take the perspective of others as an important developmental advance. Accordingly, there is much interest in learning when and how children come to the point in mental development when they can realize that other people have other frames of reference. You can see that this is both a cognitive and an interpersonal matter; the two cannot be entirely separated. Those concerned with general theories of early development have another reason to be interested. Piaget's theory and some research have suggested that the child does not transcend egocentrism until reaching the *stage of concrete operations*. The beginning of this stage is usually dated around age seven. But if it could be shown that much younger children were capable of perspective taking, then alternative theories might find a better reception.

Some well-designed studies have found that children under seven years of age are not able to express awareness of what another person with a different perspective would know (e.g., Flavell et al., 1968). Younger children did not seem to realize that somebody else might have a different frame of reference from their own. However, more recent research has come up with different conclusions.

One recent study involved middle-class children between the ages of two and six. Mossler, Marvin, and Greenberg (1976) showed the children two short videotaped stories and asked them to answer several questions about what their *mothers* knew about the stories. In the "Cookie Story," for example, a five-year-old child was shown sitting at a kitchen table. The child in the story then stood up and walked over to his or her mother. (Different versions were shown depending upon the sex of the child who was participating in the experiment at the moment.) The child in the story asked, "Mommy, can I have a cookie?" "Sure," said the mother. The experimenter spoke with the child afterward to make sure he or she understood and remembered the information in the story. All this time, the child's mother had been out of the room. Now she came back in, and they all watched the story together—but without the audio portion. The child was then asked what mother knew and did not know from having seen but not heard the story that the child had both seen and heard. If the child clearly indicated that his or her mother did *not* know what the child in the story had wanted, this was taken as a *non*egocentric response. Mother had a different frame of reference from child because only the latter had heard the crucial information. A similar procedure was followed for the other videotape story.

This study had two major findings. It was confirmed, as expected, that with advancing age there were more interpretations that were free of egocentrism. This general finding was consistent with just about everybody's theory and research. More interestingly, perhaps, it was also found that most of the four and five year olds and all of the six year olds gave *non*egocentric answers with respect to both videotape stories. The older children of those who gave nonegocentric answers were better able to explain that their mothers could not know what they knew (i.e., that they had different perspectives).

The same investigators found similar results with a different experimental procedure (Marvin, Greenberg & Mossler, 1977). It is worth knowing this study as well because it helps bridge the distance between the cognitive and interpersonal aspects of perspective taking. Again the study involved mother, child, and experimenter. All sat in a circle on the floor. There were three versions of what came next. In the first version, the experimenter closed his eyes so he couldn't see which toy mother and child would select to be their "secret." The child was then asked three questions: Do you know the secret? Does Mommy know the secret? Does the experimenter know the secret? The experimenter then said he would try to guess the secret; he chose one of the toys; and he asked if he was right or wrong. In the other versions of this experimental game, the mother and the child each hid their eyes while the other two made their secret choice.

The researchers noted that to perform correctly, the child had to observe which of the other participants did and did not hide their eyes and from this *infer* who did and did not know the secret. *Inference!* An important term here. Just seeing that the experimenter had closed his eyes did not directly demonstrate that he did not know the secret. In other words the observation, the perception that the child makes, was not enough to come to the correct conclusion; an act of inference was necessary. The fact that children as young as four years of age could give *non*egocentric responses suggested strongly that inferential thinking occurs at this early time of life.

There are two interesting inferences that the researchers themselves made. First, they suggested that whether or not such young children give evidence of perspective taking probably depends a great deal on the familiarity and security of the experimental setting and the essential simplicity or complexity of the task. The studies conducted by this group were in the home setting and were designed to minimize complexities that might distract or confuse the child. It is possible to go even further. In the real life situation of the child, there may be the *potential* to engage in inferential thought and to recognize that other people have different perspectives. However, this potential can be actualized only under very favorable conditions. This means that one observer might see inferential thinking while another does not because of a variety of circumstantial factors.

Secondly, Marvin et al. made an inference concerning the interpersonal significance of perspective taking. They noted that just about the time children show perspective-taking abilities, they also begin to play with each other in more stable and differentiated ways. "The nature of a child's interaction within a group reflects his ability to conceive of that group in terms of multiple, and often differing perspectives" (Marvin, Greenberg & Mossler, 1977, p. 513).

Studies such as these remind us again that the particular method of inquiry is closely linked to the results found. Some methods do not find perspective-taking orientations until about age seven, while other methods find this orientation earlier. This is one of the reasons why sophisticated developmentalists are not quick to emphasize specific age-relationships and are always on the lookout for different research approaches that might either validate or modify existing knowledge.

THE REPRESENTATIONAL STAGE OF MENTAL DEVELOPMENT

All areas of mental development show continued growth during the preschool years. I have explored a few of them here. Let's return now to Piaget for a broad conception of what is taking place at this time and how it relates to the overall pattern of development.

From infancy through early childhood the individual functions at a sensorimotor stage of mental activity, according to Piaget. He or she moves from one to another of six substages (*see* Chapter 6). By the end of the final substage, the child is showing much more versatility and sophistication in his or her interaction with the environment. There is more sense of purpose, of recognizing that actions have consequences (e.g., the dropped spoon falls to the floor—make a note of that Mr. Newton, please; smile and you will be smiled back to). The child knows his or her way around the house (e.g., this door leads to that room; look for sister over there if she is not here). Throughout the entire sensorimotor period, the very young child is developing a basic sense of stability, a sense that there is a world around him or her that can be depended upon to behave in certain ways.

The **representational stage** is Piaget's second stage of mental development

Around age two there is a transition to a more advanced stage of development, as Piaget sees it. This is known as the **representational stage** or the preoperational stage. It is during this stage that true symbolic functions, such as language, emerge. Although the sensorimotor infant and child have learned to use various signals and signs from the environment, this is the first time that the individual himself or herself can produce and express symbols. Language develops rapidly and play, as we will see, takes on more symbolic qualities.

One of the interesting new characteristics during this stage is the ability for new behaviors to show themselves after a considerable period of time has passed from exposure to the stimulus. In what is sometimes called *deferred imitation,* the child observes an action and hours later does it himself or herself. Performances of this type support the idea that the child has a more advanced mode of preserving experiences through time and retrieving them for later use than is generally believed.

The new ability to create and utilize internal representations does much to free the preschool child from the immediate environment. He or she now has mental processes that allow a larger psychological distance between self and the world. He or she can, in some sense, think of objects that are absent. This is the basis for later development of a far-ranging memory system and the ability to project thought into the future.

Some interesting research underlines the importance of continued intellectual development during the representational stage (Inhelder & Sinclair, 1967). Children of four and five years were presented with ten sticks of various length, arranged in a series from shortest to longest. A week later they were asked to arrange the same sticks the way they had seen them before. The children did not re-create the entire series graded in length; instead they tended to pair a long stick with a short one or divide them all into two or three groups. But the really interesting part came later. More than six months after this experiment the children were given the same memory task, and this time, most of them came much closer to reproducing the original (correct) series of stick lengths. This suggests that the growth of mental prowess can influence the reconstruction of past reality as well as the comprehension of the immediate situation. We might have expected memory to become less accurate as time went by. But apparently the children had somehow stored the ac-

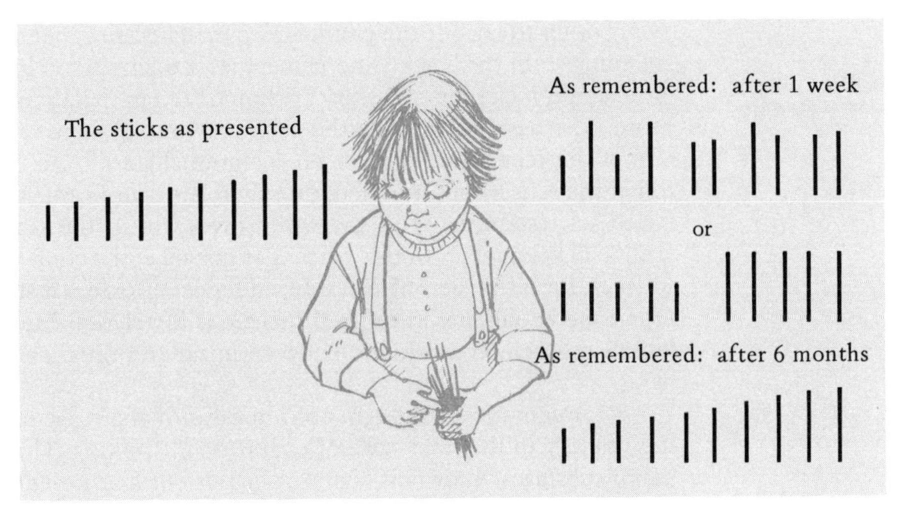

Figure 8–2. Sticking in the memory. *The following are typical responses found by Inhelder & Sinclair (1969). These responses suggest that memory improves as time goes by.*

tual sequence of the sticks even though they could not reproduce this grouping until they were in the possession of more adequate mental processes.

The child's ability to comprehend numbers provides another illustration of the growth of thought at this time. Piaget (1952) and his colleagues did such simple experiments as arranging six pennies equally spaced on a table in front of a child. The child was asked to match this array with pennies that were stacked up. Four year olds had trouble with this task. Some matched the length of the row but paid no attention to the spaces between the pennies. These children wound up with either too many or too few coins in their rows. Others focused on the spacing and ignored the length of the row; they usually continued until they either ran out of pennies or space on the table. This suggested to Piaget that the length and density of the rows had been locked into a global perception. The child at this stage had only a *figurative* or global concept of number and could not separate and coordinate the two separate dimensions of length and density.

Wait until he or she learns 1 = one = I = uno = ein!

The five year olds were a good deal more accurate in matching the row of pennies. But they used a lot of trial and error in doing so and had difficulty in explaining what they had done and why. Even the child who had successfully matched six pennies to six pennies would often say that the longer row had more! Piaget believed that this showed an intermediate substage in the development of the number concept. The child at this stage could accurately respond to a model, but did not yet grasp fundamental number properties.

Around the age of six there was a clear shift in behavior on such tasks. The six year old was a confident know-it-all. He or she sometimes did not even bother to lay out the pennies in a matched row, but simply counted out six coins from the stack. The perceptual model was no longer required; the child could make the necessary arithmetical manipulations in his or her own mind. The basis for later mathematical thinking, then, can be seen near the end of the representational or preoperational stage. And we see again, as we did at the sensorimotor period, that there are significant advances in the ability to act resourcefully in the world even within the same general level of mental development. In this particular instance, the child was eventually able to free himself or herself from dependence on external stimuli and models. He or she could combine two dimensions or relationships (in this example, length and density) in his or her own mind and thus grasp the idea of number.

Representational thought continues until about age seven or eight when it gives way to the final stage of *operational* thinking. This, in turn, has two major substages, known as *concrete operations* and *formal operations.* These will be described in later chapters. The child between the approximate ages of two and seven is *pre*operational in the sense that he or she does not yet have the ability to perform many of the cognitive manipulations that normal adults have available to them. The preschool child can represent and work on his or her experience of the world inside his or her own head to some extent. Increased freedom and flexibility of thought lies ahead, however.

CHILD'S PLAY

Much of the preschool child's time is spent in one form of play or another. What are we to make of play? It is natural to think of play as fun, recreation, a light-hearted activity. Yet many scholars take play very seriously. There is room for both interpretations. It is obvious that the child is having fun soaring back and forth on the swing or scrambling up the monkey bars at the playground. Her face is aglow with pleasure as she dashes out of concealment and makes it to home base ahead of the tag in hide-and-seek. He cracks up in laughter at his own imitation of a television personality (or a mommy or daddy). The child does have fun playing, and we should not become so serious that we forget this obvious fact.

TAKING PLAY SERIOUSLY

There are several viewpoints that consider play to be a serious and significant form of behavior. Some observers believe that the frolicking of young animals prepares them for survival-oriented behavior and that boys and girls, too, develop skills and maneuvers in play that prove useful in adapting to life. If you are a dedicated inside-out theorist, you might maintain that we have play instincts which serve the biological purpose of honing our survival functioning.

Is play a form of practice for the responsibilities of adult life? This is one theory of this complex behavior. (Photo by Talbot Lovering.)

And if you are an outside-inner, you might argue that play is one of the processes by which the young child incorporates into himself or herself some of the opportunities and expectancies provided by society (e.g., playing house, school, or soldier). Either way, play is seen as significant and acceptable behavior because it has functional value beyond itself. This might be called the *play-as-practice view.* It is all right for the child to play even if he or she does have fun! Sooner or later, the play experiences will "pay off" in adaptive, socially viable behavior.

Another serious view of play might be called the excess energy or *letting-off-steam theory.* This position means pretty much what it says. One could slip an instinct theory beneath it and propose that energies that eventually will flow through functional channels (such as working and mating) need safety valves to escape, especially in childhood. Excitement does not have a socially established pathway for release in early childhood, so the youngster may just run, leap, shout, and hurl about. Young children certainly do run, leap, shout, and hurl themselves about when excited, so this view has some support. According to this view, fun and pleasure are acceptable outcomes, but the essential purpose of play is to discharge excess energy.

The letting-off-steam theory has its limits, however. It does not explain precisely why the child discharges tension in one way instead of another. Play that is quiet and not very physical is barely touched by this view. Furthermore, only a rather simple relationship is assumed between the child's play and the psychological world in which he or she lives: inner or outer stimulation excites the child, and he or she discharges the tensions pleasurably through vigorous play.

The *psychoanalytic approach* to play offers a more subtle and far-reaching view. Yes, a child can learn skills and discharge excitation through play. But play is much more than that for the young child. It is a sort of language, a way for him or her to express thoughts and feelings naturally, spontaneously. It is as natural for a child to play needs, fears, and triumphs as it is for an articulate adult to talk them. Even more broadly, perhaps, play can be regarded as a mode of relating oneself to the world. Play is not just something that young children do; it is *how* they do many things.

This approach has already had many practical applications. In the assessment and treatment of emotional problems, the young child is often provided with a play situation. An experienced clinician (psychologist, psychiatric social worker, psychiatrist, etc.) "reads" the nature of the child's problem from the play behavior. This technique is often used for research purposes as well, to understand aspects of children's thoughts, feelings, and social development apart from any particular emotional problems. You might enjoy reading about the play technique in clinical and research use (e.g., Axline, 1947), and some of its special applications in the analysis of children's drawings (e.g., Kellogg & O'Dell, 1969). Some parents and educators—and the people who concoct merchandising programs to sell products in the preschool child "market"—have also taken pages from the psychoanalytic approach to play. It is important to distinguish here between the fundamental concept and the various uses to which it might be applied. From Freud onward (including his brilliant daughter, Anna Freud, 1959), psychoanalysts have been developing a view of early childhood that regards play as a vital and meaningful activity. This does not guarantee, however, that a particular interpretation of what a child's play "really means" will always be correct, or that people will always bring pure and altruistic motives to the study of play.

Play-work is the child's use of play to master problems and conflicts

Without forgetting that play is also fun, it can be said, from the psychoanalytic view, that play can be work, too. Through **play-work** (to coin a phrase), the young child tries to cope with the new, the challenging, the alarming. Some of the problems that a particular young child faces may be distinctive to his or her own family situation. Yet it is also likely that a child of a certain age will encounter challenges that many other children have also experienced at this time. Because a range of problems and experiences have been identified as common for young children, the observer can begin with some hunches, some clues to understanding what is taking place in the play-work of a particular child. Let's take one example from the many that can be found in the psychoanalytic literature on young children. This excerpt is from a case reported by a psychiatrist, Thomas A. Petty (1953):

> Jimmy was two and a half years old when his mother gave birth to his brother. His parents had prepared him for the event with simple direct explanations of the birth process and what it might mean to him. Just prior to the delivery a favorite game was dropping stones into a small creek while standing in the middle of a bridge that spanned it. At the time the following play was observed, he was four and a half, and the younger brother had only a minute earlier almost fallen into the fire in the living room fireplace. The play consisted of Jimmy's vigorously sweeping and batting a plastic balloon doll of Humpty Dumpty around the floor with an andiron broom. When asked who

> the doll was, he could not think of the name and identified it by saying, "He sits on the mantle. . . ." A few moments later he was balancing the doll on the fire screen and deftly caught it as it was about to fall into the fireplace. As his mother came into the room, both he and his younger brother scrambled with the dog for a position in front of her, and in the jockeying he [Jimmy] bit a hole in Humpty Dumpty, thus deflating the balloon instantaneously. (pp. 405–406).

Petty interpreted Humpty Dumpty as an effigy or stand-in for the younger brother who competed with Jimmy for the mother's attention. Jimmy's play expressed chiefly hostile feelings in this episode, but "the rescue of the doll as it fell toward the fire may be considered an attempt to master the intense feelings the older child felt in relation to the younger; and the bite that deflated Humpty Dumpty a manifestation of the failure to master those feelings under the stress of competition" (p. 407).

Sibling rivalry describes the tension and conflict among children in a family as they vie for love, status, etc.

In this example of play-work, then, a preschool boy tried to express and master the **sibling rivalry** experience (certainly a common situation when an only child finds himself partially displaced from parental attention). One can emphasize the hostility, or one can notice the attempt at rescue and mastery. Either way, the play situation serves as a sort of dramatic production in which mixed feelings can be expressed. It also illustrates the close relationship that can often be discovered between child's play and significant events within the family circle.

Interestingly, Petty reported play observations of two other children whose sibling rivalry also seemed to express itself in Humpty Dumpty dynamics. Impressed by the fact that Humpty-Dumpty-type songs were known to young children in many nations, Petty suggested "a resolution to the enigma of the Humpty Dumpty rhyme and . . . [its] popularity. The rhyme itself is a simple dramatization of the second catastrophic trauma a child experiences (if birth is the first), viz. [namely], the arrival of a sibling." In other words, "H. D." is a symbol of the painful loss of love and power. As an only child, the firstborn is perched high in everybody's affections, including his or her own. Nothing can hurt him or her. But this egg of security is dashed to the ground by the birth of another child. Things will never be the same again. All the king's horses and all the king's men cannot put H. D.'s feelings of undivided love and power together again, for the new baby cannot be returned from whence it came.

Petty's interpretation of Humpty Dumpty can be challenged, of course. It is by no means easy to prove or disprove theories such as this which depend on the close perception and experienced interpretation of child behavior as well as of literary and other cultural materials. But the example serves us well enough as an example. We see the possibility that certain themes of conflict and loss in early childhood may be common enough to show up in the songs, stories, and games that one generation passes onto the next. This suggests that a careful study of children's games (using this term in its broadest sense) would reveal more than a collection of activities that children just happen to invent or discover. Instead, we might expect to find that actual historical patterns and continuities exist, that the street and playground games of today have something important in common with the games played by children over the decades, and even over the centuries.

In fact, scholars are beginning to demonstrate that children's play does have much in common from generation to generation and from place to place. Humpty Dumpty is but one example of many. You can find hundreds of examples in the songs and games collected by Iona and Peter Opie (1969). The Opies concentrate on the accurate description of children's activities, leaving most of the interpretations to those who choose to make them. The fact that certain types of play have remained popular over the years perhaps supports the psychoanalytic principle that children express emotionally significant concerns through play. Even if one turns away from the psychoanalytic approach, the evidence suggests that child's play does have structure and continuity; it is not all made up on the spur of the moment.

The Opies declare that children at the entrance to the British Museum

> secretly playing "Fivestones" behind one of the columns as they wait to go in, little think that their pursuits may be as great antiquities as the exhibits they have been brought to see. Yet, in their everyday games, when they draw straws to see who shall take a disliked role, they show how Chaucer's Canterbury pilgrims determined which of them should tell the first tale. . . . When they jump on a player's back, and make him guess which finger they hold up ("Husky-bum, Finger or Thumb?") they perpetuate an amusement of ancient Rome. When they hit a player from behind, in the game, "Stroke the Baby," and challenge him to name who did it, they unwittingly illuminate a passage in the life of Our Lord. And when they enter the British Museum they can see Eros, clearly depicted on a vase of 400 B.C., playing the game they have just been told to abandon. (p. 6).

Hide-and-seek, tug-of-war, and blindman's buff are among the games that children believe they have discovered for themselves today, but which were played two thousand years ago on the streets of Rome. The Opies suggest that if a child of today "wafted back to any previous century he would probably find himself more at home with the games being played than with any other social custom" (p. 7). Deposited in the Middle Ages, for example, a boy or girl of the 1970s could play street football or marbles, to name just two games popular at that time. Specific rules and customs of the games have remained intact throughout the centuries. A blindfolded player, for example, is turned around three times before being allowed to chase, and the seeker is told that he or she is "warm" or "cold" in looking for a hidden object—techniques that have been passed on through many generations. There is a formidable tradition in the games delighted in by children in many areas of the world, just one more reason why child's play is a subject deserving our attention and respect.

Could store-bought and adult-supervised games destroy this ancient continuity between generations of children?

Let us waft back now to the psychoanalytic view of play. We have already seen that play is regarded as a sort of natural language by which the child expresses his or her feelings and concerns. This language often involves symbolism as, for example, when a Humpty Dumpty doll or rhyme substitutes for baby brother. Play also has the advantage of being repeatable. The child can carry out his or her play-work time and again. In the real world an event may occur only once (e.g., being scared when a friendly looking dog growls or snaps). In the play world, however, the event can be made to happen over and over again, with the child changing the script whenever he or she finds this to his or her advantage. In play *he* or *she* can bite the dog, be a dentist who removes the dog's teeth or, if the child chooses, even be the dog and terrorize some other luckless creature. Just the opportunity to repeat an event many times can be very helpful. It can use up or neutralize the emotional tension that was generated by the original reality situation. Play, then, can be regarded as a way for the child to master challenges and anxieties that arise in everyday life. This does not mean that *all* child's play is stimulated by problems and conflicts. Rather, it means that play offers the boy or girl a relatively safe and controllable theater in which either serious or lighthearted productions can be staged.

PLAYING AND THINKING

Play, like most other human activities, reflects the level and type of thought processes that are at the disposal of the participants. A three year old rolling a beach ball and an eighteen-year-old quarterback releasing a pass in game competition are both playing ball. They differ not only in physical skills but also in their thoughts and strategies. Let's consider a few of the relationships between play and thought in early childhood.

During the later phases of the sensorimotor stage (Piaget), the child becomes able to expand the focus of attention and competence beyond his or her own body into the space that surrounds him or her and the objects and people in that space. He or she can learn simple strategies and apply them to new situations. "For example, after having attained a distant object by pulling the blanket on which it had been placed, the child will generalize this discovery into using many other aids to draw closer many other objects in various situations" (Piaget, 1973, p. 114).

Different styles and qualities of play can be observed as the young child moves from the sensorimotor stage to the various progressions of the representational stage that has been described in this chapter. The very young child seems to enjoy *explorative play.* Hands, feet, or head come into contact with something that is fascinating or amusing. It is worth doing again (e.g., pulling an uncle's beard, or dropping something from a height). Explorative play, involving repetitions and variations, continues well beyond the sensorimotor stage. But it is the dominant style of play at this early period, reflecting the growing edge of the child's intelligence. As adults, we sometimes speak of "getting in touch" with another person or being "in touch with reality." For the young child, playful touching and manipulation are among the most basic ways of becoming acquainted with the world.

We should, in fact, go beyond what has already been said about the basic relationship between playing and thinking. The child's style of play does more than reflect the existing level of thought; it is probable that play also sharpens and develops thought processes. Piaget and other specialists in mental development believe that play is one of the most important means available for promoting the development of symbolic thought.

By the end of the sensorimotor stage, play and mental development have both increased in complexity. Susanna Millar (1969) reports: "Spatial arrangements often become the object of play at about eighteen months. At that age, one of my children, after an ominous period of quiet, proudly drew my attention to a beautiful arrangement she had produced: all available tomatoes carefully lined up on the doorstep, with a little bite taken out of each to make them stand up better." She adds: "From the time that hand-eye coordination is established the infant's tendency to touch everything seems almost compulsive. It is difficult for the older infant or toddler not to touch and handle everything in spite of being told not to 'fiddle' " (p. 110).

The young boy or girl is able to appreciate and enjoy making an effect on the world or other people. A splash in the water, a dab of paint on the wall, a sound produced on a musical instrument, an exclamation of surprise from a parent or another child—all of these impacts are rewarding to the child and may be the object of his or her play. We can observe a more plan-ful quality in some of the play of children who are moving from the first to the second stage of intellectual development. These children have had enough experience, for example, to know that squirmy worms can be found outside after the rain, and they will go out with the clear purpose of finding these playmates. Pleasure in making an effect or controlling the situation can also be observed in this instance, for example, by presenting the new playmate to a squeamish person who can be counted on for a shriek of panic. (In our society, little girls traditionally are supposed to shudder and turn away from squirmy, squishy, gooey creatures, but I know of no evidence that indicates that young females at first react any differently than their male cohorts. Cultural conditioning of sex-role stereotypes operates so early that styles of play in early childhood may already reflect "lady-like" or "masculine" behavior as much as they do the child's direct perceptions.)

Child's play becomes more varied and even more distinctive from one youngster to the next as the ability to represent and symbolize asserts itself. The world of make believe and "let's pretend" is now available. I recall, for example, a boy who served as unofficial alarm clock for the neighborhood by greeting each morning with lusty exclamations: "Eeeeyah, train! Eeeeyah, train!" The train that generated this enthusiastic outcry was no more than a piece of stick, its track, the bottom edge of the windowsill. A store-bought train was put aside with little interest. The locomotive engineer's heart obviously was in the stick-train.

The *pretend games* of the preoperational stage include people and animals as well as objects. Lacking companionship (or preferring one's own company), it is still possible to play almost anything. This would be out of the question if the child did not have the ability to represent and symbolize. "Now you be the Good Guy and I'll be the Bad Guy." "No, *I* am the baby, and the cat is the babysitter . . . you can be the mommy coming home." The child may also pick up books and "read" them to himself or herself, a pet, or a younger child (an activity I have noticed more in girls than in boys). This feat involves the recognition that those shapes on the pages are symbols, words that represent sounds and meanings, even if the code has not yet been revealed to the child. Enjoyment from asking questions, telling riddles, and playing other kinds of word games also reveals that the child at this level does have some ability to appreciate and manipulate symbols.

PLAYING AND SHARING

The infant who is playing with his or her own body or a nearby object eventually will be capable of complex games that involve the participation of others. There are a number of steps or stages inbetween, although these have not yet been as clearly established as the intellectual phases. Usually, however, the observer will notice that very young children do not really seem to be playing with each other even if they are playing in the same place at the same time. This phenomenon is known as **parallel play.** Two toddlers, for example, may be enjoying toys in the same sandbox. Perhaps they are even doing the same things, such as filling pails with sand and emptying them out. But neither is taking much notice of the other, let alone coordinating their activities. Interactions may arise from time to time. One child sees the other doing something interesting and imitates it, or both reach for the same toy at the same time. By contrast, older children in the same situation are likely to cooperate in small projects. Even if they have an argument and a shoving match, this behavior reflects a greater mutual awareness and range of interactions than what would be shown by a pair of two year olds.

Parallel play is the tendency of very young children to play in close proximity without interaction

Attempts at *cooperative play* often have their ups and downs during the preschool years. Fatigue, limited attention span, insecurity in a new situation, need for parental attention, these are but a few of the problems that can derail a potential extended play situation. Parents themselves can either facilitate or impede cooperative play by their attitudes and the kind of situations they provide. Typical problems associated with this type of play and suggested ways of approaching them have been discussed by a number of child psychologists. Let's sample some of the advice offered by Molly Mason Jones in her book, *Guiding Your Child From Two to Five.* Jones recognizes that property disputes often lead to breakdowns in play and in child-to-child rela-

Cooperative play involves growth in social skills, but also in cognitive structures that enable children to take other people into account more fully. (Photo by Talbot Lovering.)

tionships in general. Adults may attempt to solve the disputes by insisting that the children share or take turns. Although this approach is reasonable, it has some drawbacks as well and can be improved upon:

> As far as sharing goes, at preschool ages this is often sharing in name only. If peace reigns, it is likely to mean monopolization by one of the sharers and acquiesence by the other. Of course, if two children of preschool age spontaneously want to share a piece of play material, well and good. It is often wise to say, though, "That's fine, if you can manage it. But if it doesn't work, then David, you're the one to find something else since Roger got the blocks first."
>
> And as regards taking turns, there are two disadvantages to this technique: First, children are forever dependent upon the grownup to apply it for them. Second, it promotes constant references to the grownup in play: "It's my turn now. Tommy's had it longer than I did," and so forth. In very special circumstances, however, grown-ups dictating turns is sound. When some new activity has great appeal, then taking turns is the fairest way of letting each child take part in what he is yearning to try out, without having to wait a long time because the novelty of the material keeps holding the user's interest. (Jones, 1967, pp. 194–195).

For property disputes in general, Jones believes it is a sounder principle to "let the right to use something depend on who got it first" (pp. 194–195). Children can understand this principle and apply it themselves. The child can be told, "Sally got it first. I'll help you when she is through so that you can get it"; or "If you want to keep on using it, hold on to it. If you leave it, somebody else is likely to get it." Pursuing this topic further than we can here, Jones also suggests some ways for acquainting preschool children with the idea of "right." If successful, these approaches would not only help to smooth out some of the ruffled feelings that can arise in early attempts at co-operative play, but also offer the child useful practice in resolving disputes by words and thought rather than by violent or surrendering behavior.

While on the subject of property disputes, we might remind ourselves that cultural values such as the emphasis on competition and material success reach into the sandbox and playroom. Millar cites several studies which show that even within the preschool range there is an increase in competitive behavior from year to year (Millar, 1969). The need to achieve and accumulate goods becomes a part of the child's play before it becomes a part of his or her school and work life. By contrast, Dorothy Lee (1959) tells us that Hopi children could not be persuaded to compete in games. The boys and girls would derive great enjoyment from new games taught to them, but did not seem to care who "won." Sharing and belonging to a tight-knit group or "people" was already a deeply engrained value to these children. Individual victory over another Hopi and the aggressive competitiveness that leads to winning (or losing) was not part of their way of life. Those who emphasize the influence of the social environment on child development can find provocative information by comparing play styles of young boys and girls who grow up in cultures that have markedly different value orientations, including some subcultures within the United States.

PART OF THE FAMILY: DEVELOPMENT OF SOCIAL RELATIONSHIPS

INDEPENDENCE VERSUS PROTECTION

We have been delving into the preschool child's personal experiences and activities. It is time to bring more of the family dynamics into focus. Although the child at this age is beginning to explore a new world of relationships—to other children, to other living creatures, to the physical environment—the relationship with parents remains by far the most important. For both child and parent, there is the continuing challenge of balancing opportunities for assertion of competency and independence on the one hand with the need for guidance and protection on the other. It is difficult, perhaps impossible, to keep this balance absolutely precise. The child may, at a particular time, seek a freedom that he or she is not quite ready for (e.g., going for a walk or a tricycle ride too far away from home). Or the parents may have become so accustomed to certain developmental limitations of the child that they do not recognize when certain activities are now within his or her scope. Often the parents find themselves on both sides of the fence, feeling the need to hold back certain liberties or privileges the child seeks, while encouraging more "grown-up" behavior in other areas. As reasonable as this approach may be, we can also appreciate how such a mixed message can leave the child somewhat uncertain about his or her independence status. The child may be encouraged to dress himself or herself and clean up his or her area of the house the way adults do, but at the same time he or she may be forbidden to stay up to watch a television program that is on too late for children. This balancing act on the part of both parents and child will be going on for many years, well beyond the preschool scene. Sensitivity and flexibility will continue to be required of the parents if the child is both to feel secure in their affection and licensed to explore his or her rapidly expanding horizons.

Many parents do seem to appreciate the continuing adjustments they must make in the independence versus protection sphere during the pre-

Is it always parent-child? Or is it sometimes person-person? It is a significant development for both parent and child when they come to see each other as independent people. (Photo by Talbot Lovering.)

school years. John and Elizabeth Newson (1968) studied the family life and total "social ecology" of seven hundred four-year-old boys and girls in an English city (Nottingham). Their richly detailed book holds much that will interest both students of developmental psychology and parents. Although they discovered numerous individual differences, there seemed to be a consensus among the parents on the basic way to encourage independence without creating other problems. Whether the parents were among the white-collar or the blue-collar set, they tended to believe that "children at four are old enough to make some contribution towards looking after themselves. The attitude that 'they're still babies till they go to school' . . . is not at all typical" (p. 78). Most parents expressed the belief that it is futile and self-defeating to force the child into a highly emotional state over matters of independence. "The most frequent response to these (independence) questions . . . was in some form an *objection* to the phrase 'Do you think he should be *made* to do things'?" An administrator's wife replied to that question, "Well, *encouraged*, I think, perhaps, instead of *made*. . . . I never actually force him to do anything. But I find that he's very reasonable, and I think it's because I've been fairly—I treat him as an adult, you see, as far as possible." A laborer's wife gave a similar reply, "I don't say *make* them do it. You know, I try to *get* them to do more for themselves. But I wouldn't really *force* them to do it—I mean, that'd be idle-itis on my part. . . . But I like to tell them that they must learn to do things for themselves" (p. 81). By and large, the parents spoke of being patient with their preschool children, making their expectations for responsible behavior known, treating the child as a reasonable and reasoning human being, and rewarding responsible behavior with approval and status.

The pleasures of parenting seldom find their way into the research literature

The Newsons were alert to the pleasure-giving aspects of parenting a young child as well. Many of their respondents "confessed" the enjoyment they sometimes experienced by participating in their children's play, for example: "Say they were playing at cowboys; well, I have to hide behind the settee, and they come riding round it, and I have to (pantomimes dying in agony)—'Oh . . . you've got me!'—all that sort of thing, you know. I act daft in *that* way." Or, "We get all these pillows off the settee here, and we put them in a line all along the floor here, and then we all have to get on and go for a ride. Oh, it's smashing, isn't it, Brian?" Most mothers (about two out of three) welcomed the opportunity to join in the fun of playing with their children on an equal footing, while many of the others would also participate, but on a more "dignified" level, such as helping the child with puzzles and games. Those who observe and write about child psychology often concentrate on the problems. But why lose sight of the pleasures?

LEARNING ABOUT HUMAN RELATIONSHIPS

The child begins to learn about human relationships from his or her own experiences as part of the family. Something is learned about affection, sharing, and constancy, and something is learned about rebuff, competition, and separation. And, as social psychologists emphasize, the child learns in part by modeling himself or herself in the image of the grown-ups in his or her life. It is common and natural for both girls and boys to be influenced by, and model themselves after, both parents. In most families, however, the children are not exposed equally to mother and father. Both boys and girls, for example, are likely to see more of mother than father during the day. Furthermore, mother and father typically have different functions in the family and different styles of interaction with the children. These variations offer to the child some range of behavior that can be modeled selectively. One can try to be daddy-like in some ways and mommy-like in other ways. Most research into early infancy demonstrates the relatively greater significance of the mother who is usually more available to the baby and does more for him or her. This means that most children soon have differential perceptions of mother and father. But the differences take on additional meanings as the preschool child becomes more aware of human relationships.

The girl recognizes that she is destined to be a woman/mommy, and the boy that he is to be a man/daddy. This is an important discovery. It is like having an advance edition of one's self walking around the house. There begins a process of **identification** with the parent of the same sex. Freud and other early psychoanalytic observers tended to regard this as a single, powerful process. Developmental research during the past few years, however, indicates that identification with parents is more usefully regarded as a set of complex processes in which learning opportunities and situational factors are very important (Mischel, 1970). It is only for ease of presentation that I will continue to speak of identification as though it were a unitary process.

Identification refers to the process by which a child tries to become like an admired person

Parents often encourage same-sex identification. Boys and girls are provided with different kinds of clothes to wear, given different kinds of toys to play with, different kinds of chores to perform. Seldom is the identification with the same-sex parent complete and exclusive, however. The children will continue to learn from, or model themselves after the life-styles and behaviors of both parents (and whatever other influential grown-ups may be in their lives).

The family romance. It will not escape the child's notice that mother and father have a special relationship. In effect, they "have" or "belong" to each other. The child knows that he or she also "belongs" to the parents. But the relationship between mother and father is of another kind, even if not completely understood. Through a configuration of developmental processes, the child is often brought to the point of wanting the opposite sex parent all to himself or herself. This is known as the **Oedipal situation** or family romance. Many different accounts have been offered of the dynamics involved (e.g., Freud, 1933/1965; Erikson, 1950; Mullahey, 1948). We will con-

The **Oedipal situation** is the young child's desire to possess the opposite-sex parent

Figure 8-3. *Identification and the family romance. The young child wants to be like the same-sex parent and win the opposite-sex parent.*

centrate on a few basic points here, attempting to avoid entanglement in specialized theories:

1. If the child is to model himself or herself on the same-sex parent, then the job cannot really be completed unless the opposite-sex parent is "conquered."

2. An outsider might object that the whole idea is ridiculous. A little boy or girl cannot have a man-woman relationship with the opposite-sex parent. The child often has doubts about this himself or herself. This may be one of the reasons why the child offers to marry mommy or daddy when he/she grows up. Time-age discrepancies and other reality factors do make the youngster's case look almost hopeless. But the important point is the child's need or desire to have a special and, preferably, exclusive relationship with the opposite-sex parent.

3. The long-range goal of marrying and possessing the parent alternates with efforts to beguile and win the parent right now. The little girl tries to charm daddy to be hers. The little boy tries to take charge of mommy. In this way, both begin to learn and practice some of the social skills they may use at a later time in their lives when they are courting potential mates. And both the boy and girl may be so curious and jealous of what goes on between their parents that their curiosity or possessive impulses break into action (e.g., trying to gain admittance to the parental bedroom, or frequently interrupting conversations).

4. The quest for mommy or daddy has its scary aspects. The boy's side of the problem has had more attention, from Freud onward. There is an awesome obstacle in the path of the lad who wants mother all for himself—his father! The boy is thought to fear massive punishment from father for daring to compete with him. It is in this setting that the famous *castration-anxiety theme* comes to the fore. Some psychoanalytic writers have maintained that the boy of about age five or so fantasizes actually being castrated in punishment for his wish to possess mother and thereby terrifies himself. Other writers put a more general interpretation on the situation: the boy is afraid of being put in his place or devalued and rejected. According to this latter view, castration anxiety is usually a *symbol*. It represents the feared loss of value and worth, especially associated with masculine worth. The child must find some way of coming to terms with both (1) his wish to be like father and possess mother, and (2) his fear of being clobbered by father in one way or another for having this wish. (Wishing for something and actually engaging in public behavior are often not clearly separated in the mind of the child; on this point there seems to be general agreement.)

5. The situation is even more complex. The child does not really want to be a grown-up—at least, not all at once, not now. He wants to remain "daddy's boy," and she wants to remain "mommy's girl" at the same time they aspire to defeat father or mother. Not only do they realize, at some level, that there are overwhelming obstacles in the way of conquering the opposite-sex parents, but they also intuit that the reality, if it did come to pass, would be far beyond their capacities to manage. The "romance" is significant; but it is important as a core of thought and feeling to grow around, not as an actual goal to be achieved.

6. The child brings to this situation a growing sensitivity to sexual pleasure. Curiosity about the human body is common at this time, some of which may be expressed in undressing or peeking games. The child has always drawn certain feelings of comfort or pleasure from bodily states. Very early in life, these bodily states were diffuse, spread out. Later they became focused around the mouth area. Now, in the preschool range, the child already has had the experience of learning bowel and bladder control and paying attention to body functions associated with elimination. It is natural to "discover" the sexual organs (not that they were ever completely lost) and to be stimulated by their presence and puzzled about "what they're good for." Although a long way from being a sexually mature person, the child does have enough of a hint to bring sexually tinged feelings into the family romance.

These family romance dynamics obviously have the potential for all sorts of trouble. Parents who are immature or insecure may overly encourage the boy-for-mommy, or girl-for-daddy, maneuvers. They may seek gratification from their children that should more appropriately come from their spouse. Other parents may be so upset either by the competition or its sexual overtones that they squelch and reject the child. And, as psychoanalytic writers are fond to point out, there is nothing like a new crisis to reawaken an old one. In other words, the oedipal problems being experienced by the young child may impinge upon those once experienced by the parents themselves and contribute to anxieties and misunderstandings. The young child who is learning some important first lessons about human relationships at this time is in need of understanding and acceptance from both mother and father, not seduction or threats.

In a living and relaxed family, the oedipal dynamics can be kept in perspective. The child is not placed under intense pressure to renounce the discoveries and curiosities that fascinate him or her, nor is he or she encouraged to move prematurely into a more grown-up sphere of functioning. Some people, indeed, are haunted by unresolved problems associated with the family romance, and others never free themselves from early forms of attachment. But we are not likely to improve matters either by ignoring the complexities of early parent-child relationships or by indulging in extravagant speculations that conjur sex and neurosis everywhere we look.

DAY CARE

In middle-class United States culture, we have become accustomed to the idea of parents being on the scene with their young children both day and night. This has often amounted to the mother's responsibility, especially during the day. In fact, much of the context for observing family romance and early identification dynamics, already described, assumes a traditional pattern of parent availability. Many of the studies and theories that have come forth in child psychology over the decades assume a configuraton in which there is both a mother and a father in the house, and the mother stays there during the day. This is still the way of life for many families. Alternative life-styles are now fairly common, however, and we will probably see additional voluntary or involuntary experiments in family living as time goes on. Here I will touch on just one development, the burgeoning of day care alternatives to maternal presence.

SOME CIRCUMSTANCES LEADING TO DAY CARE

The notion, "A woman's place is in the home—period," has many advocates, especially when young children are involved. People tend to have strong feelings on this subject, pro or con, and discussions have a way of rising into the upper decibel range. There is a danger that individual differences will be swept away or lost sight of in emotionally fueled generalizatons. In fact, there are many reasons for a child to be introduced to a day care situation, and many types of day care situations. The same logic or psycho-logic does not necessarily apply to all circumstances. Consider a few of the reasons why the day care alternative is used:

1. A woman *must* earn money to support herself and her child. There is no other adequate or acceptable source of income available.
2. A woman is *willing* to work in service of a goal that is important to her.
3. A woman *prefers* working to the alternative of staying home with the child or children.
4. The *father* is the only available parent, and he must work to provide income.
5. There is something about *the child himself or herself* that makes day care seem to be the preferred alternative.

This is only a partial list of circumstances that may lead to day care placement of a preschool child. Let's examine a few of them in more detail.

Some women have the responsibility for raising one or more children without the socio-emotional or financial help of others. Perhaps you know a widow or divorcee in such a situation, or a never-married mother who resolved to keep her baby. Any of these people might prefer to be a classic stay-at-home parent, but reality is not in their favor. Another woman may be working temporarily to help her husband complete his education (and perhaps waiting for the opportunity to complete hers). Still another woman may have skills that are in considerable demand and may believe that she has an obligation to fulfill (e.g., a nurse or physician). Or this might be a woman who gets depressed around the house and feels she has a better attitude toward herself, her husband, and her child if she can be "out in the world doing something" at least part of the time. And sometimes, of course, the woman is a man. Some people who disapprove of a woman stepping out of the home when it means day care for a child apply a double standard; it is all right if it is the man who "has" to work or prefers to work. Yet the realities, the motivations, and even the morality may be similar whether it is the woman or the man who must make the decision. Finally, there are a variety of circumstances in which people might reasonably believe day care is the alternative of *choice,* not just of necessity. A child with a particular developmental disorder may benefit from participation in an appropriate day care program; another child may simply need more opportunity to interact with other boys and girls in order to be happy and to cultivate social skills, etc. As individuals, we can look judgmentally on decisions to place a child into day care, but it might be wise to also look at the precise facts in each situation.

Many a Ph.D. has been supported by a Ph.T. (Putting Husband Through); now more men are doing the same

Many young children now have day care experiences as the pattern of family life continues to change. (Courtesy of Fisher-Price Toys.)

SOME PROBLEMS ASSOCIATED WITH DAY CARE

What are the problems associated with day care? Here, again, I will sample just a few of the more significant possibilities:

1. One or both parents *feel* they are doing the wrong thing in putting the child into day care (or they have a strong disagreement among themselves). This orientation may generate bad vibrations that influence the child and heighten domestic tension.
2. Inadequate attention is given to selecting—and checking up on—the day care service. The dangers here range from exposing the child to rules and ideas that differ appreciably from parental values (and will cause friction as a result) to exposing the child to health and safety risks because of inadequate facilities or supervision.
3. The child is truly not wanted by one or both parents. In this instance, day care arrangements are only symptoms of an underlying problem. The particular day care experience may either help or worsen the child's situation. A rejected child might possibly have some constructive experiences in a sensitive day care setting.

As time goes on, we will know more about the effects of day care experiences of various types on various types of preschool children. At present, much attention is still being focused on sorting out the results of Headstart and other experiments in preparing young children for school. We must learn about the developmental implications of the total range of day care services that have come into existence, including the informal as well as the formal, the volunteer as well as the professional.

EFFECTS OF DAY CARE

Let's consider one recent study as an example of what has been learned and what remains to be learned about the effects of day care experiences. Two groups of children in the three-to-four-year-old range were involved. One set of youngsters had participated in a high-quality infant day-care program from about their ninth month. The other set consisted of boys and girls who were entering their first day care program at the age of three or four. All were enrolled in a new day care program at a major university, in which: "The program was characterized by an absence of structured activities, and . . . designed to facilitate child autonomy in using freely available diverse materials and to encourage social interaction between children. The adult-child staff ratio was maintained at 1 to 7. Since there were no assignments of children to specific areas or teachers, each teacher had ample opportunity to observe each child enrolled" (Schwarz, Strickland & Krolick, 1974, p. 503). Children in both groups were carefully matched with each other for purposes of comparison (this does not mean that they were brought into actual contact with each other, but that information about children with similar characteristics but different placement experiences was analyzed in a pair-by-pair fashion).

It was found that the children who had received day care since infancy were more aggressive and motorically active (e.g., running around, using their bodies) than the children who had been cared for at home previously. The "old-timers" in day care were also less cooperative with adults. The authors concluded: "Consideration of these results in the light of other findings with the same and other samples suggests that early day-care experience may not adversely affect adjustment with peers but may slow acquisition of some adult cultural values" (p. 506). The point for our purposes here is that differences in behavior and interaction were indeed found between young children with different histories of experience with day care. (The "newcomers" did *not* become similar to the "old-timers" after some months of participation in the day care center, suggesting that the *timing* of the experience in the child's developmental career may be quite important). Additional research should help to clarify the entire pattern of differences to be expected in later behavior when a child is given various types of day care experiences. Plato, Watson, Skinner, and some of the other theorists discussed in Chapter 7 would not have been surprised to learn that variations in environment and care procedures can lead to observable and lingering differences in behavior. The day care movement is both a response to the needs and challenges of our present generation and a testing ground for views of human nature and nurture that have been with us for centuries.

We have been discussing the question of removing the child from the home for part of the day. This is not the only form of separation that may be experienced by the young boy or girl. Parents go away on trips, or divorce each other. The child or the parent can be hospitalized. Separations of many types are experienced by young children, and the effects can at times be quite significant. I will now examine one of the most extreme forms of separation in some detail.

HAZARD: A PARENT DIES

The fact that a young child's father and mother are alive does not guarantee an ideal climate for development. Many other things can be the source of stress and insecurity. Similarly, the absence of one parent does not necessarily mean that the child will be deprived of his or her opportunity for happiness. Many competent adults have been raised by only one parent. Nevertheless, when the parent of a young child dies, there are some hazards which become more prominent. In thinking about the possible effects of parental death, we should not assume that these will be the same in every family, or that every child in the same family will be influenced in an identical way.

It would be easy to oversimplify the effects and meanings of parental bereavement. As a safeguard, let's remind ourselves of the following:

1. Death is not the only cause of separation. Some young children have little or no contact with one of their parents for other reasons.

2. The events surrounding the parent's death should be distinguished from the permanent absence that follows. Did the parent die suddenly? Or did the child see the parent gradually succumb to illness? Was the child present at the funeral? How sensitive were the other survivors to the emotional needs of the child at this time? These are a few of the questions we might ask concerning the child's experience of the parent's death as an *event*. The prolonged *absence* of a mother or father affects the child in a different way. What does *not* happen day after day can be crucial here. There is no parent of the same sex for the little boy or girl to take as model, for example, or no parent of the opposite sex to impress and win over. It is not always possible to determine how death as an event and death as an absence make their separate contributions to the child's developmental process, but the distinction is worth keeping in mind.

Some people remain memory-haunted by a parent's death in childhood; Mark Twain was one

3. A child with both parents alive may have experienced the death of a grandparent, sibling, or somebody else who was quite close. Although this child has a complete set of parents, the developmental history has been marked with one or more "significant" deaths. By contrast, some infants lose their mother or father but soon gain a complete set of parents by remarriage. A person growing up in this situation may have little or no experience of parental loss or absence. In other words, the presence or absence of parental death in a developmental history does not of itself tell the complete story of the child's relationship to bereavement and loss.

4. A child with a history of parental bereavement may show one or more of the problems that will be described below. However, this does not *prove* that parental bereavement caused his or her problem. Some people have similar difficulties although they did not suffer the loss of a parent during childhood. Furthermore, it is possible that factors other than the loss of a parent were critical in leading to unfortunate outcomes.

The death of a parent in early childhood (or at any other time) is a fact to be taken seriously when we are interested in understanding a person's developmental career. But the careful observer will not try to explain everything on the basis of this one fact.

SCOPE OF THE PROBLEM

Four decades ago in the United States, 600 mothers died for every 100,000 live births. This amounted to about 13,000 childbirth-related deaths per year. The total would be up to about 23,000 today, had the mortality rate continued at the same level. Fortunately, public health advances have reduced the maternal mortality rate to fewer than 29 per 100,000. There are now about 1,000 maternal fatalities per year despite the fact that the birth rate is appreciably higher.

The maternal death rate for whites and nonwhites differs markedly. The death rate for whites has dropped to about 20 per 100,000, while the rate for nonwhites is about 70 per 100,000. There has been a large difference between maternal death rates for whites and nonwhites ever since statistics have been collected. The white/nonwhite ratio persists, although both segments of the total population have benefited from sharp decreases in mortality. It is not by choice that I use the term, *nonwhite*. This classification is the one that has been employed for many years in the compilation of census and other population data. It is of limited value because possible differences within the nonwhite category are obscured. There are indications that a more adequate breakdown will be available in the future.

These figures suggest that the number of children who lose a parent by death depends upon a variety of conditions. The general level of public health care at a particular time is one of the major conditions; another is discriminatory practices that place some citizens (even the neonate) in a state of exceptional vulnerability. Geography can also be cited in this connection. In the United States, infants born in southern states have between two and three times greater risk of losing their mother in childbirth than infants born in the New England area. There are also appreciable differences among other nations of the world.

Warfare has been another major factor in the death of parents. Literally millions of children lost one or both parents during World War I and World War II. These included civilian as well as military casualties. During recent years the mortality of young fathers in the United States has varied with the nation's military involvements. The death rate of young men was relatively low between 1961 and 1962, thanks to a combination of rising health standards and lack of substantial military action. During the following decade, however, men in the 20–24 age range became much more vulnerable to death—and their children to parental bereavement. In 1961, for example, all-cause mortality for young men was 174 per 100,000, while in 1968 the rate had soared to 325 per 100,000. The rate will probably prove to be higher among blacks as compared with whites, and other inequalities may also come to the surface when and if final, accurate statistics become available.

There are many other factors (including personality and life-style) that influence the mortality of men and women who have young children. In general, it is the male who is most vulnerable to early death, and it is the female who most often has the responsibility for heading up a family that has been disrupted by parental death.

Developmental psychology would be limited if it ignored the social and political contexts children grow in

As sociologists might say, parental mortality is *not* equally distributed across sex, age, geographic, racial-ethnic, and socioeconomic dimensions. Furthermore, the situation is constantly shifting (military involvement is one contributing factor here). Those who are interested in the hazard of childhood bereavement must, therefore, make a continuing study of the subject. Taking another view, it might also be said that by examining patterns of childhood bereavement over time we also acquire useful insights into the hazards that society in general has been experiencing (or creating for itself).

SOME EFFECTS OF CHILDHOOD BEREAVEMENT

Bereavement at any age—including the adult years—can lead to immediate changes in behavior. There may be obvious sorrow as shown by weeping, loss of appetite, and withdrawal from customary activities. But there are other ways to respond as well, especially for young children. The youngster whose father or mother has just died may not say much about the death for awhile. Instead, he or she may resist going to bed, experience nightmares, change his or her style of play, throw himself or herself into one activity to the exclusion of all others, "act up," or become unusually difficult to live with. Clearly, adults must exercise sensitivity in observing and interpreting the bereavement behavior of a young child.

But not all the effects are short-range. Some psychologists and psychiatrists believe that the bereaved child is more likely to have serious difficulties when he or she becomes an adult (and along the road to adulthood)

Tears flow so hard they create a fountain of sorrow in this drawing by a six year old. The bird in flight often represents the soul leaving the body in folklore. (Reproduced by permission of Sandra Bertman and Equinox.)

than the child whose parents were alive during his or her entire childhood. What are these difficulties? Schizoprenia and other forms of mental illness have been related to childhood bereavement (Blum & Rosenzweig, 1944). Severe depression and suicidal tendencies have been found to be more common in men who had lost their fathers in early childhood than in men in general (Brown, 1961). Delinquent and criminal behavior has also been related to early death of a parent (Barry, 1939). Adolescents and adults with a history of parental bereavement seem to engage in more serious forms of antisocial behavior and to commit more offenses in general.

Some bereaved children experience difficulties that stay with them for a long time, although they remain free of severe mental illness, criminality, and other obvious problems. I am speaking of the "clinging child" who becomes the "clinging adult"—afraid to be left alone, unwilling to allow his or her own children to become independent, etc. And I am speaking of the person who has become convinced that he or she is a "born failure." If he or she were truly lovable and capable, then mother would not have abandoned him or her by dying. And I am speaking of the person who drives himself or herself relentlessly to become a perfect world unto himself or herself—so no other mortal will have to be depended on again.

WHAT CAN BE DONE?

Death of a parent in early childhood does not automatically cause any of the problems that have been mentioned. When these problems do begin to develop, society has many opportunities to head them off. Psychologists and other behavior scientists can help to identify the specific ways in which loss of a parent leads to developmental problems, and thus help us find specific solutions. What can the rest of us do?

1. Teachers, neighbors, and others who are in contact with the child—even years after the event of parental death—can become more aware of the problems he or she is facing. Instead of responding only to annoying misbehaviors (e.g., petty theft, truancy, pesterings for attention), attempts should be made to identify and meet the needs that have led to these behaviors.

2. Around the time of the bereavement, sensitive attention should be given to the needs of the children, even if they seem "too young" to understand death (Bowlby, 1973; Furman, 1970). I do not mean rushing in and creating a melodramatic psychological rescue scene. Far more useful is a perceptive, patient, long-term approach. It is likely to take weeks and months for the young child to work his or her way through some of the thoughts and feelings set in motion by the loss of a parent. Understanding adults should be available, people who know when to speak and when to listen and how to provide the surviving parent with the right kind of support at the right time.

3. Help in providing situations in which the child can develop real competencies and skills, and in which he or she can demonstrate these to his or her own satisfaction can be furnished. This will help to develop a core of competency-consciousness in the young child as a resource to muster against feelings of hopelessness and abandonment.

4. Refrain from saying and doing some of those well-intentioned things that tend to do more harm than good. Over-idealizing the deceased parent, for example, is a technique almost guaranteed to increase the child's guilt and slow down his or her ability to comprehend and accept the loss. Acting as though nothing has really changed is another tempting maneuver that has certain advantages, but this tests the child's concept of reality and destroys the helper's credibility. Vague and high-flown explanations of why the parent has died seldom meet the child's needs and may only serve to confuse him or her and discourage further questioning and discussion. It is usually best to respond to the child in terms of what he or she really needs in the immediate reality situation. This often comes down to such basics as affection, patience, respect for the child's own feelings and responses, and providing an opportunity for the child to work out his or her own solution to this early encounter with death. The child wants to be reassured that *all* adults will not leave, that somebody who knows and loves him or her can be counted on for support. This form of reassurance cannot be given once and for all. It will take hold when, in fact, loving adults do make themselves available to the bereaved child day after day.

SUMMARY

A new world of experience and action is opening up for the post-infant, preschool child. He or she can now walk about, move objects around and overcome physical barriers. To some extent, he or she can also predict what will happen next and think about people and things that are beyond his or her line of vision. In a sense, the young child is making mental and physical breakthroughs of the kind we usually associate with science-fantasy. It is sometimes difficult for adults to appreciate how exhilerating—and, at times, perilous and unsettling—these fresh, transfigured experiences are for the child. The experiences are common to the species, but extraordinary to the individual. *Nightmares* provide one example of the frightening turns that experience can take for the young child, amidst other experiences of success and pleasure.

Mental development is extensive during the preschool period. This is well illustrated by the child's advances in the comprehension and use of *language*. Not only is the child able to communicate through language more effectively than before, but it now becomes possible to exercise *verbal control over behavior*. *Luria* and other psychologists have shown that the very young

YOUR TURN

1. How do young children go to bed? What routines do they follow? Do they "go down easy" or "only with a fight?" What do they take to bed with them, if anything? Do they ask for stories? If so, are these reading-type stories, or making-up-type stories? Do they want a different story each night, or the same one over and again (without a word changed)? Is bedtime an occasion for quiet parent-child intimacy or something of a hassle? Does the child wake up with bad dreams? Are dreams remembered the next morning? If so, what kinds of dreams are reported? These are some of the questions you might explore yourself, sharing your findings with your classmates. See if you can gain permission to observe the going-to-bed sequence of one or more preschool children. Enrich your own observations with what the parents can add from their's. Discuss the observations from the standpoint of how the bedtime scene fits into the general family style. Conclude with your own thoughts on how a preschool child should end the day and why you have formed this opinion.

2. Adults do not always appreciate how a young child may be fearful of both objective and imagined dangers. Because the child is small, we sometimes assume the fears are small, and not very serious. But fears seem serious enough to the people who have them. What fears did you have as a preschool child? Did others have the same fears? How did grown-ups respond to them? Ask these questions (along with any additions and variations you might come up with) to some of your friends. You might make them feel more relaxed and open in answering the questions if you share some of your own childhood fears with them in turn. Think about your results and see what they can add to the discussion in this chapter.

3. Look carefully around your neighborhood (or some other neighborhood). What *places* are there in which young children can play? In particular, what places are there in which they can play as they choose, relatively free from adult control? Is there a wooded area? An alley? What hazards to play exist in this neighborhood (e.g., a frequently used railroad track, a busy intersection)? A camera or sketch pad might be useful to you. Try to observe young children actually engaged in play in this neighborhood. How do they use the available space and contours? Is there, for example, a large rock that serves as home base or starting place for various games? Think about the opportunities—or lack of them—for children in this neighborhood to play a variety of games. Is there any support here for the belief, on the part of some observers, that young children are becoming environmentally deprived in free-play areas and must now limit themselves to more restricted and adult-established games? How could play space be improved in this neighborhood? If you prepare a written report on this topic, clarify your presentation with simple maps or sketches of the environment as seen from a child's-eye view.

child is excited by the sound of human speech: words stimulate actions but do not inhibit or guide them. In later stages of development in the preschool period, the child is able to respond more to the *semantic* (meaning) properties of words, to produce words himself or herself and to guide his or her own actions thereby. Theoretical approaches that emphasize sequential changes in the quality of the child's ability to comprehend symbols do not represent the only way of interpreting the relationship between language and behavior. The work of Kendler and Kendler, for example, offers a simpler *stimulus-response* approach to the establishment of *verbal mediators.*

The growth of memory in early childhood is of particular interest because storage and retrieval of experience provides a foundation for subsequent development and functioning. A distinction can be made between *episodic* and *semantic memory* systems. The former is concerned with specific events, the latter with the individual's total system for storing and utilizing information. These two systems do not necessarily develop according to the same rules or at the same rate. Both controlled research and clinical experiences suggest that memory is best regarded as an active process, not a fixed, automatic response. Very different samples of *reconstructing* experience through memory have been offered by Schactel and Piaget. Young children are relatively deficient in strategies for retrieving information from memory storage; there is not much concern with "memory for memory's sake." But recall can be vivid and accurate when the child is personally involved in the events to be remembered. Questions raised about "normal" *childhood amnesia* lead to considerations about differences between the child's and the adult's frame of reference.

Young children generally function in an *egocentric* manner, but under certain conditions they can take the perspective of other people. Inferential thinking and *perspective taking* may not occur consistently for the preschool child, but it is possible that more of this capacity exists at an earlier age than researchers once believed.

According to Piaget, the preschool years find the child at the *representational* or *preoperational stage* of mental development. It is at this time that true *symbolic functions* emerge. Examples were given from research on memory and number development.

Play is taken seriously by developmentalists. According to one view, *play is practice* for survival-oriented behavior. Others hold a *letting-off-steam view,* while the *psychoanalytic approach* states that play is a natural language or mode of expression for the young child. It is as natural for a child to play needs, fears, and triumphs as it is for an adult to talk them. Furthermore, in his or her *play-work,* the child often attempts to master the new, the challenging, the alarming. An example of the type of situation is *sibling rivalry.* Piaget interprets play in terms of the child's existing *level of mental development.* At the representational stage, play is more planned than at earlier stages, and more enjoyment is taken in the effects of what one does. Games of "make believe" also begin to flourish. *Parallel play* remains common during the preschool years, but it is gradually enriched by *cooperative play.*

As *part of the family,* the young child challenges the wisdom of his or her elders by striking a balance between the need for protection and limit-setting and the opportunity to try out his or her independence. Many parents appear to be sensitive to this challenge. A core aspect of relationship development in early childhood pivots around what psychoanalysts describe as the *Oedipal situation.* In contemporary developmental terms this seems to be part of a broader process of *identification.* The process by which the child learns to identify or model himself or herself after the parent of the same sex—without somehow entering into a competitor's role—was described, incorporating contributions from social learning as well as psychoanalytic theory.

Much that has been concluded about child development assumes that the home situation includes a mother who stays there and a father who goes off to work. This pattern now has many more variations, some of which have built up the need for *day care* centers. Factors involved in choice of the day care alternative and some of the complications and outcomes were explored.

Quite another form of separation between child and parent is permanent loss through death. *Bereavement* in early childhood was examined with respect to scope and cause. Effects can be immediate but may also take years to express themselves fully. Suggestions for helping when the parent of a young child dies were offered.

Reference List

Axline, V. M. *Play therapy.* Boston: Houghton Mifflin Company, 1947.

Barry, H. A study of bereavement: An approach to problems in mental disease. *American Journal of Orthopsychiatry,* 1939, *9,* 355–359.

Bartlett, F. C. *Remembering: A study in experimental and social psychology.* Cambridge, England: Cambridge University Press, 1932.

Blum, G. S., & Rosenzweig, S. The incidence of sibling and parental deaths in the anamnesis of female schizophrenics. *Journal of General Psychology,* 1944, *31,* 3–13.

Bowlby, J. *Separation.* New York: Basic Books, 1973.

Brown, A. L. The development of memory: Knowing, knowing about knowing, and knowing how to know. In H. W. Reese (Ed.), *Advances in child development and behavior* (Vol. 10). New York: Academic Press, 1975, pp. 104–143.

Brown, A. L., & Murphy, M. D. Reconstruction of arbitrary versus logical sequences by preschool children. *Journal of Experimental Child Psychology,* 1975, *17,* 412–422.

Brown, F. Depression and childhood bereavement. *Journal of Mental Science,* 1961, *107,* 754–777.

Colby, B., & Cole, M. Culture, memory, and narrative. In R. Horton & R. Finnegan (Eds.), *Modes of thought: Essays on thinking in western and nonwestern societies.* London: Faber & Faber, 1973, pp. 63–91.

Erikson, E. H. *Childhood and society.* New York: W. W. Norton & Company, Inc., 1950.

Flavell, J. H., Botkin, P. T., Fry, C. L., Wright, J. W., & Jarvis, P. C. *The development of role-taking and communication skills in children.* New York: John Wiley & Sons, 1968.

Freud, A. *The psycho-analytical treatment of children.* New York: International Universities Press, 1959.

Freud, S. *Interpretation of dreams.* New York: Modern Library Inc., 1938. (Originally published, 1900.)

Freud, S. *New introductory lectures on psychoanalysis.* New York: W. W. Norton & Company, Inc., 1965. (Originally published, 1933.)

Furman, R. The child's reaction to death. In B. Schoenberg, A. Carr, D. Peretz & A. Kutscher (Eds.), *Loss and grief: Psychological management in medical practice.* New York: Columbia University Press, 1970, pp. 70–86.

Inhelder, B., & Sinclair, H. Learning cognitive structures. In P. H. Mussen, J. Langer & M. Covington (Eds.) *Trends and issues in developmental psychology.* New York: Holt, Rinehart & Winston, 1969.

James, W. *Principles of psychology.* New York: Holt & Co., 1890.

Jones, M. M. *Guiding your child from two to five.* New York: Harcourt, Brace, & World, 1967.

Katz, M. *Effects of semantic and visual stimuli on the development of self-control.* Masters thesis, Wayne State University, 1970.

Katz, M. The effects of aging on the verbal control of motor behavior. *International Journal of Aging & Human Development,* 1974, 5, 141–156.

Kellogg, R., & O'Dell, S. *Analyzing children's art.* Palo Alto, Calif.: National Press Books, 1969.

Kendler, H. H., & Kendler, T. S. Vertical and horizontal processes in problem solving. *Psychological Review,* 1962, *69,* 1–16.

Kendler, T. S. Development of mediating responses in children. In J. C. Wright & J. Kagan (Eds.), Basic cognitive processes in children. *Monograph of the Society for Research in Child Development,* 1963, Serial No. 86.

Lee, D. *Freedom and culture.* Englewood Cliffs, N.J.: Prentice-Hall, Inc., 1959.

Luria, A. R. *The role of speech in the regulation of normal and abnormal behavior.* New York: Pergamon Press, 1961.

Mack, J. E. *Nightmares and human conflict.* Boston: Little, Brown & Company, 1970.

Marvin, M. S., Greenberg, M. T., & Mossler, D. G. The development of conditional reasoning skills. *Developmental Psychology,* 1977, *13,* 527–528.

Millar, S. *The psychology of play.* Baltimore, Md.: Penguin Books, 1969.

Mischel, W. Sex-typing and socialization. In P. H. Mussen (Ed.), *Carmichael's manual child psychology* (Vol. 2). New York: John Wiley & Sons, Inc., 1970, pp. 3–72.

Mossler, D. G., Marvin, M. S., & Greenberg, M. T. Conceptual perspective-taking in 2- to 6-year-old children. *Journal of Developmental Psychology,* 1976, *12,* 85–86.

Mullahey, P. *Oedipus: Myth and complex.* New York: Hermitage Press, 1948.

Newson, J., & Newson, E. *Four years old in an urban community.* Chicago: Aldine Publishing Company, 1968.

Opie, I., & Opie, P. *Children's games in street and playground.* London: Oxford University Press, 1969.

Petty, T. A. The tragedy of Humpty Dumpty. In *The psychoanalytic study of the child* (Vol. 7). New York: International Universities Press, 1953, pp. 404–412.

Piaget, J. *The child's concept of number.* New York: Humanities Press, 1952.

Piaget, J. *On the development of memory and identity.* Worcester, Mass.: Clark University Press, 1968.

Piaget, J. *The child and reality.* New York: Grossman Publishers, 1973.

Ramsay, R. H. *No longer on the map.* New York: The Viking Press, 1972.

Reese, H. W. (Ed.). *Advances in child development and behavior* (Vol. 11). New York: Academic Press, 1976.

Rhine, L. E. *ESP in life and lab.* New York: Macmillan, Inc., 1967.

Schactel, E. *Metamorphosis.* New York: Basic Books, 1959.

Schwarz, J. C., Strickland, R. G., & Krolick, G. Infant day care: Behavioral effects at preschool age. *Developmental Psychology,* 1974, *10,* 502–506.

Smirnov, A. A., & Zinchenko, P. I. Problems in the psychology of memory. In M. Cole & I. Maltzman (Eds.), *A handbook of contemporary Soviet psychology.* New York: Basic Books, 1969, pp. 452–502.

Tenney, Y. H. *The child's conception of organization and recall: The development of cognitive strategies.* Ithaca, N.Y.: Cornell University Press, 1973.

Tulving, F. Episodic and semantic memory. In E. Tulving & W. Donaldson (Eds.), *Organization of memory.* New York: Academic Press, 1972.

Von Grunebaum, G. E., & Caillois, R. (Eds.). *The dream and human societies.* Los Angeles, Calif.: University of California Press, 1966.

Vygotsky, L. S. *Thought and language.* Cambridge, Mass.: Massachusetts Institute of Technology Press, 1962.

Yendovitskaya, T. V. Development of memory. In A. V. Zaporozhets & D. B. Elkonin (Eds.), *The psychology of preschool children.* Cambridge, Mass.: Massachusetts Institute of Technology Press, 1971, pp. 89–116.

FROM HOME TO SCHOOL

The Child Steps Out

9

CHAPTER OUTLINE

I
The beginning school child is in a phase of physical development that represents a comfortable balance between growth and stability.

Rate of growth / Transition to adult physiologic functioning / Head-body ratio / Physical coordination / Permanent teeth / Body image / Laterality / Spatial egocentrism / Body representation

II
The early school years often represent one of the peak stages of philosophical activity, if by this term we mean finding excitement and pleasure in the use of one's mind.

Maturation versus experience / Humor / Ideas about life and death / Creativity / Peer group identity / Fantasy stimulation / Perceptual learning / Quantitative attitude / Counting / Memory / Rehearsal / Intentional memory / Production deficiency / Mediational deficiency / Organizing or grouping / Metamemory

I want to climb the santol tree
That grows beside my bedroom window
And get a santol fruit.
I want to climb the tree at night
And get the moon the branches hide.
Then I shall go to bed, my pockets full,
*One with the fruit, the other with the moon.**

The young poet, seven-year-old Tomas Santos, whose imagination created the lines above (in Lewis, 1966) is one of them. So is the child who can barely put a sentence together, and the one who expects you to *know* what is on his mind without him saying a word. The "clinger" is there, and so is the "imp." Alongside the child who has grown up as the one and only is his peer with a houseful of scrambling younger and older sibs. The boy who hates girls is there and so is the girl who hates boys. A veteran of nursery schools is there; while next to him sits a child who has never seen the inside of any building except a private house or apartment. This generous assortment of little individuals, already bringing a variety of different experiences, skills, and preferences with them, will soon be expected to adapt both to each other and to the total school experience. It is a prospect well frought, as old-time novelists liked to say, with all kinds of possibilities.

We will explore a few of those possibilities here. Even at this early age, there is a high level of complexity in the physical, psychological, and social development of any one child. The range of differences from child to child also makes it difficult to portray a simple, or "typical," pattern of development, as does the variety of school and home settings that children inhabit. Furthermore, a portrait of the young school boy or girl will be misleading if it is of the still life variety. We must find a way to keep our perceptions up to date with the tempo of developmental change that takes place during these years. Let's begin, then, by catching up with *physical development.*

GROWTH AND STABILITY

The beginning schoolchild is in a phase of physical development that represents a comfortable balance between growth and stability. First, let's look at the body's side of the story, and then we'll consider some of the child's own interpretations and uses of his or her physical nature.

* From *Miracles* by Richard Lewis. "The Wish" by Tomas Santos. Copyright © 1966 by Richard Lewis. Reprinted by permission of Simon & Schuster, Inc.

PHYSICAL DEVELOPMENT

1. The *rate of growth* is now appreciably slower than it was during infancy and very early childhood. The child continues to change along many physical dimensions, but the overall process can best be described as stable growth. There will be a powerful growth spurt on the verge of adolescence. For the next few years, however, the tempo of physical change will be fairly moderate by comparison with what the body has already undergone.

2. Many of the fundamental measures of *physiological functioning are completing the transition* to adult levels. Blood pressure, pulse, respiration, and body temperature usually are closer to adult levels than to early childhood levels as the school years begin. A thorough analysis of the child's anatomy and physiology would disclose numerous examples of a shift from the temporary infant–young child status to adult status. Bone marrow, for example, is at first stuffed entirely with red cells to meet the body's great demand for production of blood cells. Around age five, this demand has slowed down enough for the red marrow to give way increasingly to fat deposits, first in the long bones, and eventually in other parts of the skeleton (Custer, 1949).

3. The child *looks more grown-up* as well. This is in large part because the head-body ratio now approaches the adult model more closely. At birth, the infant's head was quite large in relation to the rest of his or her body. The length and width of the head continued to increase appreciably for the first few years, but now arms, legs, and the long muscles are catching up. The arm span of the six year old, for example, will expand so much further that eventually it will be equivalent to adding a third arm. At this time, however, head circumference is already more than 90 percent of its final adult size (Martin & Vincent, 1960). Enough changes have already taken place in body proportions to show that the transition to adult proportions is well underway.

4. *Physical coordination* has become much improved and will continue to improve. The child can execute movements that require more refined and controlled actions than would have been possible in the preschool years (e.g., skating, using tools). The sense of balance is also more secure, as can be seen in jumping rope, climbing, and walking along narrow (and sometimes precariously high) places. Fatigue, distraction, and overeagerness, as well as attempts to over-reach present abilities, make the child of this age vulnerable to many spills and tumbles.

5. *Permanent teeth* begin to erupt at age six. This is an obvious physical sign of growing up to both children and their parents. It also affects the contours of the child's face and may at times result in the "all-I-want-for-Christmas-is-my-two-front-teeth" predicament as emergence of permanent teeth lags a bit behind the departure of their predecessors.

6. The *basic pattern of physical development* is similar for both sexes during the early school years. Individual differences such as type of body build and nutritional history seem more important than sex differences. The child who is a year or two older still tends to be more adept at various physical skills. On the whole, however, girls tend to be more advanced in bone development, more flexible in their use of musculature, and perhaps more adept at rhythmic movements. Boys tend to be a little taller and heavier and are more likely to throw their entire bodies into their actions.

MINDING THE BODY

The child must continue to adjust to the changes he or she observes or senses in his or her body. These include technical adjustments, such as taking advantage of greater strength or longer arm span. But the child also has to represent the new physical person he or she is becoming in his or her *body image* and sense of identity. Furthermore, the child is now better able to appreciate some facts about his or her physical make-up (and that of other people and objects). I will sample a few of the ways in which boys and girls at this age level go about minding the body.

The body as home base. As adults we realize that our orientation toward the physical world (including other people as physical objects) is directly related to the positioning of our own bodies. We are clearly aware that there is a front and a back to us, a top and a bottom, and the two sides we know as *right* and *left*. These are all facts that the child has to learn. By age five, most boys and girls do appreciate the "sidedness" of their bodies. But the right/left distinction is not yet reliably established (a distinction that perhaps is complicated in our society by the double meaning of *right* as correct! and the-side-that-is-not-on-the-left). About age six, the child usually has his or her own right and left side firmly in mind (except sometimes). The child's problem, however, is in relating his or her orientation in space with other people and objects. Walking straight toward you, for example, the child is likely to regard your right side as your left because it lines up with his or her left side. The six year old's uncertainty in thinking and talking about his or her body's relationship to objects in space has been demonstrated by research. Such studies have requested the child to move around in various ways and to describe his or her relationship to other people or things, or have asked the child to report whether a certain object is being moved to the right or left (Cratty, 1970). The child is less confused about laterality (right-leftness) when he or she and everything else is standing still.

An important distinction should be made here. Typically, the ability to adjust oneself "correctly" to spatial orientations develops before the ability to represent the situation verbally. The child's body seems to understand how to behave in a certain spatial orientation, even if the child cannot yet describe or explain what is taking place. The child who cannot yet tell his or her right hand from the left, for example, may dependably shake hands with the right (right) hand. There is, nevertheless, great advantage in being mentally able to represent and communicate the relationship that exists between one's own body and other points in space. The seven year old often is able to make this breakthrough. He or she knows right from left both in respect to his or her own body and in the world around him or her. In other words, the seven year old is able to use his or her body as a dependable reference point, or home base, for determining his or her relationship to the physical world. This ability becomes established even more firmly during the next year or two.

When children who are at the same age level but who have varying performances on tests of mental functioning are compared, those with better skills and more masterful performances are more adept at making right-left distinctions with their own bodies as reference point (Lacoursière-Paige, 1974). The right-left concept appears to develop as an integral part of the child's total comprehension of his or her relationship to the world, not as an isolated phenomenon.

As the child continues to mature, he or she will be able to orient himself or herself in space without having to take the momentary orientation of his or her own body so heavily into account (Wapner & Werner, 1957). The child's frame of reference will be more mental, or abstract, and, thus, more liberated from influence by specific physical cues. In the early school years, however, the child is still engaged in developing secure knowledge that his or her body has certain fixed dimensions and sides that can be used as dependable cues in orienting himself or herself within the environment.

Spatial egocentrism is being able to see a situation only from one's own point of reference

Even more difficult and complex will be the child's progression to comprehending how others orient themselves, to see the world through another person's eyes. **Spatial egocentrism** was demonstrated clearly by Piaget and Inhelder (1956) who asked young children to look at a miniature landscape of three mountains and report what a doll would "see" from various points of reference around the scene. Those who were six years old or younger usually believed that the doll's perception would be the same as their own—no matter where the doll was standing. Among other things, this suggests that for the child at this stage, the body has become a secure home base, but the developmental problem of recognizing that other bases or frames of reference also exist lies ahead. In a more recent study, Shantz and Watson (1971) asked the question, "What concepts does the young child acquire that make possible the emergence of spatial objectivity around seven years of age?" They reasoned:

> The ability to predict what another sees from various locations develops from the child's own experience in object relations. Specifically, the child's awareness of himself as an object within an organized world of objects begins with gross discriminations which follow a certain order of increasing specificity and organization. The very young child has no expectancy of objects changing in appearance with a change in his spatial position. The first step toward decreas-

> ing egocentrism would be the child's recognition that objects and their arrangements look different from different spatial locations, but no specific expectancies as to how objects appear. That is, he operates on a simple "same" versus "different" expectancy. Next, the differences expected by the child are differentiated into specific object-subject relations, but the relations are not yet organized into a total spatial framework. This latter concept is the final stage of development. In summary . . . the ability to predict another person's viewpoint follows the same developmental steps as subject-object predictions. (p. 171).

Shantz and Watson found that there was indeed a close relationship between the ability to predict where various objects would be located on the miniature landscape after moving around it, and how the landscape would look from another's (doll's) point of view. This experiment supports the general idea that children must reach a certain level of ability in comprehending and representing the relationship between their own bodies and the outer world before they can comprehend another person's orientation and viewpoint. Identifying object locations from another's viewpoint proved very difficult for most of the young children in the Shantz and Watson study.

I have lingered on this point because the ability to see the world from another person's perspective is very important in many spheres of functioning—the emotional and the motivational as well as the perceptual and the cognitive. The child's gradual achievement of understanding about his or her own body and its orientation to the world is an important early phase in the total elaboration of a world view in which there are many people, many bodies, many perspectives.

Body inventory. Knowledge of body parts, what they are called and where they are located, is another aspect of minding the body that adults may take for granted, but which each child must learn for himself or herself. Many two year olds know where head and tummy are to be found, and can say the names of these major body parts while touching them. A year or so later, the child often has become well acquainted with eyes, nose, thumb, and perhaps a few other points of special interest. More body details can be named and located as time goes by, with some parts (such as the ring finger) proving more difficult to master than others (such as the little finger). Even by age seven, however, three children in ten still have difficulty in naming and locating their eyebrows (Ilg & Ames, 1966). The young schoolchild, then, is typically just completing his or her knowledge of visible body parts.

Children enjoy learning the names of body parts, where they are, and what they do

The boy or girl of this age level does not yet have a clear idea of what bodies are like on the *inside*. There is difficulty in naming internal body parts, even more difficulty in assigning them to approximately the correct locations. Even a very bright child may imagine that people's insides resemble the stuffing in dolls or the components in mechanical toys. It is probable that children can and do become more sophisticated about their "innards" if appropriate learning experiences are available. Even city-bred youngsters, for example, learn some fundamentals of internal anatomy from watching or helping their parents prepare whole chickens for cooking. Since the supermarket-and-freezer revolution, many children no longer have the opportunity to observe that chickens and fish do have insides. Television commercials that portray the inside of the human head as a clogged drain pipe or that otherwise distort internal anatomy can hardly prove enlightening to the young child. Lack of secure knowledge concerning internal body parts is not restricted to the early school years. Many adults retain only a vague conception of those physical structures that lie within and beneath the skin.

Body representation. The child's image of the human body shows up in figure drawings and other forms of creative play. Our interpretations should not be too hasty, however. The careful observer will make sure that conclusions are based on not one, but many, figure drawings or other productions made by the same child on different occasions. The child has a variety of moods and purposes, and these can influence his or her body representations (Kellogg & O'Dell, 1969). Furthermore, the way the child represents the human body depends to some extent on the particular materials that are used (Arnheim, 1954). Drawing a two-dimensional human figure with pencil or crayon, for example, is quite different from modeling a three-dimensional sculpture with playdough. The child discovers how to make distinctive use of the available materials. We do not see a pure representation of the human body, then, but an interaction between the child's concept and the way he or she has mastered certain physical materials.

The earliest depictions of the human figure through pencil or crayon drawings are usually dominated by a large circle. Having discovered that he or she can produce this basic geometric form, the child embellishes it with markings for eyes and mouth, perhaps nose also (although not necessarily in the places we expect to find them). Why does the child "go for the head" at the very beginning of his or her artistic representations? Perhaps this is related to the anatomical fact that the young child's head is outsized (by adult standards) and dominates his or her own physique. Perhaps, again, it is more intimately associated with the significance of face-to-face contact in human communication. Or perhaps it is especially satisfying to draw the head because feelings can be expressed fairly directly by putting on a happy face or a sad face. The balloon-shaped head usually tops off a stick-figure representation of arms, hands, and legs. These items emerge from the head itself; the torso is not introduced into figure drawings until some time later.

We would be truly astounded (and so would child-artist) if this creation actually came our way down the street. But even the earliest representations of the human figure in two dimensions generally capture the principle features of the human form. The person is vertical (head up, feet down), can see and express (head), can touch (hands), and can move (legs). Not a bad beginning, at all.

Just as development moves in a cephalocaudal sequence, so the young child's drawings usually emphasize head and face. (Photo by Tania D'Avignon.)

By the time the child has reached school age, he or she includes the torso in drawings, along with more head and body detail and some attention to clothing. Special care may be taken to indicate the sex of the person who is being depicted (although not necessarily in every drawing). The boy or girl seems to have definite ideas by this time as to what comprises a "good" picture of a person; this can be taken to indicate either a more secure image of the body, the establishment of artistic standards, or both. A recent study, for example, asked children aged three to six years to complete a drawing that somebody else had begun. The drawings consisted of face-circles with eyes that were in the wrong place, either too low or too much to the side. In coming up with their various solutions, most of the children seemed to act upon the principle that a face must have the proper ingredients and look like a face, even if this means tilting the body or taking other liberties with it (Goodnow & Friedman, 1972). The older children recognized that their productions "looked funny," a comment that suggests critical thinking and a sense of artistic integrity (and possibly also a sense of humor).

In another recent study, children in the same range of ages mentioned above were given their first experiences with playdough and asked to make a number of objects, including a mommy and a daddy. As might be expected, children of various age levels approached the task differently, but all of them progressed in their style of representation as their experiences and skills with the materials increased. The youngest children, for example, went "from passive, aimless handling of the dough medium to a more active exploration of the material and finally to representation proper" (Golomb, 1972, p. 390). The three year olds typically produced an upright column (again, capturing the idea of verticality) with some attempt at facial features. It was only the older, school-age children who attempted (sometimes) to join body parts together or mold the back as well as the front of the figure. The three-dimensional possibilities of playdough modeling provide the child with a different, and perhaps more complex, way of representing the human form.

Researchers still do not know very much about the larger network of meanings that a schoolchild develops around his or her body. The occasional study on this topic tends to raise more questions than it answers. Even at the youthful age of six, for example, boys regard some types of body build as "better" than others. The muscular **mesomorph** type of physique appears to be the most popular, regardless of the actual body type of the child who makes the judgment (Lerner & Korn, 1972). Similar body-type preferences are expressed by adult males. We might expect some important differences between those who have and those who do not have the most popular type of physique in the way they regard their own bodies, and how their total self-concept develops around their body images. Yet this kind of information is lacking, as well as comparable data about girls and women. Even if much more scientific information were available, however, when it comes down to an individual boy or girl, we still need to be responsive to the distinctive ways in which this particular child minds his or her own body.

A **mesomorph** is a chunky, muscular type of physique

Measuring up. The schoolchild, with his or her heightened sense of selfhood, is often interested in comparing developmental status with other children and with adults. The child may exult, for example, when he or she goes "one tooth up" on a friend, since each baby tooth gone means another step toward grown-up status. The boy or girl may stand up against one of the parents to show, "I am up to here on you," and next year, "I will be up to there!" On her birthday, one girl I know was certain that she was taller than she had been the day before, adding, "I feel *very* six today!" The schoolchild is old enough to have a past that is usable and may take considerable delight in comparing his or her current size and physical prowess with the way he or she was "back then." The child appreciates that physical growth has already taken place, and that there is more to come. In this significant sense, then, he or she is an observer as well as a participant in the developmental process.

THE CHILD AS PHILOSOPHER

Let's shift focus now to the young schoolchild as thinker, although some aspects of mental development associated with body changes have already been sketched. Actually, we could look on the boy or girl at this age level as a philosopher and not be far off the mark. The term *philosophy* asserts a love *(philos)* of wisdom or an advanced state of knowledge *(sophos.)* Many children aged five, six, or seven are intellectually vibrant. They enjoy learning, taking on mental challenges, and extending the limits of their minds. The early school years often represent one of the peak stages of philosophical activity if by this term we mean finding excitement and pleasure in the use of one's mind. Adults do not always appreciate this side of the child's development and do not always share the youngster's pleasure in mental discoveries and gyrations. The child at this age also qualifies as a philosopher in the sense of one who pursues "ultimate" questions, including the meanings of life and death.

These statements are not intended to blur the difference between the thought processes of an early schoolchild and an intellectually mature adult. Many developmental psychologists are continuing to devote their research efforts to the identification and understanding of mental functioning throughout the lifespan. *How* different the child's thought processes are in comparison to adults (or even to older children) is a question that has not been thoroughly resolved. Some theoreticians believe that the differences are fundamental and qualitative, closely related to the child's level of psychobiological *maturation*. Others are more inclined to see child-adult differences in terms of the grown-up's more extensive *experience* with the world and greater accumulative achievements in learning. In other words, the familiar inside-out versus outside-in view of human development applies here as well. It seems most useful to seek a balanced perspective. Children and adults may think very differently in some situations, but not necessarily in every situation. We do not *have to* assume (even if we often *do* assume) that all differences in thought between a child and an adult are explained by their age differences, or that the adult's view is invariably superior. In this section, however, we are chiefly interested in becoming acquainted with the child's mental life for its own sake, not in comparisons.

Learning can be a pleasure when given half a chance. Children enjoy many of their experiences even though they know they are supposed to think school is a grind. (Photo by Talbot Lovering.)

"WANT TO HEAR SOMETHING FUNNY?"

The child's intellectual zest often expresses itself in humor. I am thinking here of humor that the child himself or herself concocts, discovers, or appreciates, as contrasted with behavior that *un*intentionally provides amusement for adults. A four year old boy, for example, unwrapped a gift and then exclaimed politely but mournfully, "Just what I never wanted!" A girl of the same age often volunteered to help her mother prepare a "chicken pox pie." A year or two later, these same children are "in on the joke" themselves. Indeed, they are actively perpetrating jests, acts, and riddles that are intended to be funny.

Few parents can escape riddles such as, "What is red and white on the outside, and grey and white on the inside?" If the respondent is stumped, the child may be delighted both by the joke and by his or her success in putting it over. In turn, the child usually enjoys the opportunity to solve riddles. One schoolboy, for example, hearing the above riddle for the first time, came up with, "An elephant—inside out," which is scarcely inferior to the standard answer ("A can of Campbell's cream of elephant soup"). The young schoolchild does not always succeed in performing jokes he or she has heard, sometimes because the point eludes him or her; he or she just knows that it is *supposed* to be funny. However, the child is now able to operate in a realm of verbal thought and expression sophisticated enough to have (and share) fun with words. Perhaps you have seen two or three children literally rolling and shaking with laughter at their own errors in tongue twisters. Substituting new words to familiar songs, poems, or phrases can also induce a state of high glee. As a punster, rhymer, and player-with-words, the child is discovering that the use of his or her mind can generate novelty and entertainment.

Spontaneous wit begins to appear more frequently in the conversations of the schoolchild. The chicken-pox-pie girl, for example, was singing her favorite selections from *The Wizard of Oz* while riding in the car with Daddy. When she paused for a moment between songs, Daddy began to sing one from the same show.

Cynthia: "No, Daddy. You can't sing that song!

Daddy: Why not?

Cynthia: Because a lady is supposed to sing it because a lady *does* sing it!

Daddy: But a man *could* sing it.

Cynthia: (After a long pause). Yes—*if* he has the voice for it (A burst of laughter all around.)

Cynthia effectively defended her role as the only vocalist in the car by mentioning Daddy's untunefulness, while making this unpleasant truth acceptable through humor. The child who develops a knack for seeing and expressing the humorous side of situations has a valuable asset both for the present and for the future.

Among children, humor can be both cruel and sociable at the same time. There are many "underground" songs, for example, that schoolchildren sing in defiance of the classroom establishment (but out of the teacher's earshot). One of the mildest I know of is:

Hi hee, hi ho!
I bit the teacher's toe!
That dirty rat—she bit me back!
Hi hee, hi ho!

The more violent songs and chants seem to proclaim a state of enmity if not an imminent revolution. Yet the singers may, in fact, be quite fond of their teachers. The rough humor of the antischool songs binds the children together, bestowing a sense of power and mutual belonging through their common (if feigned or exaggerated) grievances.

Developmental psychologists are giving renewed attention to humor as both an interesting subject in its own right and a useful approach to studying the child's mental functioning in general. Children as young as six years of age, for example, are able to appreciate jokes that depend on simple kinds of double meaning and unexpected elements being placed in relationship to each other. They do *not* see anything funny in the following dialogue:

I saw a man-eating shark in the aquarium.

That's nothing. I saw an octopus.

But they *are* amused by:

I saw a man-eating shark in the aquarium.

That's nothing. I saw a man eating herring in the restaurant.

Jokes that depend on more complex relationships between the "opener" and the "punch line" are often beyond the intellectual grasp of the six year old. By age twelve, however, the child is able to comprehend a larger variety of jokes. This suggests a growing sophistication in the use of language and logic in general (Schultz & Horibe, 1974). But even the more limited humor-appreciation of the six year old implies that considerable intellectual development has already taken place. The child must recognize a certain orderliness in the way things are and the way things happen to notice, and be amused by, the unexpected. Memory is involved, and memory requires the ability to call upon past experience in the present situation. Furthermore, the child who sees something funny or surprising (which are not identical reactions, of course) probably has some control over his or her own relationship to the situation. He or she can step back a little and consider a situation in which he or she is not necessarily the central figure. This is another aspect of the more general process by which people escape from the egocentric orientation typical of early developmental phases and achieve a sense of perspective (this emergence from spatial egocentrism has already been touched upon in this chapter).

Many intellectual achievements, such as recognition of predictability and orderliness in the world, memory, and emotional distance or objectivity, can be seen in the child's burgeoning sense of humor as well as in his or her more "serious" activities. Humor itself can be serious in that the child devotes attention and effort to it, has standards for what is not funny, and will on occasion perform chores and obligations in order to be free for humor-generating or appreciating opportunities, etc.

QUESTIONING AND CREATING

"I love animals and dogs and everything."

This is the sort of statement we would not be surprised to hear from an exuberant five-year-old girl. Yet this love letter to the universe has its probing and reflective side. She continues, "But how can I do it when dogs are dead and a hundred?" The girl's desire simply to love all that she sees about her has run up against intuitions of time, aging, and death (how can I love dogs when they are one hundred years old and have, therefore, died?). In a poem that I find deeply moving, Hilary-Anne Farley (in Lewis, 1966, p. 143) poses fundamental questions and offers her own answer:

SUN GOES UP
I love the juice, but the sun goes up; I see the stars
And the moonstar goes up,
And there always goes today. And the sun
Loves people. But one always dies.
Dogs will die very sooner
Than mummies and daddies and sisters and
brothers because
They'll not die till a hundred and
*Because I love them dearly.**

Thoughts about life and death. Not every young child is as perceptive as Hilary-Anne, nor as gifted in expressing her thoughts and feelings. But boys and girls think more often and probe more deeply into questions of life and death than their elders might suppose (Anthony, 1972). Adults may believe that they are protecting their children and enabling them to remain in a sun-filled fantasy realm by evading or glossing over the topic of death. This is really the adult's fantasy. Young children are commonly attracted to the themes of separation, loss, change, deterioration, and death (Bowlby, 1972; Maurer, 1961). I am not implying that a five or six year old grasps death-related concepts in the way that a mature adult does. But the child is very likely to be fascinated by problems related to death. The well-known childhood curiosity about "where do we come from" has its counterpart in the inquiry, "where are we going."

Later in this book I will give more systematic attention to ideas and feelings about death from childhood onward. At this point the most important considerations are:

1. The young child addresses his or her mind to fundamental questions of life and death that challenge the wisdom of his or her most intelligent elders. Leaving aside for now the kinds of answers the child comes up with, perhaps the most impressive fact is the quest itself. In pursuing questions of life-and-death magnitude, the child starts on an intellectual adventure that will influence his or her general style of mental development. Death-related thoughts are not exotic and isolated happenings; they are part of the core pattern of mental growth. Challenged both by the intellectual and emotional dimensions of death, the boy or girl pursues such other fundamental problems as time, space, cause-and-effect, and identity. Curiosity about death contributes much to curiosity about everything else (Kastenbaum & Aisenberg, 1972).

Children think about death and incorporate it into their games and conversations

2. Adults in our society often have their own "hang-ups" on the topic of death. Some past observers have gone so far as to declare that we have a national taboo on mortality (Feifel, 1959). There are signs that our attitudes are now becoming more open and flexible. The young child needs honest and sensitive communication with adults in all areas of functioning. This need is perhaps most under-appreciated in the area of death. I recommend that those who are interested either in understanding the mind of the young child or in serving as a knowledgeable and responsive "up-bringer" look into some of the relevant guidelines and discussions that are now available (Grollman, 1967; Kastenbaum, 1973).

Curiosity and self-education. The child's curiosity flickers across many topics. Questioning adults is only one of the ways in which that curiosity expresses itself. There is pleasure in taking things apart to see what they are made of, pleasure in performing experiments to see "what would happen if . . . ," and pleasure in observing older children and adults who are engaged in interesting activities. These are mostly uncontrolled learning situations, as contrasted with the planned education the young child is beginning to encounter in the classroom. It is not known precisely what or how much the child learns from the exercise of his or her own curiosity in everyday life, but it is probably a very important part of the total acquisition of knowledge and skill. There are some dangers, of course. These include the perils of experimentation with machinery, fire, drugs, and chemicals—and the handguns that some families unbelievably leave within the reach of schoolchildren. As the child comes increasingly under the influence of other children as well as parents, he or she may also pick up misinformation and distortions that can weigh heavily upon him or her (including wild stories about certain individuals or groups of people). Nevertheless, the opportunity to follow his or her own curiosity is a significant component of self-education for the young child and provides a useful balance to the more standardized instructional conditions that prevail in school. Sensitive teachers and parents are sometimes able to help the child's curiosity to flourish in the school situation as well, but there are also "casualties" whose spontaneous enthusiasm for learning is seriously dampened by the classroom atmosphere and dynamics (Kozol, 1967).

The *Death at an Early Age* Kozol refers to is the death of curiosity and intellectual zest

Creativity within the peer group. In childhood as in later life, questioning and creating are processes that seem to be closely related. In order to bring something new into existence (a work of art, an invention, a new way of doing things), it is helpful if the individual can see beyond the way things are at the moment. "Why is it done this way?" "How else could it be done?" Creativity, like questioning, may surface in virtually any area of the child's activity. Often, however, creativity bears some relationship to special interests and emotional concerns. The young schoolchild, for example, is beginning to cultivate his or her **peer-group identity.** In other words, the child recognizes that there is plenty of company at his or her own level and he or she is not just one small individual encapsulated in a world of larger and more powerful beings. Curiosity about who he or she is and what others are like is stimulated by increasing contacts with other children (even if he or she has been around many children before, now there is more capacity for mutual interaction).

Peer-group identity is that part of self-concept based on association with others of the same age and status

The emergence of peer relationships continues from childhood to adolescence, offering a choice between the social worlds of parents and friends. (Photo by Talbot Lovering.)

Some of the earliest creations at this time include secret plans and fantasies within the peer group. Several children will spin and share secrets that no outsider is to hear. Depending on a number of factors, including the kind of private space and materials available to them, there may also be the establishment of secret places and passwords. A few years later the same children may create more complex secrets, such as a code language. From the social and emotional standpoints, this sphere of creativity helps children to form deeper bonds with each other and to have an alternative to the adult world on one hand, and to solitude on the other.

In his monumental contribution, *Comparative Psychology of Mental Development*, Heinz Werner (1948/1961) reported a system of classificatory terms jointly invented by a four-year-old child and a six-year-old child. They decided on words to denote building blocks of different sizes and shapes in order to carry out a project together. One piece was known as *Big Miller*, for example, another as *Big Little Miller*, still another as *Little Little Miller*, etc. This spontaneous innovation in language probably would have delighted two of the most provocative thinkers in human history, Jonathan Swift (1726/1963) and Ludwig Wittgenstein (1958). Both of these men provided illustrations of a "thing-language" in order to illuminate the roots and possibilities of human communication. The two young children who named and passed blocks back and forth to each other were thus creating their own little chapter in the philosophy of symbols.

In *Gulliver's Travels*, people carry thing-language to an extreme: namely, in sacks over their backs

Expressive creativity. We have already looked at an example of a young child's ability to express questions about time, aging, and death in the form of poetry. Others express themselves through drawings, story inventions, and varied forms of creative play and fantasy. But there are also examples of creative expression in early childhood that appear to spring more directly from the discovery of personal abilities and the appreciation of beauty. Music provides some of the most striking examples. The "child prodigy" is an actual phenomenon. Perhaps the most famous prodigy came to the attention of his father when Leopold Mozart asked his four-year-old son what he was so busily doing with his pen. "Writing a concerto for the clavier," replied Wolfgang. "It will soon be done." This first manuscript has not survived the years, but by the time he would have been a third-grader, Mozart had created more that thirty compositions, some of them already published, performed, and acclaimed. One can marvel either at the talent demonstrated so early or the fantastic development that was yet to take place within Mozart's adolescence and foreshortened adult life. Felix Mendelssohn was another who astonished by composing technically secure and attractive music at a very early age. Then, in late adolescence, he took his place as a major composer.

Examples do not have to be drawn exclusively from the past. In this century, the late Benjamin Britten was an example of the creative musician whose gifts began to unfold in early childhood. And the late Igor Stravinsky once confessed that as a boy he excelled in discovering and performing the most alarming sounds possible (years before his orchestral sounds alarmed the entire musical establishment). Among young schoolchildren today, there are probably a number of boys and girls who are already creating music en-route to exceptional careers. Olivier Knussen, for example, composed about two hundred works for piano between the ages of five and nine. At age fifteen he conducted a major symphony orchestra and now he has entered his early twenties as a composer of recognized talent and achievement.

The child prodigy is, by definition, exceptional. We will not appreciate the full spectrum of humans developing without recognizing those who possess special gifts and attempting to comprehend their total patterns of actualization. The creative impulse may also express itself in the boys and girls who are part of our own household or neighborhood. Do we recognize, appreciate, and encourage their creativity?

Fantasy and self-regulation. Children between the ages of six and nine who introduce many fantasy elements into their play are also creative along other dimensions such as story invention (Singer, 1961, 1966). Perhaps unexpected is the fact that youngsters with a strong fantasy-creativity knack demonstrate superior ability to regulate their own actions, for example, sitting quietly on a chair for a long time. One possible explanation is that such children can entertain themselves with their own thoughts better than less creative children when there is nothing else to do, or when they must wait. The studies conducted by Singer and his colleagues also suggest that fantasy and creativity dispositions are greater in young children who have had much

Figure 9-1. *Make-believe helps children master reality. The following are areas of superiority (over peers using fantasy to a lesser degree) for six to nine year olds who introduce fantasy elements into their play. (Based on a review of recent research by Singer, 1977.)*

- Greater self-restraint (e.g., sitting quietly)
- More cooperation with other children
- More positive emotional states
- More flexible use of language
- Richer imagery
- Better recall of narrated materials
- Keener appreciation of absurdities
- Superior performances on verbal learning tasks
- Invention of more distinctive and interesting stories
- In need of less regulation from adults

one-to-one interaction with their parents, especially through story-telling and idea-sharing sessions. Such children are more likely to be the only or oldest child in the family, or to have fewer siblings than the children who have less inclination for fantasy.

The daydreaming child is often chastised as inattentive or "off in his or her own world." However, research suggests that such a child actually possesses distinct advantages over low-fantasy peers. The child who can call on his or her own mental resources for entertainment and points of reference is less at the mercy of the immediate situation. Less stimulation and control is required from the environment.

Overviewing current knowledge on fantasy and creativity among young children, Singer concluded:

> The child who engages in fantasy play is not necessarily withdrawing from "real" life or responding defensively to inner instinctual promptings. It may be more likely that he is actively exploring his environment visually and then continuing this exploration in play. Curiosity, pleasure in development of imaginative skill, and some defensiveness may all combine to foster fantasy play as a strong disposition in a given child. (Singer, 1966, p. 137).

Fostering the child's imagination. Since imagination and fantasy seem to be associated with favorable developmental and adaptational characteristics, is it possible to stimulate more of these qualities in young children? There are now a number of studies which answer this question affirmatively (e.g., Freyberg, 1973; Saltz, Dixon & Johnson, 1976; Singer & Singer, 1976). Fantasy-stimulation appears to be successful whether one uses realistic sociodramatic play as the medium or more imagination-oriented stories of adventure and exploration. Adults who are at ease with the realm of fantasy and who do not have strong needs to structure the child's play are more successful in stimulating the expression of fantasy (Gershowitz, 1974). From his extensive research in this area, Singer (1977) believes that much of the young child's make-believing is pressured to go underground during the early school years. This deprives the child of the pleasures to be experienced through expression of fantasy and imagination and also may place an unnecessary restriction on cognitive and interpersonal development. He suggests that parents and teachers relent on efforts to turn the child's mind forcefully from fantasy to reality, since the former is such a useful way of learning to cope with the latter.

OBSERVING AND LEARNING: PERCEPTUAL LEARNING

Philosophers, whether youthful or aged, must observe and learn if they are also to question, reflect and create. Eleanor Gibson (1969) has offered an ac-

Perceptual learning is the process of actively extracting information from the environment

tivist view of what is becoming known as **perceptual learning** in children. The child goes after the information he or she wants from the environment. It is not a matter of standing around and accepting a bombardment of stimuli from parents, teachers, other children, television, etc. Rather, the young child is already an active investigator who is continually improving strategies of extracting useful information from the surrounding world. It is the child himself or herself who ultimately decides what makes a stimulus a stimulus, and what features are most worth attention. Harold Stevenson, another advocate of perceptual learning in children, has capsuled this approach:

> The product of experience is not the acquisition of new responses but an increased sensitivity to the ways in which stimuli are alike and how they are different. The basic mechanism of learning is differentiation, the abstraction of distinctive characteristics, not association of stimulus and response. Behavior becomes more complex, not because the child has learned more responses but because the environment has become more differentiated. The child sees more and hears more because he has learned to discriminate more. (1972, pp. 260–261).

With increasing experience, the child tends to improve his or her ability to discover the most useful clues in a situation. Stimuli that are less relevant and less dependable are discarded; the child learns to discriminate what he or she needs from what can be set aside. Active exploration of the possibilities

An activist theory of learning emphasizes the child's inclination to go after the experiences he or she wants, rather than to wait passively for stimuli. (Photo by Talbot Lovering.)

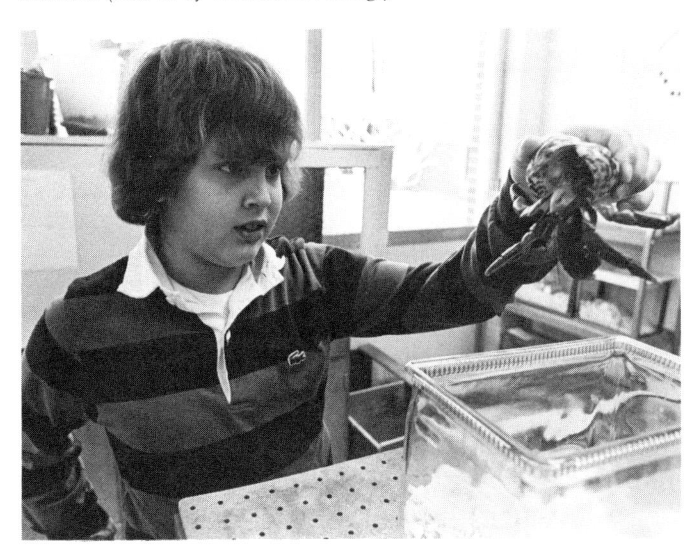

available in the environment is necessary for perceptual learning. The child acquaints himself or herself with sights, sounds, tastes, smells, shapes, objects, people. Gradually, the child discovers that objects can be similar in some ways (e.g., color) yet different in others (e.g., shape or size). The good learner can apply past experience to perceptual differentiation of new objects or situations. In other words, the child becomes better at knowing what to look for or, perhaps, how to go about looking for what might be important or interesting even if there are few clues to begin with. The child actually develops a clearer idea of what he or she is observing: the brightest star, the taller man, the wartiest toad, the older car, etc. Active experience enables him or her to carve out more precise perceptions from the total field of possible stimuli.

What is being learned? Observing, learning, thinking, and language are closely related. Because the school-age child has language and other representational abilities to work with, he or she can label what he or she is observing, and also communicate his or her experiences to others. The ability to put a word on a perception (or a perception into a word) makes it easier to learn. He or she is also more able to respond to instructions and clues offered by teachers and parents to guide his or her learning. But Gibson would have us remember that telling or showing is not an effective way of helping the child to learn, at least not as the primary technique. In one way or another, the child must participate actively—after all, it is the child's learning that is involved.

There are some theoreticians and researchers who give more emphasis to the value of rewarding or "reinforcing" children for giving correct responses. The perceptual learning approach advocated by Gibson and her disciples accepts the demonstrated fact that reinforcements can influence a child's performance in a learning situation. Nevertheless, the focus is on the boy or girl's *own* active effort to discover the most significant properties of the world not on the power of the reinforcing stimulus to affect the passive child. Even if the child is sitting still, this does not mean that he or she is passive. As Harold Stevenson notes:

> The child must look and listen, rather than merely see or hear, if what he experiences is to influence his later behavior. Observational learning is selective. Children do not incorporate all information that is available in a social situation. Characteristics such as the salience of the observed behavior, its outcome, the ease with which the information can be coded and stored, the motivation of the child, and the relation between the child and the person observed are among the many variables that influence the degree to which observational learning will occur. (1972, pp. 274–275).

One of the research examples cited by Stevenson (1972) in his review of the literature is worth our attention here. What is it that children learn from observation? More specifically, what is it that children learn from watching television and movies, in the sense that their everyday behavior shows changes related to their observations? Bandura, Ross and Ross (1961) conducted one of the first studies into this problem. In one phase of this study, the subject-child was given some materials to play with, while another person, sitting nearby, behaved either aggressively or nonaggressively. Aggressive actions of the "model" included hitting the Bobo doll with a mallet and otherwise creating a scene of active discontent. The child himself or herself was exposed to mild frustration by being told that he or she had to stop playing with the toys he or she was then enjoying, but could play with toys that were in an adjacent room. When the child went to the next room, observations were made to determine whether or not he or she showed aggressive behavior (what toys the child picked up and how he or she used them were noted).

The complete pattern of results is difficult to describe in a few words because the experiment involved several other research conditions as well. The basic finding, however, was that children who had been exposed to an aggressive adult model while they themselves were experiencing mild frustration were much more likely to behave aggressively than children who had observed a nonaggressive model. It was clear that the children's own behavior changed in an aggressive direction after they had observed somebody else behaving aggressively.

In a subsequent study, Bandura, Ross and Ross (1963) used the real-life aggression situation again, but also introduced two experimental conditions in which the children observed aggression through the motion picture format. In one film, an adult performed exactly as the aggressive model performed in the real-life situation. In the other film, a cartoon animal carried out the aggressive behavior. In this study, as in the preceding study, there was a group of children who saw no aggression at all. It was found that all three kinds of aggressive presentations led to more aggression on the part of the viewers. In fact, the children "learned" to be just as aggressive from watching an adult model knocking around the toys in the movie as from a live demonstration of the same behavior. The children "learned" to be more aggressive than they had been before by watching the cartoon animal, but not as aggressive as in the other conditions. The boys, by the way, tended to show more aggressive behavior than the girls in both studies.

We are justified in worrying about what children see on TV and at the movies

Whether physically active or sitting quietly, whether observing people in the flesh or in filmed presentations, school-age children are learning something. But what? In a general developmental sense, perhaps they are *learning how to learn*, how to select and organize certain dimensions of the total environment while giving less priority to others. But they are also *learning implicit rules that govern behavior and interaction*. This means that the sensitive adult, whether parent or teacher, will be interested both in the child's opportunity to explore the world actively and develop his or her learning skills, and in the nature of the world that is available for exploration. Aggression, for example, could hardly be removed from a child's world, even if it were a good idea to do so (which is certainly debatable). But adults do have some choice as to the type and amount of aggression their children are ex-

Aggression Displayed After Viewing Aggressive Behavior	Experimental Groups					
	Real-life Aggressive Model Seen		Human-film Aggressive Model Seen		Cartoon-film Aggressive Model Seen	Control Group
	Male Model	Female Model	Male Model	Female Model		
Amount total aggression						
Girls	65.8	57.3	87.0	79.5	80.9	36.4
Boys	76.8	131.8	114.5	85.0	117.2	72.2
Amount imitative aggression						
Girls	19.2	9.2	10.0	8.0	7.8	1.8
Boys	18.4	38.4	34.3	13.3	16.2	3.9
Amount mallet aggression						
Girls	17.1	18.7	49.2	19.5	36.8	13.1
Boys	15.5	28.8	20.5	16.3	12.5	13.5
Amount nonimitative aggression						
Girls	27.6	24.9	24.0	34.3	27.5	17.8
Boys	35.5	48.6	46.8	31.8	71.8	40.4
Amount aggressive gun play						
Girls	1.8	4.5	3.8	17.6	8.8	3.7
Boys	7.3	15.9	12.8	23.7	16.6	14.3

Figure 9-2. *Witnessing aggressive actions: effects on nursery school children.* *Exposure to any of three types of aggressive actions leads to more aggressive behavior of various types on the part of nursery-school-aged children when compared with a control group who were not exposed to aggressive scenes. (Adapted from A. Bandura, D. Ross & S.A. Ross, Imitation of Film-mediated Aggressive Models.* Journal of Abnormal & Social Psychology*, 1963, 66, 3–11. Copyright 1963 by the American Psychological Association. Reprinted by permission.)*

posed to. It is interesting to realize that some very practical guidelines might become available from basic research into learning processes. The Gibson theory and the Bandura group studies have already suggested that witnessing aggressive outbursts is more likely to "teach" children to engage in similar behavior than it is to help them discharge tensions.

THINKING WITH AND ABOUT NUMBERS

I touched on the preschool child's comprehension of numbers in Chapter 8. It is useful to follow this facet of mental development into the early school years both for its own inherent interest and for what it tells us of cognitive growth in general. John Flavell, a leading researcher and theoretician in cognitive development, suggested: "The elementary-school child has what might be called a *quantitative attitude* toward many cognitive tasks and problems. He seems to understand better than the younger child that certain problems have precise, specific, potentially quantifiable solutions, and that these solutions may be attained by logical reasoning in conjunction with well-defined *measurement* operations" (1977, p. 85). It is not just that the school-age child can think better about numbers. It is that he or she can use logical reasoning of a mathematical kind in approaching life in general. Quantitative as well as qualitative differences in the world are much better appreciated.

Making sure one has a fair share is often a first revelation of the quantitative attitude

What does the average child now understand about numbers and their place in the world? In examining the extensive research in this area, Flavell selected a few basic findings:

1. *Subitizing.* The child can look at a few objects and see how many are there without having to count them. He or she also knows that when there are *more* than a few items it is useful to take the time to count them; perception alone cannot be relied upon.

2. *Counting.* The child can now count, without counting anything in particular. The regularity of numbers is appreciated and is within the individual's control somewhat apart from specific perceptual manifestations. Most children can also do "fancy counting"—counting by two's, by ten's, and, to some extent, backwards. (Who knows how influential the reverse counting of lift-off command has been on the present generation? "10 . . . 9 . . . 8 . . . 1 . . . blast off!"). The apparently simple process of counting respects important mathematical rules. The child understands, for example, that numbers should not be skipped over or repeated. Furthermore, the child can count in accordance with task specifications (e.g., if instructed to do so, will count up to 15 and then stop; this shows an effective memory dimension in verbal-cognitive behavior as well).

3. *Cardinal and ordinal aspects.* The child understands both the place of a number in its particular series (the ordinal aspect), and the final or highest number in the set (cardinal aspect). He or she knows that 10, for example, is higher than 9, and also denotes a larger set.

Elementary school children know the basic language of numbers and can perform many fundamental operations with them. (Photo by Talbot Lovering.)

4. *Numerals.* The child can both speak and write arabic numerals. He or she appreciates that numbers are the same whether spoken or written out.

5. *Correspondences.* Unlike the preschool child, the elementary-age boy or girl is able to grasp the equivalence of sets. There are as many in this set as in that one, and both sets added together make a new set that is twice as large as either of them separately. The child may also know how to divide a set into two or more equal subsets by putting its elements into one-one correspondence with each other.

6. *Comparing set size.* By middle childhood there is usually a clear understanding of "more than," "less than," and "just as much as." Guesswork is no longer necessary, as in preschool children who sometimes achieve a correct performance through trial and error. The child now knows that "more" and "less" are *relationships,* not bound to fixed and specific numbers.

7. *Relevant and irrelevant transformations.* The increasingly sophisticated elementary-school child cannot be fooled about number through irrelevant transformations. Expand or contract that row of pennies, stack them up, or stand them on end (*see* the representational stage of mental development in Chapter 8); he or she *knows* that the number is not changed. **Number conservation** has been solidly acquired.

Number conservation is the understanding that "how many" is not changed by rearrangements

8. *Nature and generation of the set of whole numbers.* There is some grasp of the basic idea of the number series as such. The child knows that another number can always be added. Eventually there is also the recognition that "bigger" numbers behave in the same way as smaller numbers; they are generated by the same rules and subject to the same arithmetical operations. Number conservation is understood to apply to large as well as small sets.

Flavell saw a reciprocal relationship between the child's growing skill with numbers and his or her understanding of number. Learning to count properly is a skill, for example, but it provides stimulation and foundation for improved understanding of what can be done with numbers.

The general quality of the young child's thought processes is now changing. This can be gleaned from his or her more commanding grasp of the concept of number. Advances both in skills and reasoning with numbers are evident. But Flavell and others have cautioned that the specific nature of the task at hand also has much to do with a child's performance. Some tasks make it difficult for the child to apply what he or she knows; other tasks are easier because of their content or structure. Developmental assessment, whether in the realm of arithmetical reasoning or elsewhere, can never be separated from the specific measures and tasks involved.

MAKING MEMORIES—AND MAKING MEMORIES WORK

The role of memory functioning in the child's total development has been studied with renewed vigor in recent years. It is worthwhile for us to follow up on what was said about the preschooler's memory in Chapter 8. The overall view I am taking here is that what is called *memory* refers to a set of complex processes and dimensions that cannot really be separated from mental functioning and social behavior (although specific studies often do attempt to isolate memory-related variables for close attention).

It is not surprising to discover that the elementary-age child has a more adequate and useful memory than the preschooler. He or she has had the time to accumulate more experience and, therefore, more past (episodic memories) to utilize. The situation is more interesting than this, however. The general level and style of the school-age child's thinking (as opposed to the younger child's) makes it possible to do more with experience. This has already been seen, for example, in Piaget's study of reconstructive memory after a week and after six months or more. *See* Chapter 8, "Development of Memory in the Young Child.") Advanced linguistic and reasoning skills furnish the child with a more effective semantic memory as well. There are more experiences that have more meaning for the older child. This makes it easier to both store and retrieve memories.

The active, participatory quality of memory can already be seen in the school-age child. Grade-school children have been presented with stories that have certain gaps in them (Paris, 1975; Paris & Lindauer, 1976). One story, for example, told how a girl rescued an unspecified something that had been "flapping its wings" under the porch. In stories such as this, children not only recalled the essence of what was actually presented but also filled in reasonable objects and events for what was missing in order to make the stories more coherent. Usually the children believed they have actually "heard" and, therefore, remembered what was actually information their own minds had supplied to make the story complete. This type of memory that "improves" on what was actually presented is important for daily functioning by adults as well as children. Hagen, Jongeward, and Kail (1975) claimed that children could not carry on everyday conversations unless they were able to make spontaneous inferences, elaborations, integrations, and reorganizations. Flavell (1977) also reminded us that even the adult listener must often add something through his or her own understanding and memory to make complete sense of what he or she has heard. Studies such as those of Paris and Lindauer are now beginning to show that as children grow older they become better at making the kinds of inferences and elaborations that take the shape of a satisfying memory.

Rehearsal is the memory strategy of mentally practicing what one wants to retain

Other important changes take place in the *strategies* through which memory is made effective. **Rehearsal** is one of the best-known and best-studied types of strategy. In one form or another, the individual practices what he or she wants to remember. He or she might repeat a list of terms or names aloud or write them down. Spontaneous rehearsal has been shown to increase with age from kindergarten through the fifth grade (Flavell, Beach, & Chinsky, 1966). This was detected by a trained lip-reader who observed what the children were doing in the fifteen-second interval between making an observation and being asked to recall it correctly. Another study soon discovered that children who spontaneously used verbal rehearsal had better memory performances than those who did not (Keeney, Cannizo, & Flavell, 1967). But the nonrehearsing children would also use this strategy when given some instruction and demonstration. Afterward, their recall reached the same higher level of the spontaneous rehearsers. All of this sounds satis-

Reading is one thing; remembering another. Children who can use strategies such as rehearsal and grouping often have more success in remembering. (Photo by Jean Boughton.)

fying and makes sense. The children did not give away all their cognitive-motivational secrets in this study, however. Later in the same study, those who had learned how to rehearse and did so well with it were given their choice about using or not using this strategy. More than half of the learned rehearsers abandoned the technique that had improved their performances! Being *able* to do something and *wanting* to do it are not identical. Perhaps this aspect of the study has implications for the ins and outs of memory functioning in adult life and in advanced old age. Motivation and memory are both functions of the same individual.

The studies I have cited are typical of most studies into memory in that they concentrate on *intentional memory.* The child or adult learns something with the intention of recalling it later. In real life, of course, we often find ourselves called upon to recall something that we did not salt away for explicit reproduction. (This means we probably did not rehearse it or use any other deliberate memory strategy.) More research into *incidental memory* (nonintentional and nonrote memory) in children is needed to broaden our understanding of the total pattern of memory development.

Production deficiency is the inability to use a certain memory strategy

Mediational deficiency is the inability to use a certain memory strategy effectively

To understand much of the existing research, particularly that of a recent vintage, it is helpful to make a distinction between **production deficiency** and **mediational deficiency** (Flavell, 1970). When a child does not, and apparently cannot, use a particular memory strategy, he or she is said to have a production deficiency. If he or she knows the strategy but it doesn't do much good, then it is a mediational deficiency. Older children are more aware of memory tasks and of the possibility that it might be useful to have some special way of meeting them. The older child is also more likely to be able to use a particular strategy in a particular situation. For reasons such as these, there is usually more production deficiency among younger children. It is also possible that strategies for recall are cumbersome, difficult, and energy-demanding for the younger children. The somewhat older child has a smoother and easier time with these more-practiced strategies and therefore may use them more often.

Organizing or *grouping* is another type of memory strategy that finds more use, and more effective use, as the child develops through the early school years. This seems to be a more advanced kind of strategy than the repetitious rehearsal method. There are many ways to organize experiences and observations for recall, just as there are many specific ways to rehearse. Older children are more likely than younger ones to separate material they are studying into categories or sets (Moely, Olson, Holmes, & Flavell, 1969). This seems to facilitate both learning and recall.

Within the limits of present knowledge, it appears that children become more versatile in the type of memory strategies available to them, and apply these strategies more appropriately and effectively as they develop in all their mental and social spheres of competence. Older children also understand their own memory processes much better than young children (Flavell & Wellman, 1976). Older children recognize what they can and cannot do, for example, "There are too many pictures for me to remember now." While younger children overestimate how much they will be able to remember, older children are more cautious and realistic. The term ***metamemory*** has recently been proposed for the individual's understanding of his or her own memory processes (Flavell, 1977). This may point the way to a fruitful new area of inquiry.

Metamemory is one's understanding of one's own memory processes

In this section we considered some of the many dimensions of mental functioning in school boys and girls. Let's now consider some of the difficulties that children may encounter in their early classroom experiences.

HAZARD: THE SLOW LEARNER

The challenge of performing up to expectations in school is not met equally by all children. A boy or girl who begins as a "slow learner" may face several types of hazard, depending both on the underlying basis for the difficulty and on how the school and family respond. I am not concentrating here on the child with obvious and severe developmental impairments; such impairments are usually (although not always) recognized during the preschool years. Instead, I am thinking of the youngster who is labeled a *slow learner* or *different* for the first time when introduced to the classroom situation. This means thinking of more than one type of problem and more than one type of solution.

MINIMAL BRAIN DYSFUNCTION

Educators, physicians, and child clinical psychologists have been paying increased attention lately to learning disabilities that are associated with relatively mild or specific neurophysiological problems. This condition is often known as *minimal brain dysfunction* **(MBD).** It is believed that more children with MBD are entering the schoolroom today than previously. (Possible reasons will follow.) Let's first become acquainted with the kind of behavior that is typically associated with MBD. We will try to see the problem through the eyes of a perceptive elementary school teacher.

MBD is a state of mild CNS impairment that interferes with learning

The teacher notices that one of the children has an unusually *limited attention span;* he cannot keep his mind on a discussion or activity as long as most children his age. Along with this, he is more *distractible.* A noise or movement that is ignored by other children breaks his concentration. In fact, the child in general seems less able to control his behavior. Most of the children have their squirming and restless moments (as does the teacher herself). But this child is almost continuously moving around in his seat, swinging his feet, tapping his fingers, and so forth. This kind of behavior overflow or *hyperactivity* is one of the most frequent indications that MBD may be involved.

Through careful attention to the way in which this child approaches his school work, the performances themselves, and his behavior outside the classroom, the teacher may find other indications that are consistent with MBD. She notices, perhaps, that the child has difficulty in *fine visual discrimination.* He can see most things well enough. Close inspection of letters and words, however, requires a different kind of visual ability from what is usually required of a child prior to his exposure to reading and writing exercises. The child may reverse letters and words in a mirror-image way. (This is something that many boys and girls do at first, but the problem is more serious and lasts longer with the MBD child.) Similarly, the copies he attempts of geometric figures reveal more distortions than do the drawings made by most other children of his age. While some of his difficulty appears to be in fine visual discrimination, the teacher may observe that the child's *physical coordination* also contributes to the problem. This may show up in small ways in class exercises, but it can be seen more readily during recess when he has more than his share of spills and tumbles in running around and demonstrates less skill in active games.

Furthermore, the child does not seem to be a very good listener. Like his vision, his hearing may be well within the normal range for most purposes. But close listening makes demands on him that he cannot easily meet. In other words, he has problems in *fine auditory discrimination.* Sometimes he appears to be a slow learner because he has not taken in the material adequately in the first place, perhaps in connection with the visual and auditory discrimination problems that have been mentioned. At other times, it appears he has not retained the knowledge that he did acquire. The distinction between "taking in" and "giving back" problems is seldom easy to make with assurance. A special assessment of the child's functioning may be necessary to determine how much of the difficulty is attributable to perceptual uptake and how much to problems in remembering or reporting what one has in fact learned. It is fairly common, however, for children with MBD to experience some degree of *memory impairment.*

As you can imagine, any combination of difficulties such as these is likely to cause problems in mastering the required school work. Furthermore, the child may also require a greater share of the teacher's attention if his hyperactivity is not to disturb the other children. And the experienced teacher knows that she may be seeing only the tip of the problem. If the child falls behind so early in his school career, this might well contribute to many years of frustration and unadaptive behavior in the future. It is not just a matter of the child being a little slower than other boys and girls in an early grade level; it is also a matter of getting off on the right foot in the school career. This has many implications for general development and adjustment throughout the lifespan.

The teacher may at this point consult with the parents and the school psychologist. From the latter, she may learn, for example, that the child is of normal intelligence or better. (Some people are surprised when they discover that MBD does not necessarily mean fundamental intellectual deficit.) Yet the boy or girl does have a number of handicaps that interfere with learning, including problems in the perception of spatial relations and depth, and some specific impairments in the retention of information. She will try to present the situation to the parents in a manner that is as nonalarming as possible, but which conveys to them the major facts and considerations. The parents' response will be very important in determining their child's educational future. Do they feel a need to deny that anything could possibly be wrong with their child? If this denial persists ("Oh, that can't be true. . . . He'll grow out of it. . . ."), then a valuable opportunity to help the child at an early time may be lost. Do the parents, on the other hand, overreact and conclude the child is "hopeless"? This error could lead to the same delay of assistance that sometimes results from denial. Or do the parents turn their feelings inward, almost forgetting the child himself in their own distress? This reaction is often accompanied by recriminations against oneself or one's spouse, "What is wrong with us that we produced a child who is less than perfect."

Here is where good rapport between parent and teacher can make a big difference

Often the services of the school counselor or psychologist, the family physician, or another parent who has been through the same experience prove quite valuable to all concerned. Some parts of the country have well-established associations that are ready to provide understanding and help. The Massachusetts Association for Children with Learning Disabilities, for example, is a nonprofit corporation that provides a variety of services to perceptually handicapped* children and their parents. Although one cannot say that every school district and every geographical area has an abundance of services available to meet these needs, many teachers and physicians, as well as local mental health associations, are alert both to the identification of MBD and to the measures that can be taken to assist the child. Before discussing some of these measures it would be useful to examine the circumstances that often precede the recognition of MBD in the school situation. Psychiatrist Richard A. Gardner has written such a concise and informative summary that I can do no better than to quote it here:

> Minimal brain dysfunction is seen in children whose mothers have suffered with a variety of difficulties. During pregnancy, these include excessive bleeding, nutritional deficiencies, infectious diseases such as German measles, and overexposure to radiation (including diagnostic X-rays). The disorder is more commonly seen in babies who were born prematurely, in first-borns (possibly related to their mothers' longer labors), when parts other than the head appear first, and when other mechanical difficulties are present during the birth process. Toxemia of pregnancy, excessive bleeding during delivery, umbilical cord complications (such as cord around the infant's neck), and Rh incompatibility have also been implicated as causes. After birth, the disorder can result from

* A distinction is usually made between children with learning disabilities, known as *perceptually handicapped,* and children who have *sensory handicaps* such as blindness and deafness.

> any of the infectious diseases that can damage the central nervous system, such as meningitis and encephalitis. It has also been associated with a group of diseases known as the "inborn errors of metabolism" of which phenylketonuria (PKU) is probably the most well known. These genetically determined biochemical derangements interfere with the normal metabolism of substances that are necessary for normal brain cell functioning. In less specific ways genetic factors also play a role in causing brain dysfunction. There are families in which the disorder seems to be transmitted genetically. Relatives on one or both sides of the child's family may have exhibited symptoms of minimal brain dysfunction during their childhood. (1973, p. 20).

Lengthy as this list of causes may be, Gardner added that still other conditions can lead to MBD or intensify the condition (as, for example, convulsive seizures whatever their cause). The factors quoted above represent only the most common antecedents of MBD. In some cases even a very careful medical history and examination cannot establish the precise cause, but the child is considered to have MBD on the basis of the neurological and behavioral signs usually found with this condition.

Some parents need help in overcoming guilt unnecessarily attached to their child's problems

Reflection on this list of causal factors helps us to understand why MBD is becoming more visible as a problem of elementary school children. As Gardner puts it, "Children who only fifteen to twenty years ago would not have survived pregnancy and childbirth are now doing so—only to suffer the effects of their narrow escape from death" (1973, p. 21). This is probably the major reason. However, it is also likely that closer attention is now being given to the perceptual and learning processes of schoolchildren as a consequence of improved psychological knowledge and techniques. Child psychology has gained in scope and precision over the years, as might be expected, and this clinical and academic resource in turn provides teachers with more cues to guide their instruction and assessment. Furthermore, it is possible that we expect more of young children today in terms of competitive achievements. Some parents are already worried about their child gaining admission into the college or professional school of his or her choice, while the boy or girl is still learning the alphabet. This pressure, of course, has its way of influencing the school system, and vice versa. The stress of achievement orientation may come down especially hard on those who enter the situation with underlying difficulties and, thus, their problems become more visible sooner.

Medications sometimes prove very beneficial to the child with MBD. This is a situation in which it is vital to have a physician in whom the parents can put their trust. Some adults have strong feelings about the use of drugs in general or with children in particular. Alarmed at reports of drug abuse and crime, the parents may prematurely decide that their child should not be introduced to medication at such an early age. Other parents may find it all too easy to use drugs as a solution to virtually any childhood disturbance—it is more convenient to hand the boy or girl a pill than to explore and cope with the underlying problem. The knowledge and perspective of the physician is necessary here. What works for one child may fail with another; the physician himself or herself may have to do a bit of experimenting with drug and dosage before achieving optimal results, hence the folly of parents taking over the physician's role.

There is much that parents can do, however. Perhaps the best place to start is with their own attitudes. Are they able to provide the basic love and protection that all children need without either overdoing or underdoing the kinds of guidance and assistance required by this particular child? Gardner's excellent book, to which reference has already been made, is a useful source for parents in this situation; the book also includes a unique section meant to be read by boys and girls themselves (recommended for older children). Parents are also advised to contact local mental health or learning disability agencies to become acquainted with special educational programs, counseling services, and other resources that might be available to them.

OTHER CONDITIONS

Some people have spent *fifty years* in schools for the retarded although their problems were minimal

One hazard is that learning disabilities associated with MBD will not be recognized early. But another hazard is that learning disabilities will be erroneously attributed to organic problems when, in reality, the difficulties are in another sphere. One child begins his school career as a "slow learner" because his mind and feelings are absorbed elsewhere. The "elsewhere" takes in a wide variety of possibilities. There may be a new brother or sister at home who is commanding too much parental attention for the schoolchild's comfort. In this situation, instead of rising to the challenges of his new role, the child may relapse temporarily into a more baby-like orientation. On the other side of life's continuum, there may have been a recent death in the family. Even if the boy or girl is not talking much about death at home or showing any obvious reaction, this might be the major problem he or she is trying to comprehend both intellectually and emotionally. Either of these situations might have first call on the child's energies and not leave much left over for the classroom. Many types of family tension and disturbance can take their toll on the elementary school child, just as they do on the breadwinner's performance on the job.

Other problems may be linked more closely with the school situation itself. Consider, for example, the child who has had older siblings already pass through his or her current grade level at the same school. The teachers may have developed strong expectations about the next Smith, Kochansky, or Garcia. He or she may be awaited as a genius or as a dud. Children are sensitive to adult expectations and are human enough to respond in a variety of ways. The child may have, in effect, decided that he or she cannot possibly live up to the fame of an older sibling. His or her early school performance shows this resigned state of mind. But it also shows "slow learning," which might be interpreted as reflecting less intelligence, neurophysiological problems, or whatever else suggests itself to the observer.

Or the young learner may be "slow" because the instructional techniques used in this particular school conflict with those he or she encountered in a previous school (a problem that is not at all rare). His or her lack of rapid progress can also stem from hurtful experiences with other students. Classmates may have teased and ridiculed a child into a state of mind that makes him or her dread every aspect of school: "The kids are so mean to me. I'm never coming to school again!" A specific functional problem, such as a speech impediment, can make the child feel isolated or rejected. Even a regional accent that peers consider to be "weird" can have this effect. A child with physical characteristics that make him or her "look different" to peers may also be in for a rough time. These experiences do not always lead to learning difficulties, of course, but the sensitive teacher and parent will explore such factors when trying to fathom the reasons for lack of expected progress.

You can see that in some situations the problems may be temporary. The rejected child makes friends after all. The child disturbed by competition with a new baby discovers that new status and appreciation can come through school success and through being a privileged big brother or big sister. Sometimes, however, the difficult beginning sets the pattern for much of what is to follow. An attitude of defeatism takes over. There is no substitute for careful attention to the individual situation of each child, and the development of an appropriate level of interventive assistance where indicated.

In concluding this exploration of the "slow learner" I find myself thinking about the long-term effects of words and attitudes. Early in his or her elementary school career, a child may acquire a reputation that stays with him or her much of the way. Although the specific reasons why the child was a slow learner for awhile may have long since disappeared, the child by now has accepted this identity. He or she is not supposed to be quite as smart as the others, or he or she is supposed to receive "special" attention (whether for the better or the worse). We are challenged, therefore, to have our perceptions and facts in good order when drawing conclusions about the performance of a young child in school. Further, we must keep our minds open for revising these conclusions as warranted by new developments. There are problems enough ahead of the schoolchild without carrying the extra burden of a label that prejudices both the child and others.

Having now explored some of the physical and mental dimensions of the young child's life, I will conjur a scenario that places the whole child in the new school environment.

SCENARIO 6: SCHOOL BELLS FOR MICHAEL

Up until now, the school bells have sounded for other children. Today they will sound for Michael as well. It is an exciting day not only for him but for everybody. For the past two weeks all the media have given prime attention to the impending start of the new school year. There is a mood of celebration.

In fact, the opening day of school (which is the same day across the nation) has the character of a national holiday. The basic theme of the day is appreciation of the opportunity to attend school, affection toward teachers, and congratulations for those children who are entering the classroom for the first time.

The festive mood continues as the doors swing open. Michael is accompanied by a host of proud well-wishers: his parents, other relatives, and neighbors, all carrying bouquets of flowers. Next there is a ceremony in which all sorts of people make all sorts of brief speeches, emphasizing the joy and splendor of the occasion. One of the children, perhaps the smallest of the first-graders, is given a little hand bell to ring, and the children move inside to present their flowers to the teachers. They have already learned to expect the teacher to be not only an important person in their lives, but also a warm friend.

Michael is now seven years of age, as are all the other first-graders. He will be expected to learn basic skills and knowledge upon which he can build more advanced abilities. But his education will also place strong emphasis on conduct and morality. Each school day, he will greet his teachers and classmates by name, put his belongings in order, stand up when addressed, sit in an alert position for listening, reading, and writing, and perform classroom duties. The latter might include being class librarian, gardener, sanitarian, or general teacher's assistant. While he is learning the proper conduct for a first-grader, Michael is also becoming acquainted with the new roles he will have in the future. For example, when he runs out to play in the schoolyard he knows that older children will be there who take a special pleasure and responsibility in looking after him. One of the fourth-grade classes has "adopted" his first-grade class in a big brotherly and big sisterly way. The older children also help him with his lessons. Michael realizes that someday he, too, will have the opportunity to look after younger children.

Michael is also learning what is expected of him at home and in public places. When he wakes up in the morning, for example, he greets his parents cheerfully, thanks them for breakfast, and makes it his business to check that he has anything he might need to take with him to school. Throughout the day at home, he takes responsibility for his own things. Time at home is expected to be fairly active. He helps with the housework within the limits of his abilities, plays with younger children, tends the family garden, and does his homework. The rule is, "Job done—take a rest, then start another job." When in public, Michael minds his elders carefully and does not disturb others with boisterous activity and unnecessary noise-making. There are "do's" as well as "don'ts." He is encouraged to learn what goes on in his environment. How does the public transportation system operate? What kinds of businesses and industries are functioning in this area? Are there parks or buildings of special interest? In keeping with his level of development, Michael has freedom to explore the physical and social space around him. He is made to feel especially welcome in places where people are doing things for the benefit of others. And, of course, he is encouraged to become well acquainted with what his own parents do that is valuable to the community.

To introduce a small note of conflict into this scenario, let's suppose that Michael has moments when his own needs and impulses do not seem to fit in with expectations. Here he is, a first-grader, running out to the schoolyard with energy to discharge and a particular plaything in mind, perhaps a large and attractive ball. But another child has reached the ball at about the same time and in a similar mood. Michael and his classmate each want exclusive rights to the ball. A vigorous and noisy dispute follows, with the prospect of a fierce scuffle near at hand. But the scuffle does not occur. Instead, a teacher or assistant calls out in a loud and cheerful voice, "Look here, children, see how beautifully Peter and Maria are playing together, see how they have learned to play a new game!" The other boys and girls come over to observe the cooperative behavior of Peter and Maria who, in turn, showboat a little to take advantage of the praise and attention. Michael and his competitor drop the ball and join the throng. A few minutes later they are engaged in other, more approved activities without ever having been directly approached or reprimanded for their momentary deviation.

You have probably caught on that the Michael presented in this scenario is not the same boy introduced in earlier chapters (although in many respects he could be). What has been described is the early school experience of a child in the U.S.S.R. In a book that is well worth your attention, Urie Bronfenbrenner, a distinguished psychologist at Cornell University, describes and analyzes *Two Worlds of Childhood: U.S. and U.S.S.R.* (1973). This book has already stimulated considerable thought and debate among educators in this country. Bronfenbrenner's observations indicate that **altruistic behavior** (helping others; placing the needs of other people on an equal or superior footing with one's own) is systematically encouraged in the Soviet educational system. Those boys and girls who attend day care programs before entering the first grade are exposed even earlier to expectations for cooperative and altruistic behavior. Preschool children are taught to be self-reliant, to listen carefully and express themselves clearly, and to find their basic sources of pleasure through interaction with other children and adults in noncompetitive pursuits. If a child shows too much "selfish" behavior, the "upbringers" examine their own behavior to see if perhaps they have not expressed enough affection along with their teaching, or have otherwise been insensitive to the child's needs. In other words, a child who is marked by a consistent streak of self-seeking and competitive behavior represents at least a temporary failure of the system. The citizen who is the end-product of the Soviet educational system is not expected to be motivated chiefly for his or her own competitive success. Furthermore, children, even first-graders, are expected to have meaningful social functions and responsibilities. One does not wait for the day after graduation to take on serious roles in family and community life.

Altruistic behavior is action intended to help others

Bronfenbrenner is not the first or only person on the American scene who has expressed disapproval of certain emphases in our educational philosophy and practices. His observations, however, indicate that it *is* possible to do things differently and successfully. One can choose to agree or disagree with the position he has formulated, but it is difficult to ignore the information he has reported in both technical and more general publications. Bronfenbrenner believes: "Education in America, when viewed from a cross-cultural perspective, seems peculiarly one-sided; it emphasizes subject matter to the exclusion of another molar aspect of the child's development—what the Russians call *vospitanie*—the development of the child's qualities as a person—his values, motives, and patterns of social response. . . . These are matters not only of educational philosophy, as they are sometimes with us, but, as revealed in the Soviet material, of concrete educational practice both within the classroom and without—in home, neighborhood and larger community" (1973, p. xxiv). He deplores what he sees in the United States as the isolation of both school and child from the larger social environment. "Moreover, the insularity characterizing the relation of the American school to the outside world is repeated within the school system itself, where children are segregated into classrooms that have little social connection to each other or to the school as a common community, for which members take active responsibility both as individuals and groups" (1973, p. xxiv). Bronfenbrenner believes this situation is becoming even worse because of increasing social disorganization on the American scene. He arrives at the controversial conclusion that *"the schools have become one of the most potent breeding grounds of alienation in American society"* (p. xxiv). Along with reading and writing, many children, left to their own devices with an insufficiently developed sense of human relationship, learn how to be disconnected from society. This is seen as having serious implications for their lives as adults.

Is Bronfenbrenner right?

More active and personalized than traditional classroom education, the Head Start program is a contemporary experiment in the encouragement of early learning. (Courtesy of ACTION.)

In a more constructive vein, Bronfenbrenner points to a number of educational principles and some actual practices in this nation that he believes are moving in the right direction. Some guidelines have been emerging from controlled psychological research on child development. This includes suggestions for improving the effectiveness of the teacher as a *model* for conduct. Those who are particularly interested in this topic will also find rich background materials in *Social Learning and Personality Development*, an important book by Albert Bandura and Richard Walters (1963). The Head Start program and its variants are taken as a prime example of practical endeavors that can help children not only to *feel* but to *be* more actively involved in the learning process and more a part of the community.

SCENARIO 7: SCHOOL BELLS IN PEYRANE

If the "same" Michael were introduced to the Soviet and U.S. school experiences at the same time, we might expect appreciable differences in their social and academic development (assuming the accuracy of reports from Bronfenbrenner and others). This would tend to support the contentions of outside-in theorists (*see* Chapter 7). But one of the dangers in presenting a particular contrast with our own school practices is we might take this as the only important comparison that could be made. A person might usefully examine many different types of school experiences around the world and *then* come back to think about the ideas and practices currently being applied in our own nation. I cannot include such an extensive survey here. But I can include a brief description of education in a French village that had not been touched by "progressive education" at the time Laurence Wylie lived there and prepared his detailed account (1964). This scenario is typical of an important tradition in French elementary school education, but it should not be generalized to all French schools.

Children begin school at age four in Peyrane, carrying little brief cases back and forth even though they are not given homework for several years. There is no ceremonial beginning, except for boys whose long curly locks are clipped into a formal haircut with a part on the side. The teacher greets each child warmly, and consoles him or her if the child cries after mother has left (something that is not likely to happen with seven-year-old beginners in the U.S.S.R.). Within an hour or so, the child accepts the new situation and, from then on, "no more nonsense is expected of him." Even the youngest child must sit still for long periods and accept the school discipline. In some respects, this is more a nursery school or day care program than a formal class situation. Yet the child is officially launched on his or her school career. Furthermore, he or she is learning what it means to learn: "They are impressed with the fact that to learn means to copy or to repeat whatever the teacher tells them" (Wylie, 1964, p. 59). Self-expression is not encouraged, far from it. Control is the major aim of the first two years of school; later it becomes the means for further learning to take place. By age six, most children are "mature enough" to begin their formal education (i.e., they can sit quietly for long periods of time, pay attention, and inhibit any personal thoughts or feelings that might start to intrude).

The lesson plan is quite standardized. Little allowance seems to be made for individual differences in ability or temperament. In practice, however, teachers often use their own experience and good judgment. The classroom scene is more congenial and flexible than it would appear from examining the educational plan and method on paper. Nevertheless, it is likely to appear regimented and stilted to the observer who is acquainted with more participatory styles of education. Some of this impression is generated by the rote learning techniques still favored (memorizing large segments of the course materials), and by the neglect of curiosity behavior. The facts are given; the children are there to receive. Basic principles are not to be questioned. The children do seem to learn much material thoroughly, including to a very large extent information and skills that will be of practical use to them in their daily lives. But they may be at a disadvantage in such areas as formulating questions and developing imaginative approaches to new problems.

If this were Michael's school experience, he would also find himself exposed each morning to a "moral lesson" read by his teacher from a standard book stuffed with short stories and vignettes. The morals may be very similar to those taught in the U.S.S.R., for example, the value of cooperation and teamwork. However, the effect is more dubious here because: "These are not characteristic virtues of the people of Peyrane. A few lessons are so completely in conflict with the customs . . . that it seems futile to teach them" (Wylie, 1964, p. 65). Such an example was noted by Wylie. All the children were required to learn by heart and repeat aloud this sentence: "Let us be the friends and protectors of the little birds." But as Wylie observed, "In a region where a favorite dish is roasted little birds, where a husky man boasts of consuming fifty or sixty warblers at a sitting, there is little likelihood that this lesson will have much effect" (p. 65).

Meal times also have a regimented atmosphere for those children who do not have lunch at home. Absolute silence is maintained. A boy or girl may speak only if permission is granted by the teacher. Punishment for misbehavior (including lack of attention in class) often involves the *shaming process*. The child who has not controlled his or her behavior adequately or memorized his or her lesson is often exposed to a humiliation experience in the class. This technique is usually effective in changing behavior. Physical punishment is employed rarely and then usually in a token manner. The idea of shaming as a major disciplinary technique fits well with parental attitudes. The parents themselves will not tolerate "shameful" behavior on the part of their children.

Welcome opportunities for play are provided during morning and afternoon recess periods and the latter part of the lunch hour. Play equipment is almost completely lacking, but the children seem to find enough to do. The girls often sit in a sunny place and exchange secrets, or seek out and comfort anyone wandering around the schoolyard who seems to need some loving. Young boys such as Michael are more likely to run around madly, playing active games. It would not be surprising if Michael and two or three other boys picked out another classmate as a scapegoat and spent the play period chasing and tormenting him. The teachers would be on the scene, but they would not interfere unless physical mayhem appeared imminent. The children are permitted to *say* virtually anything they want to each other; verbal cruelty is tolerated. Physical attacks, however, (and risk-taking behavior of a physical kind) are definitely not tolerated. Both parties in a brawl receive punishment. It would seem that verbal techniques for controlling behavior, such as shaming and scapegoating, are practiced both by students and teachers.

THREE MICHAELS

Michael of Peyrane will soon shape up as a neat, polite, very well-behaved young man (around adults, anyway) who learns his lessons and does his share of work aroung the house and farm. The village parents are generally satisfied that their school system is producing the desired results.

Michael of the U.S.S.R. will also be well controlled and respectful toward his elders, but he will perhaps show greater spontaneity and a deeper sense of involvement in cooperative enterprises. He derives much of his sense of worth and pleasure from being part of a functional social unit.

Michael of the U.S.? If we continue to follow the main contours observed for school experience, then Mike would turn out to be quite a resourceful, competitive youngster with an advanced appreciation of how to look out for his own interests. He will appreciate individuality in others and have a flair for expressing his own. But Mike may also have a less secure sense of his relationship to the community at large and be less inclined to both extend automatic respect to his elders and accept facts and explanations without questioning.

These speculations on the three Michaels are related to observations that have been made in three different cultural settings, but they are not to be mistaken for ironclad scientific conclusions. The important point is we should recognize that the nature of the elementary school experience is intimately associated with cultural values and priorities, and it functions as a significant influence on the character of each boy and girl who steps across the threshold of the schoolhouse.

SUMMARY

As the child's horizons expand to include the school world, physical development tends to be balanced comfortably between stability and continued growth. The rate of development has slowed down appreciably, many aspects of physiological functioning are completing the transition to adult levels,

YOUR TURN

1. The house that is home to several children in the age range we have been discussing often runs short on bandages and runs long on incidents of minor injury and frustration. It has been mentioned that the improved physical coordination of young schoolchildren encourages them to attempt feats that are not yet within their mastery. Look into this yourself. Interview several boys and girls. Ask them to tell you about their latest cuts, bruises, scars, and injuries. How did they get them? Were they aware ahead of time that they might get hurt? When were they aware? With a little encouragement you may hear an extensive recital of minor (but famous) injuries suffered in the course of being an active child. Among other things, notice the attitudes the children express toward their injuries. Do they take a certain pride in them as cherished memorials to heroic actions or in the brave way they responded to the injury? Has the injury convinced them not to try the same action again? Does the injury seem "unfair" or a "punishment" to them or, by contrast, is it seen as an objective consequence of the action, something that one has to expect? Notice also if there are any differences between the boys and girls, either in the frequency and type of injuries reported or in their attitudes toward them. Thinking over your findings, what role do you believe physical adventurings and minor injuries play in the young child's experience and development? How much of the child's world of experience includes injury and adjustment? What effect, if any, does this have on the child's sense of body integrity and vulnerability? Would we understand young children better if we took these experiences into account, or are they safely ignored by parents except when serious physical injury is a possibility?

2. Draw a large outline figure of a human body. Ask boys and girls of preschool, early school, and later school ages to draw what they know or think might be inside the body. What names do they give to these inner parts and where do they locate them? What functions do they attribute to the parts they have represented? (You can ask them, "What does this do?" or "How does this work?") Discuss the drawings in terms of: (1) age differences, (2) sex differences, and (3) what areas of the human figure are more filled with internal parts than other areas, and why this might be. You may notice other features of the drawings, or the children's comments about them, that are worth discussing. This exploration could be made even more interesting if you let the children share their questions about the human body with you. What would they like to know about how the body works? Answer some of these questions (if you can). Take another outline figure and show the child how you would draw in a few of the most important internal organs, telling at the same time how these work or what they do. See if the child can then draw a more accurate internal portrait of the human figure and tell you how the parts work.

3. If you are serious about humor and its role in human development, read the scholarly review of research and theory by Paul E. McGhee (1971). This article also includes a good list of further references. Write a brief discussion and analysis based on this article and your own thoughts—you may enjoy taking the time to make your own fresh observations on humor in young schoolchildren.

4. Can you recall some of your own daydreams when you were in the early school grades? Just for your own interest, review what you can recall of your daydreams. What moods were expressed in them? What basic themes? How were they related to the situations in which you found yourself? Have those daydreams (and those themes) vanished completely from your personality today, or have they left a residue? Do you daydream more or less now? Do you respect your daydreams, or do you feel apologetic about them? You might let yourself tune in a little more to daydreams and fantasies in general, as these are topics we will continue to explore in chapters ahead.

5. Discover what resources are available in your own geographic area for helping the child (and his or her family) when there is a problem of slow learning. Where can people go for help? What services are provided? (Social service directories are available in some areas and can be very helpful not only for this but for many other purposes.) Perhaps you can attend a local meeting on this topic or make a personal appointment to interview somebody who is experienced in diagnosing and helping the "slow learner."

6. In this chapter, I offered scenarios for a child starting school in the U.S.S.R. and in a rural area of France, but I only hinted at one for the United States. Draft a scenario that fits in with your own perceptions of starting school in the United States today. How is your protagonist (boy or girl) prepared for and introduced to the school experience? What values are encouraged by the school environment? Do these values represent what the parents and community also hold significant? Be faithful to what you yourself have experienced and observed, although you may wish to supplement your personal knowledge with further readings. When you have completed a scenario that satisfies you, think comparatively. What are the "best" features of the school experiences in the three scenarios? Is it possible for any one society to offer, through its schools, a foundation of experience and learning that has all the features you believe most valuable and few, if any, of the negatives? If you are willing to make the imaginative leap, propose an ideal scenario and try it out on a few people whose opinion you value.

physical coordination has improved, and the child even looks more "grown up," to mention some of the more significant changes. The basic pattern of physical development is similar for both sexes during these years of *stable growth*.

One of the child's challenges at this time is to become better acquainted with the new physical person he or she is becoming. Improved self-knowledge of the body as "home base" includes, for example, learning both the right/left distinction on one's own person and gradually overcoming *spatial egocentrism* to recognize that other people have different coordinates and perspectives. The child's image of the human body shows up in the ability to take inventory of anatomical features and, perhaps more revealingly, in his or her figure drawings and other forms of creative play.

The child's mind seeks out other challenges at this time as well. In fact, it is not going too far to look on the five, six, or seven year old as a *philosopher*. These children exhibit a questing and intellectual vibrancy not always found among adults. The child's intellectual zest often expresses itself in *humor*. The child's spontaneous wit reveals itself more frequently, and he or she displays an appreciation for jokes, riddles, and simple paradoxes. This burgeoning sense of humor suggests a new ability to gain perspective on the world and to recognize what is predictable and what is astounding.

Questioning and *creating* are both significant aspects of the young child's mental activity. There is more curiosity about life and death (where we are going as well as where we come from) than adults usually credit to the child. Self-generated learning experiences, including many simple experiments with the world, continue to be important even as the child begins to receive more formal education in the classroom. With his or her newly cultivated *peer-group identity*, the child may also spin and share fantasies with others. Children who express much fantasy and imagination are also likely to have a superior ability to regulate their own actions. Creative expression in young children sometimes has a beauty, power, and sense of form that demands consideration as true art work. The *child prodigy* is recognized in every generation, but there are also many other creative minds at work and play during the early school years.

Perceptual learning is another fundamental aspect of the young child's mental activity. The child not only learns new bits and pieces of information but, more importantly, develops increased sensitivity to similarities and differences in his or her total environment. He or she perceives the world in a more *differentiated* manner through various types of active exploration. The child is learning about social interaction and "rules" of behavior as well as objective characteristics of the world. This can be seen, for example, in studies of *aggression* conducted under different circumstances.

General intellectual growth during the elementary school years is shown through the strengthening grasp of *numbers*. The child comes into a new set of skills in the use of numbers and displays such reasoning abilities as appreciating "more than"/"less than"/"same as" relationships.

Memory strategies improve during the school years as well. As boys and girls grow older, they become more adept at using *rehearsal, grouping,* and other strategies to assist in recall. They are also able to contribute inferences and elaborations that add to what they actually saw and heard in order to shape more complete and coherent memories.

Children who prove to be *slow learners* in their early classroom experiences sometimes have a form of *minimal brain dysfunction (MBD).* Some of the causes and manifestations of MBD were presented here, along with possible approaches for helping the child. Learning difficulties sometimes have other origins, however, such as the family or school situation. Some of these problems were explored.

The *cultural context* of the child's school experience is too important to neglect. Contrasting case histories were given of Michael as a student in the U.S.S.R., in a provincial French school, and in the United States. Implications for personality development as well as educational development were outlined.

Reference List

Anthony, S. *The discovery of death in childhood and after.* New York: Basic Books, 1972.

Arnheim, R. *Art and visual perception.* Berkeley, Calif.: University of California Press, 1954.

Bandura, A., Ross, D., & Ross, S. A. Transmission of aggression through imitation of aggressive models. *Journal of Abnormal & Social Psychology,* 1961, *63,* 575–582.

Bandura, A., Ross, D., & Ross, S. A. Imitation of film-mediated aggressive models. *Journal of Abnormal & Social Psychology,* 1963, *66,* 3–11.

Bandura, A., & Walters, R. *Social learning and personality development.* New York: Holt, Rinehart & Winston, 1963.

Bowlby, J. *Separation.* New York: Basic Books, 1972.

Bronfenbrenner, U. *Two worlds of childhood: U.S. and U.S.S.R.* New York: Pocket Books, 1973.

Cratty, B. J. *Perceptual and motor development in infants and children.* New York: Macmillan, Inc., 1970.

Custer, R. P. *An atlas of the blood and bone marrow.* Philadelphia: Saunders, 1949.

Feifel, H. (Ed.). *The meaning of death.* New York: McGraw-Hill, 1959.

Flavell, J. Developmental studies of mediated memory. In H. W. Reese & L. P. Lipsitt (Eds.), *Advances in child development and behavior* (Vol. 5). New York: Academic Press, 1970, pp. 182–213.

Flavell, J. *Cognitive development.* Englewood Cliffs, N.J.: Prentice-Hall, Inc., 1977.

Flavell, J., Beach, D. H., & Chinsky, J. M. Spontaneous verbal rehearsal in a memory task as a function of age. *Child Development,* 1966, *37,* 283–299.

Flavell, J., & Wellman, H. M. Metamemory. In R. V. Kail & J. W. Hagen (Eds.), *Memory in cognitive development.* Hillsdale, N.J.: Lawrence Erlbaum Associates, 1976.

Freyberg, J. Increasing the imaginative play of urban disadvantaged children through systematic training. In J. L. Singer (Ed.), *The child's world of make-believe.* New York: Academic Press, 1973.

Gardner, R. A. *MBD: The family book about minimal brain dysfunction.* New York: Jason Aronson, Inc., 1973.

Gershowitz, M. Fantasy behaviors of clinic-referred children in play environments. Unpublished doctoral dissertation, Michigan State University, 1974.

Gibson, E. J. *Principles of perceptual learning and development.* New York: Appleton-Century-Crofts, 1969.

Golomb, C. Evolution of the human figure in a three-dimensional medium. *Developmental Psychology,* 1972, *43,* 385–391.

Goodnow, J. J. & Friedman, S. Orientation in children's human figure drawings: An aspect of graphic language. *Developmental Psychology,* 1972, *7,* 10–16.

Grollman, E. (Ed.), *Explaining death to children.* Boston: Beacon Press, 1967.

Hagen, J. W., Jongeward, R. H., & Kail, R. V. Cognitive perspectives on the development of memory. In H. W. Reese (Ed.), *Advances in child development and behavior* (Vol. 10). New York: Academic Press, 1975.

Ilg, F. G., & Ames, L. B. *School readiness.* New York: Harper & Row, Publishers, 1966.

Kastenbaum, R. The kingdom where nobody dies. *Saturday Review/Science,* January 1973, 33–38.

Kastenbaum, R., & Aisenberg, R. B. *The psychology of death.* New York: Springer Publishing Co., Inc., 1972.

Keeney, T. J., Cannizo, S. R., & Flavell, J. Spontaneous and induced verbal rehearsal in a recall task. *Child Development,* 1967, *38,* 953–966.

Kellogg, R., & O'Dell, S. *Analyzing children's art.* Palo Alto, Calif.: National Press Books, 1969.

Kozol, J. *Death at an early age.* Boston: Houghton Mifflin Company, 1967.

Lacoursière-Paige, F. Development of right-left concept in children. *Perceptual & Motor Skills,* 1974, *38,* 111–117.

Lerner, R. M., & Korn, S. J. The development of body-build stereotypes in males. *Child Development,* 1972, *43,* 908–920.

Lewis, R. (Ed.). *Miracles.* New York: Simon & Schuster, 1966.

Martin, P. C., & Vincent, E. L. *Human development.* New York: The Ronald Press Company, 1960.

Maurer, A. The child's knowledge of non-existence. *Journal of Existential Psychiatry,* 1961, *2,* 191–212.

McGhee, P. E. Development of the humor response: A review of the literature. *Psychological Bulletin,* 1971, *76,* 328–348.

Moely, B. E., Olson, F. A., Holmes, T. G., & Flavell, J. Product deficiency in young children's clustered recall. *Developmental Psychology,* 1969, *1,* 2–34.

Paris, S. G. Integration and inference in children's comprehension and memory. In F. Restle, R. Shiffrin, J. Castellan, H. Lindman & D. Pisoni (Eds.), *Cognitive theory* (Vol. 1). Hillsdale, N.J.: Lawrence Erlbaum Associates, 1975.

Paris, S. G., & Lindauer, B. K. Constructive processes in children's comprehension and memory. In R. V. Kail & J. W. Hagen (Eds.), *Memory in cognitive development.* Hillsdale, N.J.: Lawrence Erlbaum Associates, 1976.

Piaget, J., & Inhelder, B. *The child's conception of space.* London: Routledge & Kegan Paul, 1956.

Saltz, E., Dixon, D., & Johnson, J. Training disadvantaged preschoolers on various fantasy activities: Effects on cognitive functioning and impulse control (Technical Report #8). In *Studies in intellectual development.* Detroit: Wayne State University Center for the Study of Cognitive Processes, 1976.

Santos, R. A wish. In R. Lewis (Ed.), *Miracles.* New York: Simon & Schuster, Inc., 1966.

Schultz, T. R., & Horibe, F. Development of the appreciation of verbal jokes. *Developmental Psychology,* 1974, *10,* 13–20.

Shantz, C., & Watson, J. S. Spatial abilities and spatial egocentrism in the young child. *Child Development,* 1971, *42,* 171–181.

Singer, J. L. Imagination and waiting ability in young children. *Journal of Personality,* 1961, *29,* 396–413.

Singer, J. L. *Daydreaming.* New York: Random House, Inc., 1966.

Singer, J. L. Imagination and make-believe play in early childhood: Some educational implications. *Journal of Mental Imagery,* 1977, *1,* 127–144.

Singer, J. L., & Singer, D. G. Imaginative play and pretending in early childhood. In A. Davids (Ed.), *Child personality and psychopathology.* New York: Wiley-Interscience, 1976.

Stevenson, H. *Children's learning.* New York: Appleton-Century-Crofts, 1972.

Swift, J. *Gulliver's travels.* Boston: Beacon Press, 1963. (Originally published, 1726).

Wapner, S., & Werner, H. *Perceptual development.* Worcester, Mass.: Clark University Press, 1957.

Werner, H. *Comparative psychology of mental development.* New York: Science Editions, Inc., 1961. (Originally published, 1948).

Wittgenstein, L. *Philosophical investigations.* New York: Macmillan, Inc., 1958.

Wylie, L. *Village in the Vaucluse.* New York: Harper Colophon Books, 1964.

THE CHILD PUTS IT ALL TOGETHER

10

CHAPTER OUTLINE

I

The older school child enjoys a life situation that is difficult to match. Intellectual competence is one aspect of the overall splendor of this age.

Paying attention / Cognitive structures / Conservation / Motivation / Social context / Strategies of learning / Conditionality / Lose-shift strategy / Win-stay strategy / Growth error (Bruner) / Whole-properties (articulated perception) / Concrete operations (Piaget) / Formal operations (Piaget) / Inversion / Reciprocity / Class inclusion operations / Serial ordering operations

II

Boys and girls move through several stages of moral development that are associated with the levels of mental growth.

Moral absolutist (Piaget) / Fair share (Piaget) / Situational judgment (Piaget) / Kohlberg's stages of moral development / Altruistic behavior / Social desirability

III

In recent years, accidents of various types have accounted for half of *all* deaths in school age boys with a sharp rise in mortality for girls as well.

Motor vehicle accidents / Drowning

Pick an age, any age. Select a year of life glittering with vitality, security, and confidence, an age that out of the total human lifespan is the most nearly perfect. What age have you chosen?

Age ten.

CELEBRATING THE TEN YEAR OLD

This chapter is essentially a celebration of the ten year old (broadened to accommodate those a little younger and a little older). In general, the older schoolchild enjoys a life situation that is difficult to match for its pain-pleasure, achievement-frustration, and security-insecurity balance. There are depths of thought and experience still to come, of course, and most of the achievements that keep humankind going are made at more advanced age levels. But if one were to climb out on the proverbial limb and argue for a particular year as supreme, there is much to be said for the boy or girl who has just completed the first decade of postnatal life.

To begin with, biological survival peaks around age ten. At every age level there is a probability of nonsurvival; this probability can be estimated statistically for large populations. In the United States, it is the ten year old who has the lowest probability of meeting death within the next twelve months (Dublin, 1965). Essentially, this is because the child at this age level has passed through the hazards of possible acute infection that are most often encountered very early in life, and he or she is still a long way from the chronic diseases that account for a large proportion of adult deaths. (Accident is the most common cause of death at and around age ten.) This biological security factor is important in itself. Additionally, it serves as an index of the general health and vigor of the older child. It cannot be said that he or she is without a worry or hazard in the world; this cannot be said of any human being. But this period does stand out as one of relative security.

In what other ways is the ten year old at special advantage?

1. The child is old enough to have accumulated a usable past. He or she has already had life experiences that can be drawn upon and applied to new challenges. This contrasts with the younger child who still runs into many novel situations and cannot rely on solutions and strategies previously achieved.

2. Physically, the child is capable of performing many of the actions required in daily living. Decremental changes associated with advancing age are negligible, and the physical changes that will require so many new adjustments in adolescence are still in the future.

3. Plans, fantasies, and ambitions can flourish at this age, but the child is still too young to measure or actually test them against reality, an experience that often brings crushing disappointment and frustration. The child can have the pleasure of projecting himself or herself ahead into a career or life situation that seems exciting and fulfilling. Most of these projections remain safe from decisive confrontations with reality until late adolescence or adulthood.

4. Home and family continue to provide a caring environment in which the child can both be himself or herself and try out new attitudes and skills. The child is not expected to make his or her own way. Energies are free for activities that engage the child's interest, rather than usurped by the necessity to provide income or household. The intelligent ten year old *can* engage in many basic life-support activities. However, most older children today receive from their family and society a gift of time that enables them to prepare more fully for participation in the complexities of our culture.

5. Romance and sexuality have not yet come into focus for most ten year olds. The child may have some idea of "what's going on" by observing the moods and behaviors of adolescents. And there may be some anticipatory stirrings within the ten year old himself or herself. But the exciting upheavals associated with sexual awakening have relatively little claim upon his or her energies. There is still time to consolidate the mental, social, and physical gains that have been made through the course of childhood up to this point. In psychoanalytic terminology, the child at this age is within the **latency phase,** a calm harbor where one rests and thrives before sailing off to the open sea of sexualized life.

Latency phase is the reduction of instinctual pressures in late childhood facilitating ego development

6. Social interaction is likely to be more rewarding to the ten year old than it has ever been before. He or she has come to realize something of a

Friendship becomes stronger and richer in the latency years. "Good old buddies" become a kind of second family. (Photo by Talbot Lovering.)

group identity. There is a realm that belongs just to him or her and others of that age. It is a larger realm than the family circle, but it is not as challenging and unmanageable as the whole world outside. This is the realm of good friends. The ten year old's ability to take part in fairly elaborate and rule-oriented activities provides for a great deal of mutual gratification. Strong and rich friendships are possible at this time, and some may endure and deepen over the entire lifespan.

7. The intellectual competence of the ten year old contributes mightily to the overall splendor of this age. It is obvious that he or she is "thinking better." Precisely what is involved in the older child's enhanced mental functioning is another question—or, rather, it is a series of questions that many developmental psychologists are trying to answer. We will now explore a few selected aspects of thought at and around age ten. It is possible that we will even learn a bit regarding our own thought processes as well.

"PAYING" ATTENTION

Everyday speech can reveal much, if we listen freshly. Consider the familiar phrase, *paying attention*. Why should attention be likened to a financial transaction? Where does the "payment" come from, who is the recipient, and what benefit is gained in return? These questions lead us directly into one of the core psychological developments of later childhood: the ten year old tunes in more consistently to what adults define as *reality*. This represents both a mental-maturational achievement and a willingness to become part of the larger society.

Eye contact is one useful measure of paying attention

The younger child, even the infant, *gives* attention—sometimes, to some phenomena. *Paying* attention implies that one is withdrawing a certain fund or resource to meet an obligation. In this sense, the older child is a better customer for the brand of reality that adults sponsor. He or she is more willing and able to withdraw attention from the very personal to give it to the shared or the objective. The infant and young child's fascination with the external world is closely linked with his or her own internal states of need. But the older child is ready to surrender some of his or her attention to personal needs in order to improve his or her contact with the social and physical environment. The ten year old invests more attention or perceptual energy in the world outside of himself or herself. This is truly a form of payment, like spending a nickel for one purpose rather than another. The recipient of his or her attention is some selected aspect of the world. In return, the child learns more about the world and, thereby, enters more fully that realm of physical and symbolic activity that comprises shared reality.

One of the reasons why adults are more comfortable with the thought processes of the older child is the latter has now "agreed" to take seriously some of the phenomena that grown-ups center their own thoughts around. Daddy, perhaps, was never comfortable with the imaginative play of his son or daughter at age five. "Too silly," he thought, or "Too unrealistic." He did not find it natural to follow the child's flights of fancy that suddenly interrupted games or chores in which they were both engaged. Even shifts of attention that were closely linked to reality might have annoyed him. For example, daddy has organized something approximating a baseball game on a vacant lot with a batch of young children, his own and their friends. He is aware of the game situation (who is on base, what the score is, how many are out), if nobody else is. Suddenly, one of the fielders shouts: "Wow—look at this neat bug! It's wicked cool!" Just as suddenly, the game disintegrates. The children surround the discovered creature. "It's not a bug, dumm-dumm; it's a caterpillar!" "Is it poison?" "My sister likes to eat those!" The caterpillar is clearly a part of the physical reality of the scene, but no part of daddy's game plan. He is irritated because the children are not paying attention to *his* reality. Poor dad; perhaps, poor caterpillar, too.

Alternative views of reality are possible in such a situation. The parent's view is the "correct" one because, well, because he is the grown-up and reality is what he thinks it is. But perhaps the childrens' view is more appropriate. They are focusing on the caterpillar, a tangible presence in the field, while the grown-up is caught in a set of arbitrary rules and expectations inside his own head. (Besides, there are more children than adults on the scene and so, more votes for their version of reality.) You can probably discover other alternative views as well. The most relevant point here is that reality orientation cannot be understood unless we clarify *whose* reality is being awarded precedence. In general, the older child comes to accept more and more of the adult's version. But does this necessarily prove that the child has become "smarter"? Or should movement toward the adult's version of reality be regarded primarily as a social and motivational development? The child of ten or thereabouts says, in effect, "OK, dad, I'll play your game."

HEIGHTENING IN ATTENTIONAL PROCESSES OR MORE MATURE COGNITIVE STRUCTURES?

This brings us to a problem of fact and interpretation of fact. Suppose that the ten year old differs mentally from younger children only in willingness or ability to pay attention in a more adult-like fashion. This new ability alone might be sufficient for us to characterize the ten year old as a "better" or "more dependable" thinker. In point of fact, the older child does tend to pay better attention (*see also* the discussion of perceptual learning in Chapter 9). Perhaps, then, we should accept the simplest available line of explanation, namely, the older child pays better attention because of a shift or heightening in attentional processes as he or she acquires more experience with the world.

Preference for the simplest possible explanation is known as the *law of parsimony*

But there is a strong alternative viewpoint to be considered. We have already become acquainted with some aspects of Piaget's approach to mental

The ten year old is becoming adept at paying attention selectively—concentrating, in this case, on what comes through the ears. (Photo by Talbot Lovering.)

development. Piaget, and a number of other developmentalists, believe that differences in thought and behavior are related to differences in underlying *cognitive structures* at various phases of the maturational sequence. The ten year old, for example, possesses cognitive structures that are more advanced than the *sensorimotor stage* afforded him or her a few years previously. He or she is, however, still some distance from the adult's *operational* thought structures. Why is the older child now able to pay attention in a more adult-like manner? According to this view, it is precisely because he or she now has started to develop the kind of cognitive structure that makes it possible to view the world differently in many respects. Attention is just one example.

If simplicity is our preference, then we might opt for the explanation that children just pay more attention to adult-oriented reality as they grow up. But if we wish to take a broader range of developmental facts into account and achieve a systematic view of total psychological growth from infancy onward, then we must continue to explore complexities such as Piaget (1954), Bruner, Olver, and Greenfield (1967), and other investigators have observed. The fact that there is more than one way to look at mental development helps sharpen our appreciation of the issues involved. Let's take, for example, an interesting study by Gelman (1969). She took issue with the conclusion that children make certain errors because they lack advanced cognitive structures. She proposed instead that children are often not paying attention to those aspects of the situation that would provide them with the information they needed.

At issue in Gelman's study was one of the now-classic types of experiment introduced by Piaget. The experimenter shows the child two identical beakers that are filled with the same amount of water. The child recognizes the equivalence: same amount of water in each beaker. Water is then poured from one of the beakers into another that differs in shape. When this is done, the child may say that there are different amounts of water in the two beakers because the levels of the liquid are no longer at the same height. The Piagetian explanation is such children do not yet have a concept of **conservation** because their cognitive development has not yet reached a certain level of maturation. In terms of this particular illustration, the child is unable to "conserve" the volume of the water mentally when it is poured from one beaker to another of different shape. The child's mental representation of what he or she has observed prior to the pouring lacks something. He or she can be tricked by the physical manipulation because he or she does not yet possess a secure underlying cognitive structure that maintains or conserves an observation. The same principle applies to other dimensions as well. The young child usually does not conserve the mass of a piece of playdough when it is rolled into various shapes. He or she responds as though either more or less playdough has come into existence with a change in shape although, in fact, the mass has not changed at all.

Conservation is the ability to know that something remains the same although seen in differing circumstances

Gelman's innovation was to give children the opportunity to focus on similarities and differences. The boys and girls in her study were provided with training sessions in which their attention was directed to cues that indicated whether a particular object "belonged" or "did not belong" with some others. After learning how to pay attention to such cues, the children were presented with the basic conservation tasks that they had "flunked" not long before. Many succeeded this time (about half of the responses were now correct). Not much time elapsed between the first failure to comprehend the conservation of liquid and mass and the subsequent success. *Performance* had changed. But it would seem unreasonable to argue that the cognitive structures that supposedly underly performance had had time to grow and mature between the sessions. The more logical supposition was that the children performed better because they had learned how to better direct their attention.

This type of study both informs and perplexes. The children in Gelman's study did not have to "ripen" over a long period of time in order to improve performance on what had been regarded as a task requiring the existence of relatively advanced cognitive structures. But could this mean that the training quickened the development of basic mental abilities? If so, then it is much easier to modify and accelerate human development than most observers believe. Or could this mean instead that there really are *no* complex underlying structures at all?

It is unlikely that either of these alternatives is correct. Another, perhaps more reasonable alternative, is that the ten year olds did in fact have the necessary cognitive structures available to them, but needed some appropriate learning experiences in order to translate the abilities into observable performance. Even this alternative, however, has its difficulties and has not been clearly supported by research. There is still another possibility: the children performed better after the training experiences because they had learned to use already existing abilities in a new and appropriate way, *not* because they had suddenly sprouted an advanced form of mental activity.

These are complicated questions we have been discussing. The continuing input from new research and the interplay of informed opinion in the field makes the questions even more difficult to present in a simple form. A few points have become clear, however:

1. The older child does pay better attention to the adult-oriented reality environment.
2. Improved performance by the older child in some situations can be brought about through training that concentrates on perceptual and attentional processes.
3. Perceptual learning and attention are processes that themselves probably depend on the changing basic capacities of the child. If it were *all* a matter of paying attention, then presumably the three year old could be trained to do as well as the ten year old.
4. It is important to maintain a fundamental distinction between what we can see of a child's functioning in action—his or her *performance*—and our ideas about the *cognitive structures* and *processes* that make performance possible.

ATTENTION, MOTIVATION, AND THE SOCIAL CONTEXT

Attention is related not only to the general level of mental development a child has achieved, but also to motivation and the social context. How well a child moves into the grown-up's world of thought and action depends to some extent on how much he or she finds attractive or irresistible in the larger world, how adults facilitate or interfere with his or her ventures into adult turf, and how much satisfaction is still to be discovered within his or her personal version of reality.

How are we to regard the child who does not pay attention to adult-sponsored reality (such as classroom material)? In practice, such a child is open to a variety of interpretations by others. If the social-motivational aspects of attention are emphasized, the child may be regarded as a dreamer lost in his or her own preoccupations or as a rebel who rejects the world that others accept. The child may also be seen as a victim of serious emotional conflicts whose problems interfere with the ability to tune in to stimuli outside of himself or herself. If inside-the-organism aspects are emphasized, the child may be regarded as one who is generally immature for his or her age, or as a person of less than normal intellectual capacity. Less frequently, perhaps, it may be ventured that the reality available for a particular child is too restricted, too standardized, or otherwise inappropriate for his or her needs. The "fault," if any, is not in the child but in an environment that is too barren, too oppressive, or too something else.

This range of interpretations is mentioned to add a note of caution. The fact that a particular child is not attending to school or to other aspects of his or her environment in the standard way can have more than one meaning. Inattentiveness can lead to consequences that resemble those associated with limited mental development, but the underlying dynamics may be quite different. Before leaping to a stereotyped conclusion, parents or teachers should exercise their own powers of attention over a period of time and become well acquainted with the total situation. More than one child has been labeled *slow* or *rebel* when, in fact, the boy or girl could not see or hear adequately, came from a background which made the school situation appear quite different from the way it does to most others, or was suffering acutely from hunger and lack of nutrition.

HOW DOES THE TEN YEAR OLD LEARN AND THINK?

Attention, as we have seen, is not easy to distinguish from learning, thinking, and other psychological processes. That distinction is worth making both for theoretical and applied purposes. As we explore selected aspects of learning and thinking, however, we should not lose sight of the fact that these are all components of the individual's total process of adaptation to his or her environment.

STRATEGIES OF LEARNING

"Learning to learn" is a familiar phrase in psychology. It refers, in general, to the fact that we bring to each learning task a certain "set" or "strategy." We might ask, then, how does the child, especially the older child, approach learning situations? Many of the studies that bear upon this question are summarized and evaluated in a book by Harold W. Stevenson (1972). You may wish to consult it for additional details.

Children often approach learning tasks with some kind of strategy. Their first responses in a new situation may appear random, trying this and trying that. Sooner or later, however, the child tends to hit upon a strategy. The young child is likely to base his or her strategy on simple and concrete features of the materials involved (e.g., always selecting the stimulus choice that is in a particular spatial position). But the ten year old has developed more complex strategies that are less rigidly bound to the spatial and physical properties of the stimulus materials. This has been shown clearly in what are known as oddity problems. The task is to discriminate which of several presented stimuli is the "odd" one. This requires that the child learn the rule that governs "oddity-ness" in the particular experimental situation. The superiority of the older child over the younger is revealed when the task is made more difficult. One way of making a task more difficult is to include the element of **conditionality.** A particular stimulus-choice, say, a red cylinder, may or may not be the correct (odd) answer, depending upon whether it is presented on a blue or a yellow tray. In other words, the child must learn to notice and interpret the significance of a background cue as well as attend to the basic stimulus itself. Although we are focusing here on situations contrived for experimental purposes, in daily life there are many situations in which one course of action is more appropriate depending on a particular cue, for example, the color showing on a traffic signal.

Conditionality is when one thing depends on another

In conditionality-oddness situations, the older child learns when to stop. That is, he or she stops giving a particular response when it proves to be wrong and tries another kind of response. This is known as the **lose-shift strategy.** But when a child is responding correctly, he or she remains with the successful mode and does not change his or her behavior arbitrarily, just for the sake of change. This is known, sensibly enough, as the *win-stay strategy.* This means that the learning strategy of the older child involves an integration of two approaches: lose-shift and win-stay.

A **lose-shift strategy** is trying another approach when the first does not have the desired effect

The particular strategies developed by the ten year old will depend to a large extent on the nature of the problem that confronts him or her. The important point, however, is that he or she is capable of a *planful* approach to learning that can be *adjusted* as required by the situation. As you can see, this is a very useful ability. The child is less likely to experience continuing frustration in coping with problems. He or she learns more because he or she is more planful and flexible than the younger child—and is less vulnerable to episodes of rage and regression when a problem cannot be solved at first try. We should also remember, however, that even younger children do attempt to apply strategic approaches to learning situations, but their attempts are usually less effective.

Children of outstanding intelligence tend to develop effective learning strategies more rapidly than other children. A very bright seven year old, then, may have a more effective approach to learning problems than a ten year old of lesser capacity; an exceptionally bright ten year old, similarly, may surpass many adults. An individual's level of intellectual ability cannot usually be increased, although the limits of possible tinkering with intellectual ability are still being explored. However, it is reasonable to concentrate on ways of enhancing the child's knack of learning how to learn. Stevenson puts it this way:

> The ultimate goal in any type of learning cannot be the retention of large amounts of specific information. For the most part, this information will be forgotten. What can be retained are techniques for acquiring new information,

> learning how to attend to relevant cues and ignore irrelevant cues, how to apply hypotheses and strategies and relinquish them when they are unsuccessful. Teaching children how to learn may prove to be a more lasting contribution to their education than continuously requiring them to concentrate their efforts on the accumulation of facts. (1972, p. 307).

MENTAL DEVELOPMENT AND ERRORS

It is usually assumed that greater ability and experience will be associated with fewer mistakes. In the learning strategy studies, for example, a sudden decline in rate of errors often evidences the achievement of insight. It is the same way in much of daily life. We expect the older child to know more and to make fewer errors than the younger child. The facts do not always confirm this expectation, not in daily life and not in research. Although such contradictory research evidence is of a special kind, obtained largely under controlled experimental conditions, there is enough theoretical significance here to warrant our consideration.

Heinz Werner maintained for many years that developmental psychologists should examine "process" as well as "achievement" (Werner, 1948/1961), a point that is more appreciated today than in the past. The score earned on a particular test or task, for example, does not necessarily tell us *how* the organism behaved. Two individuals who receive an identical score may have, in fact, engaged in psychological processes that were markedly different. According to this view, it is entirely conceivable that progression from a lower to a higher developmental level might be accompanied by an increase rather than a decrease in objective errors.

Evidence to support this theory has been obtained by Wapner and Werner (1957), and subsequently by other investigators. The first study involved susceptibility to a perceptual illusion known as the Titchener circles (*see* Figure 10–1). The standard stimulus is shown on the left side of the figure. It consists of an inner circle that is surrounded by a ring of five other circles. For research purposes, this stimulus is presented along with another set of circles (right side of the figure). This comparison set is changed from one presentation to the next, with the relative size of the inner circle systematically varied, sometimes smaller, sometimes larger. The respondent's task is to examine both sets of circles that are presented to him or her and to indicate whether the inner circle of the comparison stimulus is smaller, larger, or the same size as the inner circle of the standard stimulus. Many such comparisons are made by each respondent. Each response can be objectively scored as correct or incorrect; the degree of error can even be measured in millimeters. A correct score is one in which the respondent did not prove vulnerable to the visual illusion effects.

What makes this study interesting is the fact that eight and nine year olds made *fewer* errors than ten and eleven year olds. Furthermore, susceptibility to the effects of this illusion showed an increase all the way through

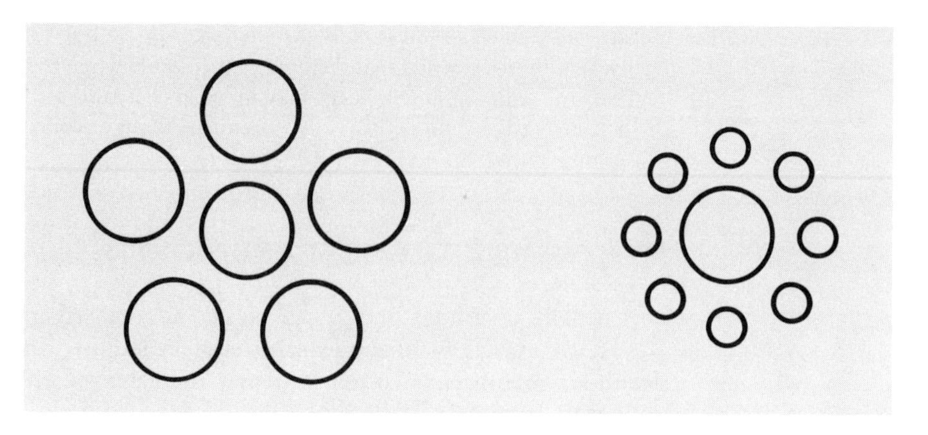

Figure 10-1. *The Titchener circles illusion. The set of circles on the left is always presented as the standard stimulus. There are as many as seventeen variations of the set of circles shown on the right in which the size of the circles differs. The respondent is asked to judge whether the inner circle in both sets is the same size (sometimes it is; sometimes it isn't), smaller, or larger. Research indicates that as the child grows up, he or she may actually be taken in more by the illusion because more attention is paid to the total design (Wapner & Werner, 1957).*

adolescence. Growing up somehow appeared to be associated with a tendency to be "fooled" more easily by the illusion. How can people become more trickable or make more errors as they grow up? Both a precise and a general answer must be given to this question. The precise answer (from Werner's developmental viewpoint) is that the older child is himself or herself more precise in the examination of the stimulus materials. He or she notices more detail. Instead of just seeing "a lot of circles" that have an overall effect on his or her judgment, the older child analyzes the detailed components of the visual display. His or her perception is more "differentiated" or "articulated." Now, it happens to be the case that this particular illusion depends on the viewer getting caught up in the details. To some extent, the more accurate and analytic the perception, the more errors will be made. For the more general aspect of this answer, let's turn to a companion study by the same researchers, using the same respondents.

The Müller-Lyer illusion is shown in Figure 10–2. As you can see, this looks quite different from the Titchener circles illusion. Directional dynamics are important. One tends to be influenced most by the overall character of the design. Respondents were asked to adjust the right side of the pattern until it appeared equal to the length of the left section (a rack and gear device was provided for this purpose). In this instance, it was the younger children who proved more susceptible to the illusion. Wapner and Werner believed that the differential results with the two types of illusion were not contradictory; in fact, together they illustrated a general developmental principle. This is the principle that we begin with perceptions that are global and un-

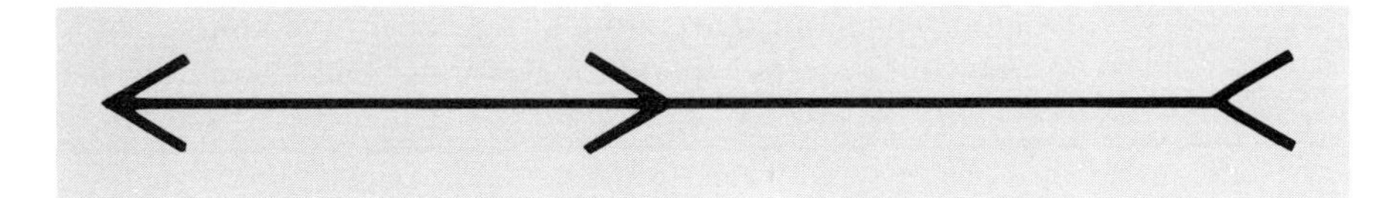

Figure 10–2. *The Müller-Lyer illusion. The line on the right side of this figure is longer—or is it? The two horizontal lines are actually the same length. In one series of studies, respondents were asked to adjust the right side of the figure until it was equal to the left side. Younger children were more susceptible to this illusion, but opposite effects were found for the Titchener circles illusion.*

Whole-properties are the characteristics of a *total* situation

differentiated. We are most apt to comprehend the **whole-properties** of the situation. It is possible to recognize that the day is "cloudy" without specifying the number, type, shape, and flow of the clouds. If there is something tricky about the whole-properties in a particular situation, then we will probably be tricked. As we develop, we become more analytic and sophisticated. Not only do our perceptions become more articulated or differentiated, but we are able to examine the relationships of parts to the whole. This latter ability, for example, is the difference between seeing a lot of people running around on the football field and seeing a total offensive and a total defensive pattern of movement interacting in the development of a play. In most circumstances, articulated perception leads to improved accuracy and more flexible adaptation to the environment. But if we happen to be in a situation that penalizes analytic perception, then we are vulnerable to error.

Growth of the same psychological process, then, may lead to more errors when an illusion depends on attention to detail, and fewer errors when an illusion depends on overall effects. The ten year old thus shows himself or herself to be more "adult" in his or her errors as well as his or her successes.

Growth errors are mistakes that occur when moving from a lower to a higher level of thinking

Experimental situations contrived by Jerome S. Bruner and his colleagues (Bruner, Olver, & Greenfield, 1967) have also revealed examples of what he terms ***growth errors.*** Older children generally perform more accurately than younger children on a variety of problems. Sometimes, however, the older child appears to be caught in transition between one level of approaching a problem and another, more advanced level that he or she has not yet mastered. In attempting to understand ratios and proportions, for example, "He must master what it takes to hold several different things in mind at once." When the child cannot manage this feat of retaining and integrating several elements in his or her mind, then the older child's performance can appear inferior to that of the younger child. It is perhaps akin to a person switching from counting on his or her fingers to utilizing a computer. At first one might make more errors in attempting to use the new system, but even when mistakes are evident there should be no mistaking the fact that the person has moved on to a significantly higher level of functioning.

The older child generally performs more competently and accurately than the younger child. But the exceptions noted here inform us that the emergence and sharpening of new cognitive abilities at times may lead to errors and even to apparent regressions. The distinction between "process" and "achievement" is well taken.

FROM CONCRETE TO FORMAL OPERATIONS

The transition from one level of thought to another is also suggested by the work of Piaget and his many followers (e.g., Piaget, 1954; Piaget & Inhelder, 1969; Elkind & Flavell, 1969). Roughly speaking, the ten year old is well established in his or her command of **concrete operations.** At the same time, however, he or she is moving toward the sphere of adult cognition, which is known as **formal operations.** We will take a step or two backwards in time to gain perspective on these developments.

Concrete operations is Piaget's stage of mental development in which rules and facts are understood but not abstract thought

Formal operations is Piaget's highest stage of mental development including abstract thought

Thought becomes more systematic. Earlier in this book I noted that the infant and very young child are considered by Piaget to function on a *sensorimotor level.* The youngster relates to the world through direct activity. Later, at about age two, he or she is able to engage in representational thought. This is usually known as the *preoperational phase,* although it is sometimes described as the intuitive stage as well. The child's behavior is more flexible and better integrated with the socio-physical environment because some of the action can take place inside or representationally, instead of being anchored entirely to physical movement. The preoperational period runs more or less parallel with the preschool days. The child becomes very adept at regulating his or her behavior and relating to the world through various types of mental images. By the time he or she is ready for the classroom experience, however, the child is also ready for a more advanced level of cognitive functioning.

The schoolchild is able to think more consistently and logically than ever before. His or her thoughts hang together more securely. Consequently he or she is also able to *behave* more systematically and to follow through with plans and intentions. It is, in fact, the *systematic* quality of the child's mental life, beginning at or around the seventh year, that gives this new phase its distinctive identity. Thoughts operate in relationship to each other. One thought can contradict, modify, or combine with another. It is in this general sense that the child is said to function on an *operational level.* We shall see in a moment why it is known more specifically as the level of *concrete operations.*

First, it should be recognized that Piaget would have us think of the child's intellectual development in a broad, encompassing manner. With or without the help of controlled studies, we can see many ways in which the child at one age differs from a younger or older child. But the specific features of mental functioning—use of language, orientation toward space and time, understanding of cause and effect, etc.—should be interpreted in terms of the

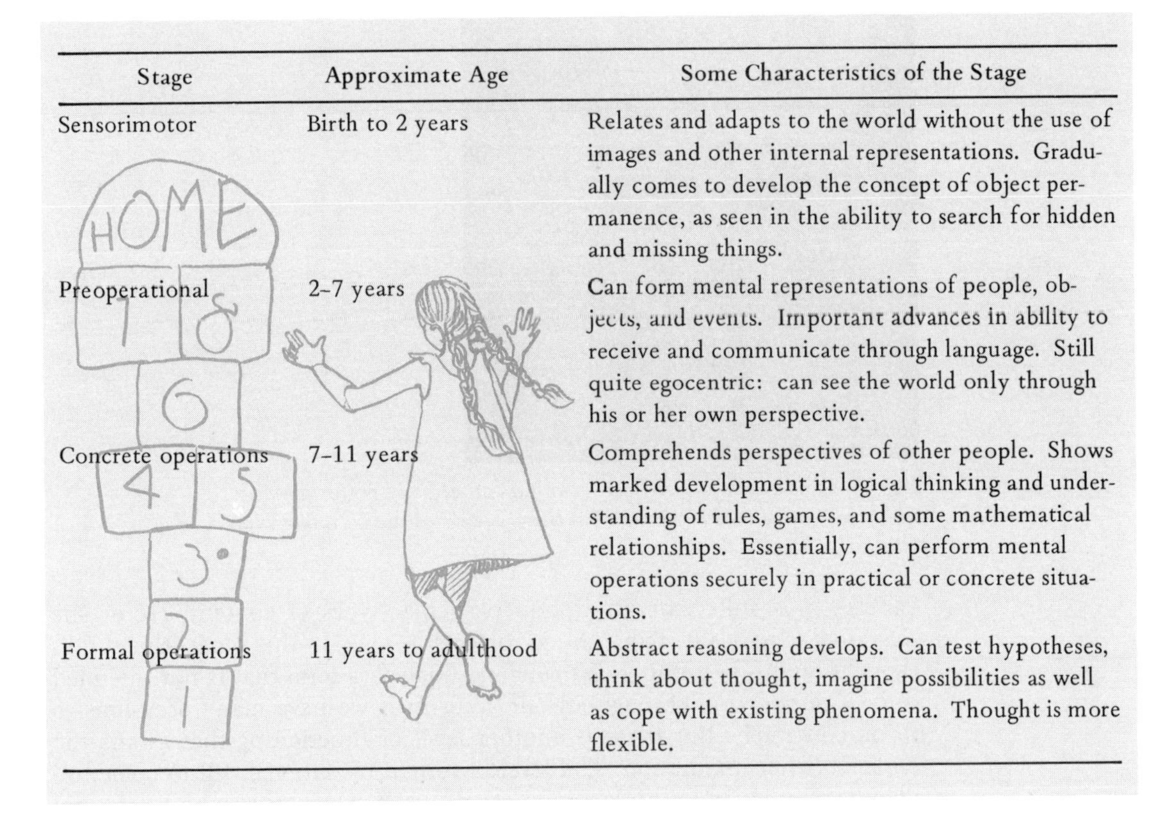

Stage	Approximate Age	Some Characteristics of the Stage
Sensorimotor	Birth to 2 years	Relates and adapts to the world without the use of images and other internal representations. Gradually comes to develop the concept of object permanence, as seen in the ability to search for hidden and missing things.
Preoperational	2–7 years	Can form mental representations of people, objects, and events. Important advances in ability to receive and communicate through language. Still quite egocentric: can see the world only through his or her own perspective.
Concrete operations	7–11 years	Comprehends perspectives of other people. Shows marked development in logical thinking and understanding of rules, games, and some mathematical relationships. Essentially, can perform mental operations securely in practical or concrete situations.
Formal operations	11 years to adulthood	Abstract reasoning develops. Can test hypotheses, think about thought, imagine possibilities as well as cope with existing phenomena. Thought is more flexible.

Figure 10-3. *The stages of cognitive development as seen by Jean Piaget.*

general developmental principles they illustrate. It is not sufficient to observe that one child comprehends the idea of reversing numbers in a series while another child does not, to take just one example. Rather, the Piagetian will discover basic rules or principles that apply to *all* adaptational processes at a particular level. Specific mental phenomena are studied as illustrations of the general case. To move, then, from preoperational to operational thought is not to "get a little smarter" or to "learn a little more," but to relate to the world and to one's own being in a radically different way.

Secondly, each time the individual moves from one level to the next, he or she must gradually master a new set of challenges and opportunities. It is perhaps something like bidding farewell to a job situation where one has become a veteran who knows all the ropes, and now entering a new and more promising situation that requires a period of exploratory activity.

Until about age eleven or so, the child is attempting to master a complex set of mental operations that lay the foundation for much of the thinking he or she will do as an adult. Although his or her level of thought is much ad-

Research indicates that the ten year old can already perform many mental operations. Piaget believes these are all of the concrete type. (Photo by Tania D'Avignon.)

vanced, it is still centered around particular types of content. He or she thinks systematically about things, objects, people. In this sense, the child's thought retains something of a *concrete* quality (a term that is not intended to be a put-down). Throughout our adult lives we have many occasions to think concretely. But there is another level of functioning that awaits the child's further maturation. The level of *formal operations*, still to come, involves the capacity to think about thought itself. The person who has formal operations at his or her command can combine and examine thoughts freely. The possibilities of mental transformation can be explored apart from specific or concrete content. Among other important uses, formal operations appear to be fundamental for the development and appreciation of systematic science, logic, and mathematics.

The ten year old, for example, can "think in reverse." If he or she puts a quarter into the piggy bank in the morning, and shakes it out in the afternoon, the child understands that he or she has engaged in a type of reversal operation (known technically as *inversion*). He or she has added a something, and then taken that something away—leaving the original sum on deposit exactly the way it was at the beginning. The concept of *reciprocity* is also understood. He or she is not deceived if the graduate student from the university rolls playdough into first one shape and then another. There is no question in the ten year old's mind that the amount of material remains the same because what now is thinner has also become longer, etc.

But the child has not yet *integrated* the ideas of inversion and reciprocity. This means he or she cannot use these concepts freely together, independent of particular content. The same holds true in many other instances. Functioning on the level of concrete operations, the boy or girl can apply logical principles to particular problems.

It is only in the next stage, that of formal operations, that "the subject becomes capable of reasoning correctly about propositions he does not believe, or at least not yet; that is, propositions that he considers pure hypotheses. He becomes capable of drawing the necessary conclusions from truths which are merely possible, which constitutes the beginning of hypothetico-deductive or formal thought" (Piaget & Inhelder, 1969, p. 132).

Class inclusion operations. Let's give a little more emphasis to what the older child *can* do with the concrete operations at his or her disposal. He or she can put classes or groupings of things together in his or her mind, and can also take them apart. To use one of Piaget's own examples, the child at this level of functioning recognizes that *boys* refers to one class or grouping that contains many specific components (two boys, ten boys, more boys than you can count, etc.). The same holds true for *girls.* The ten year old can combine *boys* and *girls* into a larger category, *children.* Further, he or she can proceed to add *children* to *adults* and arrive at an even more inclusive grouping, *people.* Each class has its component parts, and all the classes or groupings "fit into each other" and build up to larger groupings. Using his or her new reversal or inversion ability, the older child can also go back in the other direction, subtracting one class from the others, or systematically fractionating *people* into *adults* and *children,* and each of these groupings into *boys* and *girls* or *men* and *women.*

The child possessed of concrete operations can also multiply classes or groupings. To take one example:

> Before the age of about seven . . . children, given a box containing about eighteen brown and two white beads, all wooden, and asked whether there are more brown or more wooden beads, reply that there are more brown ones because only two are white. That the categories are available and observations correct is shown by the fact that the younger children, when asked the questions separately, give correct answers as to the relative proportions of brown, white, and wooden beads. However, without class inclusion operations they cannot deal with the parts and the whole at the same time, and thus they make a false generalization. (Inhelder & Piaget, 1958, p. xvi).

Class inclusion operations is the process of dividing and grouping phenomena into appropriate sets

The older child can answer the question correctly because he or she can distinguish between two properties (color and material) and multiply the possible combinations (white-wooden, brown-wooden, with the possibility of white-metal or brown-plastic also being available if indicated).

Serial ordering operations is the process of organizing phenomena in sequence by one or more of their characteristics

Relational thought. In his or her **class inclusion operations** (the type of thought I have just been describing), the older child is displaying a valuable ability to work logic on the materials of the world. Another type of ability is known as the **serial ordering operations.** How should various objects be lined up, either physically or mentally? The basic kind of logic involved here is *relational.* "This is the shortest stick, so I will put this one first, then the next shortest, and then the next shortest after that." This simple form of serial ordering is usually shown early in the concrete operations level. Later the child can understand and perform more sophisticated arrangements and also take account of the ways in which two or more independent series may be related to each other.

Even at their most complex and sophisticated, concrete thought operations lack the power of a full combinatorial logical system. The child is not yet as free and flexible as he or she will eventually become at the level of formal operations. Even so, the ten year old does have some impressive cognitive structures and is keeping many researchers around the world busy observing, testing, describing, charting, and debating fine points.

PLAYING FAIR: DEVELOPMENT OF MORAL JUDGMENT

Quality of thought is closely related to quality of life in general. The ten year old's mode of thinking shows itself in many ways. Of these, one of the most interesting and significant is in the area of moral judgment. What is the "right" thing to do in a situation? Why are certain actions "bad" or "evil"? The moral judgment of children has been studied by a number of eminent researchers including, once again, Piaget. His observations suggest that boys and girls move through several stages of moral development that are associated with the levels of mental growth already described here. Other investigators have added to this picture, tending to support his general view although differing in some details. To understand the ten year old's conception of morality and fair play, we will once more find it useful to backtrack a little.

A question currently popular among schoolchildren inquires, "Where does a five-hundred-pound gorilla sleep?" The answer, of course, is "Wherever he wants to!" The infant and very young child resemble the ponderous primate in this characteristic. Not only do they fall asleep wherever and whenever overcome by the desire, but in general they seek direct fulfillment of desires. The "good" and the "proper" amount to pretty much the same thing: instant relief or instant pleasure. Gradually the child learns that certain behaviors are no-no's. This, however, is not the same thing as having a firm sense of the rules that govern the moral dimension of behavior.

PIAGETIAN MORAL JUSTICE

Piaget's observations (1932) led him to propose that children pass through three major periods in their understanding of what might be termed *moral justice:*

1. The preschool child depends heavily on sanctions from the parent or other adult authority. Mommy and daddy determine what is "right" and

what is "wrong." Disputes about "playing fair" are to be settled by reference to parental authority. At this point, the child is a moral absolutist—right is right and wrong is wrong, no matter what the situation. If a young child were to be appointed as a trial judge, there would be little chance for a defendant to win acquital or receive a lighter sentence because of extenuating circumstances.

2. From approximately age eight through eleven, the child takes more account of the total situation in which a particular behavior (or misbehavior) occurs. There is also more sensitivity to equality of treatment. The principles of fair share and fair punishment, for example, are often invoked. Parental influence remains important in determining the moral tone of an action, but the child tends to use his or her own experiences and perceptions more fully in arriving at his or her opinion.

3. As the child moves toward the edge of puberty, his or her moral judgment shifts more clearly toward a relativistic or situational position. Whether a particular action is right or wrong depends a great deal upon the total context. Hitting another child may be a "bad" action because it is an excessive response to his or her provocation or because the victim was younger or smaller. But the same action may be morally "OK" if it is seen as a fair and measured response ("giving him what he asked for" or "standing up for my rights"). The older child is more prepared to judge each case on its own merits. He or she can take more factors into account and exercise greater flexibility because he or she now possesses the systematic mode of thought characteristic of the operational periods.

Fair play is an important concept for the ten year old. Morality centers around obeying and enforcing rules of conduct, rather than relying on the voice of authority alone. (Photo by Paul S. Conklin.)

KOHLBERG'S STAGE THEORY OF MORAL DEVELOPMENT

A continuing series of investigations by Lawrence Kohlberg (1969) and his colleagues seems to have confirmed the proposition that children regularly move through various stages of moral development. There is also some evidence that the same progression occurs for children in other nations (e.g., Kohlberg & Turiel, 1974). Because this work addresses itself to general developmental questions as well as to moral judgment as such, it is worth considering in some detail.

We have to be careful; some people will arrange everything into "stages"

Like Piaget, Kohlberg takes the idea of *developmental stage* quite seriously. The term is not used as a casual verbal decoration. When he discusses *structural moral stages in childhood and adolescence,* this phrase has precise meaning:

> 1. They are qualitatively different modes of thought rather than increased knowledge of, or internalization of, adult moral beliefs and standards.
> 2. They form an invariant order or sequence of development. . . . Movement is always forward and always step-by-step.
> 3. The stages form a clustered whole. There is a general factor of moral stage cross-cutting all dilemmas, verbal or behavioral, with which an individual is confronted.
> 4. The stages are hierarchical integrations. Subjects comprehend all stages below their own and not more than one above their one. They prefer the highest stage they comprehend. (Kohlberg & Turiel, 1974, p. 186).

You may want to take this succinct definition of developmental stages in general and determine whether phenomena in *any* area of human functioning meet the stage criteria. Extending the work of Piaget in this area, Kohlberg sees moral judgment as following the same basic rules that can be seen in the child's overall cognitive and adaptational development.

Moral reasoning is the process of judging "right" from "wrong"

We should speak, then, of **moral reasoning,** for judgments of morality are very dependent on the general level of reasoning available to the individual. Kohlberg believes his research supports the existence of at least six stages in the development of moral reasoning. These are preceded by a sort of prestage or no-stage stage that exists prior to moral reasoning as such. They may be followed by a hypothetical seventh stage for which appropriate data have not yet been obtained. The moral reasoning stages can be described as follows (discussion is based on Kohlberg, 1973):

Stage 0: No morality orientation. The child does one thing or another because he or she "wants to." There is no distinction between what is desirable or undesirable and what is "right" or "wrong" in the moral sense.

Stage 1: Punishment-obedience orientation. It is what a powerful, prestigious authority says that makes something either right or wrong. The child obeys without question to avoid punishment.

Stage 2: Instrumental hedonism and concrete reciprocity. This was originally known as the stage of "naively egoistic orientation." Decisions are made on the basis of practical considerations. The right thing is what benefits oneself and perhaps others (instrumental hedonism: doing something to bring pleasure). "I'll do something nice for you if you will do something nice for me" (concrete reciprocity) is also characteristic at this time. The absolute power of an authority no longer dominates.

Stage 3: Orientation to interpersonal relations of mutuality. This was first known as the "good-boy" stage of moral development. The motives and intentions are regarded as crucial to the morality of an action. It is not just what a person does, it is also why he or she does it. The child is "good" in order to help others and to gain social approval. There is satisfaction in being a "good person" apart from specific, concrete actions.

Stage 4: Maintenance of social order, fixed rules, and authority. The basic sense of morality at this time is drawn from the laws of the land and from the existing social order in general. The child at this time sees goodness in terms of being a law-abiding citizen and group member. The laws and regulations of the social order itself are not questioned; whatever upholds law and order must be good.

Stage 5: Social contract perspective. This stage is divided into two substages. First, the individual recognizes that there is something arbitrary in the rules and expectations of social order. Laws still should be obeyed because they comprise a social contract, an arrangement tolerated in order to live harmoniously together. Secondly, however, some laws are better than others, and one can advocate revising laws through due process. The sense of laws being relative to particular groups with morality as such being something that goes beyond simple obedience to established laws is a shift that continues within Stage 5.

Stage 6: Universal ethical principle orientation. The individual's *conscience* becomes critical here. There is an increased sense of personal moral responsibility. "I must do what I feel deeply to be right based upon all I understand about life." Moral decision-making at this stage may lead either to conventional or unconventional judgments from the social-group standpoint. This is a more inner-directed moral sense, as contrasted with the socially directed moral sense of Stage 5.

What would happen to society if *everybody* had Stage 6 morality—everyone "right" but not agreeing?

Stage 7: Contemplative nonegoistic experience (hypothetical). The possibility that a stage of this type exists was suggested by Kohlberg (1973) on the basis of philosophical and biographical information compiled through the centuries rather than on the basis of controlled research. In this stage, the question, "*Why* be moral?" is confronted, and the answer is given by each individual according to his or her understanding of life and death. It represents a movement away from the individual self, just as the previous progression of stages represented a movement away from external authority and group regulation.

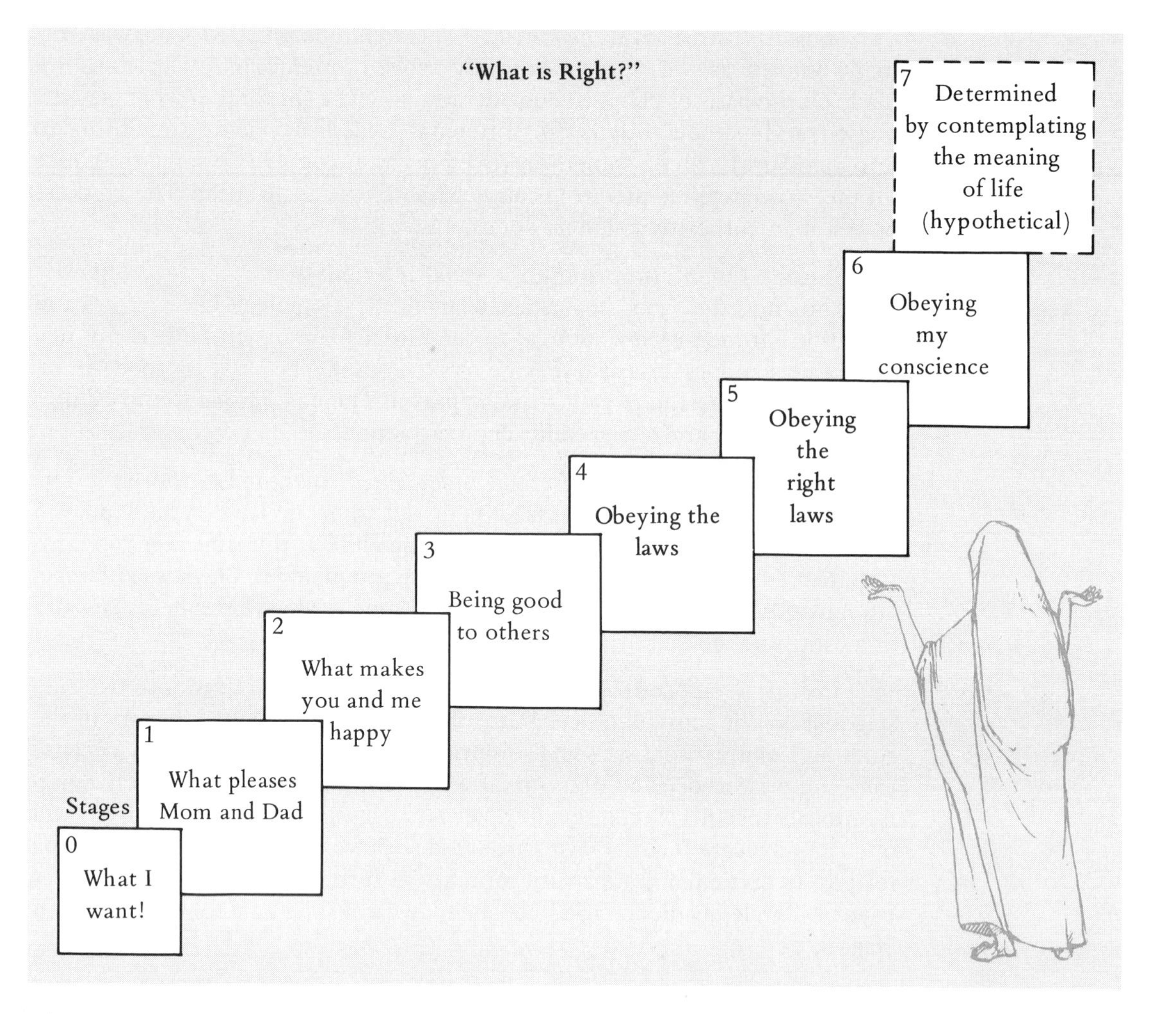

Figure 10-4. *The development of moral reasoning: Kohlberg's stage theory. (Based on the ideas and research of Lawrence Kohlberg, 1973.)*

Let's place the ten year old within this general framework of moral reasoning development and then attempt to put this entire approach into perspective.

Kohlberg's research does indicate that the more advanced stages of moral development require the type of logical thinking that is associated with formal operations (the adult sphere of thought). But this does not mean that a person who can engage in formal operations will in fact display high level moral principles in his or her actions! Development of cognitive structures is a necessary condition for the parallel development of moral judgment. Yet a person can have the intellectual resources and still be unprincipled. This finding is consistent with everyday experience. Most of us have known individuals who have all the intellectual tools, but little sense of morality. This is a matter of considerable practical concern. Under what conditions does the sequence of moral reasoning *fail* to accompany cognitive development? And what type of experience might possibly stimulate the belated growth of moral judgment in a person who is already developing mentally but not morally? These are among the questions that might be worth examining in future research.

The normal ten year old, however, shows evidence of continued development in both spheres, the mental and the moral. He or she has a relatively firm sense of fair play and a fondness for rules (although, of course, not for every rule under every condition). Because his or her own thoughts have become more systematized, he or she can appreciate games and other activities that embody rules of conduct. If his or her sense of morality sometimes catches the adult off-balance, it might be kept in mind that the ten year old still has a way to go in his or her cognitive development despite the considerable progress that has been made already. What appears to be a moral defect may represent only the present limitation imposed by his or her level of mental functioning. If, however, he or she is obviously immature in moral judgment, then it might be wise to examine the child's total life situation.

Interesting situations arise when the ten year old catches parent or teacher violating rules

Most ten year olds throughout the world show moral reasoning beyond the no-stage stage. Whether they are Chinese, Mexican, Hopi, Zuni, Belgian, Swiss, Turkish, Israeli, or Papago, ten year old boys and girls have usually reached at least Stage 1; there are many at this age who have already demonstrated Stage 2 morality (Tapp & Kohlberg, 1971). Exceptionally mature ten year olds may show the Stage 3 "good-boy–good-girl" moral reasoning pattern, although it is more typically found around age thirteen. In general, the ten year old grasps and respects an established order of things. The basis of morality no longer resides entirely in authority figures and fear of punishment, but it is not yet free from particular regulations within particular social contexts.

An important theoretical question arises here. Precisely what leads a child to move from one level of moral reasoning to the next? Is it enough to speak of maturation, or even of maturation-plus-interactive experience? Or do we need a more specific kind of answer. Kohlberg and his colleagues, especially Turiel (Kohlberg & Turiel, 1974) have offered an explanation that parallels Piaget's thoughts about movement from one general cognitive stage to the next. This involves the concept of **disequilibrium.** The individual's existing way of comprehending the world (logically or morally) is challenged. The old way of looking at things doesn't seem to do the job. In attempting to attain a new equilibrium, the mind may forge ahead to a higher stage of functioning. This explanation—progressive development through overcoming disequilibrium—has stimulated interesting speculation and research throughout a range of mental phenomena. Whether or not it is a solid explanation for the development of moral reasoning remains to be seen.

Disequilibrium is a temporary difficulty in adaptation that spurs efforts to a higher level of adaptation

If cognitive and moral development are as closely associated as the Piagetian/Kohlbergian approach suggests, then we have a powerful basis for understanding much of social as well as mental growth. But the more important a theory, the more it deserves—and usually receives—critical attention. Kurtines and Greif (1974) have raised many questions about this approach and data that can only be answered by more research and perhaps some reanalyses of existing information. Some of their criticisms concern the precise determination of what stage a person is in at a particular time. They also are not convinced that all the stages are necessarily "higher" than the ones below, especially when the top levels are reached. It may be difficult to separate out the bias of the theoretician and his culture from the conclusions actually justified by data. Perhaps we Americans are so individual-oriented that the stages transcending group-based morality reflect more of our own quirks than a true universal of human development. Furthermore, much of the research in this area has been based on responses to hypothetical situations presented in the form of questionnaires. Some work has broken out of this mold (Turiel, 1966), but one might want to see a greater variety of research methodologies directed to the same questions before coming to a firm conclusion.

Kohlberg himself has shown the capacity to take criticism into account, to reexamine his own previous work and to move ahead both to consolidate and to extend the theory. The view of moral reasoning as a developmental progression closely associated with (but not identical to) stages of cognitive growth will be worth watching in the years ahead.

In general, the ten year old's combination of mental and moral development makes him or her a fine companion for people of any age, from the young child through the aged adult. This is a time of life when, in addition to developments already mentioned, the child is likely to show a positive interest in "being good" and "doing good" for others.

What is known as *altruistic behavior* (actions intended to benefit other people) tends to increase throughout childhood (Krebs, 1970). This gives adults the opportunity to encourage the continued growth of concern for others if they so choose. Many studies (reviewed by Bryan, 1975) have shown that altruistic models encourage altruistic behavior in children. The child is more likely to be helpful to others if he or she has seen others engage in altruistic behavior either in real life or on television. Furthermore, children who enjoy a general sense of personal well-being are more likely to be altruistic (e.g., *see* Yarrow, Scott, & Waxler, 1973). It is easier to be good, then, if one feels good and has seen other people acting in positive and helping ways.

There are individual differences, then, in altruistic behavior and moral reasoning within a particular age range. Let's look a little further at individual differences through Scenario 8.

SCENARIO 8: "PLEASE LIKE ME"

Karen and Beth are both ten years old and are best friends. They walk to school and back together almost every day. Although each girl has other friends, Karen and Beth find it natural to join forces when there are plans to make, problems to solve, or hours to pass pleasantly. Despite their mutuality of interests, however, the girls are far from identical in their approaches to life. This causes them to worry about each other from time to time.

"Oh, c'mon, Karen, we're going to be late!" "That's silly, Karen!" "But Karen, we're not supposed to do that!" These are typical of the concerns Beth has about Karen. Beth thinks her friend is sometimes too careless and too much of an adventurer. True, Karen has never really been in any trouble that she knows of, but it bothers Beth to see her take chances and risks. Beth feels more comfortable doing what is obviously the right thing, and doing it so that everybody can see it. Yet Beth does take almost secret enjoyment in her friend's enterprising and carefree behavior. She is often tempted to join in herself—and sometimes does! The conflict, such as it is, exists more within Beth herself than between the two girls. Beth would like to be as free and spontaneous as Karen, but she is terribly afraid that this would displease her parents, her teachers, and perhaps the whole universe. "Terribly afraid" sounds like an exaggeration, but that is precisely how Beth feels sometimes when she is tempted to emulate her blithe friend.

For her part, Karen wishes her friend wouldn't worry so much about what people might think. Perhaps, then, she would not hesitate as much or back down when she was just about to go along with one of Karen's ideas. The time that they took that extra long bicycle ride to the other side of town was almost ruined by Beth's constant doubts and apprehensions (although Karen would not have had the courage to make the trip without her friend's company). Karen enjoys her role as the leader in extracurricular pursuits, although Beth is the one who receives most of the special notice, responsibility, and "plums" at school. Sometimes Karen gets the feeling that Beth puts her up to little tricks and adventures so she won't have to run the risks herself.

Schooling is not the only significant activity that takes place in school. Children are now becoming mature enough to pair off and make each other "best friends." (Photo by Monkmeyer Press Photo Service.)

These good friends are already quite different from each other at age ten. Will these differences continue to exist when another ten or twenty years have gone by? And what previous developments led to these different orientations in the first place? Assume that the descriptions of Karen and Beth are based on direct observation of their current behavior and thinking. I am attempting now to flashback to a time before we came into the scenario and also to project it ahead. For this purpose I will call on some research findings reported not long ago by Jacqueline Allaman, Carol Joyce, and Virginia Crandall (1972). Their study illustrates one of the tasks of developmental psychologists, the effort to discover how events that occur at one time in the lifespan influence behavior and experience at another time. Their research involved schoolchildren of various ages and also young adults. It was possible for them to arrive at preliminary conclusions regarding changes-over-time because the researchers had at their disposal information acquired during a major longitudinal study conducted at the Fels Institute in Ohio.

It was discovered that children who have a strong need to be socially desirable often have had similar background experiences. Further, people who orient themselves around the desire to please others in childhood are likely to continue in the same mode at least into young adulthood (which is as far as this study went). Perhaps Beth typifies this pattern for girls. When Beth was still an infant, her mother did not exude much warmth toward her. Throughout Beth's early development, she received more criticism than affectionate contact. Mother usually found a lot to criticize about her daughter and used a somewhat coercive and punitive childrearing style. Accordingly, Beth has learned to place much emphasis on how her behavior pleases or fails to please adult authority figures. She must be "good" in the sense of engaging only in socially desirable responses, yet her experience has been such that she does not really expect to be loved and accepted even when she is doing the right thing. Beth is a bright girl, however, and can earn both self-respect and the respect of others through intellectual achievements. Being an outstanding student, then, provides her with an area of functioning that has its own rewards. Chances are that Beth will continue to be influenced in the years ahead by her need to be socially desirable. (This need is, of course, a part of most people's make-up; we are concerned here with a need of such intensity that it tends to dominate life-style.) Her friend Karen has experienced more direct warmth and affection at home since infancy. She knows that she will face disciplinary action if her behavior wanders too much out of line, but she does not fear a basic loss of relationship with the parents for making a mistake or two.

Will they be the same in old age, or should we expect differences?

The pattern with boys is similar but not identical. Allaman and her colleagues found that "harsh parental practices" in infancy and early childhood were closely associated with subsequent development of strong social desirability tendencies in both boys and girls. Boys, however, were less often subjected to this kind of treatment by their mothers, especially when very young. Up until the beginning of elementary school, boys were usually raised in a warm and tolerant way. However, boys were more likely than girls to experience a discontinuity in parental expectations and treatment at this time. Mothers became more pushy and demanding. The boys (at least, some of them) were expected to become rapid achievers, while simultaneously receiving less direct affection.

The role of the father was not neglected in this study, although methodological limitations made the conclusions more tentative. The investigators did learn enough, though, to gain the impression that the father's influence was also quite important and, apparently, especially important for the development of his sons. The father who took a critical and unaffectionate approach toward his young son seemed to be contributing toward a high social desirability orientation, with its accompanying sense of ineffectuality and insecurity.

The investigators also gave particular attention to the possibility that early childhood experiences may not show themselves in an obvious way immediately. There could be "sleeper" effects. The impact of Beth's relationship with her parents could become more evident in her adult behavior than in her behavior at age ten. Such a possibility forces us to look at behavior and experience at any particular point in childhood in a more sensitive and complex manner and certainly invites additional research.

The fact that Karen and Beth like each other is a valuable source of comfort for both. It is a relief, even at age ten, to have somebody in one's life who fondly accepts without requiring high critical standards of performance. The relationship between these friends is not entirely free from conflict and tension, but it offers a treasured alternative and useful learning experience in a world that seems to be shaping up as relentlessly more and more adult every day.

HAZARD: FATAL ACCIDENTS

Two facts about the ten year old's vulnerability to death have already been stated: (1) in general, this is the time of life at which the probability of death is at its lowest, while (2) among the various causes of death at ten, accidents constitute the greatest source of peril. We will now consider accidents in more detail and broaden our scope to include children between the ages of five and fourteen.

SCOPE OF THE PROBLEM

In recent years, accidents of various types have accounted for about half of *all* deaths in school-age boys. Accidents are also the most frequent cause of death

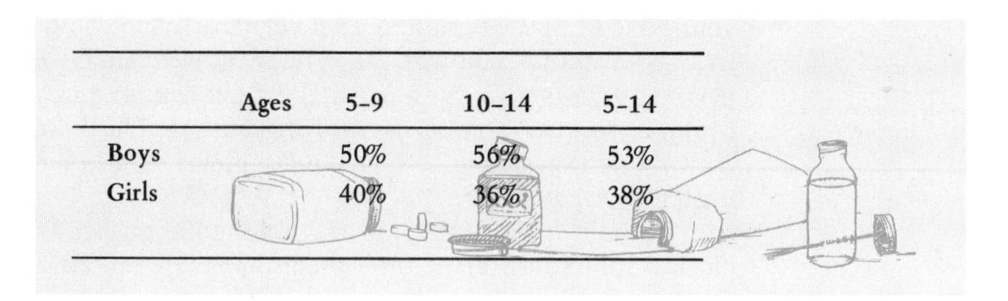

Ages	5–9	10–14	5–14
Boys	50%	56%	53%
Girls	40%	36%	38%

Figure 10-5. *The percentage of all deaths in childhood that are caused by accidents. (Based on data presented in Metropolitan Life, 1973.)*

in school-age girls, although the percentages are somewhat lower as you can see in Figure 10–5 which is based on data for 1969 (the latest year for which complete official statistics are available).

The total of 8,186 accidental fatalities in 1969 for children aged 5 to 14 means that approximately 20 out of every 100,000 children who were alive on January 1 died as a result of an accident before the year ended. This mortality rate represents an increase over the rate earlier in the decade (1960–1961). The statistics tell us not only how sizable this increase was, but also point clearly to the most important source, motor vehicle accidents. For every three schoolboys who died as a result of motor vehicle accidents in 1960 to 1961, four perished in 1968 to 1969. Mortality among the younger schoolgirls increased by about one third during this same time-period, and among older girls (ages 10–14) it increased by about two-fifths. This latter figure is exceptionally large. The nation would probably be quite alarmed if a 40 percent increase in the general mortality rate were to take place over a period of just a few years. The increasing vulnerability of older schoolgirls to motor vehicle accidents, however, does not seem to have come to general notice.

There was a general decrease in mortality due to other types of accidents for boys and for younger girls during this period, otherwise the overall rise would have been even more alarming. As it is, however, accidental fatalities (all types) rose from 11.2 percent to 12.7 percent for girls, and from 26.0 percent to 27.6 percent for boys. These increases are more serious than meets the eye, because the *percent of change* is 13 percent for girls and 6 percent for boys.

These dry statistics point up two other considerations that should not be ignored. First, in general, the death rate from accidents is much higher for boys than for girls. This also holds true for every specific type of major accident. But, secondly, the differential in vulnerability between boys and girls has started to close up. The mortality rates are becoming more similar, not because fewer boys are being killed in accidents, but because more girls are becoming victims. Clearly, there is reason to be concerned about the physical safety of our schoolchildren—and to be alert to the factors that lead to such peril.

The effects of accidental death in childhood on parents and siblings cannot be ignored either

TYPES OF ACCIDENT

In addition to motor vehicle accidents, the other major sources of mortality among schoolchildren are drowning, fires, firearms, falls, and industrial-type accidents. By far, the greatest number of fatal accidents in the water involve swimming, although about 150 children perished in 1969 as a result of boat accidents. Deaths by drowning are second only to motor vehicle accidents in boys of all ages, but most particularly those from age 10 to age 14. Perhaps the hazards from this source can be conveyed more vividly if we observe that drowning accounts for the same number of fatal accidents among boys as the *total* number of fatal accidents among girls (exclusive of the number one cause for both, motor vehicles). Fire and flames (including smoke inhalation) is the second leading accidental cause of death for girls; it is third for boys. Drowning follows closely behind fire among girls.

Particular attention should be given to deaths that are directly related to firearms. Up until recently, few girls died as a result of gunshot wounds. However, the accidental death rate among older boys is appreciable (2.4 fatalities per year per 100,000 boys in the 10–14 age range) and may still be rising. The available information does not fully reflect what has been taking place since firearms have become more common in homes and stores during the past few years. Now that more handguns are within the possible grasp of children, we can scarcely expect that the accidental death rate from this source will diminish.

Although I have been focusing here on fatal accidents, many mishaps have serious impact on the child's life even though he or she survives the incident. One recent survey has found that about 18 *million* schoolchildren suffer injuries that require medical attention or at least one day of restricted activity *each* year (Metropolitan Life, 1973). Many of these accidents take place in or around the home.

A FEW CONSIDERATIONS

A thorough understanding of fatal accidents (or accidents in general) in childhood requires attention to many factors, some of which are quite complex. One would want to become conversant with the broad topic of accident research, including the methodology, concepts, and pitfalls in arriving at conclusions. A valuable book to consult for this purpose is *Accident Research*, edited by Haddon, Suchman and Klein (1964). Although some of the material is now slightly dated, one can learn much from this volume.

Another major realm of concern is the relationship between adult behavior and expectations and fatal accidents in children. The authors of an informative statistical summary I have drawn upon here states, for example:

> A considerable proportion of the motor vehicle accident victims of school age were fatally injured while playing on or crossing streets or highways. Pedestrian

> accidents accounted for almost three out of every five deaths in motor vehicle accidents among children at ages 5–9 and for one out of every four at ages 10–14. The preponderance of pedestrian fatalities reflects not only the abrupt decrease in close parental supervision when children reach school age, but also the failure of youngsters to take basic safety precautions. (Metropolitan Life, 1973, p. 10).

The latter statement, offered as explanation, sounds plausible. However, it is a proposition that needs to be verified directly by appropriate research; the statistics by themselves do not tell us "why."

If we assume for sake of further discussion that this explanation is correct, then several series of further questions arise. Why do parents supervise their schoolchildren less adequately now than in previous years? Is this trend more powerful in some types of community or family structure as compared with others? In contemporary society, is it still *realistic* to expect close adult supervision, or should alternatives be explored? To what extent can parents be expected to teach safety precautions to their children when many adults "tempt" their own accidents through careless behavior? And to what extent are the attitudes and value-priorities of society in general crucial to determining the accidental fatality rate? Are children forced to play in the street, for example, because there is no other place for them? Does local custom condone driving at unsafe speeds in residential areas? Are environmental hazards to safety allowed to exist because of our own inertia or indifference?

Is it conceivable that adult expectations lead even more directly to childhood accidents? Consider in particular the association between "masculine" behavior and risk-taking. Historically in our society, boys are not supposed to be "namby-pamby." To win esteem in the eyes of their fathers and their peers, boys are often expected to behave in active and aggressive ways. Girls, of course, are cast in the sugar-'n-spice-'n-everything-nice model. I am calling attention, in other words, to the differential socialization of boys and girls. Encouragement of greater aggressive, independency-seeking, and risk-taking behavior among boys theoretically might be expected to lead to more accidents, including more fatalities. The available statistics are consistent with this interpretation, although they do not prove it in any rigorous way. Some of the other psychological and social complexities involved in the encouragement of hazardous behavior are explored in more detail elsewhere (Kastenbaum & Aisenberg, 1972).

For our purposes here, let's concentrate on the possible relationship between two trends that appear to exist at present in the United States. The psychosocial status of the female is undergoing intensive reevaluation from many standpoints. One implication is the schoolgirl may be experiencing more freedom and inclination to engage in types of behavior previously set aside or tolerated for boys only. The other trend is the increase in motor vehicle fatalities—especially pedestrian accidents—for girls. Perhaps this is one of the first indications that increased freedom for older girls will bring their risk factor much closer to the vulnerabilities of older boys. Next, we might fear that gunshot injuries and deaths would also show a marked increase, and perhaps drownings as well. Changes in the life-styles of society in general are reflected in the behavior of our children—and, unfortunately, may also be reflected in the mortality tables.

SUMMARY

The ten year old has biology, psychology, and social integration working beautifully in his or her favor. Biological hazards to life and health are at a minimum—although *accidents* constitute a major environmental menace. Physically, the child can perform many of the actions required in daily living. Socially, there is still the security of home and family to count on while exploring new relationships with peers and others. Psychologically, the changes associated with adolescence and sexual awakening are yet to come, and plans and fantasies can flourish without running into severe tests in reality.

Intellectual competence shows itself in a heightened ability to *"pay" attention.* The ten year old is more likely than the younger child to focus systematically on those aspects of the world that most adults interpret as reality. Two different ways to account for better attention in the ten year old are explored: (1) an approach that emphasizes learning and training experiences concerned with attention as such, and (2) an approach that emphasizes a general shift in underlying thought processes. While both aspects are probably involved, it is most likely that the increased ability to pay attention is closely related to the continued development of *cognitive structures* at this age level. The work of *Jean Piaget* is intimately associated with this approach.

A closer examination of *learning* and *thinking* indicates that the ten year old applies *more complex strategies* that are less rigidly bound to the spatial and physical properties of the stimulus materials, as compared with younger children. He or she is able, for example, to comprehend *conditionality* situations more adequately and to combine *lose-shift* with *win-stay* strategies. In general, the child at this age level shows a markedly improved ability to take a *planful* approach to learning and to *adjust* his or her approach as the situation requires.

Curiously, intellectual development is sometimes accompanied by a tendency to make *more* errors in certain situations. Termed *growth errors* by *Jerome Bruner,* these missteps may take place when the child is in the process of changing from a simpler to a more complex way of learning and thinking. This tendency brings to the fore a highly significant point made by *Heinz Werner* and his colleagues: there is a crucial difference between *performance* and the kind of *process* through which the individual operates in order to achieve his or her performance. Developmental psychologists must understand *how* the organism behaves as well as how well it performs in terms of external standards.

The most general account of mental life at this age level is once again provided by Piaget. The ten year old is seen as having moved to the stage of

YOUR TURN

1. Biographies and autobiographies offer many opportunities for the developmental psychologist to enrich his or her understanding and to gain new insights. While one might read biographies in connection with any aspect of the total lifespan, it is especially relevant here to explore retrospections on the later years of childhood. Browse through a few book-length biographies, selecting people who you particularly admire or would like to know more about. Pick out one or two books that seem interesting in general, and then give particular attention to the way in which the author recalls the childhood period. What memories dominate? What kind of enduring impressions were formed during this time? In writing his or her autobiography, did the individual portray age ten and thereabouts as a time of particular contentment and competence? Was this period, instead, a special challenge because one had to assume adult-type responsibilities while having only the experiences and resources of a child? Summarize the total biography or autobiography from a psychological-developmental standpoint, but highlight what was made of the later years of childhood.

2. Strike up conversations with a number of ten year old boys and girls. Eventually get around to the subject of their plans for adult life. What plans or hopes do they have in mind? To what extent do these future projections involve occupation, and to what extent do they center around other aspects of life, such as raising a family or living in a particular part of the world? How realistic do these projections seem to be? Are they held firmly, or are they entertained as possibilities that admittedly might change again and again before adulthood? How much time and what kind of experiences lie between the child at his or her present age and the plans and ambitions expressed?

3. Share a "situation" with a ten year old or a child of approximately that age. Select a situation that is natural and interesting to both of you (e.g., preparing a meal together, playing a game, visiting your college campus or a museum). Now pay attention to what the child pays attention to. Does the child see the situation as you do and observe the same details and features, or do other aspects catch his or her eye? To what extent is the child's attention captured (or captivated) by the same stimuli that tend to elicit the attention of adults? See how much you can notice without letting your own behavior and expectations intrude upon the child's. Is the child in any sense *less* attentive than most adults would be in this situation, or is the child attentive in different ways?

4. Here is another child-watching experience that will test your powers of observation. Bring together a ten year old and another child who is several years younger. Again, establish a comfortable, natural situation, but one in which there is to be a certain amount of activity. This might well be a seasonal activity, such as each of you building a snow creature or a sand castle. The point here is to see if you can distinguish between the end product or achievement that each child comes up with and the *process* that was used. Did the younger and the older child approach the task (or define the task) in the same way? Was the total sequence of behavior identical, or did the older child engage in a greater number of separate operations? This is the sort of observational situation that might require several run-throughs before you sharpen your perceptions to the level where you can recognize differences in process as well as achievement. What do you suppose is the relationship between the general types of process shown by the younger and the older child and the specific materials with which they were working, and the total situational context? If your curiosity gets the better of you, why not develop another situation with the same children and see if the differences carry over or show up in some other way?

5. "Tell me something *bad* or *wrong* that a person once did? Why was that a bad thing to do?" Try these questions out on a few young children. Then, take the situations they reported and discuss them with older children. Does the ten or eleven year old also see this behavior as bad or wrong? Does he or she apply the same reasoning in making his or her moral judgment? Find out what the older child would consider to be "good" behavior in the same situation. How do your observations compare with the material presented in this chapter?

concrete operations. His or her thoughts are more systematic and flexible than in the preceding stages. He or she can perform mental operations of *inversion* and *reciprocity,* for example, and apply the logic of *class inclusion operations* and *serial ordering operations.* Nevertheless, the child is not yet in command of all the cognitive processes he or she will enjoy as an adolescent and adult.

Thought and behavior come together significantly in the area of *moral development.* The work of Piaget and *Lawrence Kohlberg* strongly suggests that there are systematic phases in the development of the sense of fair play throughout childhood—and throughout the world. The ten year old is likely to have moved somewhat away from complete dependence on adult authority to define moral justice, being more concerned with the specific situations at hand and the concepts of equality and sharing. Moral judgment becomes even more relativistic and situational in the years ahead. Actions intended to benefit other people *(altruistic behavior)* also tend to increase from earlier through later childhood.

Individual differences are important throughout the lifespan. In this chapter we examined life-style differences between a pair of ten-year-old friends, Beth and Karen. We saw what differences might be expected as they continue to grow into adulthood.

Reference List

Allaman, J. D., Joyce, C. S., & Crandall, V. C. The antecedents of social desirability response tendencies of children and young adults. *Child Development,* 1972, *43,* 1135–1160.

Bruner, J. S., Olver, R. R., & Greenfield, P. M. (and nine others). *Studies in cognitive growth.* New York: John Wiley & Sons, Inc., 1967.

Bryan, J. H. Children's cooperative and helping behaviors. In E. M. Hetherington (Ed.), *Review of child development research* (Vol. 5). Chicago: University of Chicago Press, 1975, pp. 127–181.

Dublin, L. I. *Factbook on man.* New York: Macmillan, Inc., 1965.

Elkind, D., & Flavell, J. H. (Eds.). *Studies in cognitive development.* New York: Oxford University Press, 1969.

Gelman, R. Conservation acquisition: A problem of learning to attend to relevant attributes. *Journal of Experimental Child Psychology,* 1969, *7,* 167–187.

Haddon, W., Suchman, E. A., & Klein, D. (Eds.). *Accident research.* New York: Harper & Row, Publishers, 1964.

Inhelder, B., & Piaget, J. *The growth of logical thinking from childhood to adolescence.* New York: Basic Books, 1958.

Kastenbaum, R., & Aisenberg, R. B. *The psychology of death.* New York: Springer Publishing Co., Inc., 1972.

Kohlberg, L. Stage and sequence: The cognitive-developmental approach to socialization. In D. A. Goslin (Ed.), *Handbook of socialization theory and research.* Chicago: Rand McNally & Company, 1969.

Kohlberg, L. Continuities in childhood and adult moral development revisited. In P. B. Baltes & K. W. Schaie (Eds.), *Life-span developmental psychology: Personality and socialization.* New York: Academic Press, 1973, pp. 180–207.

Kohlberg, L., & Turiel, E. Overview—Cultural universals in mortality. In L. Kohlberg & E. Turiel (Eds.), *Recent research in moral development.* New York: Holt, Rinehart and Winston, 1974.

Krebs, D. L. Altruism—An examination of the concept and a review of the literature. *Psychological Bulletin,* 1970, *73,* 258–302.

Kurtines, W., & Greif, E. B. The development of moral thought: A review and evaluation of Kohlberg's approach. *Psychological Bulletin,* 1974, *81,* 453–469.

Metropolitan Life. Fatal accidents among school-age children. *Statistical Bulletin,* December 1973, *54,* 9–11.

Piaget, J. *The moral judgment of the child.* London: Routledge & Kegan Paul, 1932.

Piaget, J. *The construction of reality in the child.* New York: Basic Books, 1954.

Piaget, J., & Inhelder, B. *The psychology of the child.* New York: Basic Books, 1969.

Stevenson, H. W. *Children's learning.* New York: Appleton-Century-Crofts, 1972.

Tapp, J. L., & Kohlberg, L. Developing senses of law and legal justice. *Journal of Social Issues,* 1971, *27,* 65–91.

Turiel, E. An experimental test of the sequentiality of developmental stages in the child's moral development. *Journal of Personality & Social Psychology,* 1966, *3,* 611–618.

Wapner, S., & Werner, H. *Perceptual development.* Worcester, Mass.: Clark University Press, 1957.

Werner, H. *Comparative psychology of mental development.* New York: Science Editions, 1961. (Originally published, 1948.)

Yarrow, M. R., Scott, P. M., & Waxler, C. Z. Learning concern for others. *Developmental Psychology,* 1973, *8,* 240–260.

CHILDHOOD IN PERSPECTIVE

11

CHAPTER OUTLINE

I
What seems to be "normal" development is contingent on the particular world of childhood at a particular time.

Height / Rate of maturation / Variants / Invariants / Childhood / Sex differences / Sex-role stereotypes / Sex-role models

II
Erikson's "Eight Ages of Man" offers a framework for a more general exploration of the development of identity.

Basic trust versus basic mistrust / Autonomy versus shame and doubts / Initiative versus guilt / Industry versus inferiority / Identity versus role confusion / Intimacy versus isolation / Generativity versus stagnation / Ego integrity versus despair

III
Recent studies either illuminate or challenge our understanding of development through childhood.

Developmental stages / Fears / Moral reasoning / Generational change / Acculturation / Disequilibrium (external or adaptational; internal)

We have followed humans developing a long way, from the prenatal scene through the grade school years. What a transformation! Two biological systems, the sperm and the ovum, merge to form a new psychobiological field. Tiny, vulnerable, highly dependent upon its immediate environment, this new living entity nonetheless displays characteristics of organization and directionality. The developing embryo and fetus operate within an environment that itself is moving through phased changes. Most directly, this is the in-close environment provided by the maternal host. Less directly, but also significantly, there is the larger socio-physical environment in which the mother-to-be functions and which may exercise a variety of influences upon the prenate.

From an abstract perspective, then, we see the formation of a new, one-and-only-one-of-its-kind psychobiological field from the union of "satellites" (ovum and sperm) provided by two adults who themselves constitute independent, yet related, physical and symbolic-experiential systems. The new individual develops through interaction with the maternal host, and the relationship between the two also constitutes a field of mutual influence. At the same time, the mother remains a part of larger fields of influence (the family, the community, etc.). Fields of active and organized behavior interacting with other fields, from conception onward—such is the general picture.

A decade later we are in the presence of a person who displays much physical, mental, and social competence. The child neither looks nor behaves like the zygote (or even the neonate) with which the sequence began. The individual's relationship to his or her environment is also far different, and the same may be said about the way in which this "psychobiological field" relates to itself. This sequence of transformations is so remarkable that it should not be taken for granted simply because it happens so often.

In this chapter we will attempt to gain additional perspective on the overall process of development throughout childhood. Issues and themes that are particularly significant for this purpose will be highlighted. We will not forget that interaction is one of the keys to understanding development and functioning throughout all of childhood, indeed, throughout all of the lifespan. It is also one of the purposes of this chapter to facilitate our exploration of the years beyond childhood. Some of the basic problems and concepts encountered in understanding child development will continue to engage our attention in adolescence and adulthood.

WHAT IS BECOMING OF CHILDHOOD?

Serious attention to individual-environmental interaction requires appreciation of changes that occur in both spheres. This means that to comprehend what the *child* is becoming, it is also pertinent to understand what *childhood* is becoming. Because this approach has not yet been incorporated into the mainstream of theory and education in human development, I'll take a moment to illustrate different types of change.

Both the zygote and the maternal host move through a set of phases that are fairly predictable. At present there is no reason to believe that the basic developmental progression from zygote through advanced fetus has itself changed much over the centuries. The same proposition also seems applicable to the pattern of changes through which the maternal host supports the prenate's development. Every developing infant and every mother-to-be mutually enact a change sequence, then, but the sequence itself is relatively fixed within a certain range of variation. But one sequence or the other *can* go awry. Perhaps there is a genetic defect in the zygote, for example, or a physiological or anatomical problem in the maternal host. The total interaction may be altered or terminated in consequence. However, prenatal development serves as an appropriate example of changes within a relatively fixed system.

Could the zygote-through-fetus process have been changing subtly through the centuries?

It is the other kind of change that is most challenging here. The *world* in which the child develops has *not been the same* from century to century. Significant changes may take place within a much smaller period of time (e.g., economic disaster or the holocaust of war can intervene between the birth of one child and the birth of the next within the same family). Changes in the socio-physical system are much less predictable than the patterned changes that take place within the maternal host. Furthermore, the *limits* and *dimensions* of such change are much more difficult to establish. It cannot be assumed that the world is a fixed system, or that it invariably returns to its starting point. The total setting in which a child develops during the next generation, for example, will probably have many distinctive and unique features, and nobody who is already grown has had the kind of childhood that many will have in the next few decades. And so we are confronting a kind of change that does not guarantee where or how far it will go. And we are confronting a challenge to our ability to distinguish between the development of children as such, and the nature of childhood. This issue has been touched upon in various ways earlier in the book. Now I am considering it with the intent of improving our sensitivity to how "normal development" is contingent on the particular world of childhood at a particular time.

BIGGER, AND BIGGER, AND ??

Let's begin with two obvious physical dimensions of the child—height and weight. Have you ever examined a suit of armour in a museum, or an authentic military uniform from, say, the eighteenth century? If so, then you may have noticed the relatively small dimensions. There is reason to believe that adults today, at least in societies such as ours, are taller and heavier than adults in the past. The same holds true for children. It looks as though there is a trend over time for people to be developing larger and larger. An adult of ample, but not extraordinary, dimensions is likely to be uncomfortable in a chair or theater seat that was "just right" a few generations ago.

These observations raise a set of physical and biological questions. Has there been a systematic change in the genetic basis of human growth within the past few centuries and, if so, is this progression still in progress? Or have the *conditions* changed to favor more extensive physical growth, especially in childhood where most of the differential becomes established? Is there perhaps a more complex interaction between who is begetting with whom, the survival value of greater size and strength, and the more conducive growth conditions in society (perhaps better nutritional status, more protective treatment of young children, etc.)? A parallel line of questions could also be raised in the psychosocial sphere. These would be more difficult to study, however.

I have prepared a group of statistical tables (Figure 11–1 and Figure 11–2) that are worth consulting at this point. Like most tables, these do not provide definitive answers to complex questions, but they do enlighten. All the tables are concerned with the height and weight of males and females ranging in age from 6 through 17. The data were obtained through surveys conducted during three time-intervals within our own century: 1937 to 1939; 1954; 1963 to 1970. Look first at Figure 11–1. This reports height and weight averages for males and females respectively. Rove your eyes over the columns listing averages for the 1937 to 1939 survey and the 1954 survey. You will see, for example, that six-year-old boys were 2.4 inches taller and 2.1 pounds heavier at the later date, and girls of the same age were 0.4 inches taller and 1.1 pounds heavier. The height and weight differential continues right along the chronological age ranks.

If only these two points of measurement were available, we might conclude that children are continuing to grow larger and heavier. We could then inquire further into the possible interacting factors responsible for this progression. But we do have that third point of measurement. Compare now the data from 1954 with the data from 1963 to 1970. Sound of skidding brakes! The progression appears to have slowed almost to a halt. This conclusion comes across more clearly when we look at Figure 11–2, which compares children measured in the 1963 to 1970 survey with both of the earlier periods. The trend toward greater size has slowed down considerably (although not equally for height and weight). Whether or not it will come to a complete halt remains to be seen.

Many conclusions are based on only *one* point of measurement; many points are useful

One other dimension should be singled out: *rate* of maturation. For some time now, children seem to have been reaching their adult physiques at somewhat earlier ages, generation by generation. This trend is still reflected in the data from the most recent (1963–1970) survey. In boys, the greatest annual change in height now occurs between ages 12 and 13, in comparison with between ages 12 and 14 a decade ago, and between ages 12 and 15 about three decades ago. Girls now reach 95.3 percent of their adult height by age

Age in Years	Height in Inches 1963–1970	1954	1937–1939	Weight in Pounds 1963–1970	1954	1937–1939
BOYS						
6	46.7	46.8	46.4	47.9	49.7	47.8
7	49.0	49.2	48.5	53.8	56.7	52.6
8	51.2	51.5	50.7	60.6	63.7	58.3
9	53.3	53.6	52.7	68.1	71.0	64.3
10	55.2	55.6	54.5	73.7	79.4	70.4
11	57.4	57.4	56.3	83.9	86.6	76.9
12	60.0	59.3	58.3	94.3	93.9	84.9
13	62.9	62.2	60.7	109.7	106.3	95.1
14	65.6	65.1	63.1	124.4	120.9	106.9
15	67.5	67.3	66.1	135.3	132.4	123.4
16	68.6	68.5	67.6	142.4	141.0	133.2
17	69.1	69.2	68.4	149.5	147.8	139.2
GIRLS						
6	46.4	46.4	46.0	46.9	47.6	46.5
7	48.6	48.9	48.1	52.7	54.5	51.5
8	50.9	51.2	50.3	60.1	61.7	57.1
9	53.3	53.4	52.2	68.6	69.2	63.1
10	55.5	56.3	54.4	76.9	79.8	70.2
11	58.1	58.7	56.7	87.5	90.5	78.8
12	61.1	60.5	59.1	102.2	98.8	89.1
13	62.5	62.1	61.0	110.7	107.9	98.9
14	63.5	63.2	62.2	118.9	115.4	106.2
15	63.9	63.7	63.2	124.0	119.0	114.2
16	64.0	64.0	63.4	127.5	121.4	116.7
17	64.1	64.3	63.4	126.4	122.5	117.5

Figure 11-1. *Average heights and weights of children and adolescents: results of surveys conducted at three points in time. (Adapted from Metropolitan Life, 1973.)*

12, compared with 94.1 percent only one decade ago. And, as we will see in a later chapter, the onset of menstruation is also earlier than in past generations.

Variants are developmental phenomena or principles found only in certain situations

Thinking over these trends, we can appreciate both the variants and invariants in child development, using height and weight as the present examples. By ***variants,*** I mean those manifestations or principles of development that seem to be found only for certain kinds of people in certain kinds of situ-

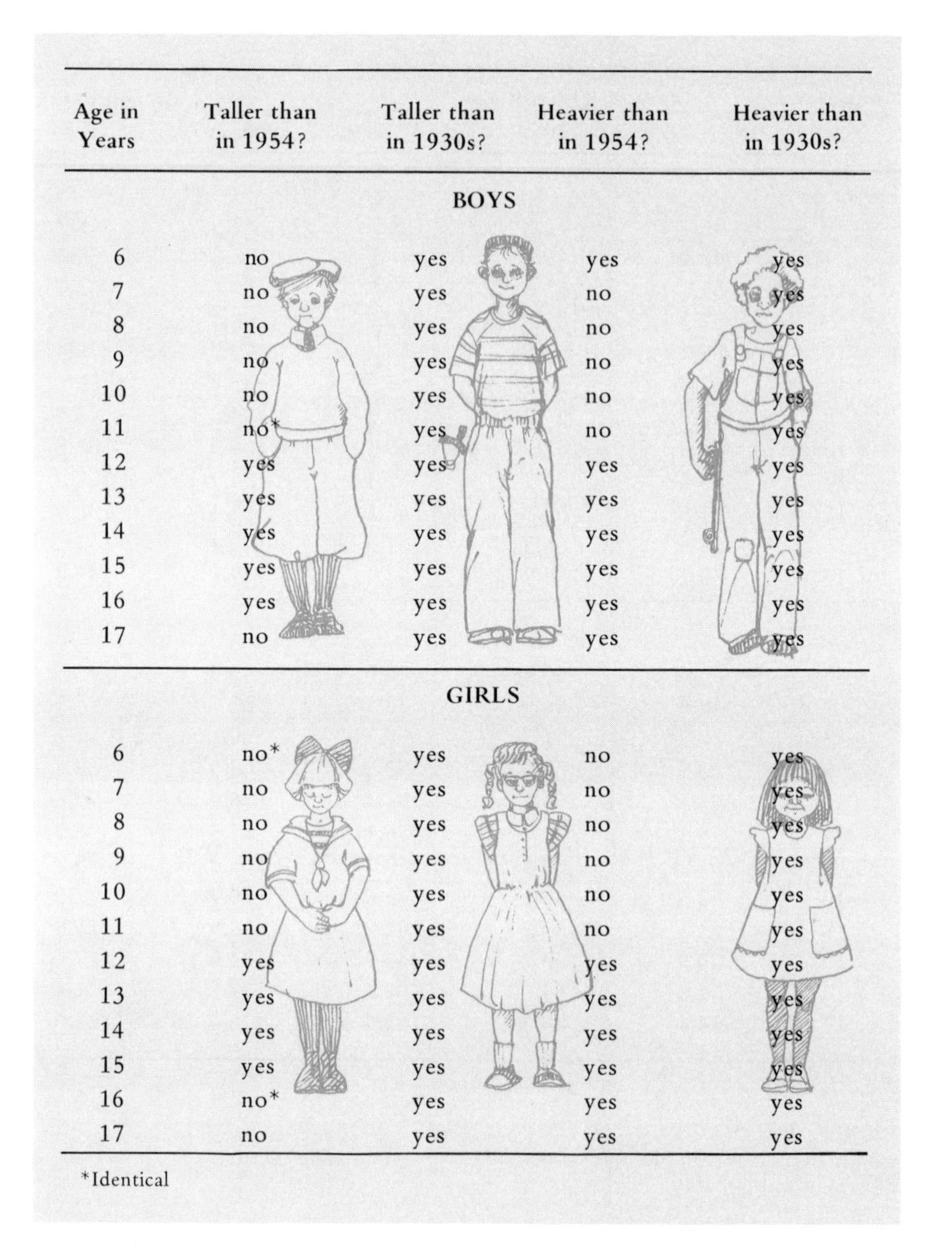

Age in Years	Taller than in 1954?	Taller than in 1930s?	Heavier than in 1954?	Heavier than in 1930s?
		BOYS		
6	no	yes	yes	yes
7	no	yes	no	yes
8	no	yes	no	yes
9	no	yes	no	yes
10	no	yes	no	yes
11	no*	yes	no	yes
12	yes	yes	yes	yes
13	yes	yes	yes	yes
14	yes	yes	yes	yes
15	yes	yes	yes	yes
16	yes	yes	yes	yes
17	no	yes	yes	yes
		GIRLS		
6	no*	yes	no	yes
7	no	yes	no	yes
8	no	yes	no	yes
9	no	yes	no	yes
10	no	yes	no	yes
11	no	yes	no	yes
12	yes	yes	yes	yes
13	yes	yes	yes	yes
14	yes	yes	yes	yes
15	yes	yes	yes	yes
16	no*	yes	yes	yes
17	no	yes	yes	yes

*Identical

Figure 11-2. *Three generations of children: still growing taller and heavier? A comparison of boys and girls studied between 1963 and 1970 with those of 1954 and 1937–1939. (Adapted from Metropolitan Life, 1973.)*

Invariants are developmental phenomena or principles found in *all* situations

ations. By ***invariants,*** I mean developmental phenomena that can be discovered no matter where we look. We cannot tell the variants from the invariants without looking, and that's much of the point here.

Some of the variants in development of height and weight can be listed with little difficulty. Actual height and weight are variants from generation to generation (considering averages only). How rapidly children at a particular age are approaching adult size is another variant. Some of the *in*variants can also be noted: size during childhood remains a good index to size as an adult; a growth spurt occurs in adolescence for both sexes; boys start out being taller than girls, but are passed by them in later childhood, and take the lead once again in adolesence. In other words, these are phenomena that have been documented repeatedly through several generations.

So far as height and weight are concerned, then, there are both similarities and differences when children of various generations are compared. Having the advantage of history's backward glance, I can speculate that physical development of many children in the past fell somewhat short of full potential. Perhaps improvements in nutritional status and other environment factors have now resulted in more children coming closer to the limits of physical development. Should significant improvements continue to take place, then we might expect a little more "talling up" in the future generations—but still within the limits of our species.

One of the reasons these objective dimensions of physical growth have been considered here is to fix in our minds the basic conception that the nature of childhood has been changing throughout human history and continues to do so. This means that what it is to be a child has been changing as well. The differences cannot be established as precisely in the psychological and social spheres as in the physical sphere, but it is becoming increasingly clear that having a childhood is quite a different proposition in various times and places.

CHILDHOOD AND ITS VARIATIONS

Psychologically oriented historians sometimes make a rather surprising statement about human development. They claim that childhood itself, as we know it, is a fairly recent "invention." What they mean is relatively few people have regarded the early years of life as a special phase of development that deserves serious attention by adults. Horace E. Scudder (1894), for example, found that children were seldom represented in literary works throughout the centuries; they just were not important or interesting enough to authors and readers. From his turn-of-the-century perspective, Scudder noted that children were starting to appear in many books, and a previously neglected form, the book written *for* the child, had recently become common. "There has been, since the day of Wordsworth, such a succession of childish figures in prose and verse that we are justified in believing childhood to have been discovered at the close of the last century. The child has now become so common that we scarcely consider how absent he is from the earlier literature. Men and women are there, lovers, maidens, and youth, but these are all with us still. The child has been added to the *dramatis personae* of modern literature" (Scudder, 1894, p. 4).

More recent scholars have confirmed and enriched this view. Philippe Aries (1962) also spoke directly of "the discovery of childhood" in his important survey of the topic. As he drew his book to a conclusion, Aries declared:

Perhaps the shorter life expectancy in the past encouraged rapid introduction into the adult world

> In the Middle Ages, at the beginning of modern times, and for a long time after that in the lower classes, children were mixed with adults as soon as they were considered capable of doing without their mothers or nannies, not long after a tardy weaning (in other words, at about the age of seven). They immediately went straight into the great community of men, sharing in the work and play of their companions, old and young alike. The movement of collective life carried along in a single torrent all ages and classes, leaving nobody any time for solitude and privacy. In these crowded, collective existences there was no room for a private sector. The family fulfilled a function; it ensured the transmission of life, property and names; but it did not penetrate very far into human sensibility. Myths such as courtly and precious love denigrated marriage, while realities such as the apprenticeship of children loosened the emotional bond between parents and children. Medieval civilization . . . had no idea of education. Nowadays our society depends, and knows that it depends, on the success of its educational system. . . . New sciences such as psychoanalysis, pediatrics and psychology devote themselves to the problems of childhood, and their findings are transmitted to parents by way of a mass of popular literature. Our world is obsessed by the physical, moral and sexual problems of childhood. (p. 411).

As Aries saw it, civilization lost the interest and insights into the importance of childhood that Plato, Aristotle, and others had offered many centuries ago. Indifference and lack of awareness, rather than new ideas, replaced the views of the ancients. The child was hardly noticed in the main course of medieval civilization. Soon after weaning, the boy or girl became the companion of the adult and functioned as best he or she could in the "big world." Only during the past four centuries has the child assumed importance and, correspondingly, have the goals and techniques of education. But even through most of this period, the life of the child did not much resemble the pattern that can be seen today in the child-conscious, increasingly urbanized United States.

Affluent parents and most physicians in eighteenth century Europe still knew and cared little about the nature of children and childhood (Lorence, 1974). In the aristocratic tradition, many barriers were placed between parent and child. The youngster was expected to respect and defer to an aloof, unbending parent. A typical diary of the time, cited by Lorence, contained passages such as, "I remember my father but little . . . he was home on Sundays and grand occasions such as birthdays" (1974, p. 3). It is striking that the

As recently as the eighteenth century, children were treated as small and imperfect adults rather than as people with special developmental needs. (Courtesy of the Library of Congress.)

makers and shapers of society, the wielders of power, considered it appropriate to offer so little of themselves to their own children. This cultural attitude had physical as well as psychological consequences: "This distancing, both physical and psychological, between parent and offspring, characterized their relationship from birth. It was the rare upper-class mother who nursed her own child, a fact of serious significance in an era when artificial feeding was unsatisfactory, and when the child's physical survival was contingent on the availability of milk" (1974, p. 3). Farming children out to wet nurses was a common practice, despite the many problems created by the lack of parental supervision and control.

Cruelty to children—and outright murder, especially of young infants—continued as part of childhood's tradition generation after generation. Then as now, there were some people who strongly objected to the mistreatment of children. Through their efforts foundling hospitals, legislation against cruelty and exploitation, and other ameliorative developments came into being. The reformers had to struggle, however, against many of the prevailing public attitudes.

In the nineteenth century, the institution of the Victorian family came into being. Some of our ideas about what a family should be like are based on the image of this family. I am talking now about the solid, close-knit, parent-respecting family of the late nineteenth century. It is against this image that contemporary family life is often compared. When a critic of our own times blasts today's family as "weak" or "failing," he or she usually has a vision of the well-ordered Victorian household in the back of his or her mind.

A **psycho-historian** combines the expertise and interests of psychologist and historian

But the **psycho-historian** takes a different view of the matter. Stephen Kern (1974) argues that the emotional tensions in the Victorian family were highly explosive because of the excessive intimacy and interdependence involved. "The family was indeed 'fortified' against the pressures of the outside world . . . but that fortification, like neurotic defense mechanisms, often stifled more than it protected or comforted" (p. 438). Many homes were run almost as military establishments. Discipline was ever-present and the dominant force. The spontaneous expression of affection was much less evident. Special problems seemed to exist in the father-child relationship, "with a special destructive element in the father-daughter relation. The father, the more assertive parent, was more likely to engage in conflict with other members of the family, particularly with the son, who generally had more rights than daughters and who sought to challenge the father's supreme authority" (p. 448). Observations of this type of family constellation probably had considerable influence on Freud's thought and his formulation of the Oedipal dynamics.

The restrictive sexual morality of the late nineteenth century bore down hard on the child, according to Kern. "The playful self-indulgence of masturbating children was especially threatening to parents who were so anxious about surpressing their own sexuality. As soon as the child was able to acquire a moral sense, the parents sought to quash child sexuality with firm moral injunctions, threats of punishment, and, if necessary, physical restraint" (p. 452). Not very surprisingly, suicide among children became increasingly common within the Victorian-type family. But even for the great majority of children who survived to establish their own families, the pressures of a Victorian childhood undoubtedly left a major impact on thought and personality.

Devices were marketed to stop children from touching themselves sexually

These brief glimpses into the history of childhood suggest that the challenges, opportunities, privileges, and hazards confronting the child have varied greatly. Important differences in child-rearing practices have been noted within our own nation and our own century (Wolfenstein, 1959). And there is no reason to suppose that time has suddenly decided to stand still today. On the contrary, we see all about us indications that the nature of childhood is continuing to change. We differ from many past generations in our emphasis on childhood. Whether or not we agree on specific theories and practices, most parents and behavioral-social scientists do agree that childhood is a special and highly significant time of life. Within this framework, though, the pattern continues to shift. This means that our ideas of childhood can become outdated quickly, without our awareness. "Things are not the way they used to be when I was growing up" is a reasonable statement, whether made by the grandparental or parental generation—or by ourselves. Some day our own children will probably be saying the same thing.

Changes in the nature of childhood should also alert us to the possibility that what it means to be an adolescent, adult, and aged adult might also differ appreciably from time to time and place to place. You can see that a kind of counterpoint is implied in the understanding of human development. One line of thought and research attempts to encompass the pattern of development from conception through the termination of the individual's life cycle. The other line attempts to chart conditions specific to growing up and growing old in a particular socio-cultural setting at a particular time in human history. It would be simpler to concentrate on one line and ignore the other. But that would seriously compromise our chances of comprehending the full reality and potentiality of human development. Having at least touched on the shifting patterns of childhood, we will be better prepared to examine what might be the variants and invariants at later points in the life cycle.

Returning to childhood, it will probably take at least several more generations of developmental research (and researchers) to determine what is variant and what is invariant in modern childhood. Repeated observations under varying conditions are necessary. We do not have to assume that children of today or tomorrow will grow up either entirely different or entirely the same as those of the past. It is the nature and extent of the shifting pattern of childhood that remains to be determined. Children today play games known to the ancient Egyptians and Romans, but also assemble space probe kits and solve "new math" problems that may challenge their parents. No matter how comprehensive and accurate anybody's knowledge of childhood and children might be at present, chances are this knowledge will require continual revision, decade after decade. One of the most obvious and interesting areas of change will now command our attention.

The technology of the computer age is part of the learning experience for today's child, whose skills in some ways make her more powerful than adults of past generations. (Courtesy of Philco.)

GIRLS, BOYS, AND CHILDREN: SEX DIFFERENCES IN DEVELOPMENT

Many of the generalizations made about the early years of development apply equally to girls and boys. One can speak, for example, about the *child's* visual development or changing pattern of sleep-wakefulness. Yet every child is either male or female. What differences does this difference make? How significant is gender in psychological, social, and physical development? And, bearing in mind the shifting patterns of childhood, how much is fixed and how much is variable in the relationship between gender and total development?

PHYSICAL AND CULTURAL VIEWPOINTS

Let's begin by recognizing two extreme positions on this topic. Neither will prove adequate, but the contrast will help us achieve a balance. It can be argued that being a female or a male involves much more than those physical differences associated with differential roles in reproduction. Females are "feminine" and males are "masculine." We think, feel, and behave in feminine or masculine styles. Because these differences in the psychosocial sphere are intimately associated with gender membership, it is evident that boys will be boys and girls will be girls regardless of the circumstances in which they develop and function. A person who maintains this view might allow that cultural factors influence the specific ways in which feminine and masculine styles express themselves, but he or she will insist that careful study always reveals intrinsic differences. Boys, for example, are usually regarded as more rambunctious, motoric, and aggressive. These traits would be taken as manifestations of "the male principle."

Interpretations of *feminine* and *masculine* are changing too fast to insist on one definition

A nine-year-old boy in Idaho confides:

It's hard to be a boy
and not get in trouble.
I feel like busting chairs
and stretching my legs
because they feel weird–
*They feel that way right now!**

Girls are often seen as more passive, receptive, and "proper." Reflecting on the female's position in life from the vantage point of her twelve years, Anu Jain finds that:

* From "Hard To Be a Boy" by G. Crookham. In N. Larrick & E. Merriam (Eds.), *Male and Female Under 18*. New York: Avon Discus Books, 1973.

Femininity is to have grace and charm,
To be polite and careful, to do no harm,
Fussing with your clothes and hair,
Beautifying yourself with very great care,
Walking in the drugstore for a cola sip,
Overhearing the neighborhood girls gossip,
Keeping a happy, lively little household,
With tidiness and coziness to have, and to hold†

The opposite view denies that physical differences in gender necessarily have any implications for personality and social role. In arguing this case, one might draw attention to the great variations that can be observed in personality among different girls and among different boys. Physical make-up cannot be all that important if boys can be so different from each other, and the same for girls. Furthermore, cross-cultural differences in the behavior of the sexes can also be interpreted to support the contention that boys and girls grow up to behave as they are *expected* to in their particular milieu. If boys and girls are expected to play, dress, think and behave differently from each other, then it is not surprising that they do. Taking this view to an extreme, it could be maintained that all the observable differences between boys and girls in the realm of personality and social function must be attributed to cultural expectations and child-rearing practices. Boys and girls are really very much the same at the start of life, and theoretically would continue to develop in a parallel manner if given the opportunity to do so.

Nine-year-old Coleen Quilliam expresses this view admirably:

Girls and boys are alike because girls and boys both
Go to school together
Live together
Sit together
Eat together
When grown up
Marry each other
Love each other
Take care of each other
Boys aren't stronger than girls. Boys with muscles are stronger than girls.
Girls with health are stronger than boys.
Boys and girls work together, but not on the same jobs and not on the same pay, because boys think girls are different, and they're not!‡

† From "Femininity" by A. Jain. In N. Larrick & E. Merriam (Eds.), *Male and Female Under 18.* New York: Avon Discus Books, 1973.
‡ From "Girls and Boys Are Alike" by C. Quilliam. In N. Larrick & E. Merriam (Eds.), *Male and Female Under 18.* New York: Avon Discus Books, 1973.

SOME RESEARCH FINDINGS

Studies indicate that both of the positions outlined have some facts to support them. There are some ways in which girls and boys tend to differ from each other, yet there is also ample evidence that the two sexes are treated differently from infancy onward. It is very difficult to say which comes first. Do parents around the world tend to respond to their young children in a sex-differential way because of actual differences in the children themselves, or are the differences exaggerated, if not created, by the differential treatment? Let's examine some of the findings that have come to light in recent years.

Sex-role stereotypes in six cultures. Children between the ages of three and eleven were observed in natural settings in "Orchard Town," New England; Nyansongo (Kenya); Taira (Okinawa); Khalapur (India); Taron (Phillipines); and Juxtlahuaca (Mexico). Beatrice Whiting and her colleagues (Whiting & Edwards, 1973) were interested in determining the validity of stereotyped opinions often held about sex differences. The researchers noted, "Females are frequently characterized as more dependent, passive, compliant, nurturant, responsible, and sociable than males, who in turn are characterized as more dominant, aggressive, and active" (p. 172). They reasoned that if the same sex differences observed in the United States and other Western-type societies were also found in a variety of other societies, then a stronger case could be made for the existence of universal (and presumably, biological) differences in personality linked to gender.

Cross-cultural research has obstacles to overcome, but is vital to understanding variants and invariants

Each child in this study was observed on the average of seventeen different times for five-minute periods over a period of between six and fourteen months. The observers included a member of the research field-team plus a bilingual assistant. After the observations were collected, the researchers focused on twelve basic types of interaction (representing more than eight thousand different specific interactions that had been recorded). It should be kept in mind that these data, although valuable in the understanding of personality expression in the three to eleven age range, cannot tell us anything directly about what pattern of development and parent-child interaction took place from infancy until age three.

The overall findings are summarized in Figure 11–3. The major stereotypes about sex differences were defined operationally, that is, with reference to specific observations of child interactions. A separate score was determined for each girl and boy, representing the proportion of interactions that fell into a certain stereotype-relevant category. These scores were then combined into group scores for girls and boys. Figure 11–3 divides the children into groups on the basis of cultural sample, sex, and age (ages 3–6, and 7–11).

Let's pick up on a few of the specific findings (you can glean the rest by inspecting the table yourself). The *dependency* stereotype would have it that girls are more dependent than boys. Seeking help, attention, or physical contact were types of behavior taken to represent dependency in this study. The younger girls did seek more help and more physical contact than the younger boys in five of the six societies. But these differences were not evident among the older girls and boys—and furthermore, the older boys tended to seek more attention than did the older girls. This mixed pattern of results is fairly typical of the findings in general. The sex-differential stereotypes show up in some societies at certain age levels within childhood, and along certain dimensions. But the differences are not massive and across the board. As we have seen in this example, girls may be more dependent in some ways, but less dependent in others.

Consider, for another example, the stereotype that girls are more *passive* than boys. Careful study of passivity versus aggressiveness required differentiation into four different kinds of interaction, as shown in Figure 11–3. It was found that older boys tended to respond more aggressively than girls of the same age to aggressive instigations by others, and that boys were also somewhat less compliant than girls to the wishes of others. However, these differences were not as great as the stereotypes had suggested. The opposite of passivity in some circumstances might better be called *initiative* rather than *aggressiveness*. For the younger children, there were no sex differences in the proportion of self-instigated actions. The older boys tended to show a little more initiative than the older girls, but not to any dramatic extent.

The whole pattern of results obtained by Whiting and her colleagues indicates that the stereotyped notions of sex-differential personality held in our own society show up to some extent in other societies as well. But the girl-boy differences often are not that marked, nor do the stereotypes hold up equally well for all realms of behavior or at all periods of childhood. The fact that some differences show up only among younger children and some only among older children is worth further thought and investigation.

Effects of cultural patterning. The investigators added some interesting reflections when they integrated their own findings with previous work done by others. They point out that:

> . . . in societies where boys take care of infants, cook, and perform other domestic chores, there are fewer sex differences between boys and girls, and this decrease is due primarily to the decrease in "masculine" behavior in boys; boys are less egoistically dominant, score proportionately lower in some forms of aggression, seek attention proportionately less frequently, and score higher on suggesting responsibly. On the other hand, the 3- to 6-year-old girls in these societies are high on assaulting and miscellaneous aggression, and both younger and older girls score low on sociability. (Whiting & Edwards, 1973, p. 186).

In none of the six cultures were the observers able to find girls who did primarily "masculine-type" tasks. In the New England community, however, the girls and boys were relatively similar in the type of experiences and responsibilities provided to them. This relative similarity in life experience seemed to

Behavior Category		Nyansongo (Kenya)	Juxtlahuaca (Mexico)	Tarong (Phillipines)	Taira (Okinawa)	Khalapur (India)	Orchard Town	All
Dependency								
1. Seeks help	Ages 3–6	+	+	+	+	+*	–	+*
	7–11	–	–	+	+	–**	+	–
2. Seeks attention	3–6	+	–	–	–	–	+	–
	7–11	–	–**	–	–	–	–	–*
3. Physical contact	3–6	+	+	–	+	+	+	+**
	7–11	+	+	+	–	+	+	+
Sociability	Ages 3–6	+	–	–	+	+	+	+
	7–11	+	–	+	+	+*	–	+
Passivity								
1. Withdrawal from aggressive instigations	Ages 3–6	+	–	+	+	–	+	+
	7–11	–	+	+	–	+	–	+
2. Counteraggression in response to an aggressive instigation	3–6	–	–	+	+	+	–	–
	7–11	–	+	–	–*	–	–*	–**
3. Compliance to dominant instigations (prosocial and egoistic)	3–6	–	–	+	+	+	–	+
	7–11	+	+	+	–	–	–	+
4. Initiative (% of acts which are self-instigated)	3–6	–	–	+	–	+	–	–
	7–11	+	–	–	–	–	+	–

Nurturance								
1. Offers help	Ages 3–6	−	+	+	+	+	−	+
	7–11	+	+	+**	−	+	+	+**
2. Gives support	3–6	−	+	+	+*	−	+	+
	7–11	+	+*	+	+	+	+	+***
Responsibility	Ages 3–6	+	+	+	+	+	+	+*
	7–11	−	+	−	+	−*	+	−
Dominance	Ages 3–6	−	−	+	−	−	−	−*
	7–11	−	+	−	−	−	−	−
Aggression								
1. Rough and tumble play	Ages 3–6	+	−	−	−	−	+	−*
	7–11	+	−	−	−	−	−	−*
2. Insults	3–6	−	−	−	−**	−	−	−**
	7–11	−	−	−	−**	−	+	−*
3. Assaults	3–6	−	−	−	−	−	+	−
	7–11	+	−	−	−	+	−	−

*$p < .05$
**$p < .01$
***$p < .001$

Figure 11-3. *Personality differences between boys and girls in the six culture study. A (+) indicates that the girls' score was higher than the boys'. A (−) indicates that the boys' score was higher than the girls'. (Reprinted from Beatrice Whiting and Carolyn Pope Edwards, A cross-cultural analysis of sex differences in the behavior of children aged three through eleven,* Journal of Social Psychology, *1973, 91, pp. 171–188.)*

be reflected in a general lack of major differences in types of interaction observed. The Orchard Town girls had relatively little experience in caring for younger children and were in an environment that encouraged academic achievement. Perhaps this is why—in comparison with girls in the other societies studied—the New England girls had the lowest scores for offering help and support ("feminine" trait), and the highest scores for seeking the attention of adults.

The case for cultural patterning of sex-differential personality is strengthened by Whiting and Edwards in the concluding statement to their study. "In both the East African societies where 'feminine' work is assigned to boys and in Orchard Town, New England, where less 'feminine' work is assigned to girls and where there is less difference in the daily routine of boys and girls, the behavior of girls and boys does not show as great differences as in other societies" (p. 188).

Sex-related stereotypes start young. Within our own culture, it has been possible to observe sex-related similarities and differences in children below the age of three. One of the key findings is that parents (and other adults) *think* of girls and boys as markedly different from each other, whether or not objective differences clearly support this distinction. Studying fetal behavior in the last trimester of pregnancy, one research team noted that the mother would interpret vigorous kicking and movement as a sign that the child was more likely to be a male (Sontag, Steele, & Lewis, 1969). Soon after birth, there are likely to be some sex-related differences both in behavior and in the way girls and boys are treated by parents. There is a tendency, for example, for baby girls to have more words addressed to them as compared with baby boys. This differential treatment might well be related to the earlier development of language sometimes observed in girls. It remains unclear, however, whether the differences should be regarded as the outcome of a parental bias in verbal behavior, or the consequence of the young girls' tendency to respond more strongly than boys to auditory stimulation (Lewis, 1972). In other words, there seem to be differences both in infant behavior and in parental interaction that are related to the sex of the child, and one cannot rule out the possible influence of biological factors or socio-behavioral factors.

By the time a child has reached the age of six months, it is likely that significant changes have already occurred in parental behavior that affect the child's sex-related development. Observational studies by Michael Lewis indicate that mothers tend to touch their baby boys more often than their baby girls. Within a few months, however, this difference vanishes. Why? Lewis speculates the following, drawing on but going beyond his actual findings:

> Comments from mothers as well as our own observation leads to the suspicion that the motive of autonomy or independence may play a role. As a function of societal stereotypes, mothers believe that boys rather than girls should be independent and encouraged to explore and master their world. This is anti-

Sex differentiation in play behavior is alive and well. More and more, however, boys and girls are feeling freer to express interest in the same toys and games. (Photo by Tania D'Avignon.)

> thetical to proximal behavior*—in fact it may be antithetical to all close interpersonal relationships. As this becomes an increasingly relevant motive, mothers start to wean their sons from physical contact with them. (Lewis, 1972, pp. 236–237).

Other studies reveal that boy-girl differences in play activities and choice of toys increase with advancing age. Some differences can be seen clearly among children as young as thirteen months, for example, the tendency of little boys to venture further away from mother and stay away longer in exploratory situations (Goldberg & Lewis, 1969). This pattern might well be the outcome of the mother's tendency to wean boys toward independence earlier. Two year olds have already begun to show preferences for playmates of their own sex (Koch, 1944), and a year later they recognize the differences between boy-type and girl-type toys. The little girl knows that she is supposed to be more interested in dolls, the little boy in trucks (although, in my own observation, this does not necessarily deter children from enjoying the "wrong" toy if they can find a way to do so without being teased or caught at it).

* *Proximal behavior* is what goes on when people are physically close to each other, whether a child holding onto mother's apron strings or a crowd of strangers jammed together in a subway car.

Role models in the media. It is very likely that differences in the behavior and interests expected of girls and boys are caused to some extent by the role models available in all the media. Girls and women are often presented quite differently than boys and men. Leonore Weitzman and her colleagues have examined this phenomenon with respect to the picture books that preschool children are likely to have brought before them. Summarizing their study, the researchers concluded:

> An examination of prize-winning picture books reveals that women are greatly underrepresented in the titles, central roles, and illustrations. Where women do appear their characterization reinforces traditional sex-role stereotypes: boys are active while girls are passive; boys lead and rescue others while girls follow and serve others. Adult men and women are equally stereotyped: men engage in a wide variety of occupations while women are presented only as wives and mothers. The effects of these rigid sex-role portraits on the self image and aspirations of the developing child are significant. (Weitzman, Eifler, Hokada, & Ross, 1972, p. 1125).

I remember a puzzled child asking a new woman in the area: "Are you a mommy or a baby-sitter?"

It is not just that males and females are depicted differently. It also seems that females are limited to a narrower band of possible attitudes, interactions, and accomplishments. Little girls see and hear about adult women as mothers and wives, with occasional fairy godmothers and underwater maidens thrown in for variety. "In contrast to the limited range in women's roles, the roles that men play are varied and interesting. They are storekeepers, housebuilders, kings, spiders, storytellers, gods, monks, fighters, fishermen, policemen, soldiers, adventurers, fathers, cooks, preachers, judges, and farmers" (Weitzman, Eifler, Hokada, & Ross, 1972, p. 1141). The researchers had difficulty finding a picture-book woman who had any occupation. As the researchers noted, "In a country where 40% of the women are in the labor force, and close to 30 million women work, it is absurd to find that women in picture books remain only mothers and wives" (p. 1141). There is little difficulty, then, in identifying some of the influences that lead young girls and boys to think of themselves as different, both at their present ages and in their eventual adult careers.

Being either male or female seems to become a dominant and well-established aspect of personality by age six or seven, with the boys taking their gender a little more seriously than girls. Money and Wang (1966), for example, have found that children of this age spontaneously draw pictures of a child of the same sex as their own even when no gender-related instructions have been given.

Relationship between earlier and later developmental patterns. An excellent source for further readings on sex differentiation in human development is the book, *Man and Woman, Boy and Girl* by John Money and Anke A. Ehrhardt (1972). Let's take a brief sample of how they view the relationship between early and later developmental patterns:

> The gender preferences of play regardless of their origin in early childhood are expanded and reinforced in the . . . games and play of later childhood. Thus they may generate or contribute to various skills or abilities differently distributed between males and females. Girls, when they play house, for example, construct what might be called a domestic nest for their baby dolls. Boys are more interested in forts and tree houses which are places of refuge and escape.
>
> Boys, more than girls in our culture, to take another example, intercept a moving ball in their games. Thereby they build up experience not only in visual-motor coordination, but also in judgments of distance, speed, and trajectory, all of which may contribute to the fact that males ultimately outnumber females in superiority of spatial, mathematical, and mechanical ability. The practice and growth of these abilities through play may take place, however, as a sequel to a prenatal, perhaps hormonally determined sex difference in directional and spatial sense. In many mammals, it is the male who establishes territorial rights by marking boundaries with odiferous substances, pheromones. The specialized glands that secrete these pheromones are regulated, in the mature animal, by androgen. Marked boundaries delineate the eating and breeding territory of an individual pair in some species, or of a troop in others.
>
> Territoriality is less prominent in the human male than in various lower species, but some signs of it are evident. Boys rather than girls are youthful explorers, fort-builders, and scouts, and boys are the ones who form gangs or troops that set up territories, dare rivals to trespass, and attack them if they do.*

THE CHANGE DIMENSION

Both the stereotypes of sex-related behavior and the research findings cited draw on a cultural tradition that is now being challenged in many spheres. During the past decade or so an increasingly effective case has been made against practices that discriminate on the basis of gender. The quest for equal opportunity for women is being pursued largely in the occupational sphere, but also in other areas of adult functioning as well. The implications are also being felt in attitudes toward children. People who are receptive to the women's-liberation message recognize that equal opportunity in adult life is best fostered by confidence-building experiences early in life.

And so it is becoming increasingly common for "dainty little girls" to participate in active sports just as the "roughhouse boys" have been doing all along. This development might, among other outcomes, provide girls with more opportunity for team work types of experience. It is also reasonable to speculate that boys in the present and future generations may come to regard girls as fit companions and equals, instead of as a species apart (although time will tell). I have had some opportunity to observe early efforts at integrating the sexes in Little-League-type baseball games. One memorable interaction occurred in a late-inning situation with bases loaded and the opposition's "big

* From *Man and Woman, Boy and Girl* by J. Money and A. A. Ehrhardt. Baltimore, Md.: The Johns Hopkins University Press, 1972, pp. 181–182.

A side effect of boys and girls working in a greater range of situations may be a more lively sharing of interests and mutual appreciation in adult life. (Photo by Talbot Lovering.)

basher" at bat. A jittery coach (male) watched the hitter drive a ball deep to center field. The outfielder stood her ground, raised her gloved hand and made the big catch (on precisely the kind of routine play other fielders had been botching all through the game). With exultant relief, the coach shouted, "Atta boy, Jenny . . . uh . . . atta *girl!*" Score one for equality! (But what if Jennifer had dropped the ball?)

The movement toward equal opportunity for boys and girls is resented in some quarters. This resistance is paralleled by obstacles that continue to be placed in the path of women who seek educational, occupational, and social opportunities of the kind usually reserved for men. Even with the limited gains that have been made up to this time, however, it is evident that many females, both child and adult, are quite capable of moving into so-called masculine activities and responsibilities. Perhaps the more subtle side of this process is the broadening of the roles and activities that are open to boys and men as well. It is becoming increasingly common, for example, to be served by male cabin attendants as well as by stewardesses in jet liners. Precisely how many girls decide to prepare themselves for once-masculine careers, and how many boys find once-feminine careers to their liking, may be less significant psychologically than the general reduction of the sense of enforced differentiation. If it is all right for girls to be brave and daring, then it is also all right for boys to be tender and nurturant.

Marriage, maternity, and wrinkles are now also accepted in flight attendants, but not without struggle

In effect, then, we have a large-scale social experiment in process. Developmental research can identify similarities and differences in girl/boy behavior, and in how boys and girls are treated by others. But, as we have seen, it is very difficult to determine how many of the observed differences (when, in fact, differences are observed) should be explained biologically, and how many evolve from the continuing interplay of feedback between child and social environment. The shift in cultural attitudes and opportunities offers a fresh perspective. Some personality and behavioral characteristics of boys and girls may persist in future generations despite the social transformations. These relatively invariant characteristics would then seem to be more definitely related to biological aspects of gender identity. Other characteristics may change easily and markedly as society changes its orientation toward sex-appropriate roles. The easy-change characteristics would then seem to depend more on social expectations and opportunities than on fixed biological factors. Girls and boys in the future may both be able to incorporate a richer medley of so-called feminine and so-called masculine traits.

Carl C. Jung, one of the guiding spirits of the psychoanalytic movement, differed from most of his colleagues in emphasizing the wholeness or integrity of the person. He believed a man is not complete if he is "only a man," nor a woman complete if she is "only a woman." Being a complete *person* requires sensitivity to and acceptance of all sides of our nature (Jung, 1963). Jung probably would have approved of the new willingness in our society to consider masculine and feminine characteristics as equally valuable. If the nature of childhood does continue to change in this direction, then it may be much easier for each girl and boy to achieve a sense of deep personal integration and not be forced to overdifferentiate along sex-stereotype dimensions.

IDENTITY AND INTERACTION

Gender identity is that aspect of total self-image centered around maleness or femininity

We have seen that girlness and boyness dimensions of personality trace their beginnings from infancy onward. A complex and continuing interaction between the child's psychobiological make-up and his or her experiences with the social and physical environment is involved in the formation of what Money and Ehrhardt refer to as ***gender identity.*** The same general principle applies as well to the child's creation of his or her *total* identity. Various components of self-identity have been touched on throughout this book (body image, for example, as well as gender-related characteristics). Now is a good time to reflect on the child's sense of identity in general, having ten years of developmental history behind us and the adolescent and adult years ahead. I have selected the framework offered by Erik H. Erikson as our basic guide here. Both a sensitive clinician and a creative scholar, Erikson offers unusual breadth and depth in his conception of human development. Although attentive to identity formation in childhood, Erikson also follows the course of life experience through adolescence and adulthood.

THE "EIGHT AGES OF MAN"

In *Childhood and Society* (1950/1963), Erikson provided insightful explorations into the relationship between individual patterns of development and the cultural milieus within which development unfolds. His scope included not only mainstream types of development in the United States, but also contrasting types in Native American cultures and other distinctive settings. He then attempted to integrate his broad-ranging observations with the psychoanalytic approach to human development. Erikson found that the traditional psychoanalytic approach remained useful, but required a number of modifications if it was to reflect faithfully the variety of culturally patterned phenomena that a keen observer should be able to discern. The framework he came up with for this purpose has since proven attractive to many developmentalists. It should be kept in mind that Erikson's "Eight Ages of Man" is a *theoretical* framework, a way of looking at some central aspects of human development. Although a number of studies have been inspired by this framework, it cannot be said at present that his ideas have been either proven or disproven in terms of strict evidential criteria. Instead, a person finds this framework more or less *useful* in describing, understanding, predicting, or influencing human development.

Erikson's stage theory is summarized in Figure 11–4. Chronological age periods are specified. These ages are given for their reference value and should not be taken too literally. You will notice that each of the stages, or developmental periods, is marked by a name that indicates its place in the total life cycle. Some special language is also involved in the early stages (e.g., *oral-sensory, muscular-anal*). These terms are intended to remind us of some of the basic processes that are in evidence at a particular time of life (examples to follow). The other feature of this table is the listing of "ego qualities." At each stage of life there is one dominant problem or crisis that confronts the individual. It is quite normal to face these problems, part of being a person-in-process. People differ, however, in the way they come to terms with each "normative crisis." One person decisively masters the dominant problem associated with his or her developmental position, another person comes to a less satisfactory resolution. The fact that, roughly speaking, a person can either win or lose at each stage is represented in Figure 11–4 by Erikson's specification of ego qualities in opposition. At the earliest stage, for example, the infant emerges either with basic trust or basic mistrust.

How a person solves his or her current challenge has important implications for the subsequent development of identity. The general principle is the forging of a strong, vibrant, growing self depends on mastery of one stage-relevant crisis after another. Early developmental periods are highly significant, then, because weak or debilitating solutions of identity challenges at the beginning of life establish handicaps for later growth. This viewpoint seems to be supported by a variety of studies into the effects of deprivation and trauma on the organism (human or subhuman) at different points in its devel-

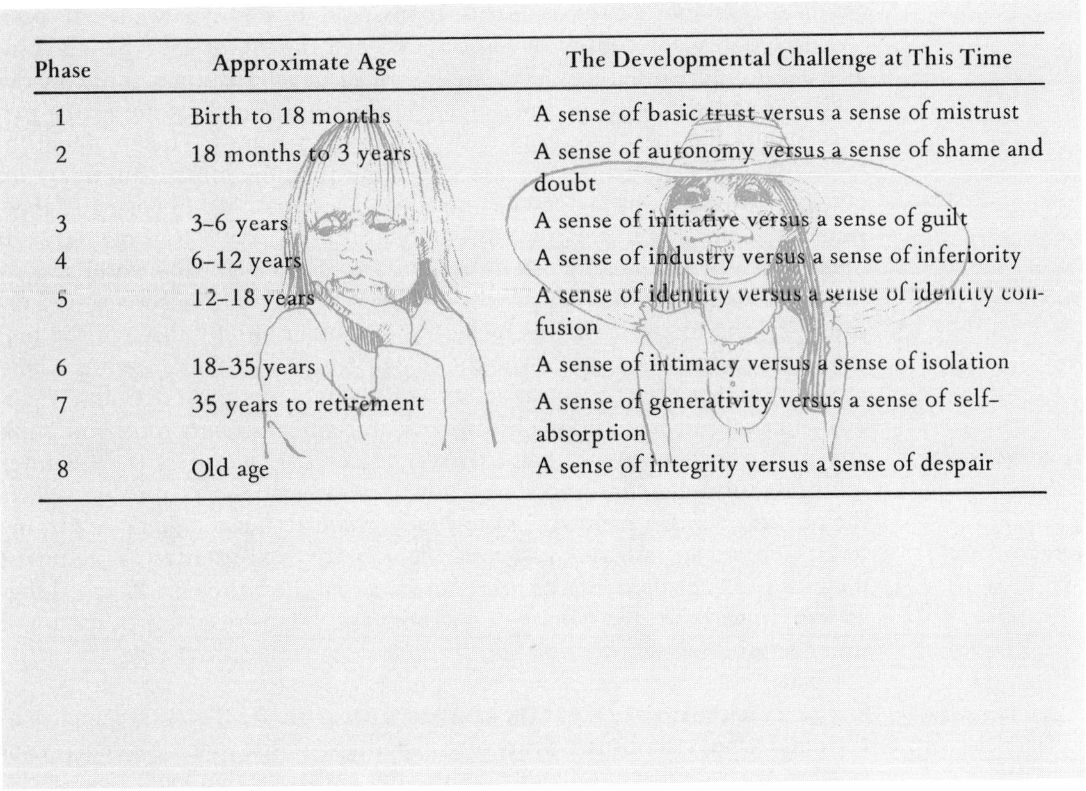

Phase	Approximate Age	The Developmental Challenge at This Time
1	Birth to 18 months	A sense of basic trust versus a sense of mistrust
2	18 months to 3 years	A sense of autonomy versus a sense of shame and doubt
3	3–6 years	A sense of initiative versus a sense of guilt
4	6–12 years	A sense of industry versus a sense of inferiority
5	12–18 years	A sense of identity versus a sense of identity confusion
6	18–35 years	A sense of intimacy versus a sense of isolation
7	35 years to retirement	A sense of generativity versus a sense of self-absorption
8	Old age	A sense of integrity versus a sense of despair

Figure 11-4. *The psychosocial stages of development as seen by Erik Erikson.*

opmental career. Other conditions being equal, early disruptions seem to have more enduring and consequential implications (see, for example, many of the contributions reported in Stone, Smith, & Murphy, 1973). The available findings tend to support Erikson's contention, even if they cannot be said to have proven the truth of his stage theory.

It should be emphasized that, in Erikson's view, the existence of problems as such is not a deterrent to development. Rather, we hone our development by confronting and, hopefully, mastering one challenge after another. It is not the challenge itself, but our failure to emerge from the challenge with a new, positive and adaptive ego quality that constitutes the real problem.

We will now consider the eight stages proposed by Erikson, with special attention to those that encompass the age span we have already covered in this book.

Stage 1: basic trust versus basic mistrust. The first developmental challenge for the infant is to develop trust in his or her own competence.

Confidence is another word for trust that would do almost as well. This positive ego quality develops most adequately when the infant has a parent (usually the mother) he or she has learned can be depended upon. Trust in the goodness and reliability of the human world, represented by the loving parent, enables the infant to think well of himself or herself. The confident infant is able to tolerate maternal absence better than the mistrustful infant, to take one important behavioral example. After a year or less of postnatal experience, the child in a favorable environment should have a "rudimentary ego identity" built around the sense that one can depend on this world and its people. This developmental period is known as the *oral-sensory period* because the infant relates to his or her environment chiefly through the oral mode. He or she has had to make the transition from intrauterine life, where needs were met directly, to the outer world where one must act more vigorously to gain satisfaction. In a broad sense, the infant *incorporates* both milk and love from the environment. **Incorporation,** a process that requires sensory ability and the knack of taking outside-good-things into oneself, is the primary way through which the young infant relates to others. According to Erikson, major failures in the oral incorporative experiences of the infant may lead to schizophrenia or other forms of psychopathology. When things go well, however, perception of outer goodness is transformed into a sense of inner certainty or self-trust.

Incorporation is making a part of the outside world an aspect of one's own personality

Stage 2: autonomy versus shame and doubt. This stage applies to the child who is about two years of age, perhaps a little older. Muscular control has been developing rapidly, as we saw earlier in the book. One of the areas in which muscular control often becomes of special interest at this time is in bowel and bladder habits; thus, this period is known as the *muscular-anal period.* Like Erikson's other early stages, this one accepts some of the

Autonomy is self-determination, independence

His first set of wheels! In the U.S., standing on my own feet is often represented by rolling along on my own wheels—a combination of developmental and culturally influenced factors. (Photo by Bobbi Carrey.)

general guidelines provided by traditional psychoanalytic theory, but is more receptive to interactive dynamics and multi-cultural variations. The child at this age has a larger realm of choice available, because he or she can do a lot more. But what *should* the child do, and how much? The prime developmental challenge is to learn "how to stand on my own feet." Parents at this time must strike the difficult balance between continuing to exercise the necessary control and discipline, and allowing the child to make and act on some of his or her own decisions. Let's sample some of Erikson's analysis of potentialities and problems at this developmental period:

> . . . if denied the gradual and well-guided experience of the autonomy of free choice (or, if, indeed, weakened by an initial loss of trust) the child will turn against himself all his urge to discriminate and to manipulate. He will overmanipulate himself, he will develop a precocious conscience. Instead of taking possession of things in order to test them by purposeful repetition, he will become obsessed by his own repetitiveness. By such obsessiveness, of course, he then learns to repossess the environment and to gain power by stubborn and minute control, where he could not find large-scale mutual regulation. Such hollow victory is the infantile model for a compulsion neurosis. It is also the infantile source of later attempts in adult life to govern by the letter, rather than by the spirit. (1950/1963, p. 252).

This passage (one of many that could have been cited) indicates how Erikson is able to suggest relationships between early attempts to master developmental challenges and the individual's later personality characteristics.

Failure at Stage 2 is apt to show itself in a heavy load of doubt and shame.

> Where shame is dependent on the consciousness of being upright and exposed, doubt . . . has much to do with a consciousness of having a front and a back—and especially a "behind." For this reverse area of the body, with its aggressive and libidinal focus in the sphincters and in the buttocks, cannot be seen by the child, and yet it can be dominated by the will of others. The "behind" is the small being's dark continent, an area of the body which can be magically dominated and . . . invaded by those who would attack one's power of autonomy and who would designate as evil those products of the bowels which were felt to be all right when they were being passed. This basic sense of doubt in whatever one has left behind forms a substratum for later and more verbal forms of compulsive doubting; this finds its adult expression in paranoiac fears concerning hidden persecutors and secret persecutions threatening from behind . . . (1950/1963, p. 254).

Stage 3: initiative versus guilt. The child who has mastered Stage 2 comes to this next developmental period with an ability to love and cooperate, and to exercise self-control. With new physical and intellectual powers at his or her disposal, the child can set more advanced types of goals. Planned and vigorous adventures become possible. There is the danger that the child will go too far with these adventures—but there is also the danger that he or she will be made to feel it is wrong to act forcefully upon the environment. Over-control by parents or other adults is the main factor underlying the latter danger, although failure by the child also contributes. Sibling rivalry, jockeying for a favored position with the parents, may provide some of the motivation for the child's initiative. This may be accompanied by the fantasy of conquering the parent for one's own self, a fantasy almost certainly doomed to defeat. As a precious fantasy is proven unattainable, the child may develop an "inner powerhouse of rage." There is little he or she can do to express this rage directly. One possible negative outcome is the conversion of disappointment and rage into exaggerated feelings of self-righteousness. This is the sort of person who, in adulthood, may try to make others pay for his or her own frustrations, using his or her own sense of guilt as a driving force to castigate others.

Children who master Stage 3 have the sense that they can grow up to be like their same-sex parents and assert initiative without running afoul of guilt-arousing experiences. In the meantime, the child is eager to discover areas of functioning in which initiative can be exercised and enjoyed, such as in cooperative play with peers and in activities that have the approval of teachers, parents, and other admired adults.

Stage 4: industry versus inferiority. The period takes in the longest age span yet considered, roughly from age six through age twelve. School and culturally valued achievement now become dominant problems to be mastered. This requires a shift in the child's sense of priorities. In Erikson's words, the child "becomes ready to apply himself to given tasks and skills, which go far beyond the mere playful expression of his organ modes or the pleasure in the function of his limbs . . . [and] becomes ready to handle the utensils, the tools, and the weapons used by the big people" (p. 259). There is an increased desire on the child's part to be productive, to accomplish things, to have something to show for his or her efforts that will win favorable recognition from others.

In the United States, the cultivation of a sense of competency at this developmental period usually involves acquisition of technical skills. The child must become ready to do the things that big people do and to handle their tools, whether mechanical or symbolic. School looms as a culture of its own, establishing goals and limits. Success in this sphere helps the child acquire new aspects of identity. He or she can hold his or her own with others. Demands and expectations can be met. The world of occupations and material productivity can be managed when the time comes.

But the child might instead despair of his or her abilities and feel unequipped to meet the challenges of school and, later, of adult competence. This sense of inferiority can be newly developed from school experiences or, in some instances, follow on the heels of failures at one or more of the earlier developmental periods. A major failure at this stage can lead to the formation of a sense of *alienation* from adult society. Even before the teen years, then, a particular girl or boy may feel a lack of belongingness. "Maybe it's your world, but it isn't mine."

Later stages. Erikson divided the first twelve years of life (more or less) into four different stages—and then conceptualized all the rest of the life cycle with just four more. This suggests that Erikson, like most other observers, saw more differentiations within the earlier years of life. What he suggested is worth summarizing here until we reach the other life-cycle periods later in this book.

Adolescence raises the dominant issue of *identity* versus *role confusion.* It is time to reevaluate identity. Who am I *now?* Who am I *becoming?* In young adulthood the key issue is *intimacy* versus *isolation.* Shall I—*can* I—commit myself fully to another person? Or am I to live my life essentially as an isolated, independent entity? Adulthood, according to Erikson's framework, encompasses approximately a fifteen-year span, considerably longer than any of the preceding developmental periods. The conflict here involves **generativity** versus *stagnation.* Am I capable of taking real satisfaction from emotional investments in my children, and what the future offers, or must I content myself by clinging to what I have in my own personal life? The final stage is also by far the longest. Maturity refers essentially to the entire second half of life, marked for convenience by the age of forty-one. *Ego integrity* versus *despair* is the ultimate challenge at this time. Can I accept this life I am living—with all its limitations and the problems yet to come—or do I find myself empty, adrift, without sustaining purpose? Ego integrity is perhaps the greatest psychological accomplishment an individual can attain in his or her developmental career; it is perhaps also the most difficult to achieve.

Generativity is the ability to encourage and enjoy others' growth and achievements

Although I will not rely exclusively on Erikson's framework throughout the remainder of this book, I will consult it regularly both for its specific insights and for the sense of continuity it recognizes in humans developing.

It is now time to return to a tighter focus on research. The grand vistas afforded by theory are only part of the landscape. The attempt to put childhood into perspective requires continuing attention to research contributions.

RECENT STUDIES THAT ILLUMINATE AND CHALLENGE OUR UNDERSTANDING OF CHILDHOOD

There is no point at which we can seal off a particular area of research and conclude, "Now we know all we can know or need to know about these phenomena." This is true of scientific work in general and certainly applies to human development. Let's take a few examples from recent research contributions to illustrate some of the ways in which particular studies either illuminate or challenge what we think we understand about development through childhood. You can decide for yourself whether studies such as these yield more in the way of new answers or new questions.

FEARS, MORAL REASONING, AND THE QUESTION OF DEVELOPMENTAL STAGES

The concept of developmental stages has proven attractive to many people. It helps to guide research, organize a wide variety of observations that otherwise might not hold together very well, and communicate ideas from researcher through educator through student and into the general society. Yet we can demand that a theory not only be attractive and socially useful, but also true. What constitutes the truth value of a statement is itself a question often raised by philosophers of science. That question goes beyond our scope of inquiry here. It is clear, however, that both of the following studies have something to suggest about developmental stages as "true" or "not-so-true" concepts for understanding the child's changing relationship to the world.

Do fears change in accordance with developmental principles? The center of gravity for developmental-stage research in recent years has been the Piaget-inspired cognitive realm. But how far across all the domains of development can the broadest principles be applied? Does emotional life, as exemplified by fears, also show the same kind of changes with advancing maturation and experience that developmentalists usually seek in perceptual and cognitive phenomena?

David Bauer (1976) interviewed children at three age levels (four to six; six to eight; ten to twelve; girls and boys equally represented). The children were asked about their greatest fears, what they might think about that was scary when going to bed, and scary dreams. They were encouraged to draw pictures of these fears as well as talk about them. Verbatim transcripts were prepared from the tape-recorded interviews and these were analyzed by a set of independent raters. The content of their fears is summarized in Figure 11–5. As you can see, about three out of four kindergarteners were afraid of monsters and ghosts, but this had dropped to about 5 percent for the sixth graders. About half of the older children referred to fears about bodily injury and physical danger while this was true for only 11 percent of the four to six year olds. On the surface there were a number of other differences among the children related to their age levels. Let's put aside for our purposes the fact that the total sample of fifty-four children is fairly small; certainly it would be useful to repeat the study with larger independent samples. Instead, let's turn to Bauer's theoretical interpretation of the findings.

In interpreting his findings, Bauer reached for a theoretical realm beyond their obvious content. He suggested, invoking Werner's orthogenetic principle (1961), that there is a definite sequence in the development of fears

Grade	Bodily Injury and Physical Danger		Monsters and Ghouls		Animals		Bedtime Fears		Frightening Dreams	
	Yes	No	Yes	No	Yes	No	Yes	No	Yes	No
K (Ages 4–6)	2	17	14	5	8	11	10	9	14	5
Second (Ages 6–8)	8	7	8	7	6	9	10	5	12	3
Sixth (Ages 10–11, 12)	11	9	1	19	2	18	7	13	9	11

Type of Fear

Figure 11-5. *The frequency of occurrence and nonoccurrence of fear types in kindergarten, second grade, and sixth grade children.* *(Reprinted with permission from the* Journal of Child Psychology & Psychiatry, *1976, 17. An exploratory study of developmental changes in children's fears, by D. H. Bauer. Pergamon Press, Ltd.)*

which reflects the child's overall mode of perceiving reality. "The observation that growth proceeds from a global state, with lack of differentiation, to one of increased differentiation of internal representations from objective reality, might explain the replacement of global fear of ghosts and monsters described by kindergarteners and second graders by more realistic and specific fears involving bodily injury and physical danger in sixth graders" (Bauer, p. 71). Bauer further suggested that the child's transition from egocentric perceptions of causality to physical cause-effect (Piaget, 1954) at about age seven or eight might account for the decline in fear of animals. The general reduction with age of frightening dreams could mean that more mature children have a greater ability to separate fantasy from reality.

Bauer realized that a somewhat different angle of explanation might also apply. Perhaps the most important factor was that older children have a more elaborated system of verbal symbols available with which to understand reality and identify specific sources of fear. Or, considering still another kind of explanation, the children's behavior in the interview situation might have been influenced by what they thought was expected of them by adult society. Sixth graders might not have wanted to reveal their fears and dreams, especially to an adult they did not know well.

Ready for still another possibility? Sex-role stereotypes have already been discussed in this chapter. Could it be that males in our society are under particular pressure to deny fearfulness (McCandless, 1967)? In this study, about three times as many of the oldest boys indicated they did *not* have frightening dreams as compared with girls of the same age, but among the younger boys and girls, the frightening dreams percentages were about the same.

This study has been brought to your attention chiefly to illustrate the variety of theoretical interpretations that can be given to the same set of data (even by the same researcher himself!). It would have been simpler and less confusing if Bauer had limited himself to one line of explanation. What he did, however, was show us that several lines of theoretical interpretation are possible here. Furthermore, one could attempt to coordinate several of the explanations into a more comprehensive overall view. It also provides an opportunity to examine the different conceptual approaches in competition with each other.

While acknowledging the different possible angles of exploration, Bauer was most persistent in championing the basic conception of developmental sequence. As children grow up, they are better able to differentiate their own internal representations from objective reality. A "pure" or extreme stage theory would not be adequate, however, because language acquisition and cultural conditioning must also be taken into account. This study suggests, then, that it is reasonable to look for evidence of broad developmental principles in domains we usually think of as emotional as well as in those that are perceptual-cognitive. But it does not enable us to distinguish empirically among alternative explanations; these we must juggle in our heads while looking over the limited data. For a different methodological approach in a different domain that also bears on developmental stages let's turn to moral reasoning.

Environment, social change, and individual development—how important in moral reasoning? We have seen that Kohlberg's work extended Piaget's approach to "stages" of cognitive-moral development. Although the research in this area has been vigorous and sustained, some questions remain that can only be answered by a richer variety of empirical approaches. White, Bushnell, and Regnemer (1978) have found a way to address several of these needs within a single study. They designed a study intended to examine cross-national, individual developmental, and socio-historical change variables at the same time. In much existing research (not just in the area of moral development), the roles of environment, environmental *change,* and the child's *individual* change over time have been confused. One cannot be sure which combination of these factors exercised specific influences to a specific extent. What looks like a difference between individual developmental patterns in the United States and the Bahamas, for example, may actually have more to do with particular social (environmental) changes that have been taking place while the children were growing up.

It takes a fairly elaborate kind of research design to include all these variables and also keep them properly insulated from each other. White and her colleagues employed a *cross-sequential design* (*see* Chapter 2 for a discussion of this research design) that combines features of the cross-sectional and longitudinal designs that are usually used in separate studies. Their participants were 426 boys and girls living on the Island of Eleuthera in the Bahamas. The Bahamas have been self-governed since July 1973, but were formerly governed by Great Britain. It is a rural environment in which English serves as the official language and public school is compulsory from age five through age fourteen. You will notice that we already have to speak about the political status of the environment in describing a piece of developmental research, something which is a departure from the norm. Each participant was interviewed on an annual basis for three years. In the second and third years of the study, *new* participants were added at the various age levels encompassed. This meant, for example, that with a **time-lag design,** it was possible to compare the moral reasoning of children who were eight years old in 1974 with that of children who were eight years old in 1975 and 1976. Of course, the performances of each subject at eight years could be compared with his or her performances at ages nine and ten.

A **time-lag design** is a research strategy in which new participants are added at regular intervals

The moral reasoning tasks themselves combined something old with something new. The standard Kohlberg "moral dilemma" questions were asked of all participants. But the researchers also developed a moral dilemma situation created to be of special relevance to the life-style of the Bahama participants. This was done to explore the possibility that cross-national differences in moral reasoning might have to do with the specific framework of life experience rather than with underlying cognitive-moral growth.

The response to the localized moral dilemma proved to be so similar to the standard task that it could be set aside as not strictly necessary. This finding lent support to the possibility that there are some basic universal aspects to moral reasoning development apart from particular environmental conditions. The lack of differences between the standard and the localized moral dilemmas in this instance did not prove the existence of universal developmental stages. *Had* big differences been found, then the concept of universal stages would have received a setback.

What of the major results? There was evidence in favor of stage-like advances in moral reasoning for the younger children and the earlier stages. But none of the participants in this study (ranging up to age seventeen) gave responses typical of the Stage-3-type moral reasoning (*see* Chapter 10 for a discussion of Kohlberg's stages of moral reasoning). The investigators suggested that this was attributable to socio-cultural influences. The Bahamian school system traditionally emphasizes obedience (a Stage-1-type morality in the Kohlberg system), and the culture in general does not support or model autonomous reasoning for anybody, including adults.

The case for socio-historical changes as an influence on moral reasoning development emerged most clearly when the time-lag comparisons were made between children who were at a particular grade level under the British or the self-governed systems. There was a trend for more Stage 2 and less Stage 1 moral reasoning as the local schools became more sophisticated and powerful. Within a short period of socio-historical time, an upward shift could be seen in the achievement of advanced moral reasoning. White and her colleagues plan to conduct continuing research to determine if the higher stages of moral reasoning appear later in life for the Bahamians, or if the cultural patterns are such that these stages never emerge.

The researchers' impression, strengthened by the study, was that environmental influence over moral reasoning, and perhaps other areas of development as well, has not been adequately appreciated, even though basic developmental stages may exist. It may be only under certain conditions, such as the demands of an industrialized, competitive, technological lifestyle, that "higher" stages have an adaptive value and are therefore generated through individual-cultural interactions. This itself is a provocative theory that will need much research before its status can be determined.

GENERATIONAL CHANGE AND SOCIAL DEVELOPMENT

Now let's take another major domain of development and source of influence. A child may grow up with an orientation that is mostly aggressive and competitive toward others, or with an orientation that is altruistic or prosocial. Most children, like most people in general, express a range of social orientations that include both the competitive and the altruistic, but the mix is highly variable. Psychoanalytic theories, and most personality theories, have emphasized the influence of early experience in the family constellation on subsequent interpersonal development. However, not very much attention has been given to the complex interaction between the entire family's place in the culture and how the individual child develops. We will consider one example of the recent and ongoing research that does concern itself with the family and its place in society. The fact that it appears in a publication entitled *Journal of Cross-Cultural Psychology* indicates an increased general appreciation for cultural influences on the individual.

Knight and Kagan (1977) studied second- and third-generation Mexican-American children enrolled in the same lower-income-level school. The children were distributed fairly equally across the fourth to sixth grades, with about an equal balance of boys and girls. Each child was shown a behavioral choice card. He or she was asked to make a decision that would determine how many toys would be received by some other child in the class and how many he or she would receive. The various possible outcomes were classified by the researchers as related either to altruistic-prosocial motivations or to rivalry motivations. The child might choose an option, for example, that would increase personal reward at the expense of the other person, or, instead, an outcome that would provide equal rewards for both.

It was found that the second-generation Mexican-American children were more generous or altruistic than those of the third generation. In turn, the third-generation Mexican-American children were more generous than a comparison group of Anglo-Americans drawn from the same school. The findings suggested that within the span of a single generational change, there was an increase in competitive values and a corresponding decrease in altruistic sharing. The longer the acculturation process had been working, the more the children resembled other children whose families had been part of the culture for many generations (the Anglo-Americans). "Even when children are sampled from a traditional Mexican-American community, which should favor preservation of traditional values or an acculturation to the barrio, the generational pattern of prosocial and competitive behavior indicates an acculturation to the majority" (Knight & Kagan, 1977, p. 280).

In a subsequent study, it was found that acculturation also seemed to be related to level of self-esteem and achievement, and in a way that invites further thought. Knight and Kagan (1978) found that Mexican-American children with greater acculturation effect earned higher achievement scores at school and appeared more "successful" within the traditional Anglo-American framework. *But* these same children tended to have *lower* self-esteem than the less accultured Mexican-Americans. While Knight and Kagan offered some speculations about the possible meaning of these divergent trends, it is the general point that commands our attention here. Not only is a family's place in the host culture relevant to the child's development, but the *way* in which it is relevant should not be thought about too simply.

One child is "mainstream American" at birth. Another child is born into a family that is transitional between two value and life-style systems. Still another grows up within an at-home environment that is far removed from the prevailing culture. We do not yet know with any one subpopulation (such as Mexican-Americans) precisely how and why these different backgrounds influence the individual. Furthermore, as illustrated by the Bahamas study, important shifts can take place in just a few years, *within* generations as well as between generations. We will not completely understand whatever might be basic and universal about early development until we understand much more about generational and acculturational factors.

Similar research could be done with other groups in the process of acculturation

THE DEVELOPMENTAL LEAP: AN EXPERIMENTAL CASE STUDY

One more research example. This time let's take a close look at one of the possible ways in which general theory and specific developmental events might be captured at the same time. Can the experimentalist make it happen? Can we see, for example, that instant when a child moves from one mode of functioning to another, more advanced mode? A recent study by Snyder and Feldman (1977) at least brings us closer than usual to such a phenomenon.

The theory first. Piaget regards intelligence as the individual's ability to survive and flourish in the world, his or her primary means of adaptation.

The various stages of cognitive development are each a particular form of *equilibrium.* Different as the sensorimotor infant and the concrete operational sixth-grader may be, both have established and are continuing to elaborate modes of equilibration with the world. But how does the child get from one mode of equilibrium to the next? Through *disequilibrium!* More precisely, a state of disequilibrium generates an adaptational strategy aimed to restore the balance. A new balance may be achieved that represents a developmental advance. This can occur area by area, function by function. It also can be seen as a more general advance from one level to the next higher level.

Disequilibrium itself can arise from two sources (Strauss, 1972; Turiel, 1969). The child may recognize a discrepancy between the way he or she has been functioning and what the situation in the world really is. This would be an instance of **external** or **adaptational disequilibrium.** (The mere existence of a discrepancy between the child's mode of functioning and the objective world does not create a disequilibrium; the child himself or herself must recognize the difference.) But there may also be **internal disequilibrium.** In this case, the child's own modes of functioning are not coordinated at the same level; there is internal disharmony of some kind. Internal disequilibrium is referred to as *level mixture* by those who believe it is attributable to an uneven level of development within the same child's functions and structures at the same time.

External disequilibrium is a lack of coordination one sees between one's functioning and reality

Internal disequilibrium is a lack of coordination within one's own modes of functioning

The Snyder-Feldman study called on 96 fifth-graders (boys and girls about equal) from an urban middle-class school. They were asked to draw a map of a miniature village landscape, a task adapted from Piaget and Inhelder (1967). (*See,* for example, the map on the left on page 411.) The level of map-drawing ability was determined through analysis of spatial arrangement, proportion, perspective, and symbolization/abstraction; the map as a whole was also given an overall measure. There were twenty possible developmental level classifications for each map drawing.

So much for the pretest measure. The more interesting part was still to come. Two different scores were derived from the children's map-drawing performances. A modal level was determined, i.e., the overall, general level of each child's map-drawing, all specifics included. Then each child's level mixture was separately determined using the percentage of scores at each of the twenty possible developmental levels.*

Next, the children were given *instruction sessions.* These were guided experiences with maps drawn at the child's own modal level, and one and two levels above. The children were led to compare the maps representing different levels. Additional training activities were carried out to emphasize the differences between various levels of map-drawing. Right after their second instruction sessions, the children drew a fresh map of the original landscape model. Five weeks later they drew still another map of the original landscape.

* Level mixture was determined by multiplying the percentage of scores at each of the twenty possible developmental levels by the number of steps separating that level from the modal level and then summing the product.

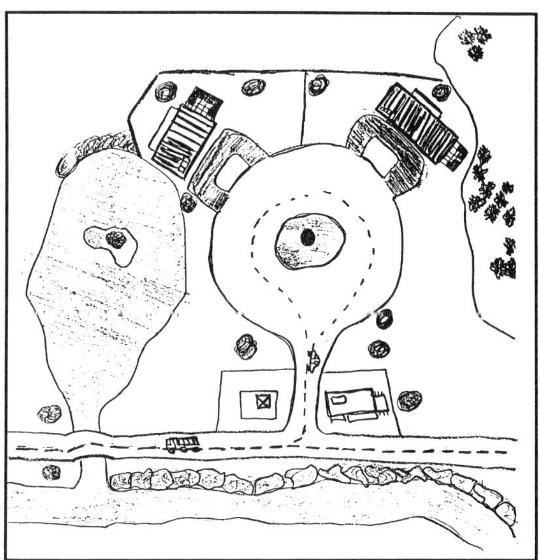

Which of these sketches of a miniature village looks more advanced, more cognitively mature? Both were made by the same child. The more differentiated and controlled rendition is on the right; it was drawn five weeks after the instruction sessions. (From Snyder and Feldman's study of disequilibrium, 1977.)

These drawings were rated by the same judge who had analyzed the first mappings. (*See*, for example, the map on the right on page 411.)

All of this research process was required in order to compare the two sources of disequilibrium, and to determine what kind of resolution would take place. The instructional sessions confronted the children with an external disequilibrium situation. Many children showed improvements in their map drawings. *All* the improvements were limited to a one-level rise above their modal levels, even if they had been exposed to greater discrepancies. This suggests that existing cognitive ability limits the size of developmental progressions. The fact that the improvement was evident on maps drawn five weeks after the instruction sessions suggests that some kind of cognitive reorganization did take place.

The differences in map-drawing performances by the same child before and after being exposed to the external disequilibrium-producing instructional sessions were marked. The experimenters seem to have induced the children to resolve their perceived disequilibrium through advancing to a higher level.

The results concerned with internal disequilibrium and level mixture also indicated patterns of developmental change, but these are too complex to summarize accurately here. While some of the results require considerable expertise to interpret, some of the features of the developmental leap to a higher level reveal themselves to the eye of any intelligent person. It is not easy by any means, but experiments such as this one indicate that we can look closely and rather directly at changes in developmental level under controlled conditions. More research of this kind in other domains of functioning would help us to understand better the link between general theory and a specific developmental event.

SUMMARY

Perspective on the first decade of human development begins with a re-appreciation of the astonishing physical and psychological transformations that occur between conception and the tenth birthday. It is also important to recognize that *childhood itself must be put into perspective.* Even relatively simple dimensions of growth such as height and weight have shown changes from generation to generation. Statistical evidence on this point was examined in this chapter. We saw that there are both *variants* and *invariants* in these physical dimensions of child development when generations are compared. More complex aspects of childhood were revealed when we explored a few of the known varieties of child-rearing and child-regarding practices over the centuries. Some experts argue—persuasively—that childhood as we know it today is a relatively recent invention. Medieval, eighteenth century European, and Victorian family patterns were briefly examined, each with specific implications for what it meant to be a child in those times. The structure and function of childhood continues to change in our own day. This means that later stages of the life cycle may also be shifting in their character from generation to generation. The person who is deeply interested in understanding human development has the challenge of distinguishing between characteristics that might be intrinsically linked with cradle-to-grave experience in the individual and characteristics of the particular time and place in which the life is lived.

An important example of the variant/invariant problem involves *sex differences* in behavior and personality. It is possible to take one extreme view or the other: (1) biological aspects of gender are so powerful that boys will be boys and girls will be girls no matter what the environmental conditions, or (2) the apparent differences between the sexes are based on differential expectations and child-rearing practices.

A review of the research findings indicated that both positions have some facts at their command. A comparative study of six different cultures was looked at. It revealed a complex pattern of similarities and differences between boys and girls in the three to eleven year old range. Some of the sex-differential stereotypes were supported, but there were also a few reversals, and a number of no-differences. Examination of the relationship between differential opportunities and experiences and differential behavior in these cultures suggested that boys and girls tend to show similar characteristics when they have undergone similar experiences. Other studies confirmed a variety of boy-girl behavioral differences from infancy onward. These seem to arise from a pattern of interaction and feedback involving both biological and psychosocial factors.

YOUR TURN

1. One of the themes this chapter has emphasized is the changing nature of childhood itself, as childhood is interpreted by various societies generation after generation. You can make a simple but useful firsthand investigation yourself. Go see one or two of the oldest people you know. If you don't know a person who is seventy, eighty, or ninety years of age, then this is a good opportunity to make such an acquaintance. After you have a comfortable relationship established with the older person, ask a few direct questions about what childhood was like in his or her youth. What was expected of children? What were children supposed to do around the house? What pleasures were they allowed? What activities, pleasures, and adventures were forbidden? Think yourself of a few other questions that will give the older person the opportunity to reveal to you the expectations, pressures, pleasures, and "feel" of what it was like to be a child a few generations ago. Now, repeat this same procedure with one or two people who are about forty or fifty years of age. Next, ask yourself and somebody else of your age the same questions: what was the essence of childhood in the generation that is your own? After this is done, compare the childhood remembrances of these three generations. What persists? What has changed over the years?

2. With your crystal ball or any reasonable facsimile, imagine what you think childhood will be like in our society for those who are starting out in life around the beginning of the twenty-first century. Take into account what you have been learning about human development, but also what you have been observing as a citizen exposed to the complexities of contemporary life and what you know of historical forces that have been operating over a longer period of time. Try out your future imaginings against those of others, and ask yourself if the child of the twenty-first century will be more or less fortunate than those of the nineteenth and twentieth?

3. Read Erikson's classic book, *Childhood and Society* (1950/1963), or as much of it as you can right now. (It will be enjoyable to return to later if you cannot complete the book at this time.) Erikson found some characteristics of the child-society relationship in Native American cultures to have valuable aspects that may not be as common in the general American culture of today. What aspects of the child-society relationship described by Erikson in Native American cultures are most appealing to you? How could they be fostered in the prevailing American culture—or would this be an impossibility?

Special attention was given to the *role models* made available to young girls and boys in our society. Fewer choices have been made available to females. American society is now reevaluating its traditional attitudes toward sex roles. Girls and women are gradually finding (or creating) more opportunity to participate in realms of activity previously limited to men. Implications for the developmental potential of both females and males were touched upon.

Erikson's "Eight Ages of Man" was taken as the framework for a more general exploration into the development of *identity*. The first four stages were presented in detail; briefer attention was given to the later stages. Erikson proposed that the successful resolution of critical problems at each developmental period enriches and strengthens the individual's growing sense of identity, while poor resolutions tend to hobble subsequent development and engender problems in adult coping and experiencing.

Recent studies that illuminate and challenge our understanding of childhood were sampled. These include an inquiry into possible developmental sequencing of *fears* which lends itself to several different interpretations, and a cross-national, cross-sequential study of *moral reasoning* that indicates the importance of socio-historical shift as well as individual developmental patterns. The influence of *generational change* and *acculturation* on social development was illustrated by a study of altruistic and competitive behavior in Mexican-American and Anglo-American school children. For a closer look at the developmental process in action, I detailed a study that tried to induce *disequilibrium* and thereby lead to a resolution at a higher level of cognitive functioning.

Reference List

Aries, P. *Centuries of childhood.* New York: Alfred A. Knopf, Inc., 1962.

Bauer, D. H. An exploratory study of developmental changes in children's fears. *Journal of Child Psychology & Psychiatry,* 1976, *17,* 69–74.

Crookham, G. Hard to be a boy. In N. Larrick & E. Merriam (Eds.), *Male and female under 18.* New York: Avon Discus Books, 1973, p. 20.

Erikson, E. H. *Childhood and society.* New York: W. W. Norton & Company, Inc., 1963. (Originally published, 1950.)

Goldberg, S., & Lewis, M. Play behavior in the year-old infant: Early sex differences. *Child Development,* 1969, *40,* 21–31.

Jacobi, J. The psychology of C. J. Jung. London: Kegan Paul, Trench, Trubner & Co., 1942.

Jain, A. Femininity. In N. Larrick & E. Merriam (Eds.), *Male and female under 18.* New York: Avon Discus Books, 1973, p. 17.

Jung, C. J. *Memories, dreams, reflections.* New York: Vintage Books, 1963.

Kern, S. Explosive intimacy: Psychodynamics of the Victorian family. *History of Childhood Quarterly,* 1974, *1,* 437–462.

Knight, G. P., & Kagan, S. Acculturation of prosocial and competitive behaviors among second- and third-generation Mexican-American children. *Journal of Cross Cultural Psychology,* 1977, *8,* 273–284.

Knight, G. P., & Kagan, S. Acculturation of second- and third-generation Mexican-

American children: Field independence, locus of control, self-esteem, and school achievement. *Journal of Cross Cultural Psychology*, 1978, *9*, 87–98.

Koch, H. L. A study of some factors conditioning the social distance between the sexes. *Journal of Social Psychology*, 1944, *20*, 79–107.

Larrick, N., & Merriam, E. (Eds.). *Male and female under 18*. New York: Avon Discus Books, 1973.

Lewis, M. Parents and children: Sex-role development. *School Review*, 1972, *80*, 229–240.

Lorence, B. Parents and children in eighteenth-century Europe. *History of Childhood Quarterly*, 1974, *2*, 1–30.

McCandless, B. R. *Children: Behavior and development*. New York: Holt, Rinehart & Winston, 1967.

Metropolitan Life. Recent growth trends in childhood and adolescence. *Statistical Bulletin*, 1973, *54*, 8–9.

Money, J., & Ehrhardt, A. A. *Man and woman, boy and girl*. Baltimore, Md.: The Johns Hopkins University Press, 1972.

Money, J., & Wang, C. Human figure drawing. I: Sex of first choice in gender-identity anomalies, Klinefelter's syndrome and precocious puberty. *Journal of Nervous & Mental Disease*, 1966, *143*, 157–162.

Piaget, J. *The construction of reality in the child*. New York: Basic Books, 1954.

Piaget, J. & Inhelder, B. *The child's conception of space*. New York: W. W. Norton & Company, Inc., 1967.

Quilliam, C. Girls and boys are alike. In N. Larrick & E. Merriam (Eds.), *Male and female under 18*. New York: Avon Discus Books, 1973, p. 23.

Scudder, H. E. *Childhood in literature and art*. Boston: Houghton Mifflin Company, 1894.

Snyder, S. S., & Feldman, D. H. Internal and external influences on cognitive developmental change. *Child Development, 1977, 48*, 937–943.

Sontag, L. W., Steele, W. G., & Lewis, M. The fetal and maternal cardiac response to environmental stress. *Human Development*, 1969, *12*, 1–9.

Stone, L. J., Smith, H. T., & Murphy, L. B. (Eds.). *The competent infant*. New York: Basic Books, 1973.

Strauss, S. Inducing cognitive development and learning: A review of short-term training experiments. I. The organismic developmental approach. *Cognition*, 1972, *1*, 329–357.

Turiel, E. Developmental processes in the child's moral thinking. In P. Mussen, J. Langer & M. Covington (Eds.), *Trends and issues in developmental psychology*. New York: Holt, Rinehart and Winston, 1969.

Weitzman, L. J., Eifler, D., Hokada, E., & Ross, C. Sex-role socialization in picture books for preschool children. *American Journal of Sociology*, 1972, *77*, 1125–1150.

Werner, H. *Comparative psychology of mental development*. New York: Sciences Press, 1961. (Originally published, 1948.)

White, C. B., Bushnell, N., & Regnemer, J. L. Moral development in Bahamian school children: A three-year examination of Kohlberg's stages of moral development. *Developmental Psychology*, 1978, *14*, 58–65.

Whiting, B., & Edwards, C. P. A cross-cultural analysis of sex differences in the behavior of children aged 3 through 11. *Journal of Social Psychology*, 1973, *91*, 171–188.

Wolfenstein, M. Changing conceptions of the infant in the United States. *Journal of Social Issues*, 1959, *15*, (1).

RIPENING
The Body Comes of Age

12

emerges.
ence / Pubescence / Ripening

ing to bodily changes and their
s.
eavier, stronger / Timing and rate / Evolutional
/ Brain-endocrine-gonad interaction / Neocortex /
ione / Epiphyses / Gonads / Primary and

M ... of a woman with both personal and social
i...
ast

IV
The ... xperiencing menarche, does experience a
resur... y and secondary sex characteristics that
runs fa... females.
Breast a...

V
Males and f... ngly different through participating in a
biological lin... is essentially similar for both.
Dimorphism /

VI
Ripening of the human's biological equipment occurs within a specific physical and emotional climate.
Age of menarche / Risk taking / Aggressive behavior / Postponement of adolescence / Maturation period for the brain

The pulse of psychobiological development throbbed away at a fantastic tempo during the earliest days, weeks, and months of life. Just think of all the changes that took place from conception through the nine months of prenatal existence. If this developmental tempo had persisted, what kind of creatures might we have become!

But the rate did slow down. Yes, the young child grew rapidly. Even so, this represented an easing of the prenatal tempo of development. Maturation unfolded at still a slower pace during middle and later childhood. This is one of the reasons why the ten year old could be portrayed as a relatively secure and stable person. He or she was still moving ahead, but was not driven relentlessly by physiological surges. He or she could cultivate and enjoy new abilities, new resources.

Now the tempo picks up once again. The years directly ahead are marked by a complex, interrelated sequence of changes. A lot happens, and it happens quickly, that is, quickly relative to developmental tempo in general. *Adolescence,* literally, becoming adult, is the term most often used to encompass all that takes place during the "in-between years." A child enters adolescence; an adult emerges. We may encounter the term *preadolescence* when our attention is focused on the earlier part of this transitional period. Roughly speaking, adolescence might be considered applicable to the teens, and preadolescence to ages eleven and twelve. There is no point in getting too fussy about the limits of these terms because their main function is to call attention to the characteristics of this period in the broadest way possible.

Puberty is the state or process of reaching physical, sexual maturity

There is, however, a related term that can be defined more precisely. **Puberty** refers to that change in body structure and function which makes it possible for the person to bear or beget offspring. The once-child is now a potential parent. When we want to focus on this significant biological change, *prepubescence* and *pubescence* are the more accurate terms to employ. The very fact that two sets of terms are involved should alert us to the complexities of human development at this period. The biological changes are really special. But the process of becoming an adult involves much more than undergoing physical transformation, hence, the value of both specific and general terms. Even within the physical realm alone, the changes related directly to readiness for parenthood can be distinguished from other marks of maturation that also manifest themselves at about this time.

Perhaps the most satisfactory term for what takes place during these years is the familiar expression, *ripening.* If nothing else, this term at least reminds us that collectively we are one branch on the tree of life, subject to some of the basic principles that apply to and define life in whatever form it expresses itself. We begin now to explore the general process of human ripening with an emphasis on the way in which our bodies come of age.

THE PRE-YEARS: BIOLOGICAL CHANGES ON THE VERGE OF PUBERTY

Preadolescence and prepubescence are significant in their own right, as well as in the way they prepare for the more advanced stages of ripening. It is during this time that the boy or girl must begin adjusting to bodily change and its psychological and social implications. New social status and opportunities arise, so do new challenges, confusions, and risks. Parents, teachers, and other children also have adjustments to make. The child experiencing his or her growth spurt may be recognized as a "big kid" by those who are being left behind for the moment. Parents may have mixed feelings about the early evidences that "my little girl" or "my little boy" has "suddenly" moved into a new phase of development. Most people are aware of the major physical changes that meet the eye at this time. These are important biologically; they are also important socially for the stimulus value they present to the child and others around him or her. We will begin with these obvious changes, and then look into some of the less visible processes also at work.

THE MORE OBVIOUS DEVELOPMENTAL CHANGES

The quickened tempo of development a little beyond the tenth year of life has already been mentioned. For some individuals, this quickening comes a little sooner, for others, a little later. Here are some other characteristic changes:

Sex differences become more obvious. Simply stated, it becomes easier to tell the boys from the girls. This general impression is based on a whole pattern of specific changes involving body proportions, style of movement, and voice. But we do not have to analyze the specific differences in order to gain the overall impression that the sexes are now starting to look and sound more different from each other—and increasingly more similar to the adults of their gender.

The face changes in size and expression. This development is something that we often seem to notice without noticing. The child looks different somehow, especially if it has been awhile since we have seen him or her, but we cannot say exactly why. It is not just our imagination. The appearance of the face, and of the head in general, frequently does change during this phase of accelerated development.

The various parts of the face have their own growth spurts. The jaw, for example, tends to change sooner and faster than the nose. With a variety of developmental clocks setting the tempo for differential growth in and around the face, the child's features (and characteristic expression) are likely to go through a series of changes. These developments in the structure of the face usually continue for several years. The "pre" (and later the adolescent) has good reason to examine what the mirror has to display; the old, familiar face is enroute to becoming something new. The shift is greater for some children than for others, and perhaps more striking for boys than for girls. At times, the changing proportions of facial features can be disturbing to the child or to others. It is useful to keep in mind that the process is a natural one and has a course to run before the face settles into its adult form.

The child becomes taller, heavier, and stronger. These three dimensions of change occur at approximately the same time and are interrelated. For many purposes it is accurate enough to think of these changes as part of a single, overall tendency toward completion of the adult stature. Those who wish to examine physical development more closely, however, will learn that the pattern is somewhat more complex, and also varies from one sex to the other. The triple spurt (height-weight-strength), as it might be called, tends to begin earlier for girls. But when the boys reach this phase of development, they tend to remain with it longer. The subsequent difference in adult height-weight-strength dimensions, then, has much to do with the male's relatively more protracted experience of growth: he starts a little later, goes a little longer.

There are appreciable differences in timing and rate of maturation. The very early maturer and the very late maturer seem conspicuous and may also feel conspicuous. It is hard not to notice that one child has started to size up and fill out, while his or her peers are still riding the slower train of development. Similarly, a child may begin to doubt that he or she will ever catch up to those on the spurt. A wide range of individual varia-

Everybody's talling up, but some are getting there a little faster. The physical differences among these boys are well within the normal range for their age. (Photo by Tania D'Avignon.)

tion is common. While there is not necessarily any reason for concern if a child is somewhat ahead of or somewhat behind his or her peers in physical development at this time, any concern that does arise usually can be relieved by medical consultation.

SOME LESS OBVIOUS CHANGES

The new developmental surge makes itself felt everywhere. Some of the more visible dimensions of change have been mentioned. It is important to understand, however, that the "pre-years" affect both structure and function at many levels of the organism. What was referred to earlier as the *psychobiological field* is becoming reshaped in its totality, taking on a different quality and style as well as greater size and heft. Here are some of the changes that do not so readily meet the eye.

The heart and most other organs participate in the general growth spurt. That vital muscular pump, the heart, enlarges to provide support for the entire body that is experiencing rapid growth. It is considered likely that most, if not all, the increase in heart size derives from the expansion of already existing cells, rather than from the creation of new cells (Tanner, 1969). Liver, kidneys, pancreas, stomach, and intestines all seem to participate in the accelerated pattern of development. The reproductive organs become larger and more adult in appearance, yet their primary phase of transformation remains in the future.

The **thymus gland** is important in early physical development and later atrophies

By contrast, some organs and tissues regress. At first it may seem peculiar that *any* components of the body should dwindle away while overall development is still in progress—especially when a spurt has begun. Nevertheless, it has been found that *lymphatic tissue* and the **thymus gland** both decrease in size as the child moves toward puberty. Lymphatic tissue is widely distributed throughout the body. It forms a channeling system that routes fluids, proteins, and other relatively large bits of matter from the spaces between cells into the blood stream. This is a vital function, for unless proteins are removed from the interstitial spaces within a few hours, the individual is likely to die (Guyton, 1967). The thymus, located in the chest, is known to be important for prenatal and early postnatal development in the formation of plasma cells and antibodies. This means that it is involved in the organism's system of defense against disease. It is not known for certain why these regressive or **involutional** (literally, rolling back into itself) changes occur during the pre-years. In the case of the thymus, it is possible that the gland has accomplished most of its job and can go into semi-retirement. Another theory with some indirect but interesting experimental support, suggests that the thymus also serves to prevent *precocious* (unusually early) sexual development (Timiras, 1972). Why lymphatic tissue should also regress at this time remains to be seen.

An **involutional** change is in the direction opposite to development

It is worth reflecting on the fact that the surge of general development is accompanied by such specialized involutions, even though their precise significance and explanation are still in doubt. We must admit exceptions to the usual assumption that development necessarily requires physical growth or expansion in all quarters. Some of us grows down so the most of us can grow up.

The brain-endocrine-gonad interaction moves into a new phase. On the research frontier, much is being learned about the specific ways in which the body transforms itself before, during, and after puberty. It is becoming increasingly clear that attention must be given to the interaction between the *central nervous system* (especially the brain), the system of ductless glands known as the **endocrines,** and the organs of sexual reproduction, or **gonads.** Efforts to explain the inner workings of physical development that are limited to any one of these components are less rewarding than the rather more difficult challenge of determining how they all function together. Since detailed familiarity with human physiology can be neither assumed nor conveyed here, we will touch on just enough of the high points to suggest what is taking place at this level in the approach to puberty. It will be necessary to introduce a few anatomical and other technical terms.

The **endocrines** are ductless glands

The **gonads** are the organs directly associated with sexual reproduction

Before the individual is able to bear or beget offspring, the body must be well prepared. This advance preparation is controlled or paced by the brain, even though the outcome is expressed through the gonads.

The brain's influence over sexual and general maturation moves along several different pathways that involve structures located inside or underneath the gray matter. However, it is probable that the top layers of brain tissue—the *neocortex,* evolution's gift to our species—exercise leadership over these inner neural workshops and mediators. In other words, the fanciest, or most advanced, brain tissue in our possession plays a role in the total sequence of maturation, although the specific operations are carried out by lower-order neural mechanisms.

The **hypothalamus** helps regulate many vital bodily processes

The **hypothalamus** is of particular interest among those neural areas and structures that influence physical growth at this time. Tucked just beneath the two hemispheres which comprise the upper section of the brain, the hypothalamus is vital to many life-sustaining functions. The regulation of body temperature, sleep and wakefulness, metabolism, and blood pressure are among its responsibilities. There are structures below the hypothalamus that are also involved in these functions. But one of the special distinctions of the hypothalamus is its ability to *integrate* specific processes so that the body as a whole remains in balance. Keeping our internal environment in a relatively steady state **(homeostasis)** is a perpetual task of the hypothalamus day and night, year after year. We cannot afford to have any of our vital systems go very far out of whack for very long. The hypothalamus also mediates much of our behavior and feeling tone: our emotions could not seethe with-

Homeostasis is the active balance an organism tries to maintain to adapt and survive

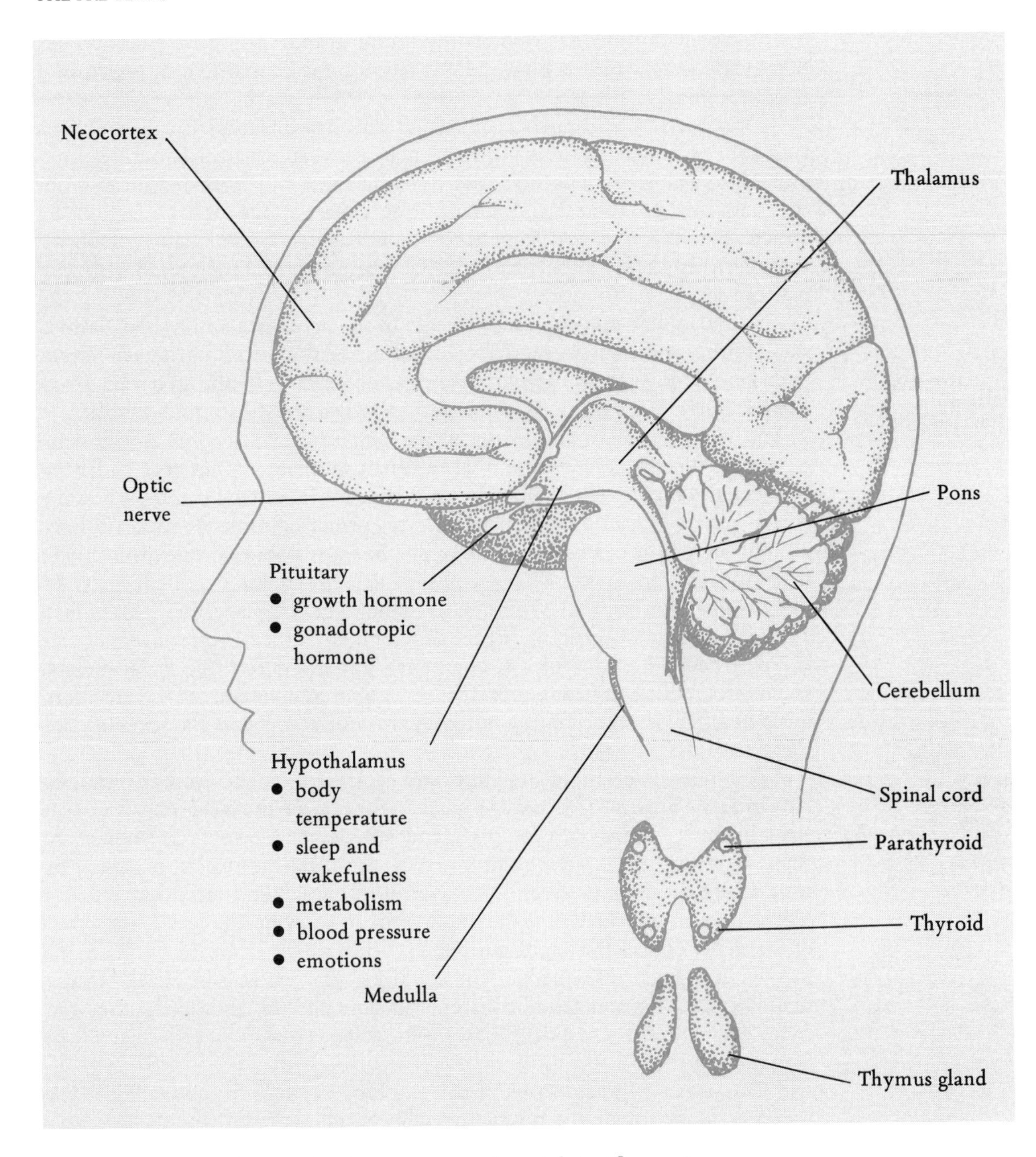

Figure 12-1. *Neural and endocrine structures of special significance in pubescence. The pituitary, sometimes known as the master gland, plays a central role in the acceleration of physical development in pubescence. A complex, systematic process operates to produce gender-specific and more general physical changes, and then to wind this pattern of development down when it has achieved its objective.*

out its collaboration. An astonishing amount and variety of vital activity takes place here, and it is no wonder it has become a favorite hunting ground for researchers.

During the growth spurt, the role of the hypothalamus is assured by its rich pattern of connections with the neocortex, subcortical structures, and endocrine glands. Signals from the hypothalamus call forth responses from the hormone producing and distributing centers. One of its neighboring structures is the *pituitary*, sometimes known as the "master gland." Transactions between the hypothalamus and the pituitary have much to do with general and sexual maturation.

Stimulated by maturational changes in the neocortex and hypothalamus, the pituitary increases its secretion of hormones that accelerate development elsewhere. One of these, known appropriately enough as the **growth hormone** (GH), leads to the growth spurt in height-weight-strength mentioned earlier. It also leads to the fusing of the *epiphyses*. This refers to the completion of the skeleton by the final joining together of bones that had been connected more loosely or flexibly until that time. Perhaps you have heard a person referred to as "ossified" because his or her opinions or life-style have become so rigid it did not seem he or she had any potential to expand his or her mind. It is during the total pre-post adolescent growth experience that actual ossification (becoming thoroughly "boned") takes place, and it is, in fact, a development that inhibits further development, at least so far as sheer size is concerned. The price of a strengthened and completed muscular-skeletal system is a reduction in the capacity for further growth. During the pre-years, these changes are accelerated, but they are not completed for several years thereafter.

The **growth hormone** stimulates the growth spurt in adolescence

Another secretion by the pituitary is known as the **gonadotropic hormone.** This means that the primary action of the hormone is on the sexual organs. There are both male and female gonadotropic hormones. As these issue forth in larger quantities, they stimulate the growth of sexual organs which, in turn, produce their own hormones. The central nervous system which, in effect, started the whole process, now begins to feel the effect of the richer concentration of hormones within the internal evironment, and its own activity is influenced. This could be regarded as a feedback circuit: neural, endocrine, and gonadic systems continually influence each other's activity even though it is possible to say in general that it "started here" and "arrived there."

The **gonadotropic hormone** stimulates the growth of the gonads

The phasing and feedback of the two broad types of pituitary hormones mentioned are important for the shift in emphasis from general to sexual maturation. The GH does most of its work first, then the gonadotropic hormones arouse the sexual organs. The internal feedback from this arousal begins to inhibit or counteract the continued production of GH. This shift in the internal balance signals the winding-down of the general growth spurt and the inception of puberty with its more complete sexual ripening.

There are more structures and types of action involved in the brain-endocrine-gonad interaction system than those described here. Perhaps enough has been presented, however, to help us move beyond the idea that sexual development follows its own rules exclusively. The timing and integration of sexual development must be understood within the context of an overall psychobiological plan that extends from the highest levels of neural control down to the most specific alterations in hormone balance—with continual interplay within and among levels. After we have extended our knowledge of the ripening process across the threshold of puberty, we will also be able to consider the role of environmental factors in the physical transition from child to adult.

INTO PUBERTY

Increased lung capacity comes at the same time many youths take up smoking

General physical development continues throughout the teen years. Lung size increases, for example, and, consequently, so does the capacity for more sustained and efficient athletic performance. The general growth spurt slows down as sexual blossoming becomes more evident, yet the process of reaching nearly complete adult stature and functioning in most physical dimensions moves along until the individual's limits are reached.

Nevertheless, it is the pattern of sexual maturation that deserves principal consideration at this time. Boys and girls both undergo significant changes in *primary and secondary sex characteristics.* In this context, *primary* refers to functions that are directly related to reproduction. *Secondary* characteristics are those we commonly associate with either masculinity or femininity. Somc of these secondary characteristics have fairly obvious indirect relationships to reproduction. The changing proportion of the hips and pelvis in females, for example, facilitates the bearing and delivery of a child. Other characteristics, such as voice timber, seem a little remote from reproductive capacity, but are linked to overall development of gender characteristics. We will now focus separately on the pubescence of each sex.

GIRL INTO WOMAN

Menarche is the first menstrual flow

It would be going much too far to say that a girl becomes a woman at the time of her **menarche** (first menstrual flow). Being a woman, in the full sense of the term, involves a sense of identity and a constellation of thoughts, attitudes, feelings, and behaviors that are seldom present in a developed form this early in life. However, menarche is a major event in the life of a female whether she is in a society such as ours or in one with a markedly different style of existence. Cultures differ appreciably in how they celebrate or respond to menarche, but the event itself is widely interpreted as a significant transformation of status. The fact that the focus here is on physical change should not dull our awareness of the personal and social implications of menarche.

1. *Pubescent changes become evident before the menarche.* Several types of development have typically established themselves in advance of the menarche itself. Breast development and the appearance of pubic hair occur at about the same time. The nipple and its surrounding *areola* become elevated. Later, the remainder of the breast fills out through the increase of fatty tissue. The budding of the breast often occurs during the eleventh year, and is of personal and social importance because it is likely to be the first visible indication that womanhood is on the way. Breast contour remains throughout the lifespan as one of the most conspicuous of the secondary sex characteristics. Cultural values and preferences for the importance of the breast in general or for a particular size or shape frequently have little to do with the biological function served.

2. *Internal changes have also taken place.* The ovaries, Fallopian tubes, uterus, and vagina begin to accelerate their developmental tempo. Stimulated by gonadotropic hormones from the pituitary, these organs renew a growth trajectory that had been temporarily slowed since early childhood. The ova ripen within their follicles. Attainment of sexual maturity does not result in the sudden production of these eggs, but, rather, in their ripening, along with that of the complete system required for fertilization and gestation.

Figure 12-2. *Some physical changes during puberty.*

Boys and Girls	Boys	Girls
Hypothalamic–pituitary activity increases, enters new phase		
Muscles enlarge, strengthen	Shoulders broaden	Hips broaden
Body hair increases (more for males)	Scalp hair decreases Voice deepens	Scalp hair increases
Breasts enlarge, change contour	Smaller and partially reversible change	Major and permanent change
Gonads enlarge and secrete	Mature spermatozoa eventually produced	Ova ripen in their follicles
	Penis becomes larger, wider	Uterus and vagina become larger, thicker
Functional capacities attained	Ejaculations occur	Menstrual flow begins, gradually becomes regular
Lung size and vital capacity increase		

A boy puts this mask over his head; a man removes it. In tribal societies, youths usually have a better idea where they stand. (Peabody Museum, Harvard University. Photo by Hillel Burger. Copyright © President & Fellows of Harvard College 1978. All rights reserved.)

3. *The first menstrual flow signifies that a new stage of maturation has been reached.* An egg has gone through the cycle of dropping into the Fallopian tubes and, after several days, reaching the end of its life cycle by being discharged through the vagina along with other cells that would have had a role in the development of a fertilized egg. Theoretically, fertilization and pregnancy *might* have occurred; this readiness to bear offspring is the most important biological characteristic attributed to the establishment of the menstrual flow. Two points should be noted, however. Menstruation may be irregular for some time after its first appearance, and the possibility that a period of sterility exists, at least in some females after menarche, has been raised by various observers (e.g., Bernard, 1971). Furthermore, we have already seen that menarche itself is but one step—if a significant one—in a longer process of sexual development that has already made considerable headway.

4. *Sexual maturation continues beyond the menarche.* Secondary sex charcteristics, as well as the primary characteristics, still have a way to go. The more important dimensions of change are shown in Figure 12–2. It is evident that some types of development are similar for both sexes, while others are more gender specific.

BOY INTO MAN

There is no easily defined male equivalent to the menarche. In the ordinary course of development, neither the boy nor others in his social world have one particular experience that establishes manhood in quite the same way the first menstrual flow signals womanhood. Perhaps this is one of the reasons some males appear to be so intent on "proving" their masculinity. Cultural anthropologists have provided us with many illustrations of *rites of passage* through which boys earn adult male status (e.g., Whiting, Kluckhohn, & Anthony, 1958). While females often have initiation rites as well, there is a tendency for males' rites to be more elaborate. This is possibly an attempt to establish by ritual what biology leaves somewhat ambiguous.

Apart from the missing menarche, however, the pubescent male does experience a resurgence of growth in both primary and secondary sex characteristics. This runs fairly parallel to the pattern in girls.

1. *Breast as well as pubic hair development occurs in males.* The increase in body hair in the genital area and elsewhere is a well-known indication of pubescence in both sexes. Less often recognized, however, is the fact that the male breast also undergoes changes roughly similar to those of the female. Growth of the breast does not go as far in the male, though, and often the size will actually decrease a year or so after it reaches its peak. Occasionally the development of the male breast confuses and disturbs its host, who may have no idea that this is normal and not adverse to the attainment of manhood.

2. *Production of testicular hormones accelerates development.* Just as in the female, the stimulation of gonadotropic hormones from the pituitary not only touches off renewed development in the sexual organs, but also leads to production of substances within the gonads themselves that further spark the surge. The ovaries churn out such hormones in females; the male equivalent is the testes. The testicular hormones stimulate growth of the primary sex organs, but also stimulate such secondary characteristics as deepening voice. This phenomenon has been known, if not in physiological detail, for many years. When the distinctive sound of the castrato voice was in favor, boys sometimes were castrated to prevent the normal developmental change from occurring.

3. *The reproductive system becomes larger and more functional.* Growth usually occurs first in the testes in their sac-like container, the scrotum. About a year later, the developmental acceleration encompasses the penis, which becomes larger and wider. Internal glands in the reproductive system enlarge and begin to form and secrete a variety of substances. Mature *spermatozoa,* capable of reaching and fertilizing an ovum, gradually come into production. The odds of any single sperm making its way to a receptive ovum are very small. Begetting a child usually requires not only the production of a mature type of sperm cell, but also of a large number of cells. It is estimated that this capacity is established by age fifteen in most males (Timiras, 1972).

SEX DIFFERENCES AND SIMILARITIES AT PUBESCENCE

Since early childhood, the pattern of physical development had emphasized changes that were similar for both sexes. The pre-years, however, introduce a new trend that continues throughout adolescence. Many of the changes now have the effect of sculpturing sex differences, a tendency known as **dimorphism.** The new dimorphism is experienced by both males and females, although the precise effects are different. Males and females become increasingly different, then, through participating in a biological line of development that is essentially similar for both.

Dimorphism is the process of becoming physically distinct as male or female

Another perspective for examining similarities and differences is through comparison of the menarche with—what? The onset of the menstrual flow is a conspicuous indication of functional, as distinguished from anatomical, maturation in the female. For males, onset of ejaculation (the discharge of seminal fluid, including spermatozoa) is probably the most conspicuous and important functional indication. In their own ways, each of these phenomena signify entrance into adult sexuality and the ability to replenish the human race. But the specific differences are so great, physically and psychosocially, that these two coming-of-age events cannot be equated. Biologically, for example, menstruation and ejaculation have very different roles to play in sexuality and reproduction. Ejaculation is one of the earlier

Characteristic	Menstruation	Ejaculation
Time of onset	Relatively easy to determine	More difficult to determine
Subsequent occurrences	Settles into monthly rhythm	No set rhythm, variable
Circumstances of occurrence	Basically under internal physiological control	Psychological and behavioral factors more directly involved
Role in reproductive cycle	Signifies the conclusion of a fertility cycle without fertilization	Signifies the possibility of fertilization occurring if a receptive ovum is available
Duration of event	Measured in days	Measured in seconds
Feeling tone	Not especially exciting, pleasure-giving	Can be exciting, pleasure-giving

Figure 12-3. *A comparison of menstrual flow and ejaculation as signals of new functional capacity.*

events in the reproduction sequence, while menstruation represents the termination of a cycle in which fertilization of the ovum might have—but did not—occur. Methodologically, it is easier (although not that easy) to determine the onset and subsequent rhythm of menstruation than it is to learn at what age the first ejaculation occurred. Psychologically, there are important differences as well. Ejaculation is often preceded by excitement and sensual pleasure; by contrast, menstruation is not usually experienced as "sexy" or thrilling. Some of the similarities and differences between these phenomena are summarized in Figure 12-3. To the extent that males and females center their emerging sense of sexual identity around these phenomena, then, we might expect an increasing divergence of feeling and attitude between the sexes.

Another point, obvious though it is, should be mentioned here. Attainment of reproductive capacity has very different implications for the male and the female. Apart from whatever personal meanings sexual intercourse may have to a pair of adolescents, the possibility of becoming pregnant is a *natural* concern to one of the partners only. Even if we assume essential similarities in maturity and personality, the distinctive biological perspectives would introduce radically different orientations.

THE PHYSICAL AND EMOTIONAL CLIMATE OF PUBESCENCE

Ripening of the human biological equipment occurs within a specific physical and emotional climate. This environmental context ranges from the obvious differences between blazing tropical sun and chilling polar winds to the subtle influence of cultural values. We will now look at a few examples of possible relationships between the inner and outer climates of sexual maturation.

CHANGES IN THE AGE OF MENARCHE

Much information has been collected over the years—over the centuries, in fact—regarding the age at which menarche occurs. Observations have come down to us from as long ago as the fourth century B.C., an indication that the topic has been considered important by many people in many times and places. The rich accumulation of data encourages evaluation of possible environmental influences. As interesting as it would be to make a parallel inquiry into the onset of male pubescence, comparable data are not available.

Both ancient and modern data were surveyed recently by Carol Jean Diers (1974). She found evidence of a very interesting historical pattern, although there are some gaps remaining in our knowledge of certain periods of time. For well over a thousand years, 400 B.C. to approximately A.D. 700—menarche usually occurred between the thirteenth and fourteenth year. Adequate information is lacking for approximately three centuries. Picking up the trail again in the eleventh century, the average age still held at about the same level. As Europe moved from the medieval period toward modern times, the age of menarche seemed to be delayed. The statistical evidence suggests a slowdown in the rate of maturation especially during the end of the eighteenth century and during the early decades of the nineteenth century. By the 1800s, most females in Western society did not reach menarche until about age seventeen.

Too bad we don't have data going so far back on other developmental phenomena

What makes the historical pattern of special interest is that sexual maturation has accelerated since the early days of industrialized, urbanized society. Age at menarche has dropped steadily since about 1830, and has now returned to approximately the same level reported more than two thousand years ago. Why?

The most popular line of explanation has been physical environment, especially the supposed effect of a hot climate to induce early maturation. The available evidence, however, does not support this theory. J. M. Tanner (1971), perhaps the leading investigator of physical development at adolescence, holds that improved nutrition and lessening of disease in infancy and early childhood probably are major factors. He also believes it is possible that changes in psychosexual stimulation may have been influential, but does not consider this possibility to have been adequately tested as yet.

We have already seen (*see* Chapter 11) that people have been growing taller and taller through the generations, but that this tendency is perhaps approaching biological limits. Possibly, the earlier onset of menarche is also approaching the biological limits in our generation, with similar factors at work for both dimensions of change. Furthermore, if nutrition, disease, and the psychological influences we have upon each other are factors responsible for broad changes in maturation across the generations and the centuries, then these factors should also be influential in the development of every individual as he or she grows up today. Much remains to be learned on this subject, but we have at least strong hints that our overall style of life as a culture has something to do with the maturational pattern in and around adolescence.

VIOLENCE AND RISK

Improved nutrition and improved control of disease have had another effect, one which we tend to take for granted. Just a few years ago, each wave of adolescents suffered an appreciable number of casualities from polio, a condition that crippled, paralyzed, and sometimes killed its victims. In one of the most dramatic of recent medical breakthroughs, polio has become a preventable and, hence, less common condition. The frequency of other diseases that sapped the strength or claimed the lives of children and adolescents has also been appreciably reduced over the past century. This means that fewer adolescent lives in our society are at risk from diseases that are beyond medical understanding and management.

Other cultural dynamics, however, continue to pose threats to the continued health and well-being of the strong, sound individuals who are approaching the completion of physical growth today. Motor vehicle accidents and suicide are leading causes of death in this age range. Problems associated with the use of alcohol and a variety of drugs have produced outcomes ranging from temporary to permanent in duration, and from moderate to lethal in severity. Venereal disease, although relatively amenable to successful diagnosis and treatment, continues to occur frequently enough to alarm health authorities. Cigarette smoking, a well-established hazard to health, is still taken up by many people around adolescence, partially as a sign of being "grown-up."

Injuries suffered in competitive sports take their toll each year, and nobody knows how many abortion attempts, conducted in the absence of medical expertise, result in unnecessary pain and injury. Violence involving young males is a phenomenon known in many cultures. In our own time, the weaponry employed includes the lethal and all-too-available handgun, as well as the clenched fist. And what is there left to say about war? Young males of many nations have been decimated by combat and the other life-threatening perils associated with organized strife. Vietnam is a name still fresh to the current generation of adolescents; past generations have also had their encounters, even if the Argonne, Iwo Jima, and Korea seem remote to many readers today.

This recital of hazards confronting the young has its place in a chapter devoted to physical development in adolescence. In many ways, our society has become a safer place for healthy development, but the variety and incidence of hazards is much greater than what is strictly "necessary." It is improbable that a single explanation would prove adequate for the multiple hazards. The following are some of the possible partial explanations, reduced to capsule form.

Adolescent judgment does not match opportunity for experience. As physical development becomes well advanced, the individual has more freedom and access to engage in a broad range of adult activities, some of which involve risk factors (e.g., operating a motor vehicle). The young will have more than their share of mishaps because they have not yet acquired the judgment that goes along with experience.

Society encourages risk-taking behavior. Society encourages risk-taking behavior and even certain kinds of violent behavior in the young, especially in males. Society's motivation can be seen as a need (for whatever reason) to establish "manhood" tests of courage, or as a vicarious outlet for the aggressions and frustrations felt by older generations. Anybody who has heard the violent oaths shouted by parents as they urge their offspring to brutal and and bloody deeds on the field of athletic competition must ponder both of these possibilities seriously.

How would you prove or disprove this proposition?

Another possible cultural dynamic is one that may appear absurd on its surface: *perhaps society really does want to do injury to its young.* This unflattering theory is based on the thought that the established generations may not welcome the competition represented by the energetic newcomers. If society has mixed feelings about the young, might not these feelings be expressed in mixed signals, some which foster, and some which jeopardize, health and development? So unwelcome is this theory that one seldom hears it mentioned.

Two implications are worth keeping in mind for future reference. Is it possible that a desire to limit competition for power and privilege is responsible for some of society's actions toward both the young and the old? Perhaps we encourage excessive risk-taking among the young for the same reason we nudge the elderly from roles of influence; this possibility will be discussed in more detail in later chapters. Additionally, it may be time to think about the consequences of increased opportunity for girls and young women in our society. Will females now be expected to prove their "masculinity" and take the kind of risks customarily expected of males? If so, what will be the effects on their subsequent development and their maternal roles? Or will the hard edge of competition and the subtle encouragement of risk-taking and injury be reduced somewhat as it becomes increasingly common for females to participate in vigorous sports and other realms of action and risk?

A less extreme view than the one just discussed is the opinion that society encourages risk-taking behavior to sharpen the competitive edge of youngsters who need confidence and aggressiveness to succeed in contemporary life. Certainly, this is an argument often heard in support of highly disciplined team-sports. But the matter can also be put the other way around: were contemporary life itself less competitive, then fewer such pressures would be needed and exerted.

The adolescent's very nature increases hazardous behavior. There might be something in the nature of the adolescent, apart from particular cultural influences and opportunities, that has the effect of increasing certain hazardous behaviors. Here, again, there are many pathways through which explanations can be sought. Perhaps sexual tensions impel release in violence and other risk taking. Maybe there really is an "instinct" that ripens along with physical maturation and demands expression in aggressive actions. There are no hard data to support either of these theories, but, as previously mentioned, some heightened expression of aggressive behavior has often been noted in early adolescence in many different cultures. This at least raises the possibility of a biological inclination in this direction.

Adulthood itself is more hazardous than childhood. The prevalent *adult* life-style itself incorporates many risks and hazards that the culture has adopted (e.g., excessive use of drugs, alcohol, and cigarettes). This view leads to probably the simplest of explanations: as the child moves through adolescence into adulthood, he or she must negotiate increasingly hazardous territory. It would be no wonder, then, that some adolescents become casualities fairly early to destructive elements in our cultural life-style. It may be more the wonder that so many bypass or overcome these perils.

Adolescents have a greater opportunity than children to experiment with the hazards and follies, as well as the pleasures and privileges, of adult life. (Photo by Talbot Lovering.)

Our general point here is that as young males and females approach general and sexual maturation, they enter into new modes of interaction with the attitudes and life-styles of their culture. This phenomenon probably occurs in every cultural setting, whether similar to ours or far different. An enormous breadth of knowledge would be necessary to understand the specific interactions between cultural milieu and the biological development and well-being of the adolescent. I can only recommend an alert and inquisitive attitude on the part of the reader. In our so-called youth-oriented society, there are often many unnecessary dangers and deprivations placed in the pathway of continued development, even though other factors do favor healthy physical growth.

WHY ADOLESCENCE? A BIOLOGICAL PERSPECTIVE

Physical ripening gives the adolescent years many of their distinctive characteristics. The basic functions of this resurgent growth are not difficult to recognize: the body must attain its adult stature and functional capacity, and the ability to reproduce and replenish the species must be established. Yet questions remain about the existence of adolescence in the particular form it is known to us. Let's take one of the problems that have been raised and see what kind of answer can be given from a broad biological perspective.

Most other creatures do *not* spend a high proportion of their total lifespan in a state of relative immaturity. Even such advanced animals as the monkey develop more rapidly than human beings during the prenatal period and do not dawdle so long in attaining general and sexual maturity. Why do we wait so long? The "why" is frowned upon by some scientists who believe it implies purpose and intention in the universe, when nothing of this kind should be assumed. Nevertheless, I will ask why here and call on J. M. Tanner for an informed response:

> The cause of the postponement of puberty in the primates appears to be traceable to a mechanism in the hypothalamus. It seems to be the brain which initiates the events of the adolescent spurt; but how the brain can tell when the requisite pre-adolescent development is complete we have no idea. One thinks of a series of clocks, the running-down of one being the signal for the starting of another: but the clocks in the brain are highly dependent on the growth and development of the body and very little dependent on chronological time. The increased time necessary for the maturing of the primate brain has been sandwiched in between weaning and puberty; the maturation of the hypothalamus has been put back so that maturation of the association areas of the cortex . . . can take place first. The literature on precocious puberty makes it clear that the adolescent mechanism, once started, is readily capable of going all the way to spermatogenesis and ovulation, even at the age of 5 or 6.
>
> Some, at least, of the biological reasons for this postponement are not far to seek: for example, it is relatively useless to produce young while the brain is

> still immature, as the advantage the brain gives in their protection will not then occur. Also it is probably advantageous for learning, and particularly learning to co-operate in family or group life, to take place while the individual remains relatively docile and before he comes into sexual competition with adult males. The actual existence of groups containing numbers of these tractable but moderately able individuals might in itself be advantageous. At any rate *the postponement of adolescence is a highly distinctive evolutionary trait of primates, and a necessary one, if advantage is to be taken of the long maturation period for the brain.**†

The biological perspective, then, makes reference to intellectual maturation, learning ability, and social interaction as well as to specific sequences of physiological activity. What we become as individual personalities while "waiting" for adulthood to strike is so important that our own biology conspires to keep us waiting long enough to become more competent human beings. It is this side of the story that will be pursued in the next chapter.

SUMMARY

The tempo of psychobiological development picks up during the years that directly precede puberty. Both general and sexual maturation now proceed at a rapid pace. During the "pre-years," a number of obvious developmental changes make themselves known. These include a sharpening of differences between the sexes, changes in the size and expression of the face, and gains in height, weight, and strength. Appreciable individual differences in the timing and rate of physical maturation can be noted. Some of the more significant changes are less obvious, for example, the growth spurt in the heart and most other internal organs, and paradoxically, the decrease in lymphatic tissue and the thymus gland. Particular attention was given here to *brain-endocrine-gonad interaction* where understanding of the mechanisms underlying growth at adolescence can most fruitfully be sought. A sequence of action that moves from the *neocortex* to the *pituitary*, and then to the *hypothalamus* and the *gonads* was outlined. This pattern of interaction becomes increasingly more complex as, for example, when hormones produced by the gonads start to inhibit the original spurt in general development.

Moving into puberty itself, children cross a line that is partially biological, partially psychosocial. In females, *menarche* traditionally represents the transition from girlhood to womanhood. Actually, much development takes place in the months preceding the first menstrual flow, and continued development occurs afterward. The transition is less clearly marked in males whose general pattern of sexual maturation nevertheless has much in com-

* Italics added.

† From *Growth in Adolescence* (2nd ed.) by J. M. Tanner. Oxford, England: Blackwell Scientific Publications, 1969, pp. 238–239.

YOUR TURN

You have been reading about other people's physical development before and during adolescence. What about the body you know best—your own? Try to recall your total experience of growing up physically from late childhood through adolescence. Think of the little experiences and observations that made you aware of the process. Were there any decisive moments that made you feel you had really changed? What aspects of your development were most impressive to yourself? Did certain developments puzzle or alarm you? Are there certain facts you could have used back then? Were you prepared for what happened to your own body? How did others respond as you changed?

With questions such as these in mind, try to reconstruct your physical development. This could be done in straightforward narrative form, starting from a time in late childhood when you were more or less at a plateau and moving ahead through adolescence. Another, or supplementary, approach would be to separate the physical changes as such from their personal and social meanings. This could be done simply by dividing a page down the center and making entries such as in the example at the bottom of the page.

In looking over your adolescence, can you see any specific ways in which your health and safety was safeguarded by society? Can you see any ways in which potential hazards were added?

After you have taken the opportunity to think back on your own development, perhaps you will be ready to enter into informal discussion with several of your peers and share at least some of the experiences you had and what you now make of those experiences. This exercise will also help you to warm up to the topics that will be considered in the next chapter when the psychological and social aspects of adolescence are examined in more detail.

Physical Development	Age	Personal and Social Response
Start of hair in moustache and beard area	13½	Proud, but too cool to show it. Shaved secretly once. Thought of growing beard.

mon with what is experienced by females. Similarities and differences in the sexual development of males and females were surveyed and discussed.

The physical and emotional climate in which human ripening occurs exerts a number of probable effects. Changes in the age of menarche have been recorded over the centuries. At least some experts believe that factors within cultural control, such as nutrition and prevention of disease, have been influential in these changes. Another way in which environmental (including socio-cultural) factors influence development can be seen in the many trials and hazards that confront each generation of adolescents. After reminding ourselves of some of the risk taking and violence that takes its toll around adolescence, we looked into possible explanations. These ranged from the mostly biological to the mostly psychological, and all are worth further attention.

Finally, we recognized that humans spend a greater proportion of the lifespan in an immature state than any other organism, and wondered why. The biological perspective offered in response to this question emphasized opportunity for intellectual maturation and social learning, as well as purely anatomical and physiological considerations.

Reference List

Bernard, H. W. *Adolescent development.* Scranton, Pa.: Intext Educational Publishers, 1971.

Diers, C. J. Historical trends in the age at menarche and menopause. *Psychology Reports,* 1974, *34,* 931–937.

Guyton, A. C. *Textbook of medical physiology* (3rd ed.). Philadelphia: W. B. Saunders Company, 1967.

Tanner, J. M. *Growth in adolescence* (2nd ed.). Oxford, England: Blackwell Scientific Publications, 1969.

Tanner, J. M. Sequence, tempo, and individual variation in the growth and development of boys and girls aged 12 to 16. *Daedalus,* 1971, *100,* 907–930.

Timiras, P. S. *Developmental physiology and aging.* New York: Macmillan, Inc., 1972.

Whiting, M. W. N., Kluckhohn, R. C., & Anthony, A. The function of male initiation ceremonies at puberty. In E. Maccoby, T. M. Newcomb & E. L. Hartley (Eds.), *Readings in social psychology.* New York: Henry Holt & Co., 1958, pp. 359–370.

WHO IS THE ADOLESCENT?

13

CHAPTER OUTLINE

Four major theoreticians present systematic conceptions of adolescence on the psychosocial level.

Evolutionary idealism (Hall)—Storm and Stress / Adaptation through cognitive structures (Piaget)—concrete operations, formal operations, combinatorial logic / Identity integration versus identity confusion (Erikson) / Cultural relativism (Opler)—metropolitan youth culture, role discontinuity

Adolescence is a time of life that invites fantasies—by all of society, not just by those who are themselves moving through their teen years. Consider some of the remarks we often hear about adolescence:

> It is a time of *daydreaming*, of wandering about in a vague mental mist of one's own making.
>
> But it is a time of *awakening*, vibrating to new urges and possibilities.

This is only one set of contradictory-sounding propositions. Take another set:

> The teen-ager is the ultimate *conformist*. All across the nation, every adolescent is talking, dressing, and stylizing him or herself to be just like every other adolescent.
>
> But the adolescent is the very embodiment of the *revolutionary!* Idealistic and intolerant of the status quo, the teen-ager is given to extreme "causes" and wild innovations.

And still another set:

> It's really tough to be an adolescent in our society. They won't let you do your own thing. You just have to wait and suffer until your day finally comes. You're wracked by inner *storm* and attacked by outer *stress*.
>
> But adolescents don't realize how lucky they are. They can *hang loose*, goof off, freak out, or whatever. Wait until they come up against the real pressures of life. That's when they'll find out what it's like to shape up, settle down, produce, and adjust—whether they happen to like it or not.

Each of these statements seems to catch a bit of reality. However, none of these propositions adequately characterizes the nature of adolescence, not separately and not together. The stereotypes cannot be dismissed without a hearing. Both daydreaming and the awakening to new urges and possibilities, for example, can be observed in adolescence. And society can choose to emphasize either conformity or revolutionary tendencies through selective perception and response. It is no trick at all to put together a few observations that suggest adolescence is a time of looking alike and thinking alike; nor is it difficult to find examples of establishment-shaking or taunting behavior that could be interpreted as revolutionary. It is so easy to urge one view or another, we are well advised to pay attention to the dynamics of teen-watching as well as to adolescent thought and behavior per se. Perhaps it is the insecure adult who is on the lookout for any sign of challenge from the adolescent generation, quick to shout "revolution!" every time he or she detects a wave of

Later we'll see how easy it is to form stereotypes of old age as well energy or innovation. And perhaps it is the adult who has settled into his or her own comfortable routine of life who is most adept at noticing adolescent conformity, but not the ways in which *his* or *her* personal life-style has come to resemble the standard pattern for his or her own age and socioeconomic echelon.

One step further. Take hundreds of young people and gather them together in the same place for a good part of the day, five days a week. Expose them chiefly to each other, and rather little to the broad spectrum of life from young children through adults and the aged. Is it at all surprising that these people begin to exercise great influence over each other? Furthermore, since they come under not only strong peer influence but also the same environmental inputs, it is not especially mysterious when they show a certain conformity or mutual resemblance.

The school situation is also one in which control and discipline are usually *intended* to prevail. The adolescent is expected to be in certain places at certain times, doing certain things. There are administrative and educational reasons for exercising this kind of control. Students are expected to gain competency in a variety of subjects considered important by society. The school system, as we know it, requires careful scheduling, utilization of resources, and consonance of values with parents and community. Additionally, it could be argued that the high-school situation is a helpful preparation for the 9-to-5 careers that many will enter in a few years. The person is socialized to do his or her own work and to interact appropriately with others in a situation that sets limits and makes demands not always of one's own choosing.

Yet from the adolescent's standpoint, he or she is up against an external, authoritarian framework. Each individual has his or her own balance of likes and dislikes regarding the school situation, but the daily experience for most adolescents is to enter a prescribed, established, adult-run world for many of the daylight hours. This can be contrasted with other types of social arrangements in which adolescents and adults live and work side-by-side instead of becoming fixed in a formal, educational-power relationship. Our high schools may encourage and further some of the student's interests, as well as stimulate the development of new interests. However, it is likely that other interests are considered inappropriate, at least during school hours. This would seem to ensure some tension and resistance to the high school, the most available representation of adult authority. The potential for rebellion, then, has something to do with the total situation, not just with characteristics that are peculiar to teen-agers.

Furthermore, our society encourages independence of mind, if not in a thoroughly consistent manner. Some teachers and some class materials encourage adolescents to think for themselves, to develop their powers of critical and imaginative thought. These heightened intellectual powers can be turned in more than one direction. The adolescent may be regarded as a menace if he or she tries these new independent thinking abilities out on traditions cherished by the established generations. It is possible, for example, for a young person to become involved in those aspects of the Revolutionary War that are often emphasized in this country. Should this revolutionary spirit be applied to contemporary society, however, the adolescent might find himself or herself on the outs with teachers, parents, and the community. Respect for independence and "Yankee ingenuity," then, are among the components of our cultural values that could encourage adolescents to cultivate and act on new thoughts—only to run into opposition from the same society that glorifies these values.

Let's continue to examine the structure of the situation in which the adolescent develops as well as those characteristics that seem to belong more directly to the adolescent himself or herself. If our culture is giving mixed signals as to what it wants and expects from adolescents, then this should be considered an integral part of the developmental challenge.

We will now consider several theories of adolescence. Each differs in the way both society and individual are taken into account. And each offers a different answer to the question: *who is the adolescent?*

INTERPRETATIONS OF ADOLESCENCE

EVOLUTIONARY IDEALISM (HALL)

The first major psychological theory of adolescence came about in an interesting way. Darwin's theory of evolution had churned beyond biology to almost every sphere of scientific and intellectual activity. Some people found the "monkey theory" distasteful or alarming. Others accepted the evolutionary view and attempted to pursue its implications. G. Stanley Hall belonged to the latter group. He was one of the "firstest" of men: the first person to receive a Ph.D. in psychology in the United States, the founder and first president of the American Psychological Association, the first president of Clark University (Worcester, Massachusetts), the originator of the child study movement, and the first to write a major treatise on the psychology of adolescence (1904). Before all of this, however, he had been a student of theology. Throughout his educational and scientific activities, Hall remained fascinated by the possibility of human perfectibility. We could be better than we are, he thought, and the key to improvement of our species might be found through an integration of evolutionary theory and the new sciences of human thought and behavior.

Knowing something about Hall's own background and interests helps us to put his theory of adolescence into perspective. This was a man equally devoted to the development of the human sciences and to the further development of the human race. A true scholar, he could draw upon the past, but also remain up to date in a variety of fields of study. When a person of this stature decided to "explain" adolescence, then, it was no wonder that many people listened, and some agreed.

Essentially, Hall saw adolescence as the most critical time of life for the advancement of our species. The child is too young, and the adult too old and rigid to carry human evolution forward. The adolescent is just right. Society must take advantage of the potentialities of the adolescent if the potentialities of the human race are to be realized. This approach guaranteed that a great deal of attention would be given to the adolescent and to various means of reaching, shaping, or influencing him or her. Although Hall's particular theory was rejected by later generations of psychologists, his emphasis on adolescence has been carried forward to the present day.

Recapitulation is going through the same stages in one's life as did the species when evolving

Hall believed that the individual passes through four developmental stages: infancy, childhood, youth, and adolescence. More remarkable, however, was his contention that the first three of these stages **recapitulate** earlier and more primitive evolutionary levels. A person does not become a thoroughly human human until adolescence. Centuries ago, our ancestors were more comfortable on all fours, like most other animals, and preoccupied with the development of basic survival skills. Just look at infants and very young children, invited Hall, they crawl about, too, and are chiefly sensorimotor creatures. At a later point in the evolution of the human race, our ancestors lived in caves, venturing forth to hunt and fish. Hall saw parallels to this activity in the behavior of children within the four- to eight-year-old range. Still later in human cultural history, people lived together in communities, but were "barbarians." Youths between the ages of eight and twelve are reenacting this phase of human evolution, thought Hall. They are ready and able to learn the various skills necessary to function as group members. But it is a simple, nonreflective, nonprogressive way of life. Children at this age have reached the level of evolutionary development that was achieved by most societies prior to the establishment of "great civilizations." The boy or girl at this age can be a most competent savage or tribesperson, but cannot be mistaken for a fully actualized individual.

Adolescence itself, as seen by Hall, encompasses a broader time range than this term usually suggests today. The individual enters adolescence at puberty and remains in this stage until about age twenty-five, or whenever full adulthood is reached in a particular case. The adolescent years are crucial for the species as well as for the individual. This life-stage hearkens back to that time in human history when our race was starting to break away from the stagnant, simple, barbarian mode of functioning. Humanity had to shake itself loose from a state of relatively low level equilibrium and seek a higher and more creative mode of existence. In the same way, each individual adolescent has the challenge of moving through a phase of disorganization and conflict in order to achieve a higher level. Hall dramatized this transition process by characterizing it as a "second birth."

This was a powerful conception. Hall was proposing a point-by-point parallel between evolution of the species and development of the individual. This would appeal to people who feared that the new scientific breakthroughs were fragmenting their general understanding of the world. Hall could alleviate these anxieties by demonstrating that there is an order in things that runs through biology, zoology, archeology, history, and individual lives. Furthermore, his theory acknowledged certain features of adolescent experience that are themselves rather dramatic, as we will see in a moment. Finally, Hall's approach, which we might call **evolutionary idealism,** had important implications for what society should do with and for its adolescents.

Evolutionary idealism is Hall's theory that distinctively human development begins in adolescence

The teen-ager who feels confused, frightened, and vulnerable—but also bold, excited, and confident—is a good advertisement for Hall's theory of adolescent experience. Going beyond the simple and secure world of childhood may be likened to a passage through stormy seas to an unknown destination. In fact, the phrase *storm and stress* came to represent the nature of adolescent experience in Hall's view. The *Sturm und Drang* (the German origins of this phrase) keeps the adolescent off balance much of the time. These storms of the soul represent the young person's attempts to feel and act upon strong but contradictory impulses. He or she wants privacy and solitude, for example, but also craves human contact and interaction. Enormous enthusiasms alternate with a sense of boredom and inertia. On and on goes the list of extremes between which the adolescent swings. One of the most significant impulses of the adolescent, however, is to envision a better world. If a person is ever to be *idealistic,* now is the time for it. Even this tendency has its opposite. The adolescent can turn abruptly from idealism to cynicism and cruelty.

But Hall did not stop with merely describing what he took to be the distinctive characteristics of adolescent experience. He also had an explanation and a plan of action. By the time a person is well into adolescence, he or she

Adolescents may declare their own psychosocial moratorium and hang loose between bouts of storm and stress. (Photo by Talbot Lovering.)

has also neared the end of genetic control over his or her functioning. We might say today that he or she is "running out of program." The next step in development is up to the individual and his or her society. Genetics and evolution have played the major role in bringing the person to this point in life. How far he or she develops now, and in what particular manner, depends more on self-directedness and environmental influence.

There is something curious about this theory. During the past several centuries, many thinkers have taken the position that our developmental careers are propelled chiefly by intrinsic factors. Others have emphasized the role of society and environment (*see* Chapter 7). Hall took a firm position on both sides of the fence! He saw early development as bio-genetic. Obviously, he did not have reams of suggestions to make about child rearing and early schooling. But when it came to adolescence, Hall had lots to say. He hopped over to the other side of the fence and maintained that environmental influence was almost everything. This shift in emphasis from bio-genetic to environmental influence makes sense so long as one accepts Hall's total theory of development, but it remains an unusual maneuver that is open to the charge of inconsistency.

From the evolutionary idealism standpoint, then, distinctively human development really begins in adolescence, specifically, during the later years of this period. Should society ignore adolescents or make the wrong moves, then it will have spawned a dangerously maladjusted generation—and, at the same time, missed the opportunity to advance evolutionary progress. Instead, society should expend its resources generously in the education of adolescents. Young men and women are more open to experience now than they will ever be again. The streak of idealism in adolescents should be appreciated and engaged. Adolescents should be given the opportunity to carry their idealism into the future, instead of encouraged to let it go as adult realities intrude.

Recent studies suggest that openness to experience increases for some in middle and late adulthood

There are two points of particular importance here. Hall would have had society concentrating its resources on the more *elite* young men and women. Those of mediocre potential would receive enough goodwill and education to attain a satisfactory adjustment to adult life, but only the best minds and bodies of each generation would have the opportunity during adolescence to participate in an all-out program of self and social actualization. The other point concerns the *goal* of this activity. The success and happiness of the individual person was desired by Hall, but this was secondary to the continued advancement of the human race. It was the super-person that Hall had in mind. Find and nurture the best adolescents and the human race has an opportunity to reach beyond its own present levels of achievement and fulfillment.

Whatever else might be said about Hall's theory, it cannot be said that he ignored the adolescent! The vigor and breadth of his approach helped create a cultural atmosphere in which what teen-agers think, feel, and do is considered to be especially important. Should the adolescent feel grateful for all of this attention? That is quite another question. Not everybody would be comfortable with the idea of being selected by society or fate to carry forward the course of evolution, nor would those excluded from the elite cadre necessarily appreciate their second-class status. There is more to be said about Hall's theory and its serious flaws. Now, however, it is time to expand our horizons with an alternative view.

THE ADAPTIVE MIND (PIAGET)

Some of Jean Piaget's contributions to developmental psychology have been described in previous chapters. In this section we will try to see the adolescent through his eyes. This must be a partial and selective portrayal, for Piaget has elaborated a broad-ranging, complex, and detailed account of the first two decades of life. Piaget and his commentators are a library unto themselves.

For Piaget, the adolescent does not suddenly spring into prominence as for Hall. Mental activity, in its most inclusive sense, can be observed very early in life. Even the youngest infant is establishing his or her own form of knowing relationships with the world. Long before IQ tests can be given, it is possible to see the distinctive marks of human intelligence. Piaget, of course, does not find much of value in psychometric techniques anyhow. He and his students concentrate on careful observations in naturalistic and contrived situations, supplemented by questions when children have reached the age of verbal communication. The boy or girl who has now reached adolescence already has a personal history of developmental achievement that cannot be ignored by either society or science. There is the clear sense *continuity as well as transformation.* The child's particular level of mental functioning today represents the outcome of past developments, but is also the origin of developments yet to come. Some researchers have concentrated on delineating the interrelationship between one level of thought and another within the same children.

Whether a preschooler or a teen-ager, the individual is using his or her mind as a primary resource for adaptation. Intellectual activity, then, is not to be regarded as an isolated part-function of the whole person. It is through operations of knowing that the person experiences, behaves, and survives. For a balanced view of development, we would have to appreciate both the persistent role of mental activity in total adaptation to life and the significant changes in the level of mental activity as the person moves from infancy through adolescence.

Piaget's theory and research indicate that the adolescent takes possession of the highest level of mental activity yet recognized by cognitive developmentalists. He or she enters the realm of pure or abstract thought, capable of far-ranging and flexible mental operations. The adolescent can still be practical. He or she retains the intellectual know-how that came into being with the stage of *concrete operations* (*see* Chapter 10). It was this level of devel-

There are fourteen-year-old Davids on both sides of this chessboard, each well-equipped with formal thought operations that help them appreciate better the intricacies of life and intellectual games. (Photo by R. Warren Johnson.)

opment that enabled him or her at an earlier age to comprehend rules of various sorts and to understand how things work. Much of school learning depends on understanding rules, whether of mathematics, grammar, science, or interpersonal relations. The ten year old seems especially suited for coping with opportunities and challenges that are embedded in concrete experiences. It is the adolescent, however, who takes the final step up. In what is known as the stage of *formal operations,* the individual can think about the possible as well as the actual. Thought itself becomes an object of thought. This transformation in *cognitive structure* is one of the keys to the emergence of new personality characteristics in adolescence. Piaget and his colleagues acknowledge the significance of the feeling-and-emotions side of human development, but their emphasis is placed on the distinctive cognitive structures that are necessary for certain kinds of emotions, motivations, and actions to emerge.

David Elkind (1970), a leading American protégé of Piaget, has illustrated some of the relationships between cognitive development and new ways of experiencing life in adolescence. He took as his starting point the adolescent's new ability to think about thought:

> For the first time, the adolescent can take himself as an object, evaluate himself from the perspective of other people with respect to personality, intelligence and appearance. The adolescent's self-consciousness . . . is simply a manifestation of this new capacity for introspection. Now that the adolescent can . . . look at himself from the outside he becomes concerned about the reactions of others to himself. Many adolescents undertake a regime of physical or intellectual exercise because in examining themselves they find a discrepancy

> between what they are and what they wish to be, between the real and the ideal self. For the child, this discrepancy is seldom conscious but in adolesence, the capacity to introspect and examine the self from the standpoint of others brings it home in full force. (p. 78).

As can be seen, the emergence of new cognitive structures has the potential for bringing new doubts and concerns. Elkind has cited the example of a child with a physical handicap, such as a deformed arm, who does not become depressed about his condition until he reaches the self-consciousness of adolescence. We see in this instance an illustration of one of the most important implications of the Piagetian approach. The adolescent may behave emotionally. But we miss much of the point if it is only the emotional aspect of the behavior that registers upon us. Significant new intellectual powers are also at work.

Take another example of the relationship between cognitive structure and behavior in adolescence. Parents have become accustomed to the child who did what he or she was told (usually), and accepted their word as final. Now they notice a change that is not entirely welcome. The child-unto-adolescent is "talking back." Parental opinions are being questioned and challenged. The adolescent may not only question parental directives, but may also go out of his or her way to find out what the parents think and believe on a variety of topics. This altered relationship is accepted better by some parents than others. One parent might appreciate the opportunity to discuss serious matters with an offspring who is now ready to explore them. But another parent might resent the questioning either as a challenge to authority, or as an opening-up of topics that the parent feels uneasy about.

Some accounts of the adolescent's relationship to his or her parents emphasize the emotional and interpersonal dynamics associated with the apparent challenges to authority. The adolescent is seen as a rebel, whether this be interpreted as a favorable or an unfavorable role. But from the cognitive development viewpoint, the emphasis is on the new mental processes that seem to make talking back not only possible but perhaps even essential.

The adolescent's new mental processes include a heightened ability to envision the future. He or she becomes aware both of the problems and opportunities that might lie ahead. Unlike the adult who has developed a relatively settled view of futurity, the adolescent has many possibilities to consider and sort out. These include practical decisions. Should I go to college? If so, where? And with what objectives? Should I marry? If so, what kind of a person? And now, or when I am a lot older? Philosophical questions also require attention. What is the meaning of life, if I'm really going to die one of these days? Do I want the kind of life my parents are having, or some other kind . . . what really is the best way to live? The future is becoming real to the adolescent in a way that was unknown to the child.

Closely associated with the adolescent's future-scanning abilities is his or her new gift of possibility construction. He or she can imagine ideal conditions that contrast sharply with the world that meets his or her eyes and ears everyday. He or she can imagine things being precisely opposite to the current situation or exaggerated to the point of fantasy or parody. And he or she can ask himself or herself, "what if . . ." and proceed with a vigorous spin-off of possible consequences.

A mind that is becoming liberated in this manner has considerable potential for creativity and perspective. But some of the insights, imaginings, and perspectives can be alarming to the adolescent and to society as well. Both the pleasures and the perils of liberated thought are freshly experienced as the adolescent begins to exploit his or her new powers on the stage of formal mental operations. These related abilities to scan the future and to construct visions of the world as it is not, or as it might be, contribute much to the adolescent's distinctive experiences.

A more advanced ability to manipulate symbols also becomes part of the adolescent's repertoire. The ten year old could use certain symbols adequately. He or she knew, for example, that the symbol, 10¢, could be translated into a dime, a palpable object that one can feel in the palm of the hand. But the adolescent goes well beyond that level. In Elkind's words:

> The capacity to symbolize symbols makes the adolescent's thought much more flexible than that of the child. Words carry much more meaning because they now can take on double meanings, they can mean both things and other symbols. It is for this reason that children seldom understand metaphor, double entendre, and cartoons. It also explains why adolescents are able to produce many more concepts to verbal stimuli than are children. (1970, p. 76).

Combinatorial logic is the ability to analyze logically and organize components several ways

Still another new ability is the adolescent's improved facility with **combinatorial logic.** The ten year old may be able to reason his or her way through straightforward logical problems. But the adolescent can take complicated problems apart and put them back together again—often in several different ways. He or she is more likely than the child to see a situation in its full complexity, instead of trying to base his or her response or solution on just one or two elements of the situation. Since much of "real life" requires coping with many factors interacting at the same time, the adolescent will find ample opportunity to apply this new mental versatility.

In trying out these formidable new intellectual abilities, the adolescent finds himself or herself with many more decisions to make. He or she can see further ahead and appreciate how things might be *if* this, or *if* that. Some decisions can be made, unmade, and remade many times over; others tend to be more binding. Tuned in to decision-making, then, the adolescent naturally takes on a new relationship to parents and peers. Even if he or she wanted to, the adolescent would now find it difficult to accept decisions made by external authority without question. He or she has the ability and the need to know "why?" The adolescent is noticing for the first time that there are other possible ways of handling situations than the particular way presented to him or her by authority. The adolescent is not a born rebel, however, who is out

"If I dress as others dress and do as others do, then maybe I'll be on the right track." In adolescence, the peer group becomes an important support for the individual. (Photo by David Kelley.)

to challenge or undermine adult authority because of deep instinctual or dynamic reasons. Instead, the adolescent is a person challenged for the first time by the scope and complexity of life. He or she is not likely to trust parental decisions and views as completely as a child, but neither is he or she completely sold on the alternatives he or she has been coming up with.

Elkind has suggested that the adolescent is often in a quandary about decision-making. He or she is having trouble making his or her own decisons, but does not want anybody else to take over. This is seen as leading into a heightened dependency on peers. He or she will do what other teen-agers do and thereby patch over the underlying challenge of arriving at his or her own solutions. Interestingly, the adolescent may also become even more dependent on his or her parents; the teen-ager needs some kind of clear standard from them against which to test strength and opinions.

If pressed for a specific answer to the question of linkage between cognitive structure and adolescent experience, some Piagetians would emphasize the significance of "the capacity to construct ideals and to reason about contrary-to-fact propositions." Elkind has pointed to the adolescent's tendency to construct idealized individuals, families, and societies. The adolescent "crush" is usually short-lived because the real other person has been replaced by an idealized mental image. In general, the ideal is compared with the existing realities, with the realities coming off a distant second. This leads to a tendency to disparage the adult world. Grown-ups have "sold out" or failed to measure up to the high standards of the adolescent's idealism. In a critical vein, Elkind declared:

> These ideals . . . are almost entirely intellectual and the young person has little conception of how they might be made into realities and even less interest in working toward their fulfillment. The very same adolescent who professes his concern for the poor spends his money on clothes and records, not on charity. The very fact that ideals can be conceived, he believes, means that they can be effortlessly realized without any sacrifice on his part. (1970, p. 79).

Elkind's criticism of some adolescent attitudes, however, does not detract from the generally high respect the teen-ager receives from the Piagetian approach. The adolescent can *think!* He or she is no longer dominated by the appearance of things. Appreciation for the reach and agility of the adolescent mind comes readily to the student who has followed the course of mental development closely from infancy onward. Looking at the teen-ager through Piaget's eyes, then, we see a person who has new powers of cognitive adaptation with which to establish a new equilibrium with the world.

Gaining increased respect for the adolescent as a thinker would help parents, teachers, and society in general to comprehend the experiences and behaviors typical of this age. Yet Piaget has not seized on the adolescent as the chief protagonist in a drama of human perfectibility. Unlike Hall, Piaget emphasizes educational and social inputs earlier in childhood. It is true that intellect flowers in adolescence, but wise cultivation of mental growth throughout childhood is the basic advice he would offer.

THE IDENTITY SEEKER (ERIKSON)

Who is the adolescent? That is just the question the adolescent is so involved in wondering about! Erik Erikson's interpretation of adolescence emphasizes the quest for personal identity. This quest should itself be interpreted within the context of Erikson's overall view of human development through the lifespan.

Erikson sees the adolescent as a person who has already passed through a series of significant developmental stages, each of which was marked by a central challenge, crisis, or "task" (*see* the discussion of Erikson's "Eight Ages of Man" in Chapter 11). The earliest, for example, centered around the infant's attempt to develop a sense of *basic trust.* Success at this stage made it easier for the individual to confront the next developmental challenge. Failure at an earlier level makes it more difficult to cope with the challenges that are typical of the next level. This means that some people come to adolescence with a winning record, while others are still struggling with developmental problems that represent earlier life stages. These dynamics will continue beyond adolescence as well. The person who masters the challenges of adolescence will be ready to take on the new responsibilities and opportunities of adulthood.

We might compare two people as they stand on the threshold of adolescence. One infant experienced the world as good, giving, dependable, supportive. This sense of goodness became internalized. The outcome was a sense of *basic trust* in self as well as world. As a two year old, this same child had the confidence to develop a sense of *autonomy.* Increasing control was gained over both mind and body through a nontraumatic achievement of muscular-anal discipline, and successful experiences in early attempts to explore the environment and make some decisions on his own. Able to love and cooperate, the child could move easily to the third stage of psychosexual development. At the conclusion of this stage, there was a delightful sense of *initiative.* The child felt good about acting on the world and exploring his own potentialities. Entering the school system also meant entering the fourth stage of development. The new demands of learning and peer competition were met and were, in fact, enjoyed. "I can do it!" the child felt, having established a sense of *industry* or competence.

The other person had a more difficult time of it right from the beginning. Experiences of frustration and deprivation during infancy led to a sense of *basic mistrust.* One couldn't count on things going right; neither the world nor one's own self were really dependable. The new challenges of meeting parental demands, gaining bowel and bladder control, and exploring the environment proved stressful for this child who lacked a basic foundation in self-confidence. Even at age two, this person was already working hard to make up for a lack of security. Despite his efforts, stage two concluded with a sense of *shame and doubt.* This orientation, in turn, increased his difficulties in establishing an adequate love-and-identity relationship with the parents. *Guilt* became a dominant sense. It was wrong, somehow, to act spontaneously on the world; this might threaten whatever precarious security had been attained. Preoccupied with inner problems and not confident in his own resources, this child entered school at a disadvantage. Although the new surroundings and opportunities might have been welcome, there was also the expectation of failure. Before long, a sense of *inferiority* had taken over. "I can't make it here . . . either!"

Is it likely that these two people will experience adolescence in the same way? Differences in the developmental careers up to this point strongly suggest that they will have markedly different orientations toward the new challenges and opportunities that lie before them. In fact, I might go so far as to say that only the first-mentioned person is truly entering adolescence. The other person remains burdened with adaptive concerns that date back to earlier developmental situations. However, the outside world will at first see them as essentially the same, just two more teen-agers. Stereotypes about teen-agers will be applied to both, and both will be expected to perform and conform.

Most of us carry unresolved problems from childhood, but the burden is heavier for some

When problems do become apparent, it is quite possible that "adolescence" will be blamed for them. In other words, our society is more apt to recognize psychosocial problems at adolescence than at earlier points in the lifespan and to assume that the problems are new and peculiar to adolescence. Actually, the problems might indeed involve new dimensions. But some of the negative attention given to teen-agers probably reflects limited recognition of problems that many children *bring with them* into adolescence. If we have not noticed these problems before, then this says something about our own perceptions and values.

Even before examining Erikson's interpretation of adolescence per se, we can see that his lifespan developmental perspective can be instructive. We are in a better position to appreciate whatever may really be distinctive about adolescence when we understand the preceding developmental stages and the various ways in which people traverse them.

Distinctive about the adolescent phase of life is the need to establish an encompassing sense of identity. Who I was as a child and who I will be as an adult are spheres of identity that must be integrated with who I am now. The child's sense of identity was first nurtured within the intimate network of family relationships. "I am the good child who does what mommy and daddy want me to do," or "I am the one who the cat likes best." Through the school years new components are added. The child begins to define himself or herself in relationships with other children and in relationships with a broader spectrum of adults and activities. "I am the second-best jumper in my class," or "I am the one who gets invited to play when the kid they want to play with them can't play."

The adolescent has significant new components to add to his or her identity picture, but also more of an urge to establish a clear and integrated self-image. The identity integration task is complex because, as mentioned, it now takes into account past, present, and future selves. Furthermore, the task must be pursued while the individual is in the midst of many new developments, including physiological transformations (*see* Chapter 12). At this developmental stage there is once again the possibility for either a successful or an unsuccessful outcome. Every young person has the task of identity integration, but some falter and leave adolescence with a sense of **identity confusion.**

In **identity confusion,** the person has not integrated components of the self into one, coherent identity

Self-reflection is one of the major themes of the adolescent's life, and appropriately so. (Photo by Talbot Lovering.)

Every adolescent must also work toward identity integration within a particular cultural context. Our own culture, in Erikson's view, tends to make the task more difficult. "As technological advances put more and more time between early school life and the young person's final access to specialized work, the stage of adolescing becomes an even more marked and conscious period and, as it has always been in some cultures in some periods, almost a way of life between childhood and adulthood" (Erikson, 1968, p. 128).

It is typical for adolescents to attempt a kind of group solution to the identity problem. An adolescent subculture is likely to be established. This subculture changes in its specific characteristics as its membership and the cultural atmosphere change. However, the purpose of the subculture remains constant: to convey the impression that a final, solid identity has been achieved. Actually, it is only a temporary, try-out sort of identity that the group has generated, decorated with the fads of the day.

Erikson, then, has pointed to three types of force at work in identity formation: the individual adolescent, the peer group that is attempting to establish a new society with its own norms and characteristics, and the larger culture which exercises significant influence over the success of both individual adolescents and the entire generation. Although identity integration is a developmental task of the individual, we have to go beyond the individual himself or herself to understand identity formation in the adolescent. Identity confusion, for example, can be regarded as a phenomenon of the culture as well as of the individual. It is so common in our society today, suggests Erikson, that it should not necessarily be considered abnormal (although in some instances the state is such a severe one that the adolescent becomes disabled and dangerously vulnerable).

The relationship between individual, peer group, and the larger culture can be sharpened by introducing concepts that Erikson has been emphasizing in his more recent writings. He believes:

> Identity formation normatively has its dark and negative side, which throughout life can remain an unruly part of the total identity. Every person and every group harbors a *negative identity* as the sum of all those identifications and identity fragments which the individual had to submerge in himself as undesirable or irreconcilable or which his group has taught him to perceive as the mark of fatal "difference" in sex role or race, in class or religion. (Erikson, 1975, p. 20).

This negative identity can be unleashed either in the individual or in the society during a time or crisis. The adolescent is especially vulnerable because his or her total identity has not yet taken shape, therefore making the escape of negative identity components more difficult to control. Furthermore, the energies and passions of the youth can stir up quite a storm when they are released.

It is crucial for the adolescent that his or her culture offer a positive and coherent view of the world. There has to be something "out there" for the adolescent to integrate with his or her own thoughts, feelings, and ideals. The adolescent carefully tests out the prevailing belief systems and enthusiasms of the adult world he or she is starting to enter. Does this world look stable or shaky? Is it too confining for his or her own individuality, or is it too unstructured and chaotic? Is it really consistent and logical within itself, or riddled with contradictions and hypocrisy? In a way, the adolescent replays some of the earlier developmental themes, but on a more conscious and resourceful level. We see the tension between trust versus mistrust, initiative versus shame and doubt, etc. reappear for a time. Adolescence is thus a time when the emerging individual and the larger culture are checking each other out with unusual rigor and intensity. The stakes are high. The individual seeks fulfillment, security, and the opportunity for continued growth. Society wants to be reassured that the new generation will be motivated and equipped to maintain essential traditions, beliefs, and modes of survival. And, as Erikson has noted, the possible break-out of negative identity components in both individual and society is always lurking in the background.

A particular culture at a particular moment in its history might be offering either a relatively serene or a relatively stressful bridge for the adolescent to cross. But within any particular culture, there are also differences with respect to *which* adolescents are in the most favorable situation. In our own culture, "Adolescence . . . is least 'stormy' in that segment of youth which is gifted and well trained in the pursuit of expanding technological trends, and thus able to identify with new roles of competency and invention. . . ." (Erikson, 1968, p. 130). Occupational success and technological mastery are highly valued in our society, and so successful identity integration in adolescents is much dependent on the particular individual's efforts toward succcss in school and work. The person who carries a sense of inferiority from preadolescence is poorly equipped to establish this aspect of total identity. This means, for example, that racist and sexist discriminatory practices which make it difficult for a child to establish a firm sense of personal competence will also intensify the adolescent's difficulties in creating a positive self-image and in forming constructive links with adult society.

Erikson perceives a danger for both the individual adolescent and the larger peer group when the adult culture does not provide an acceptable and coherent framework. The need to have faith in someone, somethings, or some ideal is very strong at this time of life. Having difficulty in establishing a personal identity, the adolescent might fall prey to extremist movements and totalitarian appeals. Wearing a uniform and marching in step, chanting simplistic slogans, and drawing strength from numbers—these are among the appeals of a totalitarian alternative to the individual's quest for identity. The "group think/group do" need not be this obvious, but the same functions can be served. The cultural climate and what it offers the adolescent is definitely on Erikson's mind, and he would have us keep alert to this problem as well.

On a more intimate level, the search for a new and more encompassing identity often involves what the sociologists term heterosexual attachments;

Teen-agers discover much about themselves as they develop intimate and loving relationships with each other. Sexual awakening is only part, although an important part, of this process. (Photo by Talbot Lovering.)

most others call it "falling in love." Erikson suggests that the teen-ager's first few serious affairs of the heart have perhaps more to do with self-discovery than with full sexuality. He does not minimize the sexual awakening of early adolescence. But it is the opportunity for two people to explore their identities together that is often the most compelling theme. Teen-agers teach themselves and each other valuable lessons in who they are through intimate shared experiences. According to Erikson, ". . . many a youth would rather converse, and settle matters of mutual identification, than embrace" (Erikson, 1950, p. 228). (No, he does not provide statistics on this point!) Erikson has proposed another major concept that is relevant to identity formation in adolescence, the **psychosocial moratorium.**

The **psychosocial moratorium** is time when one does not have to make firm decisions: adolescence

> A moratorium is a period of delay granted to somebody who is not ready to meet an obligation or forced on somebody who should give himself time. By psychosocial moratorium, then, we mean a delay of adult commitments, and yet it is not only a delay. It is a period that is characterized by a selective permissiveness on the part of society and of provocative playfulness on the part of youth, and yet it also often leads to deep, if transitory, commitment on the part of youth, and ends in a more or less ceremonial confirmation of commitment on the part of society. (1968, p. 157).

This sanctioned intermediary period between childhood and adulthood makes sense to Erikson. Some of the adolescent's experiments in quest of identity may be foolish, ill-fated, or apparently antisocial. The same may even be said of some of the behaviors and temporary commitments a person engages in a few years after adolescence. Given the opportunity to survive them, the adolescent-grown-older may look back on these experiments and shake his or her own head in wonder. Society can help by not being too quick to label an adolescent and drive him or her into a permanent niche on the basis of such experiments. The adolescent can help himself or herself by taking full advantage of the psychosocial moratorium, avoiding a premature hardening of identity.

THE METROPOLITAN (OPLER)

Cultural relativism is the idea that values depend on the particular society

The **metropolitan youth culture** is adolescents serving as their own major influence

The cultural context of adolescence in general receives more attention from Erikson than from Hall and Piaget. Erikson is credited with having renovated traditional psychoanalytic accounts of human development so that more justice is done to cultural as well as individual dynamics. Yet there are some observers who believe Erikson has not gone far enough in this direction. Marvin Opler, a noted anthropologist, is one of these people. He has proposed an alternative view of adolescence that emphasizes **cultural relativism.** The nature of society is seen as having a decisive influence on what kind of person the adolescent becomes. And the kind of person that most adolescents become in our own society must be understood in terms of the **metropolitan youth culture.**

Opler's point of departure from Freud and Erikson is a fundamental one. He notes that psychoanalysts regard the individual's psychological characteristics as emerging directly from basic physiological processes. The psycho-sexual stages of development assume a rather fixed relationship between the individual's physiological status and his or her psychic needs, motives, conflicts, and capacities. Culture's influence comes in mostly after the physiological-psychological basics have appeared. Even the culture-conscious Erikson starts with the biological transformations of early adolescence.

Opler has a different idea on the subject:

> . . . we prefer to view this important physical and physiological revolution as having, in a more central sense, immediate social relationships. The psychological potential derives less from the constant of a universal physiological change than from the cultural helps or hindrances which enable or prevent the adolescent from growing into the texture of society. Thus, rather than assuming that physical events give rise to characterological ones, we say instead that cultural settings generate the entire process of what adolescents are to become, including their characterological assets or liabilities. (Opler, 1971, p. 153).

He grants that the "physical revolution" in and around pubescence is much the same in all societies. But Opler believes that the remarkable differences in adolescent character that anthropologists have observed around the world point to the need for a more central appreciation of cultural relativity. If it were mostly a matter of biology, then why would the nature of adolescence seem so different in different places? He challenges the psychoanalytically inclined: "While Erikson moves farther away from simple biological determinism, like Freud he tends to retain certain universal psychic consequences in the response to physical growth and transformation. Yet the variability about the nature of adolescence around the world does not reveal one single healthy or one single form of disturbed reaction to the relativity fixed and simple set of physiological changes" (1971, p. 153).

While Opler draws his observations from many different societies, it is his conclusions about our own culture that are most relevant here. In the United States and other technologically advanced cultures, there has been a major change in the role of adolescents and young adults. There was a time when many young people came of age within small local groups. The adolescent would naturally take on increased responsibility on the farm, in the family business, raising the younger children, etc. He or she would be growing up within a familiar, compact, face-to-face culture. Contacts with the larger society would be sporadic. Under such circumstances the adolescent would tend to think of himself or herself primarily as a member of a particular family living in and part of a particular town. Family and town would usually include people representing various ages and conditions of the total human lifespan, so the adolescent was not made keenly conscious of his or her age-related identity. He or she was one of the "young 'uns," but this was secondary to his or her basic role in family and community.

Today, however, society has changed in a way that has also changed the nature of adolescence. The teen-ager is much more likely to live in a metropolitan area. Even those who do not actually live in a metropolis tend to be influenced heavily through television and other mass media. As the individual grows up today, adjustment must be made not just to the immediate social world that has been known since childhood, but also to the large, impersonal, remote but powerful world of the mass culture. Opler believes, ". . . youth in modern societies sense that they are constantly demeaned by lock-step mass educational techniques, competitive and categorizing examinations which threaten their quest for individual identity, conditions of increasingly meaningless mass employment, and of course deferment of social sanctions for sexual and other gratifications which adults say must be postponed until the educational and employment labyrinths are conquered" (1971, p. 154).

The stress generated by mass culture strikes at many citizens, but it is particularly hard on the adolescent. Youth answers by attempting to establish its own society-within-a-society, what Opler calls the metropolitan youth culture. This means that the adolescent is now very much aware of his or her age-related identity, and it is in consort with others who are in the same predicament that he or she attempts to come up with a viable alternative. Adolescence has become clearly set aside as a special time of life. From the moment a person sees himself or herself as a member of the adolescent generation, affinities and relationships will likely turn more to other adolescents than to society in general.

Some similarity between the views of Opler and Erikson can be seen. This is hardly surprising, since both are contemporary students of human be-

The metropolitan youth culture may serve to insulate the adolescent against grinding pressures from society at large. (Photo by Tania Mychajlyshyn-D'Avignon.)

havior who live in the same society. But a different emphasis can be detected in Opler. He seems to take the metropolitan youth culture more seriously as an attempted solution to problems experienced by the adolescent. Opler calls attention to some of the realistic hazards and hardships that many adolescents experience in common. The young are apt to be very directly affected by the culture's involvement in wars, for example. Inequalities in status and opportunities that seem built into our culture stare many adolescents directly in the face. The young person, then, may have good reason to oppose the mass culture to which he or she is exposed at this influential time of life. New forms of slang, music, dance, dress, and hair style are not mere fads, in Opler's view, even though these new forms may themselves change rapidly as others move into the adolescent generation. All these fads serve as symbolic expressions of youth's need to maintain its distance from a mass culture that looms before them in a demanding and oppressive manner.

The protest of youth at times expresses itself in substantial forms. Opler points to examples of political activism, to the Freedom Riders who contributed much to the on-going civil liberties movement, to antiwar lobbying, and to the rejection of the sexual component of the psychosocial moratorium. From Opler, one comes away with the impression that adolescents are not just playing games and biding their time. The metropolitan youth culture, as he interprets it, has some effectiveness in establishing its own view of life and translating it into action. His personal sympathies seem to be mostly on the side of youth.

Opler recognizes a broad range of difficulties that the adolescent can experience. He discusses various types of psychiatric disturbance at this time of life, as well as the prevalence of suicidality. These are topics that concern Erikson as well. As we might expect by now, Opler's explanations tend to emphasize cultural factors. Agreeing with some earlier work on delinquency by Aichhorn (1964), he argues that our culture increases the adolescent's difficulties by its emphasis on competition. Our social structure separates the young from economic and educational independence, yet at the same time, it demands achievement from them. And our social structure also does little to help the adolescent bridge the distance between his or her present situation and full adult status. The youth is often in a situation of **role discontinuity:** he or she can't go on being a child, but he or she is not yet welcome to step into adequate new roles. "We . . . constantly stress in our culture 'what is beyond his ability' for the adolescent, whereas in other cultures . . . work and social involvement are not seen by youth to be a special and separate precinct beyond their reach" (Opler, 1971, p. 169). Preliterate societies seem to do a better job than ours of integrating adolescents. If we seek to improve on this count, suggests Opler, perhaps we should turn for guidance to the metropolitan youth culture itself.

In **role discontinuity,** the person has left one major social involvement, but has not yet firmly settled into another

INTERPRETING THE INTERPRETATIONS

Four views of adolescence have been presented. Now let's run these through again, very briefly, and see how they stack up with or against each other.

Hall was inclined to see the adolescent as superhuman material. Although the individual passes through a period of severe storm and stress at adolescence, he or she is also more susceptible than at any other time in life to cultural influence. Associated with this concept is the belief that individual development recapitulates the development of the species, that is, the adolescent is reenacting a transitional phase of evolution. This makes him or her the best candidate to advance the species. Society should select the elite of each adolescent generation and offer them top incentive and opportunity to develop both themselves and the entire human race beyond society's existing level.

Piaget emphasizes the adolescent as a thinker. Since infancy, the individual has been adapting to the environment through his or her available cognitive structures. In adolescence, the individual finally comes into possession of flexible, far-ranging thought processes; this is known as the stage of formal operations. An appreciation of the adolescent's cognitive needs and abilities helps to place his or her emotional and interpersonal functioning into perspective. While Piaget would encourage better communication between adolescent and society, he sees childhood as the more critical area for cultivation of mental growth.

Erikson interprets the adolescent as a person in quest of a new and more encompassing identity. Experiencing the dramatic biological changes of pubescence and moving into new realms of activity and opportunity, the adolescent is no longer the child he or she used to be, but is also not yet an adult. The challenge of establishing a revised identity (along with its risk of identity confusion) is the latest in a continuing series of developmental tasks that began in the earliest days of infancy. The psychosocial moratorium, the time of adolescence, is a useful way to cushion the young person's transition and provide latitude for the identity search.

Opler argues that other views have neglected the fundamental power of social structure to influence adolescent development. Although pubescent changes are universal, the nature of adolescence differs from society to society. In advanced technological societies such as our own, adolescents are now forming themselves into metropolitan youth cultures. These age-conscious, age-segregated echelons represent the adolescent's attempt to defend against the frustrating and oppressive presence of mass culture. The problem and the response are both to be taken seriously.

These four interpretations do not have equal status in the field of human development today. Hall's theory, by far the oldest, is also the most severely criticized. His concept of the individual recapitulating species' development (at least in the ambitious way he uses it) has been on the reject pile for years. Hall's relative disregard for the earliest years of life has also proven inadequate in the face of many subsequent studies and clinical observations. Furthermore, there is something about Hall's advocacy of evolutionary progression and elitism that has not won favor with many subsequent developmentalists (although here we are perhaps more in the realm of attitudes and values than straightforward factual questions).

Theories come and go in popularity. Hall may yet have another day in the sun

By contrast, the work of Piaget and Erikson is highly visible today. These men are among the most respected and influential contributors to the understanding of human behavior. The wisdom of retrospection is not yet available to place their contributions into clear perspective. Erikson's views of adolescence are probably more disseminated than those of any other current thinker; they are hashed about in the media as well as among scientists, teachers, and clinicians. Piaget's work is tied more closely to an extensive and continuing series of experiments and close-knit analyses and classifications. This tends to guarantee a smaller if no less fascinated readership and following. It takes more effort to develop the implications of Piaget's contributions for the understanding of adolescence per se, although there is little doubt that significant implications exist.

Opler's interpretation of adolescence represents only one facet of his overall approach to the relationship between the individual and society. He seems to be a leading advocate, in our generation, of the outside-in tradition (*see* Chapter 7), combining expertise in both modern anthropology and psychiatry. Response to Opler's theory depends somewhat upon the individual's general position with respect to the role of inner and outer forces in shaping human development. Many of his specific contributions to knowledge have been well received, but developmentalists do not seem to have come around yet to evaluating his total approach to adolescence.

All four theorists agree that adolescence is a very special time of life. The physiological transformations are acknowledged as introducing new dimensions into the developmental sequence. All see adolescence as a period of heightened intellectual activity. The most detailed and data-based exploration of intellectual activity in adolescence is Piaget's. But years earlier, Hall had also singled out the mind of the adolescent as a major cultural resource. Erikson's discussion of identity formation assumes a high level of cognitive structure and versatility, as does Opler's description of the critical and imaginative thinking sometimes embodied in the metropolitan youth culture.

There is also general agreement that the interface between the individual and society is particularly complex and eventful at adolescence. Much can go wrong. All of the theorists, in their own ways, suggest that the teenager tends to be idealistic, impressionable, and ready to make strong if not necessarily binding commitments. Depending on characteristics of the adolescent himself or herself and what the culture has to offer at the moment, the energies and talents of the young can either be channeled into fruitful pursuits, or thwarted, wasted, and misdirected. None of the theorists seems to believe that the cultures in which they themselves live serve as the best possible model; all would have our society in the United States modify its relationship to the adolescent generation.

What about Hall's storm-and-stress characterization of the adolescent experience? The consensus seems to be that this phenomenon can still be observed, but that it does not inevitably accompany puberty. Opler indicates that "S & S" is peculiar to some cultures at some periods, and not to others. Erikson also acknowledges cultural differences that either induce or alleviate stormy disturbances in adolescence. Furthermore, in Erikson's lifespan theory, individual differences *prior* to adolescence can have much to do with the emotional climate of the teen years. Disturbances at puberty are not denied by contemporary theorists, then, but are considered to be conditional rather than universal.

I have the impression that Hall's observations linger even more pungently in current thinking than what might seem to be the case. Here, for example, Erikson describes the to-and-fro motion of adolescent dynamics around the crisis of identity, the ambivalence toward intimacy, conformity, authority, etc. And there is Piaget or Elkind speaking of the unstable idealism that see-saws into cruelty and cynicism. Standing behind both of them, gravely nodding his beard in approval, is the spirit of G. Stanley Hall who had voiced these same ideas earlier.

Some of the apparent contradictions in adolescent experience and behavior that were mentioned at the start of this chapter drift back to mind: daydreaming, but awakening; conforming but revolutionary; stressed, but still insulated from some of the grinding pressures of adult life. These are part-truths. The hasty student or theoretician might attempt to develop a complete view of adolescence by selecting only those parts that go together easily. The more systematic thinker—and the adolescent himself or herself—will devote the time and effort to integrate these part-truths into a more encompassing and authentic whole. The adolescent has the biological (Tanner, *see* Chapter 12) equipment, and the cognitive structures (Piaget) to create an adaptive new identity (Erikson) that a wise society (Hall, Opler) might encourage and welcome for the good of all.

SUMMARY

Many stereotypes about adolescence are prevalent in our society. Each of these seems to embody a part-truth. For a more systematic conception of the nature of adolescence on the psychosocial level, I consulted four major theoreticians.

Early in this century, *G. Stanley Hall* proposed a view within the tradition of *evolutionary idealism*. He believed development of the individual from infancy onward recapitulates the development of the species. Adolescence is a period of special significance because it represents the transition from the rigid simplicities of tribal existence to the unlimited possibilities of advanced civilization. The adolescent is open to significant new learning, and culture should seize upon this opportunity. The *elite* of each generation of youth can become the "next step up" in human evolution, taking society along with them. Although chiefly concerned with the evolutionary future of the human race, Hall also described the inner life of the individual adolescent in vigorous detail. His concept that the teen years are wracked by *storm and stress* has remained influential even when other aspects of his theory fell into discredit.

Jean Piaget offers a detailed account of mental development from infancy through adolescence. The individual always adapts to the environment through his or her available *cognitive structures.* Adolescence is special because this is the time of life when the person finally comes into the intellectual estate that previous stages could only anticipate. The adolescent enjoys the ability to employ *formal operations* of thought. This brings a new depth, reach, and flexibility; the possible as well as the actual can be conceived. Many of the emotional and interpersonal characteristics of the adolescent can be understood more adequately if their basis in the new mental powers is appreciated.

Erik Erikson zeroes in on the adolescent's *quest for identity.* The individual experiences himself or herself as being some place between the who-I-used-to-be and the who-I-am-going-to-be. Creation of an encompassing and unifying self-image takes on high priority; this is the leading *developmental task* or *normative crisis* at this period of life. Failure can produce *identity confusion.* Some individuals come into adolescence well prepared by virtue of their successful negotiations of previous stages; others remain haunted and distracted by developmental problems that were not adequately resolved. Erikson favors the full utilization of the *psychological moratorium* so the adolescent has time to make useful personal experiments without being pressed to make definitive commitments.

YOUR TURN

1. Education is one of the major expense items for local governments in the United States. It is probable that in your own city education is the single largest item of expenditure. Suppose that you were called upon for expert advice in the distribution of educational funds between grade school children and adolescents. There is only so much money to go around. You know that the psychoanalytic approach to development and much of developmental psychology in general lays emphasis on the significance of early experiences—the earlier the experience, the more its significance. But now you have been exposed to the evolutionary idealism of G. S. Hall, as well as to other views that emphasize the critical importance of adolescence. What kind of advice would you offer on the distribution of the community's educational funds between younger children and adolescents on the basis of these differing views? Prepare an official consultant's report to the Board of Education and justify your recommendations as closely as you can. And, now that you are in the position of recommending how everybody's money should be spent, do you feel the need for any additional type of information on human development before, during, or after adolescence in order to have confidence in your position? What kinds of information would be helpful in making practical decisions that affect the amount and type of resources we, as a society, make available to younger children and to adolescents?

2. David Elkind has been quoted in this chapter to the effect that the idealism of adolescents is "almost entirely intellectual" and is often accompanied by "little conception of how they might be made into realities and even less interest in working toward their fulfillment." This is a view that was around in previous generations as well, before Elkind, Piaget, and other current researchers. Do you think this is, on balance, a fair statement of adolescent idealism? Critically review the Elkind-Piaget position here in light of your own observations of idealism in adolescents and in adults as well. See if you can keep personal bias under control in evaluating this proposition.

Marvin Opler represents a viewpoint that emphasizes the role of culture in shaping the nature of adolescence. Development does not proceed directly from the physiological to the psychological realm. The total act of growing up takes place within a cultural context that influences the individual powerfully throughout the lifespan, and especially at adolescence. An important example of the impact of *social structure* on adolescent development can be seen in the United States and other countries where advanced technology has resulted in a markedly revised life-style. Teen-agers now tend to participate in what might be called a *metropolitan youth culture.* This age-conscious peer society should be taken seriously as an alternative to the frequently oppressive influence of the *mass culture* on the adolescent.

Differences among these theoretical positions should not be ignored. However, some significant points of agreement can also be noted. The most general area of agreement is that adolescence is, indeed, a critical phase of human development in which the individual's emerging personal resources enter into a new relationship with the established culture.

Reference List

Aichhorn, A. *Delinquency and child guidance.* New York: International Universities Press, 1964.

Elkind, D. *Children and adolescents: Interpretive essays on Jean Piaget.* New York: Oxford University Press, 1970.

Erikson, E. H. *Childhood and society.* New York: W. W. Norton & Company, Inc., 1950.

Erikson, E. H. *Identity: Youth and crisis.* New York: W. W. Norton & Company, Inc., 1968.

Erikson, E. H. *Life history and the historical moment.* New York: W. W. Norton & Company, Inc., 1975.

Hall, G. S. *Adolesence.* New York: Appleton, 1904.

Opler, M. K. Adolescence in cross-cultural perspective. In J. G. Howells (Ed.), *Modern perspectives in adolescent psychiatry.* New York: Brunner/Mazel, Inc., 1971, pp. 152–179.

INTIMATE THOUGHTS AND RELATIONSHIPS IN ADOLESCENCE

14

CHAPTER OUTLINE

I
Adolescents speak for themselves about their more intimate thoughts and relationships.

"Normal" boy / Ego ideal / Continuity of values / Dating / Anger / Drug use / Political activities / Continuous-growth, surgent-growth, and tumultuous-growth patterns / Career goals / Trustworthiness

II
Most adolescents are seeking relationships in which affection, sex, and mutual respect are integrated.

Mutuality of love / Situational ethic / Serial monogamy / Egocentrism / Decentering

III
The adolescent is old enough to think about death more systematically and resourcefully.

Suicide

Adolescence has been considered from the standpoint of the biological transformations involved (Chapter 12) and from the standpoint of several of the major psychosocial theories (Chapter 13). Although each of these theories draws on a base of actual observations, we have not yet delved seriously into the extensive research literature that has become available over the years.

How extensive is the research literature on adolescence? Thumb through most journals and textbooks concerned with human behavior. Who is the favorite "subject" of research? If the adolescent is not at the very top of the list, he or she is certainly close. An excellent laboratory species is this *Homo teenagerius:* well equipped to fill out questionnaires, follow instructions, and participate in either simple or complex experimental operations. Best of all, perhaps, the adolescent is so available! Why seek more exotic species, when all the researcher has to do is lock the classroom door and distribute his or her scales, inventories, and etceteras? More than one psychologist has observed that psychology seems to be a science of lab-bred rodents and undergraduates, rather than a thorough-going science of human behavior.

Much of the research that involves adolescents is opportunistic in the sense that these are the people most readily available. This means that when we pursue almost any theme or problem in psychology we are likely to find that young men and women have been tapped to provide much of the data. But when researchers concentrate on adolescents because they are available, it is not the same thing as concentrating on them because the researcher is especially interested in them. While much psychosocial research happens to involve adolescents, then, not all such studies contribute directly to our understanding of thought and behavior at this time of life.

In this chapter we will examine a few selected studies that have in common a strong interest in illuminating the nature of adolescence. These studies also tend to share one broad methodological principle, although the specific techniques vary, let the adolescent speak for himself or herself! Some of the material that follows here bears closely upon the theories that have already been presented; other material perhaps suggests areas of concern that theory-builders have slighted. All the observations center around the adolescent's more intimate thoughts and relationships. The adolescent, after all, is much more than the object of somebody else's theory; he or she is a real person capable of love and terror, fulfillment or frustration.

MOVING THROUGH ADOLESCENCE WITH THE "NORMAL" BOY

Pick out a "normal" boy, get to know him, and stay in contact as he moves along through adolescence. This sensible research strategy was employed by

Daniel and Judith Baskin Offer (1969, 1975) in a study useful enough to deserve a detailed summary here. Their method provided the opportunity for the teen-agers to disclose their life views and experiences in a natural manner as the experiences themselves unfolded.

The participants were seventy-three boys who represented one kind of mainstream developmental pattern in the United States: middle-class, midwestern, suburban. The time frame is a recent one. Preliminary data were gathered in 1962, and the interviews started the next year. This means that we cannot look to this particular study for data about girls, low income or very high income adolescents, or those belonging to ethnic subcultures. Individuals who were obviously disturbed or deviant on the verge of adolescence were also excluded. Additionally, it would be prudent on our part to limit conclusions to adolescent development as it occurred during one span of time in our culture. These are significant limitations. But as long as we are clear about what any particular study can and cannot contribute to knowledge, we are in a reasonable position to derive valuable information from it. And, in this case, what has been made available for our consideration is a careful, follow-along tour through adolescence with boys who were about as "typical" as any one population sample can be in this variegated nation of ours.

Offer and Offer reported their findings most extensively in two books. *The Psychological World of the Teen-ager* (1969) traced the beginnings of the study and followed the participants through four years of high school. *From Teenage to Young Manhood* (1975) followed most of these same people to age twenty-two. Offer and Offer were able to continue their contacts with sixty-one of the original participants. The fourteen year olds who were recruited for this study came from two high schools, after a larger population of relatively problem-free youths had been identified at those institutions. A questionnaire administered to students at both schools and information from schools and parents were used to increase the chances that all participants in the study would be "typical," "average," "modal," or "normal" (the authors used these terms interchangeably). Offer and Offer were mainly interested in securing a population of young adolescents who would be quite different from the troubled young person that psychiatrists most often see in offices, clinics, and institutions. They were aware that it is tempting for mental health people to generalize about adolescents (or any other subgroup) on the basis of the unhappy or crisis-ridden people who come to their attention on a clinical basis. This study was designed in part to correct a view of adolescence that would emerge from problem cases only.

Each participant provided some information about himself through a self-image questionnaire devised by the authors. The adolescent was then interviewed periodically for a total of nine individual sessions throughout the high school years. The semistructured interviews were considered to be the most important avenues to understanding the adolescent's experiences, but were supplemented by self-ratings, psychological testing, and information from parents and teachers.

THE HIGH SCHOOL YEARS

The **ego ideal** is a person another has taken as a model

School and the people met at school become significant in the adolescent's network of thoughts and experiences. The teacher, for example, has the potential for becoming what psychoanalysts sometimes call the **ego ideal.** The boys might have found qualities in their teachers that they would have liked to bring out in themselves as the grew up. However, this seldom happened. The adolescents rarely spoke about their teachers with any enthusiasm. Offer developed the impression that neither the teachers nor the students felt comfortable when a relationship looked as though it might become closer than the stereotyped roles established for each. Attitudes toward the teachers were generally negative. Few considered their instructors to be exemplary either as teachers or as individuals to be emulated. Frequent contact was not enough to bring the teacher into the adolescent's world of intimacy.

Offer noticed, however, that the boys did seem to adjust to the classroom and total school situation even though they felt a sense of constriction and overcontrol. The school environment seemed tense for both teacher and student. The students usually scurried from class to class and stood in line for almost everything, forced into a stereotyped mode of functioning. "Although the school does appear to stultify some of the student's individuality, it does not affect him as adversely as it might; he has learned to live with it" (Offer, 1969, p. 45). Interestingly, the athletic coaches were those faculty members most often acknowledged as being interested in the students as individuals, and as going out of their way to do something intended as helpful. But it was also the coach who was likely to coerce and influence the boys in ways that were less than helpful, for example, bullying the fellow who did not want to go out for the team, or making locker-room speeches against interracial dating (all but two of the participants in this study were white).

What were their experiences at home? Almost all of the boys were part of intact families and reared by their natural parents, both of whom lived at home with them. The boys and their parents gave remarkably similar accounts of their relationships with each other. Essentially, the teen-agers *shared* their parents' values. They wanted the same kind of things out of life. "The goals they were striving toward were very appropriate in the sense that they fitted very nicely into the cultural milieu in which they were living, and were almost always in accordance with the parents' expectations" (Offer, 1969, pp. 60–61). Middle-class values seemed to be alive and well for another generation!

The finding that teen-agers and their parents had core values in common deviates from much of the talk that is heard about the "generation gap" and the protest and rebellion of youth. But one doesn't have to be completely disappointed on this score. Both the boys and their parents reported some intergenerational friction despite their shared values. The pattern that emerged from these reports had familiar elements along with the unfamiliar. There were impulses and behaviors that could be interpreted as rebellious, but the peak time for teen-parent conflict usually occurred *before* the high school years. The "age of protest" tended to come around ages twelve and thirteen. Some disagreements lingered past that time, and new bones of contention did arise later (e.g., access to the family car around age sixteen). Nevertheless, the boys and their parents agreed that interpersonal friction had been greater in the very early teens than during the high school years. The boys remembered that they seemed to be getting into scraps with their parents over little things. The twelve or thirteen year old would refuse to take out the garbage or make his bed, or perhaps get home a little later than his parents expected. Chances were that he affected a hair style and preferred garb that made his parents wince and grumble as well.

Parent and teen could agree, then, on basic life values and still annoy each other considerably. It was at about this time in the boys' developmental careers that conformity in clothes and customs with the peer group became prominent. The boys did start to look, talk, and behave more or less alike, especially in group situations. Signs of rebellious behavior at home, and conforming behavior with age-peers seemed to go hand in hand. Delinquent activities were not unknown. The boys admitted feeling an urge to resist the authority of parents and other adults from time to time. Usually, however, they could find ways of expressing this urge without courting serious trouble. There were some indications that in their later teen years, the felt need for aggressive activity was channeled into sports by many of the boys.

Both family and school seemed able to give a little when the adolescent boy felt the need to assert himself through minor protests and rule-breakings. Adults might look the other way, for example, when a group of boys met secretly to smoke cigarettes. Offer noted that this seemed to illustrate Erikson's concept of a psychosocial moratorium: the adolescent can "play around" with various behaviors that challenge existing rules, as long as this remains within limits of tolerance. Not all adults encountered abided by this moratorium, however, and from time to time a boy found himself in trouble even though he was rebelling in the same way as many of his peers.

Adult affection for the teen-agers seemed to remain strong despite occasional tribulations. The parents characterized their sons as "good," "kind," "enthusiastic," "responsible," "conscientious," etc. Offer believed the parents were praising qualities in their sons that they also valued in themselves; this was another illustration of the *continuity* of values between generations. Most parents observed that the conditions for coming of age had changed quite a bit since their own youth. Often they felt that the current generation of teen-agers had more intense pressures on them, but also more opportunities. The liberties and hazards associated with the availability of automobiles were frequently commented upon. In general, the parents saw their sons as being essentially the same as teen-agers in the past, although having a different kind of world to encounter in some respects. Within this population of adolescent boys, then, the age of protest did not seem to overturn a relationship of affection and respect with the parents.

"He's basically a good kid; after all, he takes after his father."

It is a fairly safe bet that one topic frequently on the mind of the adolescent male is the adolescent female. Offer acknowledged, however, that obtaining reliable information about the sexual thoughts, feelings, and behaviors at this time of life is difficult. He attempted to learn gradually about the participants' heterosexual behavior as a natural aspect of his general interest in how the teen-agers experienced life. "Our data seem to indicate that it is possible to be interested in the adolescents as functional human beings and include sexuality as part of their functioning" (Offer, 1969, p. 80).

The general impression from this study was that the typical adolescent boy approached relationships with the opposite sex rather slowly. During the first year or two of high school, many of the boys did not think that dating was especially important. They felt they did not understand girls very well and tended to feel uncomfortable with them. Often they communicated anxiety about being pushed somehow (perhaps by their parents) into dating situations that they were not yet prepared to manage. The interviewers noticed that conversations about girls frequently shifted to relationships with the mother, whether favorable or unfavorable. Most of the boys had only occasional dates during their first two years in high school and felt that was quite enough. There was little sense of pressure from other boys to date more intensively.

Relationships with other boys of their age seem to be more vital to young adolescents than romantic pairings-off with girls. (Photo by Jean Boughton.)

Attitudes toward girls had changed appreciably by the senior year. Most of the boys were dating by this time—and, what's more, enjoying it! Even those who had not yet entered the arena admitted they were interested, but were still looking for the courage to advance. Asking a girl for a date was an anxiety-provoking experience at first. The boys had to work through this anxiety, motivated by their heightened interest and curiosity. Often enough, the first few dates came in large part from the desire to have an experience one could share with friends who were also starting to go out with girls. According to Offer:

> They shared their experiences with their peers almost immediately after they brought the girl home. The minute dissection that goes on among the boys telling each other what they did, right or wrong, is extremely helpful. . . . They try to do better next time, not so much because they enjoy kissing or petting, but so they can tell their boy friends. As their anxiety diminishes in the relationships with girls, they begin to enjoy the encounter more, and eventually can look forward to a date simply because they like the girl, and want to share their experiences with her, and her alone. (Offer, 1969, p. 82).

There was little evidence to support contentions that sexual activities are rampant among high school students. As late as the junior year, for example, only about 30 percent of the teen-agers in this study could be classified as active daters (going steady or having more than an occasional date). And only about half of the boys within this limited group reported experiencing heavy petting. Looking at the total group, approximately 90 percent had *not* experienced sexual intercourse by the end of the junior year. To the extent that the Offer data are dependable on this point, then, it would appear that the modal adolescent male was not a bold sexual adventurer. And, as Offer noted, his data agreed with results of another study conducted at about the same time, indicating that sexual behavior was relatively conservative for high school males (Reiss, 1961).

Despite what their limited experience might suggest, the teen-agers did not usually think "going all the way" was wrong. Instead, they were inclined to feel that it was better to wait a few years. "Too much sexual closeness was frightening. 'We just are not ready for it' was repeated over and over. Statements about a certain maturity level being necessary for intercourse are replacing fear of pregnancy responses as the latter lose their realistic value" (Offer, 1969, p. 84).

Daydreaming about girls was popular with most of the boys. Interestingly, these fantasies usually revolved around people they did not know personally, or with whom an actual involvement was quite unlikely.

Taking an overview of the findings on adolescent development during the high school years, Offer made mostly positive observations. The teen-agers seemed to take a realistic view of themselves and the world. They were careful, as in the realm of heterosexual behavior, not to move hastily beyond their own zones of comfort and experience, but to advance gradually. It was more difficult for most of the boys to control their aggressive and, occasionally, hostile impulses than their sexual feelings. However, most of them realized what they had to contend with and usually found channels of expression that would not disrupt significant relationships or invite serious consequences.

> Direct angry outbursts were one . . . way of ventilating aggressive impulses, but these were either kept within socially controlled limits or indulged in briefly and then later smoothed over. The anger was most often directed toward siblings who formed handy targets. For some students, aggression could be sublimated into school work, but we saw few who attacked a political theory or a mathematical problem with fervor. The family car was a likelier recipient of this extra-enthusiastic attention. (Offer, 1969, p. 213).

In general, the teen-age boys were optimistic about their futures, yet very much involved in their immediate life situations and with each other. They were active, gregarious youth, not much given to solitary pursuits. They seemed to enjoy other people and to keep their own lives in perspective. Often their senses of humor came to the rescue. The self-awareness that other observers have noticed in adolescence was also seen by Offer, usually as an adaptational strength.

Nevertheless, many of the boys went through periods of feeling lonely and isolated. During these "blue" periods, the individual was inclined to question his existence in general, the purpose of life. According to Offer, these periods usually did not persist for any significant length of time. (We will look at this kind of experience from another perspective later in this chapter.) When the teen felt bad, he was likely to acknowledge his distress. This open communication with himself had the advantage of making it unnecessary to invest energy and efforts in elaborate psychological defenses. However, anxiety tended to show up in physiological as well as psychological symptoms.

Before moving on to the later years of adolescence, it might be worth remarking on one or two of the characteristics that were *not* found prominent in these boys. There was little sense of a strong commitment to social change. Flaming idealism, or any deep concern with ideological issues, was seldom encountered. Offer suggested that the relatively affluent and stable backgrounds enjoyed by these boys might have been a factor. However, political scientists know that leadership for idealistic movements often comes from individuals who have enjoyed the advantages of wealth and social status. Whatever the reason might be, it is interesting to see how the typical brand of idealism found in this study tends to confirm the observations of Piaget and Elkind (*see* Chapter 13). Idealism usually did not come out of its own accord. Offer reached for it by asking questions, such as what would they do with a million dollars. Almost all the boys indicated they would use the money for personal purposes: save it, spend it, invest it. One could hear the voice of the middle class speaking through them; they had no burning quest to improve or reform the world. Similarly, the fact that most of them did have opinions on major issues of the time (e.g., the war in Vietnam and civil rights) showed no carry-over in the form of intentions to do anything about these problems themselves.

THE LATER YEARS OF ADOLESCENCE

Most of the teen-agers who entered the study during their freshman years in high school continued to participate until age twenty-two (61 of 73). Some interesting changes took place during that time, including a subtle shift in the relationship to both parents. Previously, most of the participants had seemed to be closer to their mothers. The father was dependable; he was all right in his own way. But mother was the one who really understood them. After high school, however, the sons seemed to feel more warmth for their fathers. It was as though the teen-agers had begun to experience themselves as more mature and therefore more like adult males such as their fathers. At the same time, the relationships with their mothers had a way of becoming more argumentative. The sons seemed to be trying to shift some emotional investment from their mothers to their girl friends. From their perspective, the parents were still inclined to be satisfied with the manner in which the sons were developing. They saw them as doing well in their major spheres of activity and took pleasure from the adolescent's own sense of satisfaction.

By this time, about three-fourths of the teen-agers were living away from home. Most were attending college, just as they had planned; 10 percent were working full-time; and another 10 percent served in the armed forces (several of the students were also working full- or part-time). Career choices included business, law, medicine, counseling, the physical and social sciences, and military service.

Young men have something of a reputation for getting into trouble. How had these people fared? Although drug abuse was receiving much national attention during the time period the study was conducted in, none of the participants had become an addict. About one in four had tried marijuana and continued to use it occasionally. There were a few experiences with LSD and amphetamines. But there were no indications that drug abuse had become a problem of serious proportions. Some of the participants expressed strong disapproval of drug usage; others volunteered that problems with alcohol were more common. Drinking with friends had become a weekend recreational activity for several of the participants. The researchers, however, did not collect systematic data in this area. This was also a period of time in which a number of people were involved in social and political activities that included sit-ins and confrontations of various kinds. Although several of the participants reported having participated in legal on-campus demonstrations, none had been arrested for protest activities.

Particular girls become particularly important to the older *adolescent male. It's especially nice when the girl has the same idea! (Photo by Talbot Lovering.)*

The intervening years did not seem to produce new values that might separate the adolescents from their parents and their values. Paraphrasing Offer and Offer (1975), the generation gap was hard to find. The sons could find things to complain about in dealings with their parents: there were still hassles about long hair, and the company they kept, for example. But most of the adolescents appeared able to view themselves as independent people without having to overturn their parents' basic values.

Intimacy with young women did increase appreciably. This trend could be seen not only in the statistics of how many did what how often, but also in the respondents' expressed attitude toward the females in their lives. During the high school years, the boys rarely spoke of the personal qualities of their girl friends; often they would not even mention the name of their favorite. Now the adolescent males were ready to commit themselves more fully to particular women. In Offer and Offer's interpretation, this represented a progression beyond the use of heterosexual relationships to separate from mother. The young men had feelings for specific young women, and this new quality came across in the way they spoke about them.

An example of how, with increasing maturity, we are better able to appreciate individuality

Most of the teen-agers were dating regularly or going steady as they moved through the post-high school years, but it was not a quick-marrying class. All remained single at the end of the first post-high-school year, and only eight were definitely known to be married by age twenty-two.

Sexual activity had increased, as might have been expected. Yet even by the end of the third post-high-school year, the group was evenly divided between those who had and those who had not experienced sexual intercourse. Many of the young men cited environmental reasons to explain why they had or had not been sexually active; it had to do with whether or not they had an apartment, a car, the opportunity, etc. It was as though they did not want to take full responsibility themselves. Those who had not yet become sexually active seemed more defensive or shy about this fact than had been the case back in high school. "Perhaps, having heard about the existence of a sexual revolution that their generation was said to be enjoying, they were wondering with some degree of anxiety if they were missing it" (Offer & Offer, 1975, p. 33).

The investigators also brought to our attention the discrepancy between biological readiness and the actual participation in intensive sexual activities. Most of the adolescent males in this study had been biologically ready for sex for several years before having their first experiences. There was no reason to believe that this delay had any negative effects on the general developmental process. If anything, holding off intensive sexual involvement seemed to provide time to work through other problems, with control of aggressive impulses being one of the more prominent.

Friendship with other males remained important during the post-high-school years. Many of the close friendships established during high school were replaced by new relationships as patterns of interest changed. Most often, boys became buddies in high school on the basis of similar interests in sports. Later friendships were built more on participation in the same work or intellectual activities and personality characteristics that were mutually suited.

Offer and Offer made a special effort to explore the boys' "identity crises" during their post-high-school years, including the use of a self-descriptive scale based on Erikson's theory (Hess, Henry, & Sims, 1968). There were signs that an identify-refining process was still at work, but no indication that many of the young men were experiencing a crisis in this area. The participants seemed to have maintained a fairly stable view of who they were from the first post-high-school year to age twenty-two. Between the ages of nineteen and twenty-two, the group as a whole increased its total identity score, which suggests a continued movement toward a strong, encompassing sense of identity.

No "identity crisis," "rampant sexuality," or "generation gap"

DIFFERENT PATTERNS OF DEVELOPMENT

Let's now follow Offer and Offer's description of the group as a whole to their focus on different patterns of development as shown by different adolescents. Contemporary research seems to increasingly highlight the *variety* of ways in which people can and do develop. It is difficult to know how much of this shift should be attributed to the technical nature of the research being pursued, and how much should be attributed to a new orientation on the part of social and behavioral scientists, a willingness to accept and value individual variations. In any event, we will continue to find examples of emphasis on varied patterns of development as we go through the lifespan.

Offer and Offer extracted from their data three types of growth pattern. Although these patterns were described in connection with the results of the follow-up study (1975), they were based on the entire eight-year journey through adolescence.

Continuous growth is psychosocial growth that is smooth, gradual, with little conflict

Continuous growth was a pattern shown by 23 percent of the total group. The adolescents displaying this pattern were usually favored by genetic and environmental circumstances. Their families had stayed together, their health was excellent, and they seem to have started out life with good coping abilities. Affection and respect between the generations was strong throughout all the years of the study. They made their parents feel good, and vice versa. To specify all the characteristics of people in this group would be merely to repeat a list of desirable or mentally healthy qualities. Of particular interest, perhaps, was the relationship between their fantasy lives and their lives of action. These adolescents seemed to enjoy the best of both worlds: "They could dream about being the best in the class academically, sexually, or athletically, although their reactions would be guided by a pragmatic and realistic appraisal of their own abilities and of external circumstances. Thus, they were prevented from meeting with repeated disappointments" (Offer & Offer, 1975, p. 41).

The continuous-growth adolescents were described as relatively happy human beings. Such people, whose developmental careers proceed so smoothly, do not often come to the attention of counselors and psychotherapists; as a matter of fact, in this group, none did. Offer and Offer noted that this is one of the reasons why theories of adolescent development tend to be built on experiences with people who are experiencing some obvious difficulties.

Surgent growth is psychosocial growth in fits and starts

Surgent growth was a pattern shown by 35 percent of the total group. These individuals functioned well much of the time. In their backgrounds, however, were incidents of trauma and loss (e.g., deaths in the family, long-term illness, separations). The adolescents themselves would go along well for awhile, achieving and integrating new skills and experiences. But then they would enter a period of being stuck in their tracks, or even regressing. They tended to grow, then, in fits and starts, with stretches of discomfort and lack of progress in between.

One of the big differences between the surgent developer and the continuous developer showed up when something unexpected and unwelcome intruded into life. The surgent adolescent could usually cope with what Hartmann has labeled, the "average expectable environment" (Hartmann, 1958). But should a crisis arise, the individual's anxiety was likely to interfere with adaptive response. His usual style of reacting to his own anxiety was to tighten up, control, keep everything under wraps. This response was often accompanied or followed by depression. If the stress continued, the youth might explode in anger or blame others for his difficulties.

The surgents tended to be short on self-confidence. They relied much on the good opinions and support of other people. While they could form and maintain close interpersonal relationships, this seemed to require more effort from them than it did from the continuous-growth adolescents. Vocational goals were kept in mind, but were pursued less systematically and with less enthusiasm. Some of the surgents also seemed to fear their own emerging sexual impulses. Additionally, as a group these teen-agers were not as reflective or introspective as the other participants in the study.

It bears repeating, however, that individuals with the surgent pattern of growth often did show an overall adjustment that was comparable with the "smoothies" of the first group. "The adjustment was achieved, though, with less self-examination and a more controlled drive or surge toward development, with suppression of emotionality" (Offer & Offer, 1975, p. 44).

Tumultuous growth is psychosocial growth occurring around a core of conflict and distress

A third pattern, **tumultuous growth,** was shown by 21 percent of the total group. This group was approximately the same size as the continuous-growth subsample. These were the people in turmoil. More problems were evident in their genetic and environmental backgrounds than with the other groups (e.g., strife between their parents, history of mental illness in the family). Life was difficult in many ways for them; they had troubled relationships with their parents, frequent self-doubts, and a greater likelihood of performing below par in school or at work. From the outside, these were the adolescents who seemed unpredictable and the source of many problems to others. But from the inside, these were people who frequently experienced pain and insecurity.

Seeing this tumultuous growth pattern in perspective is not easy. One is tempted to emphasize the problems and the turmoil. Yet, in Offer and Offer's study, the turmoil actually seemed to be a significant part of the growth process. These adolescents were still struggling to move ahead. Often they were sensitive people. They had more open access to their emotions than the surgents, although the latter seemed to have a relatively easier time in making it through adolescence. The "storm and stress" mood swing was more evident here than in the other groups. But this also meant that they could experience both intense pleasure and intense despair, as compared with the surgent's policy of keeping emotions under tight control.

Although they could not be described as generally happy with themselves, the adolescents in this group did show the capacity to move ahead, and to have their share of academic and vocational success. Not surprisingly, perhaps, there were more individuals with definite psychiatric symptoms in this group than in any of the others.

Offer and Offer reflected on the differences they had observed. Was it better for a person to be contented with himself? If so, then, the continuous-growth subgroup would seem to have the most advantageous developmental pattern. But what if deep maturity and sensitivity could only be achieved through coming to grips with conflict? In that case, the tumultuous-growth subgroup might ultimately have the more advantageous pattern. Questions such as these help us to sort out values from facts. Three patterns of adolescent development can be discerned from the data patiently collected over the eight-year period. But there is nothing directly in the data (so far as I can tell) that clearly establishes that one pattern is somehow "superior" to the others. Throughout our total exploration of lifespan development, there often will arise the temptation to select a "best" pattern, one that appeals most to us as individuals or for some particular purpose. And, as individuals, perhaps we are entitled to give in to these temptations. But are we as scientists and respectors of fact?

ADOLESCENT GIRLS IN SELF-PORTRAIT

There is no single research effort entirely comparable to the work of Offer and Offer on the development of the girl into the young woman through adolescence. One recent contribution, however, is a rich descriptive sampling from the views of 920 girls ranging in age from twelve to eighteen. Gisela Konopka, who had previously studied *The Adolescent Girl in Conflict* (1966), assisted by an all-female interviewing staff, tape-recorded in-depth discussions with a wide variety of adolescent girls in eleven states. A special effort was made to insure the inclusion of girls from all the racial and ethnic groups residing in a particular area. In addition to the interviews, Konopka hosted group discussions among some of the girls, read their diaries, poetry, and stories, and administered a questionnaire to 600 rural teen-agers based on preliminary analysis of the interview data. The findings and interpretations offered in *Young Girls* (1976) admittedly reflect both the results of the study as such and Konopka's accumulated experience in working with adolescents over the years. The study was also cross-sectional, so it is not possible to follow any particular respondent through the years of adolescence on an observe-as-she-develops basis. Despite flaws that can be found with the study as "pure" research, however, it provides a useful self-portrait of the adolescent girl who seems to feel at ease in expressing herself.

LIFE GOALS

I want to go it alone—yet be helped—but not too much! This was the typical orientation toward the future. The older girls had more crystallized plans, but all were actively thinking ahead and welcomed the chance to share their outlooks. Konopka underlined: "*Practically all girls contemplated some job or career after high school and accepted the fact that they should be prepared for some kind of gainful employment.* Only ten years ago, marriage was frequently regarded as an escape from work. This concept rarely surfaced in our interviews" (Konopka, 1976, p. 15).

Although there seems to have been a shift in life goals within recent years for teen-age girls, there was still the question of how to combine work with marriage and having children, as well as the question of choosing among an ever-increasing range of work possibilities. This latter question was perhaps more important than for most girls in previous generations because employment was now felt to be a truly significant part of their lives. Most had traditional careers in mind, at least at this time in their lives. Nevertheless, among the girls there was a broad range of career possibilities envisioned.

It is no longer unusual for an adolescent girl to be thinking of a career as well as marriage; nor does she feel she has to exclude one if she chooses the other. (Courtesy of ACTION.)

The majority of the girls wanted to be married, but did not want this to be the only important aspect of their lives. They all wanted to have "done something" before settling into marriage, and none expressed the wish to wed while still in high school. A person should not settle down too early, they felt, there was so much in the world to learn and experience first. The wish to be married was expressed by almost all the girls, whatever their geographic, economic, racial, or ethnic background. Marriage was thought of as an equal partnership.

While marriage was an important life goal, they were well aware that unhappiness and divorce sometimes are outcomes, and those who were themselves children of divorce had particular concern about this. The fear of becoming a beaten and abused wife was expressed by some of the girls.

Marriage also meant children for most of the girls. But those who did not think children would fit into their life plans were frank to discuss this outlook. Often this was also associated with the feeling that the world was too miserable a place to bring a baby into. There was no inclination to have children purposely born out of wedlock in protest of degrading marriage laws—a phenomenon that Konopka had observed when she was an adolescent herself in the 1920s. Most of the girls had in mind a family of two children, which is smaller than those mentioned in a study a decade earlier (Lee, 1971). But the number of children was not as important to them as the *integration* of work, marriage, and motherhood. They expressed the desire to know more about child development so they could be good parents.

Alert guidance can help girls take advantage of expanding career opportunities

The girls as a group did not seem to have had much exposure to the full range of career opportunities available to them, nor to guidance and counseling. Choice of careers was influenced mostly by life experiences and by the adults they knew.

SEXUALITY

The respondents were not asked specifically about their own sex lives, but most made spontaneous comments and also discussed their general thoughts and feelings about love and sex. Sex was spoken of very openly. This did not mean, however, that the girls were participants in any kind of "sexual revolution," or that they advocated casual and promiscuous sex.

Most of the girls did not think of premarital sex as sinful, whether or not they wanted it for themselves. Premarital sex was acceptable to most of the girls if there was love, understanding, and commitment on the part of both people. This was clearly distinguished from casual sex which was considered demeaning.

Sex experiences occurring early in life—before age fourteen—were often reported to be disagreeable, and they included many incidents of force or coercion. Sex experiences later in adolescence were not frightening or considered unusual. There was, in general, tolerance for people with different sexual patterns than oneself. They did not look down on girls who had illegitimate pregnancies, for example, but were aware of all the problems that situation could present.

Konopka was impressed by the inadequate information and learning experiences that had been made available to many of the girls. Often they had had their first menses without preparation and were subject to doubts, fears, and myths. While sex information was more available at the time of the study (1976) than in past years, Konopka felt it may have been given too late to be of the most value, and too often it came from friends who did not necessarily know any more than the girl herself.

RELATIONSHIPS

Many of the girls yearned for a relationship with an adult who would really listen to them and take their own thoughts and feelings seriously. Similar qualities were sought in friends. The most prized quality in another person was trustworthiness. Whether the other person was adult or peer, it was the same kind of relationship that most girls wanted: sharing, trusting, confiding.

They drew an important distinction between *friends* and *close friends.* Most of the very closest friends were of their same age, and most of the friendships had started through contacts made in school. Church groups, teen centers, and neighborhoods seldom brought friends-to-be together.

Most girls had both female and male friends. The boys were regarded "as *friends,* as *people.* Girls often talked about the same activities with boyfriends as with girlfriends and in a very relaxed manner. Sex may have been or may not have been part of the relationship, but it was not the exclusive goal" (Konopka, 1976, p. 89).

SOCIOPOLITICAL CONCERNS

The adolescent girls, in general, conveyed a sense of alienation and powerlessness with respect to most larger sociopolitical concerns. They "rarely showed an awareness of belonging to something larger than themselves" (1976, p. 94). The sense of disenchantment with the whole political enterprise may have been influenced heavily by Watergate events and may have reflected a response typical of many people throughout the country.

When they did think of large-scale issues, sympathy with the oppressed and the suffering was often expressed. They did not have many solutions in mind, but took the kind of idealistic view that has long been associated with adolescence (*see* Chapter 13), and tended to be simplistic in their analyses and "cures." They were acutely aware of the many prejudices that exist in our society, having seen many examples themselves. While some expressed harsh prejudices against other groups of people, there was an overall desire to overcome discrimination. Often they felt that they had to struggle against entrenched patterns of adult prejudice while trying to establish their own discrimination-free modes of life.

But how many *adults* are *not* simplistic in their analyses and cures?

Although their opinions of the women's liberation movement were solicited, the girls did not have much to say about this. Few felt that they wanted to be part of it, or that it had much relevance to their own lives. Nevertheless, it was evident to the researchers that the major concerns and themes of the women's movement were very much in their minds, even though the movement itself had not reached them explicitly. They were concerned about the *issues* (e.g., equal wages for equal work), but were not cause-oriented.

In general, the adolescent girls expressed a strong individualistic orientation toward social and political issues. They had their perceptions of social injustice and their desires (for universal peace and the elimination of prejudice, among top priorities), but not much sense of how to translate ideals into reality.

THE MATURE AND THE LESS MATURE ADOLESCENT

Both the major studies we have considered in the chapter included adolescents who were at various levels of maturity for their age. A closer examination of psychosocial maturity in adolescence will improve our understanding of individual experience and behavior during these years.

THE CONCEPT OF PSYCHOSOCIAL MATURITY

The term itself communicates readily, but a more specific definition is useful. Josselson, Greenberger, and McConochie (1977a & b) think of psychosocial maturity as having three basic dimensions: *individual adequacy* (the ability to function effectively on one's own), *interpersonal adequacy* (the ability to interact satisfactorily with others), and *social adequacy* (the ability to contribute to social cohesion).

Individual adequacy includes self-reliance as displayed in a sense of control and initiative and in the absence of an excessive need for social validation. It also includes being skillful in work, establishing reasonable standards of competence, and enjoying work. The adolescent with a clear self-concept, who thinks about his or her life goals, and who has good self-esteem and an internalized set of values is psychosocially mature in this sphere.

Interpersonal adequacy includes communication skills, such as empathy, and knowledge of the major roles one has to play. The psychosocially mature adolescent knows what is expected of him or her in key situations and can manage circumstances in which he or she is caught between the conflicting expectations of more than one role. Interpersonal adequacy also includes what Josselson and her colleagues speak of as "enlightened trust." This means that the person can let himself or herself depend on others within reasonable limits and has an overall view of human nature that is not too simplistic. People are not seen as all good or all bad. One does not trust everybody to the fullest extent, nor does one turn away from trusting relationships as impossible.

Social adequacy includes a sense of social commitment. The adolescent feels himself or herself to be part of the larger community, connected to people outside the immediate interpersonal circle. Furthermore, there is a willingness to modify personal goals in favor of social goals, to form alliances, and to pursue social goals that will take many years to achieve. The psychosocially mature adolescent is also open to sociopolitical change and tolerant of individual and cultural differences (including an awareness of the costs and benefits of tolerance and the practical willingness to interact with people who differ from the norm).

STUDYING PSYCHOSOCIAL MATURITY: HOW MORE AND LESS MATURE ADOLESCENTS DIFFER

Josselson, Greenberger, and McConochie administered a Psychosocial Maturity Inventory (PSM) to 192 eleventh-graders at a relatively small public secondary school. The males and females who scored as the most and the least mature (twelve boys and twelve girls in each group) were interviewed in depth by experienced clinicians. A "developmental-**phenomenological** portrait" was written for each of these participants based on interview transcriptions. It was thought that this method best preserved the richness and individuality of each adolescent. The portraits of relatively mature and immature adolescents could be seen against the background of PSM responses

Phenomenological refers to one's private, inner experience

for the total group. Results for boys (1977a) and girls (1977b) were reported separately, but we will consider them together.

All the adolescents had some important things in common. Sports, cars, and girls were of interest to most of the males whether high or low on the PSM; girls at both levels enjoyed close friendships, held down part-time jobs, and were interested in boys. But the differences were also impressive.

The high-maturity boys were more varied among themselves than their low-maturity peers. They were hard to put into neat little categories. The high-maturity males were "not as stuck on hypermasculine pursuits or as interested in swinging with the crowd" (Josselson, Greenberger & McConochie, 1977a, p. 41). Instead, each had his own pattern of diversification. One of the most obvious characteristics of these boys was their future orientation, their concern with who they would be. It could not be said that they were overly ambitious in the sense of seeking unrealistically high achievement. Rather, they were much tuned-in to the future and addressed considerable thought, feeling, and active preparation in that direction. Although enjoying friendships, they did not rely as heavily as low-maturity males on peer approval for a sense of self-worth. They showed more self-confidence and more ability to look at themselves realistically, including flaws and problems.

There was the overall impression that the high-maturity boys lived in a bigger, more complex, and more differentiated world. They seemed to find a lot more to life than the low-maturity boys. They were more individualistic and more relaxed about being themselves rather than conforming to pressures from the group. Friendships were more intimate. The researchers believed that girls remained "fairly mysterious" for the high-maturity boys who were curious and a bit romantic, but who were not deeply invested in emotional relationships at that time. The low-maturity boys seemed to have much concern about the control and expression of their impulses. This was of less concern to the high-maturity boys. They were described as mostly "straight," but not "square" or overly inhibited. They had their share of impulses, but felt it was to their advantage to keep themselves out of dangerous situations.

Perceptive and self-aware, the high-maturity boys also showed greater empathy toward their parents. They could see mother and father as individual people in their own right. Furthermore, they could also take the parents' perspective at times and see the world as it looked from the adults' standpoint. This was something the low-maturity boys never did, at least not within the context of the study. The parents of the high-maturity males, in turn, seemed to have encouraged their sons to make their own choices in life. They were expected to follow some general rules and guidelines, but were not forced into many specific situations and commitments. The boys had autonomy within limits. In general, the high-maturity boys seemed to have a better life balance than those of lower maturity.

The high-maturity girls were exceptionally articulate in expressing themselves. Their interview data were richest of all and showed the least self-consciousness. While the low-maturity girls were chiefly interested in material

Caring about the environment and taking on a share of community responsibility is one sign of psychosocial maturation. (The Boston Herald American. *Photo by Mike Andrews.)*

things and good times, these adolescents took life and themselves more seriously (which is not to say that they did not also enjoy themselves). They differed from high-maturity boys in that they did not draw as much self-esteem from occupational goals. Instead of being quite so future oriented, they were focused more on discovering who they were, right at that moment, as individuals. As the investigators reported, the high-maturity girls often began their statements with "I am the kind of person who . . ."

Like the high-maturity boys, these girls showed much individuality and found satisfaction as much in their own emerging interests as in group-oriented activities. Most of them dated somewhat older boys, an association that helped to open up their awareness of a world beyond family and the high-school scene. Yet, intimate girlfriends were still of greater importance than boyfriends to many of the high-maturity girls.

Peer pressure did not seem to be as compelling for the high-maturity girls as for the low-maturity girls. They were able to see people more objectively, as individuals in their own right, just as the high-maturity boys did. The low-maturity girls tended to have egocentric views of others, seeing them only in relationship to themselves.

The investigators were careful to point out that high psychosocial maturity was not identical to perfect mental health. Some of the high-maturity people were struggling with problems that might be with them for many years. Many of the high-maturity girls talked about having to work through conflicts inside themselves to reach their present state, and about having other conflicts still to resolve. There was the impression that many of the high-maturity boys had not yet come into a full encounter with needs and impulses.

Studies such as these make it clear that adolescents, even those at the same age-grade level, differ appreciably in level of psychosocial maturation. We might well expect them to perform differentially on such measures as the Kohlberg Moral Dilemmas Test and on perspective-taking and egocentricity tasks. Individual differences related to psychosocial maturity at adolescence also prepare us to realize that similar differences can be found throughout the adult age-range as well. Any profile of *the* adolescent which is so generalized as to overlook maturational differences within this age-range is probably also so unrealistic that it would be more harmful than useful.

Important maturational differences can be found over the entire adult age range

THE CONTINUITY OF VALUE ORIENTATIONS BETWEEN PARENTS AND CHILD IN ADOLESCENCE

One of the adolescent's most intimate relationships, of course, is the one that exists from infancy. We have already come upon some impressions of parent-child relationships in adolescence. Let's consider a few studies that looked at this topic more explicitly.

The continuity of value orientations between parents and their teen-age sons was one of the interesting findings by Offer and Offer. Troll and her colleagues (1969) took a different approach but came up with a similar result. Fifty college students who had been selected for their high degree of political involvement were compared with another fifty who had no such involvement. These students were carefully matched on a variety of characteristics (e.g., ethnic and religious background, age, neighborhood) so that political activity was the major difference between the subgroups. Men and women were included in roughly equal proportions. This study focused on the degree of resemblance between the college students and their parents on a variety of personality traits and values. Of particular interest to us was Troll and her colleague's conclusion that the salient values of the students were also the salient values of their parents, whether or not the students were activists. This suggests that the youth who "makes waves" is not necessarily rebelling from parental values but may, in fact, be actualizing them.

A related study was conducted by Kalish and Johnson (1972) who examined similarities and differences in values held by three generations of women. Their participants were fifty-three women with an average age of about twenty, and the mothers and maternal grandmothers of each young woman. There was an overall trend for all the family members to hold similar values, although the strength of this similarity varied with the generations compared and the particular values involved.

Perhaps more than the Troll et al. and Offer and Offer studies, the Kalish and Johnson study indicates the importance of specifying what particular value is in question when discussing similarities and differences. All three studies seem to support an earlier observation by Kenneth Kenniston (1967) that even though so-called militant youth may be acting on different values

Most adolescents share the basic values held by their parents and remain close despite times of discord. (Photo by George E. Finch, III.)

than those of the older generation taken as a whole, they are probably representing the basic values of their own parents. This impression holds up for nonmilitant youth as well. One of the best ways to predict what a college-age individual holds dear to the heart is to know the values held by his or her parents.

Rebellion and aggression were themes that Offer and Offer heard with some frequency from their participants. Yet these reports did not quite fit the stereotype of adolescence as a period dominated by forceful attack on the establishment. Indeed, much of the data suggested a sharing of the prevalent values. Additionally, there were no indications of widespread identity crisis or rejection by their parents.

These results are consistent with those reported by Albert Bandura and R. H. Walters in *Adolescent Aggression* (1959) and related writings. Bandura and Walters also worked with middle-class adolescent males and also found that much of the parent-child friction had usually been worked through by the start of the high-school years. The boys in Bandura and Walters's study showed increased maturity in adolescence, and their parents respected them for it. Instead of becoming uptight and imposing more controls, the parents showed even greater trust in their sons, allowing them to make more decisions for themselves. The parents had expected their children to become more independent around adolescence, and this was just what they did! Increased activity and closer ties with other adolescents did not alienate them from their parents, either. In general, then, Bandura and Walters interpreted their results as challenging the idea that adolescence is a time of great emotional turmoil.

Bandura and Walters did, however, study a subgroup of boys who were described as "hyperaggressive" and "antisocial." These individuals seemed to fit well with the storm-and-stress stereotypes. But Bandura and Walters added an important observation here: the "defiant oppositional pattern of behavior" had not suddenly bubbled up in adolescence; it had been there for many years previously. The difference was that at adolescence, the boys had grown big enough and strong enough to shake off the earlier parental controls and raise a sizeable fuss.

David Friesen (1972) investigated the value orientations of one thousand adolescents in two large Canadian cities. Both sexes were adequately represented in this sample. Additionally, three levels of economic status were included, and within each of these, three different ethno-cultural backgrounds were represented (the latter point was not specified by the investigator in his published report). Friesen's major findings were consistent with what Offer and Offer and others had had to say:

> Most of the value indicators were directly related to the ethno-cultural background of the students involved in the study. The current popular line of reasoning that youth culture is separate and distinct from the parent culture gains very little support from these data. The current popular suggestion that communication between the generations is breaking down, and that the generation gap is all-pervasive is not supported strongly by these findings. The fact that youth's aspirations, activities, and attitudes are related both to socioeconomic background and, independently of this, to ethno-cultural background, suggests that forces in society, other than the youth culture, continue to share significantly in the development of value structures in modern youth. (Friesen, 1972, p. 271).

Continuity of values between parent and adolescent seems to be a *general* finding, then, not limited to white, middle-class males. And it is a finding that reminds us that plausible and fascinating theories (e.g., *see* those of Hall and Erikson in Chapter 13) are no substitute for systematic empirical research.

SEX STEREOTYPES AND COMPETITIVE BEHAVIOR

Perhaps it is not surprising that studies have tended to focus on aggressive behaviors in adolescent males. In our society, tradition has it that the male takes the initiative, cultivates major ambitions, and acts vigorously on the environment. This view is now being challenged in many quarters, yet it has been with us for a long time and might be expected to exert an influence.

Matina S. Horner (1972) has been exploring the relationship between cultural stereotypes of masculine/feminine and the woman's quest for achievement. She has noted that previous researchers (e.g., Broverman, Vogel, Broverman, Clarkson, & Rosenkrantz, 1970) have found young men and women still think of themselves in terms of these stereotypes. A person who competes, who achieves intellectually, and who takes on independence and leadership qualities is being "masculine," whether that person is male or female. Horner believes that young women often experience a conflict between being feminine and being successful. This can mean that the adolescent female may actually shy away from being "too good" in certain spheres (a shying-away process that might have started even earlier but which becomes more pronounced at this time). She has framed a set of hypotheses about what kind of person would be most likely to *avoid success*, and under what conditions. Such a person would usually be a woman, and especially a high-ability, high-achievement oriented woman, as distinguished from women in general, who finds herself in competitive situations.

Horner's studies indicate that many highly competent young women have, in college, experienced conflicts between their self-images as feminine and their pursuits of achievement.

> Among women, the anticipation of success especially against a male competitor poses a threat to the sense of femininity and self-esteem and serves as a potential basis for becoming socially rejected—in other words, the anticipation of success is anxiety provoking and as such inhibits otherwise positive achievement-directed motivation and behavior. In order to feel or appear more feminine, women, especially those high in fear of success, disguise their abilities and withdraw from the mainstream of thought, activism, and achievement in our society. (Horner, 1972, p. 173).

As Horner also noted, this is a high price to pay for reduction of anxiety; both the individual and society lose what the young woman might have been.

Sex differences in aggressive behavior have been illustrated in just one realm, the striving for competitive success. But our society appears to expect and approve of very little in the way of aggressive behavior from women in any realm. "Boys will be boys" typifies the lenient attitude toward rule-breaking escapades by young males; there is no such expression or attitude available to condone similar behavior by young females.

The "tomboy" syndrome may be close, but it doesn't usually carry into adolescence

From the observations of Horner and some others, it appears that a significant psychological conflict is "normal" (in the sense of being common) for adolescent females. This may be conflict over self-image: can a woman act vigorously and pursue achievement while still feeling herself to be feminine? Does this kind of finding confirm theories that maintain adolescence is "normally" a time of turmoil? Or is it more reasonable to think that when boys and girls move into adolescence, they just become more vulnerable to problems that exist throughout society? Studies to date have not yet provided answers to these questions. But if women in general were not penalized for being active, aggressive, and competitive, then adolescent females probably would be spared at least some turmoil.

It may be difficult for the individual to judge whether certain thoughts, feelings, and experiences are "normal" when they are being encountered for the first time. Sexual fantasy and behavior provide important examples. Although some data on sexual development have already been reported here, there is more to be said on the subject.

LOVING INTIMACY IN ADOLESCENCE

Every adolescent generation is accompanied by its own sound track. My own memory dredges up a droopy pop tune entitled, "They Tried to Tell Us We're Too Young." What a difference from the Rolling Stones's invitation, "Let's Spend the Night Together!" Expression of sexual interest has become more direct in recent years. Perhaps these two songs symbolize some of the changes. The older of the two is a whining sort of protest against the establishment, coupled with talk of "falling in love"; the more recent makes explicit requests or demands for sex. Yet it is doubtful that human nature has changed this much in a few years. The "in" teen-ager might in one generation sing of *love* and mean sex while, in another generation, flourish *sex* while actually longing for affectionate intimacy. Some fantasies and some actual relationships may approximate either "raw sex" or a "platonic friendship." In the long view, however, I believe most adolescents are seeking relationships in which affection, sex, and mutual respect are integrated.

SOURCES OF INFORMATION

Much of the data I will be drawing on here has emerged from three recent studies. Background data on these studies will be summarized now, in order to proceed with a minimum of methodological interruptions when we consider the facts themselves.

Robert C. Sorenson, a social psychologist, conducted an intensive and systematic questionnaire study of adolescents in the United States during the early 1970s. There are several basic points about his study to keep in mind. It is up to date; it encompasses a broad age spectrum (ages thirteen to nineteen); it includes a multitude of clearly stated questions; and it makes a determined effort to represent teen-age thoughts and experiences throughout the nation. This means that male and female, white and nonwhite, city and rural, prosperous and impoverished—all these respondent and demographic conditions are represented within the study population in a manner that is reasonably consistent with their actual representation in the nation's total population. The final sample is only 411 adolescents, but it has been obtained in such a way as to stand for the approximately 27 million adolescents identified by the 1970 census. National **probability samples,** as you probably know, are used frequently to estimate such things as political preferences and consumer needs. Sorenson's study has many formal characteristics in common with such surveys, except that he asks another kind of question. The

Probability samples are small numbers of people, events, etc. selected to represent the whole

methodology as well as the findings are reported by Sorenson in extensive detail in his book, *Adolescent Sexuality in Contemporary America* (1973). Although the probability sampling approach, like any other research strategy, is not beyond criticism, this book is one of the best available resources on how adolescents in our own time and place characterize their sexual thoughts and relationships.

A study by another social psychologist, this time an Englishman in his homeland, makes a useful supplement to Sorenson's research. Michael Schofield (1971) conducted personal interviews with a representative sampling of fifteen to nineteen year olds in Great Britain. The opportunity to make further cross-national comparisons is provided by Eleanore B. Luckey and Gilbert D. Nass's (1969) study of unmarried undergraduates from universities in Canada, England, Norway, Germany, and the United States.

I will draw mostly on the Sorenson study, but with supplementation from Schofield, and Luckey and Nass.

WHAT DO LOVE AND SEX MEAN TO THE ADOLESCENT?

Each person has his or her own way of looking at love and sex. Enough of a general pattern emerged from Sorenson's study, however, to offer a portrait of the "typical" adolescent's views, so long as we realize that there is also a range of variation. Let's begin with a few interesting nonfindings.

Most adolescents do *not* consider sex to be the most important thing in their lives. Important—yes; *most* important—no! Furthermore, having sexual adventures with a variety of partners is not considered of great importance or value. One way of indicating the relative importance of sex to adolescents can be seen in Figure 14–1. The figure outlines what activities are most and least important to adolescents. As a sidelight, it is interesting to see that "getting along with my parents" is a major concern of the younger adolescents, both the males and the females. This would seem to provide further confirmation for the findings of Offer and Offer (1969, 1975) and Bandura and Walters (1959).

Most adolescents do *not* think of sexual activity as immoral. However, this attitude does not necessarily mean that the younger generation is interested in using sex to challenge tradition, to fly in the face of "the establishment." There is *no* evidence of a desire to make a revolutionary cause out of sex (*see also:* Konopka, 1976). Society might choose to feel shocked by what adolescents think and do (or by what society imagines adolescents are thinking and doing), but the *intention* to shock or challenge is usually not there.

Could society enjoy being outraged by what it imagines youths are up to?

Also missing is any substantial support for the notion that teen-agers are constantly seeking a perfect romantic love. It is *not* typical for them to wander about, looking for that one-and-only ideal mate with whom to live "hap-

Respondents	Three Items Most Often Picked as *Very* Important	Three Items Most Often Picked as *Least* Important
All boys	Having fun Becoming independent so that I can make it on my own Learning about myself	Getting loaded and hanging out Having sex with a number of different girls Trying to change the system
Boys 13–15	Preparing myself to earn a good living when I get older Having fun Getting along with my parents	Getting loaded and hanging out Having sex with a number of different girls Trying to change the system/Doing creative or artistic things
Boys 16–19	Learning about myself Becoming independent so that I can make it on my own Preparing myself to accomplish meaningful things	Getting loaded and hanging out Having sex with a number of different girls Trying to change the system
All girls	Learning about myself Having fun Preparing myself to earn a good living when I get older	Having sex with a number of different boys Getting loaded and hanging out Making out with boys
Girls 13–15	Learning about myself Preparing myself to earn a good living when I get older Getting along with my parents	Having sex with a number of different boys Getting loaded and hanging out Trying to change the system
Girls 16–19	Learning about myself Becoming independent so that I can make it on my own Preparing myself to accomplish meaningful things	Having sex with a number of different boys Getting loaded and hanging out Making out with boys

Figure 14-1. *Activities considered most and least important by adolescents. (From* Adolescent Sexuality in Contemporary America *by Robert C. Sorensen, Copyright © 1972, 1973 by Robert C. Sorensen, by permission of Harry N. Abrams, Inc.)*

pily ever after." The foreverness of a love relationship is *not* emphasized by most adolescents, nor is the desire to possess or be possessed.

Having considered some of the nonfindings, it is easier to see now what values actually are held in high esteem by the typical adolescent.

Love and sexuality are both valued, and recognized as different but interrelated. Adolescents emphasize the *mutuality* of love: both partners trying to understand each other, share joys and problems, do their part to make the relationship valuable. Sorenson found:

> . . . the intensity of love is more important to adolescents than duration and . . . possession of one another neither demonstrates love nor assures it. The ability to relate to each other is the key ingredient: one does not love another because she is intelligent or he is handsome, but because of what the two mean to each other. If people mean much to each other, they will not feel compelled to test each other by demanding commitments or promises to marry. (1973, p. 365).

The **situational ethic** is a philosophical view emphasizing the entire context, not just the act

Sex itself is seen as a natural part of life. This means that it is not "wrong" or "bad" in general, but it can be harmful or out of place in a particular relationship at a particular time. Adolescents tend to apply what is known these days as the **situational ethic** to sexual behavior. Sex may or may not be right, depending on the entire situation including, of course, the people who are involved. It is seen as wrong, for example, to force a person into sex, just as it is wrong to force a person into anything or take unfair advantage. Sex tends to be seen as "right," for example, when two people really like each other and make each other feel happy through physical contact. Basically, sexual situations are seen as very personal, and up to the people involved to work out for themselves.

Many teen-agers are seeking tangible, satisfying relationships in a society they regard as mechanistic and insensitive (on this point, then, Sorenson's findings agree with Opler's views; *see* Chapter 13). They see sex as a pleasurable means of establishing and maintaining communication, something that is directly within the power of the individuals themselves to give and receive. Sorenson became convinced through his data that adolescents value personal intimacy and sharing more than any other aspect of sexual activity (although it is "simply a lot of fun," too).

Marriage remains an accepted institution. Most of the teen-agers studied (85 percent of all the boys and 92 percent of all the girls) thought they would eventually marry. But some of the traditional goals of marriage do not seem to come across strongly. The adolescents are much more interested in what is happening in their lives now and what their prospects are in the next few years, than in the long-term companionship, security, and obligations implied by marriage. Although moving away from their own families, most do not seem ready to replace living within the family relationship with another fixed and obligatory relationship. They are not eager to commit themselves to any long-term situation, at least, not just yet.

Most adolescents agree that getting married is just a legal technicality if two people love each other and are living together. Even those who believe they will marry someday often express concern that the obligatory nature of this relationship might interfere with their love and compatibility. One of the factors that wins many adolescents over to the side of a lifelong marriage is the belief that children need two parents in the house. Although the present generation of teen-agers is well acquainted with the prevalence of divorce, or perhaps, *because* of that fact, most are resolved not to marry until they are confident that they have a relationship likely to last for the rest of their lives.

The emerging profile of love and sex values also represents a more equitable redistribution of rights, opportunities, and responsibilities between the sexes. The old double standard of sexual morality is rejected by many teenagers. Furthermore, both sexes seem more comfortable in shifting around roles and tasks instead of being fixed into the traditional stereotypes. This does not involve much tendency toward unisex, despite superficial signs to the contrary. By and large, males and females both seemed secure and satisfied with their respective sex identities, but neither felt obliged to stay within the limited compass of "pretty little girl," or "big, tough guy."

SEXUAL BEHAVIOR IN ADOLESCENCE

Focus shifts now from values and attitudes to the actual sexual behaviors reported by adolescents. As might be expected, there is a broad range of sexual activity. About one boy in five and one girl in four report no sexual experiences at all. But there are also beginners, adventurers, those who have a continuing sexual relationship with one partner, and those whose sexual experiences are few and far between. A summary is given in Figure 14–2.

It is not surprising that there is more sexual activity among the older adolescents. But this reminds us that the age span within the teen-age group is large enough to be associated with significant differences in many spheres (e.g., occupational plans, relationships with parents, sexual activities). "Adventurers" (more males than females in this classification) usually begin their sexual experiences earlier than others. About 60 percent of the adventurers have had intercourse by age fourteen, suggesting that the age of first intercourse might be a predictor of subsequent sexual activity.

The five nation study (Luckey & Nass, 1969) found that the first experience of sexual petting occurred at about the same age for all the countries surveyed. (For adolescents in the U.S.A., this was 16.3 and 17.3 average years of age for males and females respectively.) There was only a slight tendency for males to have earlier petting experiences. There was not much difference either in the average age of first intercourse for males and females in all five nations.

An interesting comparison is the period of time intervening between the first petting experience and the first intercourse. Canadian and American girls reported the shortest "delay" times. This was followed by the American teen-age male, and English adolescents of both sexes. Teen-agers of both sexes in Norway and Germany reported having waited considerably longer—well over two years between the first intensive petting experience and the first

Type of Sexual Behavior	Total	Boys	Girls	13-15	16-19	White	Nonwhite
Virgins	48%	41%	55%	63%	36%	55%	49%
Sexually inexperienced	22%	20%	25%	39%	9%	25%	23%
Beginners (virgins who have experienced sexual petting)	17%	14%	19%	12%	21%	20%	9%
Nonvirgins (intercourse one or more times)	52%	59%	45%	37%	64%	45%	51%
Serial monogamists (nonvirgins having a sexual relationship with one person)	21%	15%	28%	9%	31%	19%	14%
Sexual adventurers (nonvirgins moving from one partner to another)	15%	24%	6%	10%	18%	11%	18%
Inactive nonvirgins (nonvirgins who have not had intercourse for more than one year)	12%	13%	10%	15%	10%	11%	14%
Nonvirgins who have had intercourse during the preceding month	31%	30%	33%	15%	45%	24%	31%

Figure 14-2. *American adolescents: sexual behavior groups.* *(From* Adolescent Sexuality in Contemporary America *by Robert C. Sorensen, Copyright © 1972, 1973 by Robert C. Sorensen, by permission of Harry N. Abrams, Inc.)*

coitus. Luckey and Nass's conclusions, although subject to qualifications, help to place the sexual activities of American adolescents into a cross-cultural perspective:

> The English student has more sexual activity, begins younger, has more partners, has more one-night stands, more sadomasochistic experiences, and is more likely to have been influenced by alcohol at the time of the first coital experience. North American students are less experienced and generally more conservative; the Canadian youth is somewhat less liberal than his counterpart in the U.S. The Norwegian students tends to be less the "swinger" than the English. Premarital sexual experience does not start so early and is restricted to fewer partners. Although the picture of German students is not so clear, they generally occupy a place between the liberal English and Norwegian samples and the North American. (1969, p. 303).

Schofield's interview study of British adolescents (1971) is consistent with the other studies in emphasizing differences in sexual activity between the younger and the older teen-agers. Schofield found more differences between the sexes, however, in how fast they moved toward advanced sexual relations. At age seventeen, the males became considerably more active, but the introduction into sexual intercourse seemed more gradual for the females. Schofield noted that premarital intercourse among adolescents was not uncommon in Great Britain, but that it was by no means rampant; there were many virgins even by nineteen.

Schofield found some sex differences in response to the first intercourse. The males expressed pleasure more frequently, for example: "Rather pleased with myself. Swaggering in fact. It's only natural, isn't it?" Females were more likely to say, "Hell, what's all the fuss about!" But there were, again, many individual variations. The first sexual intercourse usually did not weaken or strain the relationship even if it did not prove as gratifying as hoped.

Serial monogamy is a pattern in which each partner has only one love at a time and new partners replace old ones

Returning to the basic American study, Sorenson gave particular attention to the **serial monogamy** patterns, that is, an adolescent couple with an intimate, sustained relationship from which either party may depart when he or she desires. He considered this, in fact, "the new sexual relationship." According to Sorenson's data, promiscuity is not being embraced by the current generation of teen-agers, nor does the commune-type of arrangement

"Just the two of us"—but maybe not forever. Many adolescents prefer an intimate attachment relationship, but may have several over a period of time. (Photo by Talbot Lovering.)

find many adherents. Instead, many adolescents find comfort and satisfaction in what might be termed a *temporary marriage.* The intimacy and sharing therein can meet the needs of both individuals at a particular time in their lives without commiting them to future obligations. Sorenson noted:

> Sexual monogamists view other adolescent sexual behavior tolerantly and with an air of slight superiority. They usually have little dissatisfaction with their own sex lives, and they know it. They usually feel no envy toward the adventurers. They feel no missionary zeal about their relationships, because they are not inviting other adolescents to join them. Their combination of personal freedom and sexual satisfaction are such that they do not believe other young people's sexual behavior interferes with their own or offers any attractions that are not available to them. (1973, p. 244).

What will be the outcome through adult life of the present trend toward serial monogamy remains to be seen.

MUTUAL RECONSTRUCTION OF REALITY: IS THIS WHAT FRIENDS ARE FOR?

Let's now bring intimate thoughts and feelings together, at least in one respect. Many studies have agreed on the significance of peer friendships in adolescence. Other studies have emphasized the adolescent's continued cognitive growth (e.g., Piaget, *see* Chapter 13). Is it more than just coincidence that the cultivation of intimate relationships with people outside the family and the development of high level cognitive abilities come at about the same time? Consider one possible interface between the two lines of development: that between social interaction and decentering.

Piaget (1928 and elsewhere) and Elkind (1967) have pointed out that the adolescent now has the ability to take account of other people's thoughts as well as to reflect more systematically on his or her own. Yet—and here many psychoanalytic observers agree—there often seems to be an absorption in one's own private world. Perhaps, as Elkind suggests, the adolescent is paradoxically taking the other person's point of view too much into account. He or she assumes that The Other Person is looking at and thinking about him or her in the same terms that occupy the adolescent. If the adolescent is particularly concerned about physical appearance, why, then, The Other Person must be intent on this as well.

This orientation suggests some of the possible basic, core functions of both larger peer group arrangements and more intimate friendships. If each Other Person looks and behaves pretty much as I do, then I do not have much to fear from others. I can define myself through this social feedback and, in turn, help to define all the others. It is true that this conformity among peers may be fairly superficial. Beneath the uniform of implicit dress code, gesture, and vernacular that characterizes a particular adolescent clique, there may be a great deal of individuality. But the outward appearance may be enough to prove helpful, at least to reduce the sense of being regarded critically by The Other Person. At the extreme, The Other Person is myself, reflected reassuringly.

But there is much more to the cognitive-social interaction relationship in adolescence. Elkind has suggested that the adolescent often imagines himself or herself to be the focus of attention. The audience is imaginary in the sense that in most actual situations the other people have something besides admiration or criticism of the adolescent on their minds. There is also a second form of adolescent egocentrism that Elkind called the *personal fable.* This is an exaggeration of one's own uniqueness. Nobody has ever felt this way before, not ever! Adolescents probably differ greatly in their personal fables (an inviting area for research), but the general tendency is seen by Elkind and some others as an outcropping of the level of cognitive functioning at this time of life, coupled with the adolescent's general situation.

What happens to this sense of uniqueness when the youth joins an impersonal work force?

Looft (1971) has suggested that social interaction in adolescence serves as an important condition for decentering. Intimate relationships are especially well suited for having one's own private fable and sensitivities permeated by what The Other Person is really thinking. The heart-to-heart talks, the trusting relationships, the yearning for a good listener that comes up again and again in research, all this suggests a need that might be as *cognitive* as emotional. Adolescents need adolescents (and others as well) to help them shed the sometimes excessive self-absorption typical at this time of life. Gradually the adolescent discovers through intimate relationships that most other people most of the time are not preoccupied with judging him or her. They have their own interests and problems. Furthermore, the adolescent's personal experiences, no matter how treasured and vivid, are not entirely unique. Other people have sometimes felt this way too, and sometimes they've found other solutions and opportunities that did not occur to the adolescent.

It may be that the relatively higher priority given to very close friendships, as distinguished from sexual relationships, in early adolescence reflects the need to define oneself more clearly. Recalling one of the studies reported earlier in this chapter (Josselson, Greenberger, & McConochie, 1977a, 1977b), the more psychosocially mature adolescent may reach this state of need sooner than others. Long and deep sharings (interspersed with personal, internal dialogues) may be important steps in decentering from a heightened egocentric state and thus being able to make more adequate use of one's new formidable cognitive abilities. It remains to be learned whether the low-mature adolescent decenters later in adult life through accumulated life experience, or tends to remain caught in an egocentric orientation which becomes progressively more unadaptive. It also remains to be learned whether the "loner" in adolescence misses out on decentering and other areas of cognitive development because of the lack of appropriate social interaction.

The interface between social interaction and the refinement of cognitive functioning in adolescence is an area well worth further theory and research.

DEATH AS THOUGHT AND REALITY

The adolescent is old enough to love, to enter into sexual relationships, to become a parent. In these matters the individual has some control over what does and what does not happen. Furthermore, thoughts of love and sex are important to most adolescents whatever their actual relationships might be at the time.

The adolescent is also old enough to die. This possibility has been true, of course, since conception and birth. But the new cognitive structures that become evident during adolescence equip the individual to think about death more systematically and resourcefully than ever before. Observations made in the relatively new field of life and death psychology suggest that death is a very important subject to most young people, although far from a comfortable one. There is also the impression that not only in adolescence, but through adult life, we have more influence over the chances of our continued survival than what has often been assumed to be the case. Let's now consider one of the ways in which the mortality rate in adolescence is influenced by individual actions, and, then, a few of the thoughts and feelings about death that tend to be experienced in adolescence.

HAZARD: SUICIDE IN YOUTH

Suicide is one of the leading causes of death in adolescence. Furthermore, it seems to be on the increase in recent years (Cantor, 1975). Precisely how many young men and women attempt suicide and how many actually die as a result remains open to question. There are errors, omissions, and distortions in both the attempted and completed suicide rates at all age levels; these mostly have the effect of underestimating the true prevalence. A conservative estimate is 4,000 deaths and approximately 400,000 attempts in the U.S. each year. But there is reason to suspect that both rates are considerably higher. Mishara (1975), for example, found that 65 percent of the college students he and his colleagues interviewed had thought of attempting suicide with sufficient intensity to have contemplated specific means of self-destruction. (The person who has devised a specific plan of suicide is generally considered to be a more serious risk than the person who has not.) About 25 percent of those with suicidal thoughts mentioned killing themselves through automobile accidents. This is of special interest, because automobile fatalities are, in fact, the number one cause of death in the fifteen to nineteen year old group. Furthermore, these "accidents" are not represented in the suicide statistics. The completed suicide rate in adolescence would look much larger, then, if self-destruction disguised as automobile accidents was included. Thinking about suicide is different from actually making an attempt. Mishara found that 15 percent of his respondents reported at least one actual past suicide attempt, and some had made multiple attempts over the preceding five-year period.

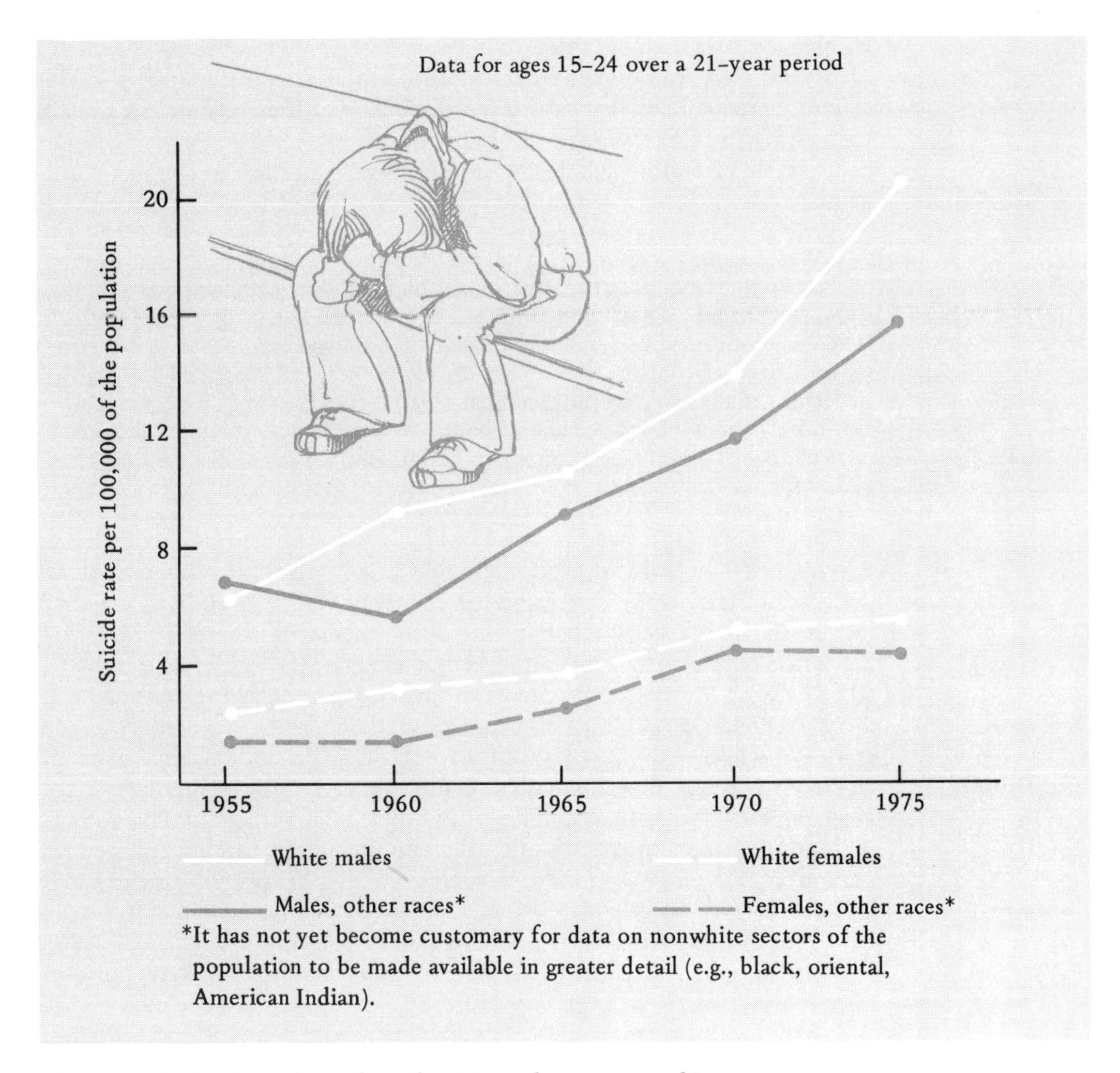

Figure 14-3. *Youths and suicide in the U.S.A.: the recent trend is disturbing. (Data provided by the Division of Vital Statistics, United States Public Health Service.)*

Although the true incidence of suicidal thoughts, attempts, and fatalities is hard to pin down, it is high enough to alarm counselors, "hot line" and crisis intervention workers, and researchers. Why is suicidality such a problem in adolescence? Psychiatrist Herbert Hendin (1975) thinks he may have the answer; this is based on his diagnostic and therapeutic experience with fifty college students in New York City. These were young men and women who had made serious attempts to kill themselves, for example, a person who had survived jumping from a seven-story building. Hendin believes that their troubled relationship with their parents was a major factor in the suicidality.

He characterizes many of these young men and women as having "death as a life-style." They were linked in a sort of psychological "death knot" with one or both parents. In many ways these students had been attempting to exist without pleasure or enthusiasm. Unusual as it may seem, they seemed more comfortable when they felt deadened, depressed, and stressed by the environment.

> Why seek out depression? The death bond these students had formed with their families required that they remain lifeless. Depression served the triple function of prolonging the tie with parents, shielding the student from the intensity of his own fury, and warding off the excitement of the new life that was opening up before him. For some, depression was sought in unhappy, coerced affairs based on the continual experience of rejection. Serious suicide attempts and profound suicidal preoccupations were, however, often required to extinguish both their anger and rebellion and to prolong the parental tie. Suicide for such students was the radical answer to radical emotions. (Hendin, 1975, p. 218).

Many of the students had made previous suicidal attempts or gestures designed to earn parental attention before literally taking their lives into their own hands with a lethal attempt. One of the coeds, for example, had slashed her wrists several times in the hope that her parents would at least notice (they hadn't). Many of the students also had lost one of their parents through death (a higher proportion of bereavement than for the population in general). But whether the parents were alive or dead, the adolescents seemed to be convinced that they themselves could not dare to be emotionally alive. Preoccupied with deadness and death and their relationships to their parents, these students became overtly suicidal when life—in the form of college stimulation, graduation, or serious involvements with other people—threatened to unravel the psychological death knot.

Hendin's intensive study of one population of seriously suicidal college students in one metropolitan area deserves attention. It is unlikely, however, to bear upon every, or even most suicidal attempts in adolescence. At the time of this writing, for example, I know of specific young people who have become increasingly suicidal because of the generally disheartening economic and career situation that exists at the moment. Still other adolescents and young adults move to the verge of suicide after what they interpret to be failure experiences indicating that they "aren't good enough," or "will never amount to anything." It is not very useful to insist on a single explanation for suicidality in adolescence, or at any other point in the lifespan.

Worth repeating: people who talk about suicide sometimes do attempt it

Sensitivity and a balanced perspective are needed in responding to suicidality in adolescence. One too-prevalent attitude dismisses talk of suicide as just talk. "People who talk about it, don't do it." This is not true. Many serious and even fatal attempts have followed verbal communications. It is true, however, that some people think about and talk about suicide and never make an attempt. It is also true that some people make attempts that are not likely to prove lethal. But every discussion of suicide and every life-threatening gesture is a potential call for understanding and help. "We never thought he'd really do it, I mean, sure he talked about it sometimes. We just never paid any attention, you know. . . ."

On the other hand, thoughts of suicide—like thoughts of death in general—do not necessarily mean that a person is "sick." A person who is squeamish about any death talk might overreact when somebody else brings up the subject. Adolescents simply have a lot of thinking to do about many subjects, including death as well as sex. Sharing and testing the new thoughts with other people can be a valuable aid to development. Rejection of suicide or death talk is likely to have the opposite effect, as in the parallel rejection of sex as a polite subject for conversation. Let's take a step or two further in exploring the adolescent's thinking about death in general.

SCENARIO 9: "ONE BUN—HOLD THE BURGER!"

"That's the best thing to do the night of an exam," Mike had advised, "Go to a movie!" So it was off to the movies for Mike and Linda. But which theater? The Strand was showing a couple of good old semiclassics, *Conqueror Worm* and *The Mummy's Revenge*. The Fleetwood was pairing *Jaws* with *Earthquake*. They quickly agreed that *Worm* and *Mummy* were kid stuff (besides, they had already seen that program twice, joking and giggling much of the time).

Jaws and *Quake* proved to be sterner stuff. There was not much time to sit back and joke. Mike didn't even bother to tease Linda when she screamed aloud once; he had felt like screaming himself.

After the movies, they didn't have much to say to each other right away. Back in the car, Mike asked Linda if she was hungry.

"I don't feel so good. All quivered up inside."

"That's because you're hungry."

"No, it's not. But we can go eat if you're hungry."

"Aw, you're hungry too, you just don't know it."

At the hamburger place, Linda stood so she wouldn't have to look at the serving counter. She tried not to watch people eating, but there was no avoiding the mingled fragrances of burgers and fries. This made her feel even worse. It took a moment for Linda to realize that Mike had been talking to her.

"Hey, anybody home?" he rapped playfully on the top of her head. "Come back to *this* world. What do you want, cheeseburger or hamburger?"

"Nothing."

"One nothing-burger coming up!"

"I'm not hungry, I told you."

"Okay, okay."

Linda sat across from Mike with her elbows on the table, her face between her hands. Mike paused between bites of his hamburger and glared accusingly at her. She giggled. "You *are* funny sometimes, Mike! Like now." Mike looked so troubled by that remark that she was sorry she had made it. "I mean, you sit there eating all that bloody animal flesh and acting like I'm spoiling your whole, entire evening because I don't want any!"

"You must be sick in your head *and* your gut! This 'bloody animal flesh' is a hamburger, and a good one, just like the one you had with me after the movies last week. Geez!"

"Well, that was last week. I feel different tonight."

"*Jaws* was too much for your delicate sensibilities, huh?"

"Oh, it's more than that. I don't know. I think I'm going to be a vegetarian from now on—don't laugh! I've been thinking about it for awhile, and about a lot of stuff like that."

Mike didn't laugh. He became reflective himself. "Yeah, it is weird. Some of that stuff that goes on. Blood on the screen is one thing, and people getting torn apart . . . monsters crawling out of tombs. You scream or you laugh. Sometimes you don't know which to do."

"I didn't know you ever thought about things like that."

"Sure, I think about a lot of things. That doesn't mean I go around talking about them or not eating hamburgers."

Linda began to relax a little, and helped herself to one of Mike's fries. "Why don't you be a vegetarian, too?"

"No, I really like meat. But I don't like a lot of blood. I mean, not real blood. And I really don't like the idea of animals getting killed so we can eat them, but that's life, you know."

"But you've been hunting, haven't you?" Linda asked as she reached for his cola. "You've shot at squirrels, rabbits, I don't know what else."

"Yeah, but that's different, too. You know—well, I can't explain it all; it's just that it's all different somehow."

Linda sprinkled more salt on the fries, which now were on her side of the table. "But I think it's all the same somehow, Mike. You know what I think sometimes—everybody is meat!"

"That's a weird thought."

"I guess so, but isn't it true? We are flesh and we eat flesh, and then the worms eat us, and the birds eat the worms and . . ." Linda suddenly noticed that Mike was starting to look a little green. "Well, I don't mean to spoil your meal. But I just don't like all the killing that goes on this world . . . although I guess I like *some* of it, or I wouldn't go to that kind of movie. I just wish—I just wish that nobody ever had to die." During that last statement, Linda's voice had taken on a softer, little girl's tone.

Mike found himself thinking about how bad Linda felt not long ago when her old half-blind cat had been hit by that car backing up. And then he began to drift off inside his own head, seeing images of death and absence that made him feel empty, yet stirred up inside.

"Hey, anybody home?" Linda smiled, rapping on his knuckles. "I'll make you very, very happy!"

"How is that?" Mike beamed.

"I'll let you buy me fries, and a cola, and even a hamburger, so you won't have to sulk the rest of the night."

"A hamburger? Ycch! How gross! Where are your sensibilities?" Mike approached the counter and carefully ordered fries, cola, and one toasted bun with tomato, lettuce, ketchup, mustard, and relish—but no burger.

Later that evening, Linda rested her head against Mike's shoulder. "Thanks for not laughing at me—too much."

"I guess we're both pretty funny sometimes. But it sure feels good to hold you. It sure feels good to be warm and alive."

SUMMARY

Intimate thoughts and relationships in adolescence are placed in a new light by recent studies. One major study (Offer, 1969; Offer & Offer, 1975) followed a panel of "normal" middle-class boys from their first year in high school until age twenty-two. Contrary to some expectations, the personal values of these boys remained similar to those of their parents. *Continuity of values* between parents and adolescent was also found by other researchers whose participants included girls as well as boys and a variety of economic and ethnic backgrounds. This challenges the stereotype that adolescents in general are seeking to overthrow parental values. And, although social ideals and opinions were not lacking among the adolescents studied, few showed a deep personal commitment to changing "the way things are" in the world.

The peak season for intergenerational friction and *rebellious behavior* came on the threshold of adolescence for most of the youths. Both adolescents and their parents agreed that mutual irritation and hassles were most common around ages twelve and thirteen. Although areas of disagreement remained throughout the high-school years, in general the bonds of affection and approval between generations also remained strong. Control of aggressive tendencies, nevertheless, was of concern to many teen-agers, usually more so than control of sexual impulses. Relationships with teachers were generally disappointing; seldom did a high school student report being inspired by the adult models encountered in the school situation.

The well-advertised *identity crisis* was *not* common. Although many of the adolescents were still refining and revising their sense of personal identity, the typical individual maintained an essentially stable and realistic sense of who he was. In fact, the adolescents in general came across as realistic, active, gregarious, practical-minded young people with a sense of perspective and humor.

YOUR TURN

1. Three patterns of development through adolescence were observed by Offer and Offer. Think ahead twenty years. Imagine three adolescents, each of whom typified a particular pattern: the continuous-growth, surgent-growth, or tumultuous-growth pattern. Now each of these people is forty years of age. What differences, if any, would you expect to find in their life-styles, based on their backgrounds and characteristics around age twenty? Would they differ in types of career? In "success"? Status in the community? Personal satisfaction in life? In how they raise their children (if any)? Might they possibly differ in health status as well? The forty year olds (and sixty and eighty year olds) we will be considering later in this book were once adolescents themselves. This will be a useful opportunity for you to test your abilities to interrelate points on the lifespan. I suggest you prepare scenarios of about five hundred words each for these adolescents, bringing them forward to age forty. Be ready to describe and defend your basis for suggesting that their future life courses will be in the direction you have projected. This exercise will also sharpen your appreciation for the difficulties involved in predicting or understanding continuities and changes in life-style through the years.

2. An alternative method of exploring this same question might be enjoyable if class time can be set aside for this purpose. Get together with two of your friends in class, or a couple of people you would like to know better. Discuss among yourselves the possible future trajectories of the three "types" of adolescents. Then, each of you select one of the people to impersonate, and figure out more of the details of this particular character for yourself. Back in the classroom, the three of you offer an improvised dramatization. The now-forty year olds meet each other again after not having seen each other for awhile (the circumstances of the meeting are for you to determine). Through this interaction, you can convey both the current status of life experiences for the old classmates, and also something of the changes that have taken place in each person's life since high school days. As you get into your character you will find it easy to improvise answers and responses that stay in character.

If time permits, two different dramatizations by different sets of students would make for interesting comparisons. A classroom discussion of the points raised by the dramatization can combine methodological, content, and theoretical issues. And, although the Offer–Offer study was limited to males, the various patterns can also be enacted with females or with a mixture.

Even within a subpopulation of adolescents with many demographic characteristics in common, several different patterns of development were found. A *continuous-growth* pattern was found for those most favored by genetic and environmental circumstances. *Surgent-growth* adolescents had more incidents of trauma and loss in their backgrounds. They did not engage in as much self-reflection or emotional expression as those in the first group. Adolescents with a *tumultuous-growth* pattern had more background problems than those in the other groups. They were sensitive and struggling people. A general theory built on experience with only one of these patterns would fail to represent the actual diversity of personality development during the teen years.

Konopka's study of adolescent girls found that almost all the girls included careers in their life goals and were concerned about the integration of occupation, marriage, and motherhood. Sex was spoken of openly, but there was no indication of a "sexual revolution" or of a casual or promiscuous attitude toward sexual behavior. The teen-age girls looked for much the same personal qualities in their friends of either sex as they did in adults: trustworthiness and a willingness to listen.

Some adolescents are more psychosocially mature than others. A clear pattern of differences emerged in one study (Josselson, Greenberg, & McConochie, 1977a, 1977b) of males and females who scored high and low on a measure of psychosocial maturation. Among other differences, the high-maturity adolescents were more individualistic and resistant to peer pressure, better able to see others as people in their own rights, and more interested in a wide variety of activities.

In the quest for *loving intimacy,* there is a strong tendency today for adolescents to emphasize *mutuality,* a nonpossessive, give-and-take relationship that meets the present needs of both partners. Although most adolescents plan to marry some day, long-term obligations and commitments are not favored at the moment. The emerging trend, if any, is for a type of relationship known as *serial monogamy,* rather than promiscuity or commune arrangements. Sexuality is considered normal and natural by most adolescents. Whether it is "right" or "wrong" for a particular couple at a particular time depends on the situation more than on traditional moral standards. (Yet there does *not* seem to be much need to use sex as a revolutionary cause against "the establishment".) The actual patterns of sexual activity reported by adolescents are quite varied. But there is little support for the contention that sexual activity is rampant throughout adolescence, either in the U.S.A. or in other nations.

There is a bit of research and some interesting theory to suggest that social interaction in adolescence, especially among peers, plays an important role in cognitive as well as emotional development. It is possible that there is a mutual reconstruction of reality whereby adolescents help each other to *decenter* from excessive *egocentricity.*

The new cognitive structures and the new life situation of adolescence provide the occasion for the individual to revise his or her relationship to death. This is illustrated by that extreme alteration known as *suicide*. The hazard of suicide during the adolescent years is probably greater than the available statistics indicate, and even these show that self-destruction is a major cause of mortality in this age range. Apart from suicidality, however, it is common for adolescents to think about death from many standpoints, both direct and indirect. Love and death remain central themes from adolescence onward, although in shifting patterns as life experiences change.

Reference List

Bandura, A., & Walters, R. H. *Adolescent aggression.* New York: The Ronald Press Company, 1959.

Broverman, I. K., Vogel, S. R., Broverman, D. M., Clarkson, F. E., & Rosenkrantz, P. Sex role stereotypes and clinical judgments of mental health. *Journal of Consulting & Clinical Psychology,* 1970, *34,* 1–7.

Cantor, P. The effects of youthful suicide on the family. *Psychiatric Opinion,* 1975, *12,* 6–13.

Elkind, D. *Children and adolescents: Interpretive essays on Jean Piaget.* New York: Oxford University Press, 1967.

Friesen, D. Value orientations of modern youth: A comparative study. *Adolescence,* 1972, *7,* 265–288.

Hartmann, E. *Ego psychology and the problems of adaptation.* New York: International Universities Press, 1958.

Hendin, H. Student suicide: Death as a life-style. *Journal of Nervous & Mental Disease,* 1975, *160,* 204–219.

Hess, R. Henry, W. E., & Sims, J. H. The identity scale. In *The actor* (unpublished). University of Chicago, Department of Human Development, 1968.

Horner, M. S. Toward an understanding of achievement-related conflicts in women. *Journal of Social Issues,* 1972, *28,* 157–175.

Josselson, R., Greenberger, E., & McConochie, D. Phenomenological aspects of psychosocial maturity in adolescence. Part I: Boys. *Journal of Youth & Adolescence,* 1977, *6,* 25–55. (a)

Josselson, R., Greenberger, E., & McConochie, D. Phenomenological aspects of psychosocial maturity in adolescence. Part II: Girls. *Journal of Youth & Adolescence,* 1977, *6,* 145–167. (b)

Kalish, R. A., & Johnson, A. I. Value similarities and differences in three generations of women. *Journal of Marriage & Family,* 1972, *11,* 49–54.

Keniston, K. The sources of student dissent. *Journal of Social Issues,* 1967, *22,* 108–137.

Konopka, G. *The adolescent girl in conflict.* Englewood Cliffs, N.J.: Prentice-Hall, Inc., 1966.

Konopka, G. *Young girls: A portrait of adolescence.* Englewood Cliffs, N.J.: Prentice-Hall, Inc., 1976.

Lee, S. High school senior girls and the world of work—Occupational knowledge, attitudes, and plans (Research & Development Series #42). Washington, D.C.: U.S. Government Printing Office, 1971.

Luckey, E. B., & Nass, G. D. A comparison of sexual attitudes and behavior in an international sample. *Journal of Marriage & Family*, 1969, *31*, 364–379.

Mishara, B. L. The extent of adolescent suicidality. *Psychiatric Opinion*, 1975, *12*, 32–37.

Offer, D. *The psychological world of the teen-ager*. New York: Basic Books, 1969.

Offer, D., & Offer, J. B. *From teenage to young manhood*. New York: Basic Books, 1975.

Piaget, J. *Judgment and reasoning in the child*. New York: Harcourt, Brace, 1928.

Reiss, I. L. Sexual codes in teen-age culture. *Annals of the American Academy of Political & Social Sciences*, 1961, *337*, 53–63.

Reiss, I. L. How and why America's sex standards are changing. In J. H. Gagnon & W. Simon (Eds.), *The sexual scene*. Chicago: Aldine Publishing Company, 1970.

Schofield, M. Normal sexuality in adolescence. In J. G. Howells (Ed.), *Modern perspectives in adolescent psychiatry*. New York: Brunner/Mazel, Inc., 1971.

Sorenson, R. C. *Adolescent sexuality in contemporary America*. New York: World Publishing, 1973.

Troll, L. E., Neugarten, B. L., & Kraines, R. J. Similarities in values and other personality characteristics in college students and their parents. *Merrill-Palmer Quarterly of Behavior and Development*, 1969, *15*, 323–336.

PRIME TIME

The Early Adult Years

15

CHAPTER OUTLINE

I
The young adult may encounter both challenges to his or her ability to cope and stresses that affect life satisfaction during this "prime time."

Life satisfaction / Subjective age / Future-time perspective / Rate of life change events / Stress / Personality types (Neugarten)

II
The young adult may be characterized by his or her physical and mental abilities, and social competence.

Physical resiliency / Brain weight / Vision—accommodation and adaptation / Audition—pitch discrimination and acuity / Muscle strength / Development versus aging / Omnibus measures (IQ tests) / Determinants of mental functioning—experience and acculturation versus decrease in physiological functioning / Crystallized intelligence / Fluid intelligence / Factor analysis

When is life most exciting and rewarding? In general, our society appears to think the early adult years are the most exciting. The twenties and thirties are prime time. It is as though human development sets these years as its goal or destination. Younger people look ahead with keen anticipation; older people look back with nostalgia on those relatively few years within the total lifespan.

This image of young adulthood must be put into perspective. The fact that our society pictures these years as prime time is significant in itself, whatever the reality of individual experience might show. Theoretically, we could live in a society that holds *all* phases of the lifespan to be equally valuable. Were this the case, then we would not have to contend with the intense loading of a few years with so much hope and expectation. The pressure on young adults to achieve everything, to experience everything, to "squeeze everything in" would be much less than it is today. Furthermore, in moving from young adulthood to later adulthood there would be much less sense of loss, less fear that "the good stuff" is being left behind. The intense valuation of just a few years from the total lifespan, then, tends to: (1) impart a strong sense of directionality and anticipation to the young, (2) increase the pressure on young adults to make the most of their hours in the sun, and (3) gnaw away at people as advancing age inevitably carries them beyond the assumed developmental crest.

We will now examine some of the expectation and some of the reality associated with the prime-time interpretation of the young adult years.

LIFE SATISFACTION: ANTICIPATED AND ACTUAL

VALUES, EXPECTATIONS, AND THE SENSE OF FUTURITY

What thoughts and attitudes do adolescents express about the lives they have had, the lives they are now leading, and the lives they are heading toward? In my own studies of normal adolescents, I found a strong sense of *directionality*, of moving rapidly toward the future (Kastenbaum, 1961, 1965). But this sense of forward movement often was *not* accompanied by a worked-out view of the future as such. Typically, the teen-ager is thinking only a few years ahead, is projecting relatively few events and experiences (although these are psychologically important events and experiences), and is anticipating little coherence or "fine structure" in the future. To this pattern can be added the impression that the "typical" adolescent is uncomfortable with his or her

How does this compare with your own view of the future?

past. The past is something one escapes from, moving at an ever-accelerating tempo toward an exciting, promising future when one really becomes one's own self. *Values* are seen as concentrated in but a few years of life, the years just up ahead. When attempts are made to determine how people on the brink of young adulthood see the rest of their lifespan, a shrinking back from the prospect and an association of middle and later life with vague, and negative, boring, or threatening experiences is usually seen (Kastenbaum, 1959).

This constellation of thoughts and feelings generates much pressure on the person as he or she enters prime time. "I have to make it—and right away!" is often a self-imposed command that adds to pressures coming more directly from the outside. But it also embodies such a negative expectation of the rest of the lifespan that we are left wondering how a person could achieve life satisfaction later, carrying the burden of impaired values and lacking positive conceptions and plans. Obviously, some people do change their attitudes toward the life beyond prime time. But the apparent fact that most young people have attitudes that *need* to be changed if they are to enjoy the rest of their lives remains something to contend with.

In a later series of studies, my colleagues and I approached this problem area in a different way. Adults of various ages were asked to respond to a procedure known as "The Ages of Me" (Kastenbaum, Derbin, Sabatini, & Artt, 1972). This procedure consists of a series of questions regarding the individual's own perceptions of how old he or she is from a number of dimensions and standpoints. Sample items are: "I *feel* as though I were about age _____"; "I *look* to other people as though I were about age _____"; "My *body* is like that of a person who is _____ years of age"; "I *do* most things as though I were about age _____"; etc. One of the most general findings derived from these studies was that adults at all age levels assign different ages to themselves depending on the specific question asked; in other words, we see ourselves as being younger in some ways, older in others. Another general finding was that the older we are, the larger the difference between our chronological age, based on the date on our birth certificate, and the ages we privately assign to ourselves. Our private ages become even more private. And the *direction* of the difference between chronological age and personal or private age? It is just as we might have expected: people tend to see themselves generally as younger than their chronological age.

This tendency has the effect of keeping people (subjectively) closer to prime time than the calendar allows. The forty year old who still "feels twenty-five" keeps part of his or her self-image afloat in the culturally valued decades of life. By the way, no implication should be drawn here that such a person is deceiving himself or herself or others. Rather, a person who is actually feeling fit and vigorous at, say, forty or fifty, is thinking in terms of a culturally-understood metaphor for well-being in saying that he or she is twenty-five, or "young." To say or think we are *young* is one way of expressing a favorable morale and level of functioning. (Whether or not this *should* be the case is another matter.)

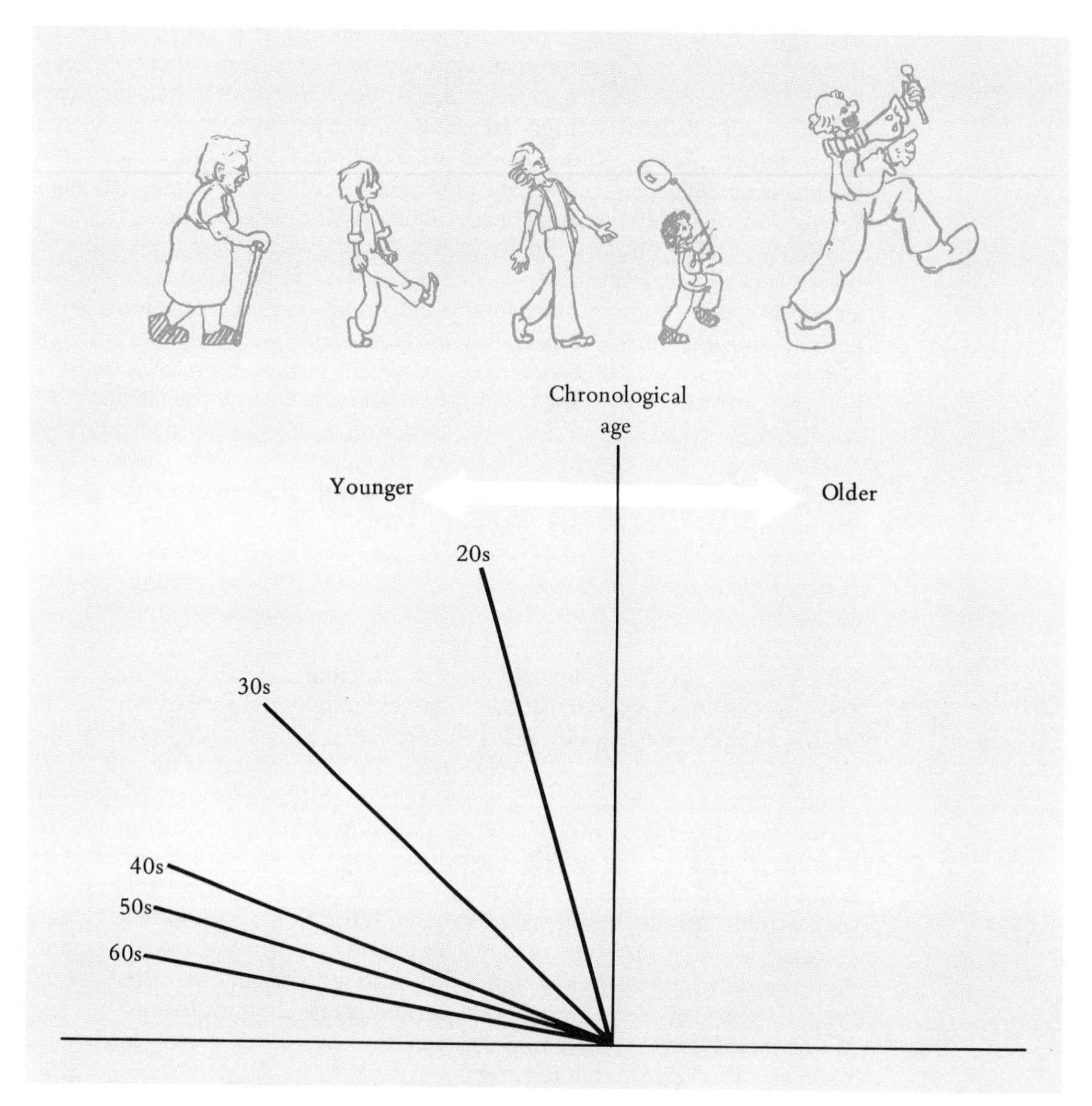

Figure 15-1. How close is personal age to chronological age? *The person who feels neither younger nor older than his or her chronological age would appear straight up in this representation. But most people feel themselves to be younger, with the lean already showing itself for those in their twenties and becoming more pronounced with each succeeding decade. Of the people studied, no age group has leaned toward the older side of the spectrum on any of the items assessing personal age. (Data presented by Kastenbaum, Derbin, Sabatini & Artt, 1972.)*

The implications for life satisfaction become even clearer when we notice how people answer the question, "I would *prefer* to be age ____." Less than 2 percent of adults ranging in age from the early twenties through the eighties would prefer to be older than their present age. In fact, the percentage dips below 1 percent as soon as the twenties are left behind. Most young adults prefer to *remain* at their present age indefinitely or "for a long, long time"; most adults over thirty would prefer to be somewhat younger. There is certainly no inclination to rush through prime time in the way that some preadolescents and adolescents want to rush *into* prime time. Furthermore, even those who are twenty-five years of age or thereabouts are more likely to prefer being a younger age than their own, although the big shift toward preference for a younger age comes later.

Observations such as these suggest that the strong sense of the future as a zone of time in which satisfaction and self-esteem can be found diminishes soon after the early adult years, and may even erode while the person is in the middle of prime time. Many young people do not project themselves into the more distant years of their own lifespans, and what they do foresee ahead is often disagreeable to them.

Future time perspective is an orientation toward the future

Notice that this limitation in **future time perspective** is not likely to be the result of a deficit in reasoning ability or imagination. The work of Piaget and many others has established that adolescents and young adults are quite capable of abstract and broad-ranging thought (within the limits of their particular level of mental gifts). The restricted and skitterish approach to long-range futurity probably has more to do with the individual's perception of social values: in most respects, our culture considers the young person more valuable and interesting than the old.

LIFE SATISFACTION AT VARIOUS ADULT AGES

I have been concentrating on *anticipations* of the quality of life in the future. Now let's turn to some studies that examine the kind of life satisfaction people *actually* report during and after prime time.

One researcher asked adults of various ages about their sense of happiness, both as seen by themselves and as they think people of other ages see them. Cameron (1972) found that there were no significant differences between the young, the middle-aged, and the old with respect to their own "fun and happiness," although the young adults claimed to have a greater *desire* for pleasure than did the others. Who was considered "happiest" by other groups? It was not the young, but the middle-aged! Cameron's own conclusion was that happiness is not determined by, or closely related to, age as such; it is relatively constant across the lifespan.

This view has support from other studies as well. A number of researchers have found that life satisfaction has at least as much to do with the individual's own personality or life-style as it does with his or her age. The kind of activities in which the person is engaged and other facets of the total life situation also seem more important than age. It might be simpler to say that people of a certain age are happier or unhappier than others, but the truth is more complex and also more interesting.

Take an example from old age. When young people project themselves into their own remote futures, or try to imagine what old people experience (e.g., *see* Kastenbaum & Durkee, 1964), they usually associate growing old with markedly decreasing life satisfaction. But this was not the result of an important study by Neugarten, Havighurst, and Tobin (1968). They followed-up on a group of men and women who had earlier participated in a pioneering study of the psychosocial aspects of aging (Cumming & Henry, 1961). The participants were in their seventies at the time of the follow-up, and their current life experiences could be compared with those they had reported about six years before. They "should" have felt less life satisfaction at the time of the follow-up, because they were then even older and more vulnerable to illness, bereavement, etc. But there was no such overall decrease in life satisfaction. If pressed to make a generalization, I would have to conclude that life satisfaction is at least well maintained over the lifespan for the typical person.

But the Neugarten et al. study helps to emphasize the point that there is no "typical" person! At the very least, one has to recognize several types of personality. Neugarten and her colleagues classified their participants into four general personality categories.

1. The *integrated* person was the man or woman who was able to accept life, both the life inside his or her own thoughts and feelings and the life in the world outside. He or she was flexible and open to new experiences. He or she was also described as "mellow" and "mature." The integrated person proved to be high on life satisfaction.
2. The *armored* person was the one who put much energy into defending himself or herself from real or supposed dangers. This person was likely to be an ambitious, achievement-oriented striver—probably the sort of individual who more recently has been classified as a "Type A," high cardiovascular-risk person (Friedman & Rosenman, 1959). The armored person had a medium level of life satisfaction in general, although those who were very successfully armored had high levels of satisfaction.
3. The *passive-dependent* person, as the name suggests, was the man or woman who leaned on others for support, comfort, and guidance. Some of these people maintained a medium level of life satisfaction; these were the ones who were able to reach out to other people and, in fact, had other people around within reach. But some had become apathetic, or rather, *more* apathetic with age. Neugarten and her colleagues felt that these were people who had always been quite passive, but now

found it more easy to give into this leaning. The apathetic, passive-dependent person had a low level of life satisfaction.

4. The *unintegrated* person was generally disorganized in thought and behavior. This person did little in the community and had few coping resources available. As might be expected, his or her life satisfaction was low.

When *all* the participants in this follow-up study were considered as a group, only about 15 percent showed a low level of life satisfaction, while about 58 percent showed a high level and 27 percent a medium level. These findings have certain limitations that we will discuss when we consider the later years of life more extensively in other chapters, but they strongly challenge the view that old age is necessarily a time of low and steadily decreasing life satisfaction.

Even more perspective can be gained when we examine the results of a study encompassing people who ranged in age from high school seniors to those nearing retirement. This major research project headed by Lowenthal, Thurnher, and Chiriboga (1975) involved urban, white, middle- and lower-middle-class participants who were interviewed at great length and in great detail. The findings are complex. I will focus here on results that are especially relevant to the question of life satisfaction and the prime time decades although some related variables will be considered as well.

Although the Lowenthal et al. study attempted to compare four specific stages of life, it turned out that neither age nor stage provided the most useful basis for understanding what a person was experiencing. Knowing whether a person was male or female was usually a more valuable piece of information than knowing the stage, whether this was a high school senior, a newlywed (prime timer), a middle-aged adult, or a preretiree.

The men tended to have a more positive image of themselves than the women across the four life stages studied. The women tended to be more self-critical. You can see that just this one comparison makes it possible to avoid some misinterpretations of the relationship between age and self-concept. More women survive into the later years of life. Studying old people in general, then, might give the impression of lower self-concept and higher self-critical tendencies. But this could chiefly be a reflection of the fact that the proportion of women is greater in the more advanced age ranges, not proof that people necessarily become more self-critical with the age. And it would also be hasty to conclude that men are somehow more successful or more enviable. Their higher self-image may reflect a tendency among males in our society to ignore or deny the less flattering aspects of their own personalities. Perhaps the women are more depressed or aware of inferiority, but perhaps they are simply more realistic about their limitations than are the men.

Again, it was sex rather than age that tended to distinguish people when Lowenthal et al. examined their friendship patterns. The men liked those people who shared their interests and activities the best. The women considered the ability to communicate and share feelings to be a more important cornerstone of a friendship relationship. They also offered more detailed and subtle characterizations of their friends than did the men.

The sense of subjective well-being, or life satisfaction, also came through somewhat differently for the men and the women. The women seemed to express more complex states of feeling. There was also the impression that they had more tolerance for the ambiguities and frustrations of life. The men seemed more distressed by "the ups and downs of emotional life." Most of the people in this study—in this case, regardless of their age or sex—considered themselves to be relatively happy. But the happiest women (by self-report) were those who had had a rich mixture of both positive and negative experiences, especially during the recent past. The men who described themselves in the happiest terms usually had had predominantly positive experiences in the recent past. Even when face to face with equally *happy* men and women, then, it was likely that the balance of experiences that led to that state was somewhat different.

Yet all these findings (and others not reported here) do not completely obscure the influence of age or stage. Consider *stress*, for example. The newlyweds reported the most encounters with stressful situations within the past decade. In fact, they reported approximately two and one half times more such experiences than the two older groups of adults. The high school students also reported more stress in their lives than the older adults, but not

A beautiful moment—but also a time of tension. Newlyweds report more stressful life situations than do older adults. (Photo by Stanley J. Kislowski.)

nearly as much as the people a few years ahead of them in prime time. There were also age/stage differences in type as well as amount of stress encountered. Those in the two youngest groups reported that most of their problems derived from the sphere of educational experience and achievement. In other words, something about going to school had upset, angered or depressed them. They were also upset by changes of residence. Moving from one place to another seemed to challenge their sense of security. Even if the move was just a few blocks away, it sometimes made the high school student or young adult feel that their lives and friendship patterns were coming apart. Interestingly, more stress was felt in relationships with friends than family. Marriage was seen in a favorable light, usually associated with increased life satisfaction, seldom with great stress and dissatisfaction.

Middle-aged and older adults reported fewer stress experiences than the young, as has already been mentioned. There were sex differences in the type of stress encountered most often by older adults. The men were more likely to have challenges and concerns related to their occupations; the women were more likely to be concerned about health (other people's as well as their own) and interpersonal relationships.

Findings such as these make it difficult to maintain the simple view that young adults necessarily are the happiest people of all. Personality and sex (themselves closely related) seem to be more important than age in determining life satisfaction. The influence of age or life-stage on psychological well-being can best be understood when it is put in the context of the individual's own style of functioning and the type of environment he or she encounters.

Perhaps the aura of nostalgia makes the "old days"—of being young—seem better than they felt then

CHANGE, STRESS, AND UNCERTAINTY

Other types of research have also emphasized the many problems that can detract from the real and supposed splendors of early adulthood. Holmes and Rahe (1967) devised a method to study the number and type of events that take place in a person's life during a specified period of time. They pointed out that the more significant events a person experiences in his or her life, the more readjustments required. The frequency and density of life events, then, should be related to the amount of stress and change with which the individual must contend. Many studies of this type have been carried out, both by the original researchers and by others. In general, there has been support for Holmes and Rahe's starting proposition. The clustering of many life events in a short period of time has been related to a variety of stress reactions: more complications during pregnancy (Nuckolls, Cassel, & Kaplan, 1972), chronic asthma (Araujo, Van Arsel, Holmes, & Dudley, 1973), psychiatric hospitalization (Fontana, Marcus, Noel, & Rakusin, 1972), suicide attempts (Paykel, 1974), physical illness of various kinds among college students (Anderson, 1975) and physical illness of various kinds among members of a number of occupational groups (Holmes & Holmes, 1970).

It is difficult to ignore the accumulating evidence that a high rate of life change affects the individual psychologically and physically. In fact, influences have also been discovered on such basic physiological processes as the secretion of epinephrine, serum cholesterol, and catecholamines (Theorell, 1974).

This pattern of research findings suggests that the young, emerging adult may be experiencing more stress than we usually appreciate. There is the transition from the home to independent living, from student to worker, and often, from a single adult to a person with strong obligations to a mate and to young children. Uncertainty and a sense of lacking control over life events further contribute to stress. The young adult has many important decisions to make for himself or herself; often it is difficult to know if the right decision has been made. Events sometimes spin out of control, well-made plans are upset through no fault of one's own. The young adult may find too many demands pressing on him or her because of changes in life situation. Research indicates that such a condition leads to more physical illness (e.g., *see* Hinkle, Pinsky, Bross, & Plummer, 1956) which, in turn, can lead to difficulties in holding or advancing on the job. One negative life change may stimulate another, and then another.

Studies of this kind do not detract from the obvious fact that many people have physical, mental, and social resources of a particularly promising sort available to them during the early adult years. These can, indeed, be years of pleasure, discovery, and achievement. But it is possible for any young adult to encounter a barrage of life events that challenge his or her coping abilities and detract from life satisfaction. For some people, the stress of change and uncertainty at this time of life makes young adulthood anything but prime.

There is at least some fantasy, then, in our culture's stereotype that young adulthood is the time of peak satisfaction. The results that are starting to come in from well-planned studies suggest that we should consider the *types* of satisfaction a person is most likely to receive at a particular stage in life, as well as who that particular person *is* at any stage. Furthermore, sharper recognition of the difference between an individual's *objective* place among life's stages and how this person *subjectively* classifies himself or herself can also help us toward a broader and more realistic view of life satisfaction.

THE ABLE YOUNG ADULT: PHYSICAL, MENTAL, AND SOCIAL COMPETENCE

As the individual enters prime time he or she receives increasing social recognition as a responsible, able person. The voting franchise is afforded just for having reached a specified age. The young adult may also seek election for almost any political office in the land; a demand for experience as well as ability is made only of presidential aspirants for whom thirty-five is the magic age. The young adult is now able to enter into contracts and various legal arrangements on his or her own say-so, rather than needing the authority of older adults. And, yes, he or she can imbibe alcoholic beverages if he or she chooses, view the "X-est" of pornographic movies, and either embrace or resist the whole spectrum of opportunities and temptations that are in the province of "grown-ups" in our society.

In some respects the young adult achieves more than equality. Job opportunities tend to favor the young, although there are important exceptions, especially where experience is considered critical. Whether or not the young adult likes the array and level of job opportunities available, there is not yet the bias against employment that will be increasingly encountered by the same individual if seeking work at forty, fifty, or sixty. This tends to hold true for promotion as well. In many companies it is easier to move up during the first few years. By contrast, there is often pressure to move older employees *out.* Special training opportunities (e.g., training as a retail store manager) are more likely to be extended to young adults. This seems to be based on the assumption that young adults are more open to experience, better learners, and more energetic. Perhaps there is also the less visible assumption that young adults can be more easily shaped or controlled, because they have less life experience and security with which to counter the influences exerted on them.

Employment opportunities in our society are usually best in prime time even though the individual will later have more experience. (Photo by Talbot Lovering.)

In actuality, the experience of any particular young adult in seeking employment can vary from very fortunate to very disappointing. The economic climate of the times is, of course, a major factor, although not the only one. (The residues of sexism and racism are among the influences that may work against an individual.) But the young adult tends to have the age bias going in his or her favor. When pressure is put on elderly employees to retire early, or when people aged forty or over are discriminated against as they seek jobs, it is often with the claim that "we must make room for the young."

Many have suggested that the young and the elderly should ally to combat job discrimination

This claim can be viewed as an appreciation of the distinctive vitality, freshness, and potential of the young adult. Or it can be viewed as a cover-up for unwillingness to pay higher wages and fringe benefits to people with more experience. Both younger and older adults are discriminated against or exploited in various employment situations for factors related to their age and level of experience (although they are not necessarily discriminated against for the abilities they possess as individuals or as a group in a certain age bracket).

Here I will briefly consider some of the abilities that characterize the young adult. These will not be limited to those involved in employability and work performance. Focus is on the relatively healthy, unimpaired person—the able young adult.

PHYSICAL ABILITY

The body is a major resource for the young man or woman. The balance between vigor and impairment or deficit is quite favorable. The intact functioning and resiliency of the body provides for a certain competitive edge and a certain margin of acceptable or "forgiveable" error and misuse. In other words, the healthy young adult can draw on his or her physical resources as a large component in overall happiness and success. There are implicit dangers here, however, which become real and present dangers in the lives of some people during prime time. Because the body can bounce back so well, there may be a tendency to push it too far. Accumulated mistreatment of our physical endowment in young adulthood may show up soon, or it may catch up some years later. Wise use of our own bodies when we are young adults is perhaps one of the more significant determinants of satisfaction and achievement throughout the lifespan. Destructive habits that we can "get away with" when young may become increasingly difficult to change and may show increasing impact on our body's ability to sustain the type of life-style we desire. Poor nutrition and sleep habits or an excessive dependence on alcohol or drugs are some of the tendencies that subtract from prime performance during prime time—and which may be even more handicapping in the years to come.

The early adult years are a period when most of the physical growth processes (apart from those which maintain homeostasis or internal adapta-

What do you want your body to do for you? The options are usually at their peak during prime time. (Sun Valley Photo.)

tion) have just about run their course. The twenty-five or thirty year old does not have to contend with the anatomical and physiological transformations that are experienced around the time of pubescence. Furthermore, marked negative changes in functioning, loosely spoken of as *aging*, are usually not in evidence. It is a treasurable high plateau between the territories of surgent growth and decline or senescence. Yet biology does not hold still for an instant. Let's scan a few indices of physical status and its direction during early adulthood.

Hold your own brain in one hand, that is, the brain you had or will have at age twenty. Now, in the other hand, hold your brain at age thirty. Can you feel the difference? Brain weight, a measure widely used to approximate the amount of intact neural tissue available to the organism, decreases even within the prime time years (Kovenchevsky, 1961). It is a small decline, but it appears to be a real one. Sex differences appear in this realm, as elsewhere. The brains of females tend to increase in weight and then remain relatively stable a little longer than the brains of males. At a particular age in or around early adulthood, then, males are somewhat more likely to have topped out in maximum brain weight and be undergoing decline. The picture is made more complex by the fact that the male brain tends to be heavier than the female brain throughout this period and for most of the lifespan. It is clear enough, however, that our brains do not continue to grow heavy with our weighty thoughts in young adulthood. We already seem to be passing the crest on this indirect measure of the "stuff" that supports our intellectual activity. A caution: the findings summarized here are best applied to age groups in general. There may well be significant individual differences not only at any particular age, but in the pattern of change in brain weight over the years.

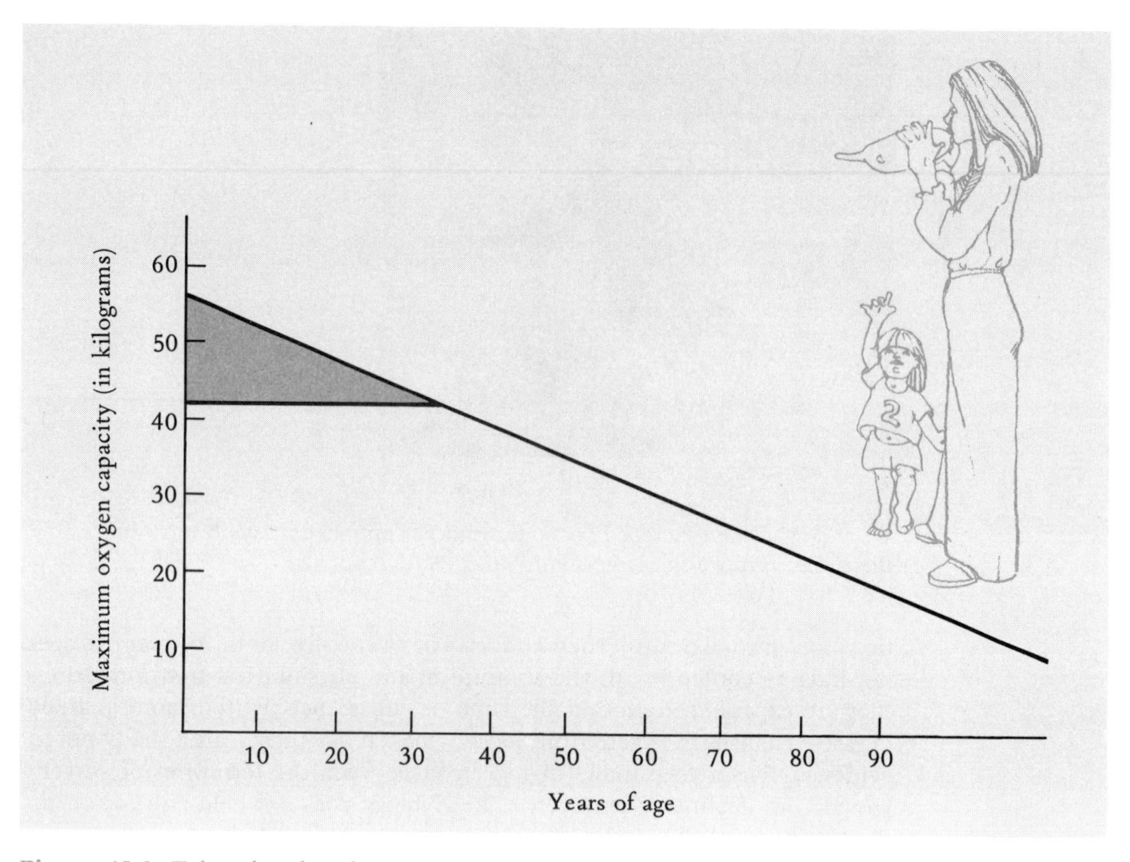

Figure 15-2. *Take a deep breath: an example of early physical decline. A decline in maximum oxygen capacity occurs from the early years onward. By the time a person reaches prime time, there has already been discernible loss (shaded area). (Based on data presented by Dehn & Bruce, 1972, from results of seventeen studies, all with males.)*

Accommodation is the eye's response to changes in distance to object viewed

Adaptation is the response to changes in lighting

Vision and audition are generally considered to be our most advanced or sophisticated sensory modes. Certainly they have been the most intensively studied. While the young adult is bright-eyed and sharp-eared, he or she is already starting to undergo decremental changes. Both vision and audition are complex processes. The timing and rate of both growth and decline differ from one dimension of vision or audition to another. Perhaps the earliest-fading sensory ability is **accommodation.** This refers to the ability of the eyes to respond appropriately to changes in distance to the object viewed. Accommodation reaches its peak in the grade-school days, so it is on the decline well before the individual approaches prime time. **Adaptation,** the ability to adjust to changes in light, begins its downward course at just about the time the individual enters the early adult years. Other aspects of visual performance tend to show little change during these years unless affected by environmental injury or health problems. Nevertheless, the overall pattern of decline in visual performance has begun to assert itself.

Pitch discrimination is the ability to distinguish between higher and lower tones

Pitch discrimination, the ability to distinguish between one tone and another tone that is higher or lower, starts to decline in the mid-twenties. This means that when a person finally has enough money to purchase a good stereo system, there may already be some loss in ability to discern the high tones. The ability to hear high frequency sounds starts to diminish in adolescence and continues thereafter. There is loss at lower frequencies as well, but to a lesser degree. Auditory acuity in general shows some impairment, although it is still a relatively small impairment.

The slight negative changes that occur in some dimensions of vision and audition during the early adult years often have little effect on the person's daily life and may go unnoticed. However, those who make special demands on their sensory systems are more likely to notice and begrudge the differences. A musician will not be pleased with even the slightest decrement in his or her pitch discrimination, nor will a thirty-five-year-old baseball player care for the new problems he is facing in identifying the type of pitch that is hurtling toward him in the split-second available for that purpose.

By contrast, *muscle strength* appears to increase at least into the late twenties. Large muscles such as the biceps may become even more massive over the next two decades, although gradually their efficiency begins to fade. For strength and coordination, these are indeed prime years.

We have looked at only a few indices of physical status in early adulthood, yet these suggest the larger pattern that takes shape when a more comprehensive review is made. In short, physical development plays itself out either a little before or during the twenties, with a few small areas of continued growth thereafter. Both the tremendous surge of early development and the renewed surge of growth related to puberty are over. From now on—and beginning within the prime years themselves—there are signs of another process at work. The twenty-five or thirty year old has the apparently competing processes of development and aging both operating on his or her physical person. It is a distinctive, perhaps unique developmental meeting place. The forces of growth are relatively weak now, having fulfilled their mission, and the forces of aging have not yet accelerated and taken command. There is a kind of quiet balancing between growth and decline tendencies, but with the shift already edging gradually toward the latter. Whether or not it is most useful to think of development and aging as competing processes is a matter we will consider again when we have moved further along the lifespan.

MENTAL ABILITY

The most familiar way of measuring mental ability is through the so-called IQ tests such as the Wechsler Adult Intelligence Scale, the Stanford-Binet, and

Omnibus measures are tests that measure several types of ability

the Army Alpha. Researchers speak of these as **omnibus measures.** They mean that each test actually assesses a variety of different mental abilities that are summed up in a total score. It is quite possible for two people to achieve the same total score, but for very different combinations of mental ability. IQ tests have a number of limitations, but they remain useful when interpreted with recognition of their advantages and disadvantages.

Performance on IQ tests appears to reach its peak in the early adult years, roughly between ages twenty and twenty-five. This conclusion is based on careful attention to many studies and to their methodological similarities and differences (Horn, 1970). Furthermore, this high level of performance holds stable, accompanying the person through his or her thirties and well into the forties.

We would know more about possible changes in mental ability, however, if *specific* components of intellectual performance could be identified and their distinctive paths of development traced (the way this has been done, for example, with different components of visual and auditory functioning). Fortunately, there has recently been some useful research accomplished in this area, although much remains to be learned. Horn (1970) and other leading researchers have speculated that there are at least two sets of determinants of the intellectual performances that are tapped by IQ tests. One set of determinants facilitates continued improvement over the years, while the other initiates a pattern of decline. When tested, an individual might show either growth or loss in mental functioning—it all depends on the specific mental

The sort of keen-mindedness that IQ tests measure reaches a peak at this time of life and stays there for at least a decade. (Courtesy of ACTION.)

abilities that are examined by a particular test (e.g., ability to define words, ability to visualize spatially) and their relationship to these two types of determinants.

The set of determinants that facilitates improvement is in the realm of learning and cultural experience. The young adult knows more than the adolescent who, in turn, knows more than the child. As the individual acquires more experience with life, his or her mental performances improve in some ways because more knowledge has been gained. The individual also has had more opportunity to master such cultural tools as language and mathematics. Were this the *only* process at work, then, we might anticipate continued improvement in mental performances throughout the entire lifespan.

But the other set of determinants carries the negative sign. It is thought to derive from the physiological changes that are associated with the aging process. We have already seen that early signs of aging appear even during the prime time years. The assumption is that those mental functions that are most dependent on intact, well-toned physiological functioning will show increasing impairment and inefficiency with advancing age.

The possible existence of two very different processes at the same time suggests that the intellectual functioning of the young adult is not adequately represented by IQ test scores. If both a growth and a decline process are taking place simultaneously, these might well cancel each other out, or the stronger process might obscure the existence of the less dominant process. According to this view, intellectual functioning in early adulthood may not truly be at its peak, nor be as stable as researchers once believed. It becomes necessary to look very closely at the different types of intellectual processes that go into the determination of the overall IQ score.

There is, in fact, a fairly impressive body of research that is consistent with this view. I must now introduce two concepts that are necessary to understand the findings and theory. Both derive from the long-term, systematic research of Raymond B. Cattell (1968). One concept is that of **crystallized intelligence.** This refers to the kind of knowledge and mental skills that are extremely dependent on exposure to the cultural environment. We would not know how to speak or write English, for example, if we had never been exposed to this language in some way. On IQ-type tests, these mental skills show up on tasks requiring general information, the ability to define words and use them correctly, and the ability to make effective use of familiar concepts. It can be said that crystallized intelligence (usually symbolized as Gc) is the distillation of experience, what we bring with us from previous situations that can find applicability in present and future situations. The pattern of evidence suggests that Gc does continue to *increase* throughout the early adult years and for some time thereafter.

Crystallized intelligence is culturally acquired knowledge and mental skill

It is a different story for **fluid intelligence** (Gf). The specific types of mental ability included in this category are thought to be highly related to the state of the body, especially the central nervous system; in other words, they are vulnerable to the effects of aging. On IQ-type tests, these abilities

Fluid intelligence is the ability to cope with new situations and to think abstractly and creatively

show up in tasks of abstract reasoning, fluency of ideas, spatial visualization, and the like. Such tasks have in common the fact that previous knowledge and experience are not sufficient to perform well. One must either cope with new problems and materials, or do something new with familiar materials. The person is less able to fall back (successfully) on previous experience alone. Fluid intelligence appears to decline fairly rapidly with age. This decrement seems to be present from adolescence onward. In other words, even while general mental performance is still moving toward its peak, one major component has started to drop off.

Factor analysis identifies and groups the commonalities within complex data

Some of the most persuasive research on this topic has made use of a sort of purifying statistical technique known as **factor analysis.** Essentially, this technique permits the extraction of a few basic components that underlie many specific items and tasks. With the tool of factor analysis (and some related multivariate statistical techniques) it is possible either to devise new measuring instruments that focus on the particular abilities one wants to examine, or to make better use of exisiting instruments (such as IQ-type tests). A significant study of this type was directed by Horn and Cattell (1967) who examined twenty different "primary mental abilities" in people ranging in age from fourteen to sixty-one. Figure 15–3 offers the striking results.

Notice that three developmental trajectories are presented. The omnibus measure represents the overall performance scores in which no distinction is made among specific components. These scores indicate that mental performance holds steady from mid-adolescence through the prime time years, and then actually rises a little in middle age. The Gc and Gf trajectories are dramatically different from each other, and also from the omnibus trajectory. Crystallized intelligence advances rapidly through the early adult years and keeps climbing. By contrast, fluid intelligence slides early and continues its descent without letup. Observe where the climb of Gc and the descent of Gf intersect. This takes place in the twenties. It is another way of visualizing the emerging pattern of the early adult years as a time when both developmental and aging processes (or incremental and decremental trajectories) are starting to contend for control of the organism.

The complete picture of mental development and decline in and around the early adult years is more complex than what has been presented here. There are some indications, for example, that people who start out with a higher level of intellectual functioning show more increase in crystallized intelligence and less decrease in fluid intelligence than people with more modest endowments. But the role of cultural opportunity is also important. The development of Gc, and perhaps also the maintenance of Gf, is related to the type of education, stimulation, and opportunity available to an individual. This means that intellectual development may not reach its potential or may fade too rapidly in a barren or threatening environment. For the person with an intense interest in this problem, there are detailed discussions and analyses available (*see,* e.g., Schaie, 1973; Baltes & Labouvie, 1973). Whether or not the rather extreme patterns of differential mental growth and decline re-

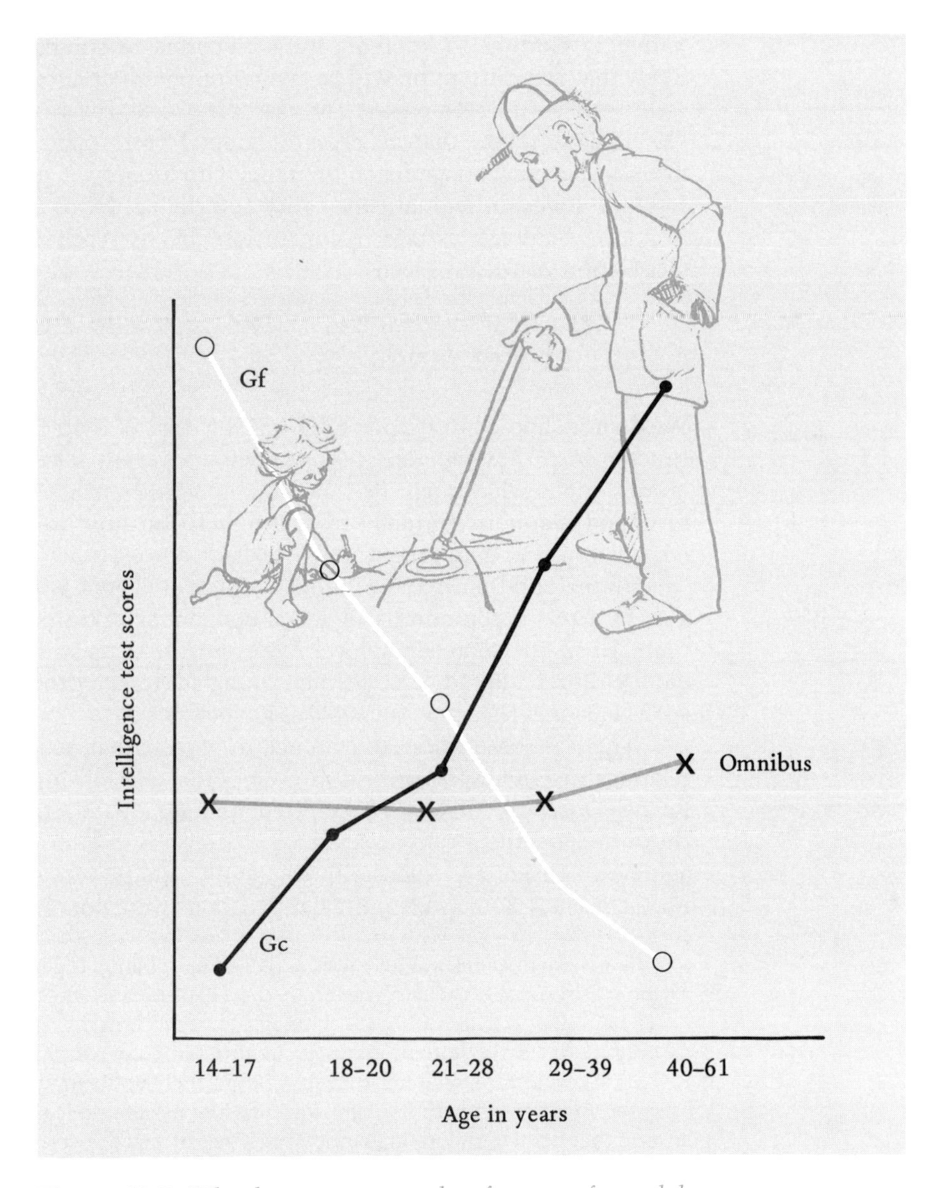

Figure 15-3. *What happens to mental performance from adolescence to late adulthood? Fluid intelligence (Gf) declines with adult age, while crystallized intelligence (Gc) increases. And the mixture of both on omnibus measures of mental functioning shows relatively little change. (From Organization of Data on Life-span Development of Human Abilities, by J. L. Horn. In L. R. Goulet & P. B. Baltes (Eds.),* Life-span Developmental Psychology: Research and Theory. *New York: Academic Press, 1970.)*

ported by Horn and Cattell are fully supported by subsequent research, it is likely that more attention will be given to individual differences in mental as well as physical development. For example, Costa, Fozard, McCrae and Busse (1976) have found that all types of mental performance (Gc and Gf) are lower in people who approach the tasks with anxiety. It may be as important to know about individual personality and the nature of the situation as it is to know about age or stage (a conclusion already reached in the area of life satisfaction, as we have seen).

SOCIAL COMPETENCE

We do not choose our friends solely on the basis of their physical and mental functioning. There is much more to being a person. There is, for an important example, what might be called *social competence.* This is the ability to form and maintain mutually rewarding relationships, to take one's share of responsibility, and, in general, to be a person among other people.

Young children may be enjoyed and appreciated although they are not able to meet all the criteria of social competence. We make allowances for their age and developmental level. We continue to make some allowances for older children and adolescents. But young adults are expected to be fully responsible, or at least as responsible as anybody else.

There is some research evidence to suggest that people who form good relationships in adolescence form even better relationships a few years later. Rachel Dunaway Cox (1970) studied undergraduate student leaders while they were on their campuses and a decade or so later. Some had already achieved considerable success in the world, others were still seeking a place for themselves. Some were quite happy, others were not.

> But none of them was an unattractive human being. Indeed they were even more winsome in their thirties than they had been as bright undergraduates. The qualities that probably led their classmates to honor them now show forth even more clearly. In their maturity, qualities of color, warmth, generosity, and honesty were expressed more freely than in their undergraduate days. The uncertainty of youth with its slight wariness and its sometimes edgy self-assertiveness that denied underlying vulnerability has dropped away. Now, in maturity, they were easier, more confident, less defensive. They were wiser, too, and much more humble, for almost every one of them had come face to face with ultimate issues of human existence and had confronted limitations in himself as well as in the human condition. Easy optimism was no longer possible for them, but at the same time they had developed or discovered new inner resources. (Cox, 1970, p. 278).

This impression associates growth in social competence with ample mental endowment. Yet even those participants in the study who had obvious limitations, impairments, and vulnerabilities "often inspired the greatest respect and warmth, for despite their difficulties they showed extraordinary tenacity" (p. 278). In general, the participants in this study had passed the early years of their development with many of the "advantages": intact families, adequate income, good opportunities for a variety of social interactions and experiences. It is not surprising, then, but perhaps a little reassuring that continued growth in social competence does flourish under such circumstances.

Erik Erikson believes that it is necessary for the individual to solve the challenges of one developmental stage to move on successfully to the next one. In young adulthood, the person who has resolved the major identity questions of adolescence is now ready for deeper, more intimate relationships with others. A relationship that is truly personal, an in-depth opening up and sharing, depends on having attained a firm sense of one's own self (Erikson, 1968). The individual who is still much occupied with an adolescent-type search for selfhood is not quite up to full appreciation of other selves. This view is similar to Abraham Maslow's (1954) distinction between relationships that are based on meeting one's needs, and those more uncommon and more mature relationships that flower when a person has become secure enough to appreciate others on their own terms. This does not mean that adolescents necessarily lack meaningful personal relationships. Rather, it indicates that there is an even more mature level of love and concern that a person can grow into after the developmental personality needs of adolescence have been met.

Throughout this book there has been emphasis on the dynamics of the *situation* as well as on the developmental process itself. Erikson (1975) has

Social competence has roots in childhood experiences with people who care about each other. The "family outing" is one of many experiences from which such competence is developed. (Photo by Talbot Lovering.)

suggested that the transition to adulthood has been complicated today by unusually strong distrust of the parental generation's competence. Adolescents usually find much to criticize in the world they are inheriting, but today's adolescents—around the world—have had more opportunity to learn of social inequalities and shortcomings. They have both more ammunition and more targets for criticism. Furthermore, Erikson has described adolescence today as being more of an independent stage or style of life than it was in the past. Previously, individuals were adolescents long enough to prepare themselves for the responsibilities of mature adult life. But now there is much peer group identity and power (*see*, e.g., Opler's views described in Chapter 13). The increasing separation of reproducing and child-bearing responsibilities from sexual behavior gives adolescents an earlier entry into adult-type relationships, but without the accompanying system of adult accountability and obligation. These are just two of the social-climate phenomena that are thought to give the transition from adolescence to early adulthood its special character today.

We might speculate together about the outcome of these phenomena when an adolescent does make the transition. In our very recent past, there were indications of an unusual intensity of political and "antiestablishment" concern on the part of adolescents. Many of these young men and women, including charismatic leaders, appear to have since become part of "the establishment" themselves. The eighteen year old who took a cause or philosophy of life seriously enough to risk many consequences, may now, only a decade later, be a pillar of community respectability (and perhaps a much more resourceful person for having undergone this sequence of experiences). Perhaps it is this person, who has made the most strenuous effort to develop an alternative life-style or to reform society, who has the greatest adjustments to make when he or she eventually confronts the daily responsibilities and obligations of adult life. It is not as simple to function in the day-to-day world without "moral compromises" or without trade-offs between advantages and disadvantages as it once appeared from the outside.

Erikson has spoken of the transition from the adolescent's *ideological experimentation* to a new process of *ethical consolidation* (Erikson, 1975). Social competency in the young adult seems to require a broader, less narcissistic (bound-to-the-self) conception of what is desirable and what is possible in life. The adult may give the appearance of seeming "less moral" than the adolescent, but this is not necessarily the case. The young adult's relationships with other people are based on a broader, more flexible outlook than the adolescent's. This broader outlook does not wear its moral concern with flamboyant colors and gestures. Yet this is an area in which development can go awry, isn't it? The adolescent may become disillusioned and discouraged, but without necessarily developing a new level of social-moral concern. Idealism tips over into cynicism. This particular sequence, however, would not be accepted by Erikson and other developmental theorists as the normal progression. It would be, rather, an instance of an adolescent growing older and more bruised by life, but without actually becoming an adult. Much remains to be learned about the subtle moral and social transformations between adolescence and adulthood—and why these transformations are less than 100 percent certain. The brief experience sketched in the following scenario may suggest something about these transformations.

SCENARIO 10: CH'ENG-JEN

Even the late afternoon sunlight was coming in at just the right angle. Mike paused in his duties as the day's designated houseperson to appreciate the physical ambiance of what Tanya and he had together. What a neat little place it was, and what good vibes it gave off!

But now it was the phone vibrating to life. Mike was very pleased at first to hear Doug's voice. They had been close back in high school, but hadn't seen much of each other for awhile. As the conversation went on, however, Mike's mind began to wander. He thought of what he still wanted to get done before Tanya came home. In a light way, he introduced the fact that he was becoming a pretty good cook. This didn't go over at all with Doug, though. The more he listened to Doug (and a listener was all Doug seemed to want), the more Mike realized that his old pal was still where he had been years ago. Except now, the quips were a little more forced, and everything seemed to be a going-through-the-old-routines rather than a really live interaction. It was good to talk over old times, but Mike felt he and Doug were just not on the same wavelength anymore. Doug seemed to recognize this too, and his voice was taking on a wheedling, almost accusatory tone. Mike finally wrapped up the conversation with a vague promise of getting together some time soon and having a "blast."

Actually, after he was off the phone, Mike wished that either he could have been more enthusiastic in talking with Doug, or more honest in expressing his feeling that they seemed to be going different ways. Mike continued to rummage in his mind about Doug and other friends of a few years back, those he still felt close to and those he had drifted away from. At the same time, he also continued putting finishing touches on the Chinese-style dinner he was preparing. Maybe there were a few too many can-openings and package-shreddings strewn around to qualify the dinner as an advanced-type achievement, but he was confident of Tanya's approval. She would, at least, congratulate him for getting the cans open!

Having supper together in their own place was still special and kind of thrilling for Mike, and he felt it was for Tanya, too. It seemed as though everyday they felt closer together, and yet every day was different; they were becoming different together while still staying themselves. "Ah, mystery of conjugal bliss," he hissed in his best imitation of what he imagined to be an oriental sage. "Confucius say . . ."

Mike's mood suddenly took a more serious turn. "Yes, Confucius did say. He spoke of finding your Way. But not the same Way for every person. And not a *place* or a *destination*, more a way of traveling through life. The Way is learning . . . becoming. Not scrambling ahead just to get ahead, whatever that means. And not sitting down and letting the grass grow all over you either." Wasn't there a word for it? Yes, and he found it readily came to mind: *ch'eng-jen*. "That's Tanya and me, both of us separately, and both of us together. We're something already, we have something. That would be *ch'eng*. A fulfillment, a completion. And we are still becoming . . . becoming something more that grows out of what we are. That would be *jen*. Chinese for adulthood, Chinese for a grown-up. Not *all* grown-up, but grown-up and still growing . . . through a long life together, I hope. Sorry you're not coming along, Doug."

Tanya at the door. The embrace. The appreciation. "That's a Chinese dinner like I've never seen before. Ummm. Does it have a name?"

"I think *ch'eng-jen* will do."

HAZARD: THE ONE-WAY RIDE

Gangster movies of the old Edward-G.-Robinson-and-James-Cagney type often featured menacing threats of a "one-way ride." Today—and for many years in the recent past—the one-way ride is a menace to young adults in real life. In an earlier chapter attention was given to the hazard of fatal accidents (*see* Chapter 10). It was seen that accidents of all types posed the number one life-threat to children. Let's return now to this hazard but focus our attention totally on the automobile.

A similar story could be told about the motorcycle as well

The young adult is still relatively free from life-threatening diseases, although no age level is completely immune. Furthermore, certain types of accident are especially uncommon among those in prime time. Fires, for example, account for about 7,000 deaths each year in the United States (Metropolitan Life, 1973), but children and elderly people are those most vulnerable to this form of accidental death. Young adults, perhaps because of their greater agility, have a relatively low rate of death by fire.

Automobile accidents, however, take a great toll. About three out of four fatal accidents in the twenty to twenty-four year old group involve automobiles, by far the largest number occurring when the person is driving or riding in the vehicle (Metropolitan Life, 1975). Neither cancer nor heart disease, the two most common physical conditions that threaten life at this age level, are responsible for nearly as many deaths as the motor vehicle accident. For every person who dies of cancer during their early twenties, ten are killed in automobile accidents; for every heart disease death, twenty-five are slain in vehicles.

It may be important to add that the motor vehicle death rate begins to decline markedly after this peak between ages twenty to twenty-four. Perhaps this is related to the possibility of temporary vulnerabilities at various points of the lifespan (Kastenbaum & Aisenberg, 1976). A person, in other words, may be "accident prone" in certain respects and for a certain, limited part of the lifespan. If he or she survives this particularly dangerous period, then the vulnerability may decrease considerably.

Whatever the reason behind the number of fatal accidents, it is difficult to overlook the fact that most young-adult deaths involve a lethal intersection of person and machine (and often alcohol as well). We have all heard the sermons on highway safety and the many suggestions offered to reduce the death toll. I have simply stated the facts here.

SUMMARY

The early adult years, roughly the twenties and early thirties, are considered by our society to comprise "prime time." This leads to heightened anticipation on the part of adolescents and heightened retrospection on the part of older adults. Furthermore, the social emphasis on early adulthood tends to increase pressure on people who are actually in prime time to accomplish and enjoy while they can.

Studies of *time perspective* and *subjective age* reveal the special value placed on young adulthood. However, empirical studies of *life satisfaction* do not support the simple conclusion that early adulthood is necessarily the best time of life. People may continue to derive considerable satisfaction from life through middle age and old age as well. The importance of individual *personality* or *life-style* has been emphasized by research findings. Often it is more useful to know *who* the person is than *how old* he or she is if we wish to understand level of life satisfaction. Important differences in life satisfaction also appear to be related to *male and female* roles and experiences. The interaction between personality, male-female experience, and age or stage of life requires close examination. Furthermore, the high frequency of *significant events* in early adulthood can lead to increased *stress*, with both psychological and physical consequences.

In general, the young adult is an able person. In this chapter, the abilities and challenges of this period were examined in the *physical*, *mental*, and *social* spheres. While many physical abilities are around their peak efficiency at this time of life, the developmental or growth process has slowed down appreciably. Some physical functions are already showing decrements. Young adulthood can be seen as a time of *balance* between the ebbing tide of development and the coming influence of aging.

Assessment of mental development is difficult for many reasons. It has to be considered with respect to the specific concepts and measuring techniques involved as well as individual and situational differences. There is increasing evidence, however, that certain aspects of mental functioning continue to improve into and beyond early adulthood; these aspects are called *crystallized*

YOUR TURN

1. Make your own survey of life satisfaction before, during, and after prime time. It does not have to involve a long list of questions. Simply ask questions such as: (1) "What gives you the most satisfaction in life now, these days"; (2) "What gave you the most satisfaction a few years ago"; and (3) "What do you expect will be your biggest satisfactions in life a few years from now?" Inquire of people who are in early or middle adolescence, in their twenties or thirties, and in their forties or beyond. You might conclude by asking them to indicate which of these periods (a few years ago, now, a few years from now) they feel will prove to be the most satisfying. Be ready to accept the answer that they will be equally satisfying! Be ready to identify some of the factors associated with the answers, for example, do males and females tend to emphasize different types of satisfaction? And, while you are at it, why not ask yourself?

2. Erikson's influential statements about the adolescent-adulthood transition were up to date just a few years ago. And yet the social climate seems to have shifted again (e.g., less political activism and cause-advocating among young adults). Read carefully some of Erikson's analyses (e.g., *Identity: Youth and Crisis,* 1968) and then provide your own follow-up and modifications based on your understanding of what has been happening more recently. How much of what Erikson suggested still appears applicable? Do changes in the social situation contradict or undercut the persuasiveness of his earlier observations?

3. Examine the concept of *adulthood* in other cultural settings and compare them with our own. Or examine the *roots* of the concept of adulthood in our own culture. One place to begin is with the Spring 1976 issue of *Daedalus* which offers a variety of views on the nature of adulthood.

intelligence. Another major aspect, *fluid intelligence,* seems to wane at about the same time. The total picture of mental development and decline in early adulthood is a complex one that has not yet been entirely clarified.

Social competence appears to increase from adolescence through early adulthood. Rough edges are smoothed out; there is a greater sense of security with other people. Erikson and other developmental theorists believe that this is the time of life when it is possible to form more *intimate* relationships with other people because the individual has now well established his or her own sense of identity. People who have not come out of adolescence with a firm sense of identity will have difficulty in developing truly intimate, sharing relationships. Consideration was also given here to *situational dynamics* which influence the transition from adolescence to adulthood (e.g., a climate of political activism or its opposite).

The *motor vehicle accident* was spotlighted as a significant hazard, especially during the twenty-to-twenty-four-year-old range.

It was suggested that the view of a "Way" of life that derives from the teachings of Confucius still has applicability today, integrating as it does the idea of having attained a certain level of "grown-upness" while continuing on a life-long journey toward further growth. This differs from the notion that prime time, or any other specific place along the lifespan, is *the* only or preferred destination for human development. The young adult may be ready to appreciate that the journey itself is the most significant part of the story.

Reference List

Anderson, C. College schedule of recent experience. In M. Marx, T. Garrity & F. Bowers, The influence of recent life experience on the health of college freshmen (Appendix I). *Journal of Psychosomatic Research,* 1975, *19,* 94–98.

Araujo, G., VanArsel, P., Holmes, T., & Dudley, D. Life change, coping ability, and chronic intrinsic asthma. *Journal of Psychosomatic Research,* 1973, *17,* 359–363.

Baltes, P. B., & Labouvie, G. V. Adult development of intellectual performance: Description, explanation, and modification. In C. Eisdorfer & M. P. Lawton (Eds.), *The psychology of adult development and aging.* Washington, D.C.: American Psychological Association, 1973, pp. 157–219.

Cameron, P. Stereotypes about generational fun and happiness versus self-appraised fun and happiness. *The Gerontologists,* 1972, *12,* 120–123.

Cattell, R. B. Are I.Q. tests intelligent? *Psychology Today,* 1968, *1,* 56–62.

Costa, P. T., Fozard, J. L., McCrae, R. R., & Busse, R. Relations of age and personality dimensions to cognitive ability factors. *Journal of Gerontology,* 1976, *31,* 663–669.

Cox, R. D. *Youth into maturity.* New York: Mental Health Materials Center, 1970.

Cumming, E., & Henry, W. E. *Growing old.* New York: Basic Books, 1961.

Daedalus (magazine, whole issue). Spring 1976, *105.*

Dehn, M. M., & Bruce, R. A. Longitudinal variations in maximal oxygen intake with age and activity. *Journal of Applied Physiology,* 1972, *33,* 805–807.

Erikson, E. H. *Identity: Youth and crisis.* New York: W. W. Norton & Company, Inc., 1968.

Erikson, E. H. *Life history and the historical moment.* New York: W. W. Norton & Company, Inc., 1975.

Fontana, A., Marcus, J., Noel, B., & Rakusin, J. Prehospitalization coping styles of psychiatric patients: The goal directedness of life events. *Journal of Nervous & Mental Disease,* 1972, *155,* 311–321.

Friedman, M., & Rosenman, R. Association of specific overt behavior patterns with blood and cardiovascular findings. *Journal of the American Medical Association,* 1959, *169,* 1286–1296.

Hinkle, L., Pinsky, R., Bross, I., & Plummer, N. The distribution of sickness disability in a homogeneous group of "healthy adult men." *American Journal of Hygiene,* 1956, *64,* 220–242.

Holmes, T. S. & Holmes, T. H. Short-term intrusions into the life-style routine. *Journal of Psychosomatic Research,* 1970, *14,* 121–132.

Holmes, T., & Rahe, R. The social readjustment rating scale. *Journal of Psychosomatic Research,* 1967, *11,* 213–218.

Horn, J. L. Organization of data on life-span development of human abilities. In L. R. Goulet & P. B. Baltes (Eds.), *Life-span developmental psychology: Research and theory.* New York: Academic Press, 1970, pp. 424–467.

Horn, J. L., & Cattell, R. B. Age differences in fluid and crystallized intelligence. *Acta Psychologica,* 1967, *26,* 107–129.

Kastenbaum, R. Time and death in adolescence. In H. Feifel (Ed.), *The meaning of death.* New York: McGraw-Hill, 1959, pp. 99–113.

Kastenbaum, R. The dimensions of future time perspective: An experimental analysis. *Journal of General Psychology,* 1961, *65,* 203–218.

Kastenbaum, R. The direction of time perspective. I: The influence of affective set. *Journal of General Psychology,* 1965, *73,* 189–201.

Kastenbaum, R., & Aisenberg, R. B. *The psychology of death.* New York: Springer Publishing Co., Inc., 1976.

Kastenbaum, R., Derbin, V., Sabatini, P., & Artt, S. "The ages of me": Toward personal and interpersonal definitions of functional aging. *International Journal of Aging & Human Development,* 1972, *3,* 197–211.

Kastenbaum, R., & Durkee, N. Young people view the elderly. In R. Kastenbaum (Ed.), *New thoughts on old age.* New York: Springer Publishing Co., Inc., 1964, pp. 237–249.

Kovenchevsky, V. *Physiological and pathological aging.* New York: Hafner, 1961.

Lowenthal, M. F., Thurnher, M., & Chiriboga, D. *Four stages of life.* San Francisco: Jossey-Bass, Inc., Publishers, 1975.

Maslow, A. H. *Motivation and personality.* New York: Harper & Row, Publishers, 1954.

Metropolitan Life. Mortality from accidents in the United States and Canada. *Statistical Bulletin,* 1973, *54,* 9–11.

Metropolitan Life. Accident mortality at the preschool ages. *Statistical Bulletin,* 1975, *56,* 7–9.

Neugarten, B., Havighurst, B., & Tobin, S. Personality and patterns of aging. In B. Neugarten (Ed.), *Middle age and aging.* Chicago: University of Chicago Press, 1968, pp. 173–177.

Nuckolls, K., Cassel, J., & Kaplan, B. Psychological assets, life crisis, and the prognosis of pregnancy. *American Journal of Epidemiology,* 1972, *95,* 431–439.

Paykel, E. Life stress and psychiatric disorder: Application of the clinical approach. In B. S. Dohrenwend & B. P. Dohrenwend (Eds.), *Stressful life events: Their nature and effects.* New York: Wiley-Interscience, 1974, pp. 135–150.

Schaie, K. W. Developmental processes and aging. In C. Eisdorfer & M. P. Lawton (Eds.), *The psychology of adult development and aging.* Washington, D.C.: American Psychological Association, 1973, pp. 151–156.

Theorell, T. Life events before and after the onset of a premature myocardial infarction. In B. S. Dohrenwend & B. P. Dohrenwend (Eds.), *Stressful life events: Their nature and effects.* New York: Wiley-Interscience, 1974, pp. 101–118.

NEITHER YOUNG NOR OLD

16

CHAPTER OUTLINE

I

The voices of theory and research are remarkably quiet about what it is to be a grown-up.

Reduction to the routine, holding on / Social integration, holding society together / Pluralistic model of development (conditional, selective)

II

Because so much time and energy is spent on the job, there is much that could be learned by discovering the developmental implications of the occupational experience.

Time absorbed by work / Opportunities / The job situation

III

Several studies of developmental trajectories improve our understanding of the relationship between who we are at one time of our lives and who we become later.

Gifted children (Terman) / Change and continuity / Normative aging in Boston veterans (Costa and McRae) / San Francisco citizens (Lowenthal / Complexity or simplicity, male or female

IV

Development continues throughout the adult years but we have much to learn about this process.

Turning the corner / Alcoholism as a hazard / Interiority (Neugarten) / Mid-life challenges (Gould, Levinson)

There is guarded optimism that this might be a decisive session of The Commission To Do Something About Middle Age. After all the studies and all the committee work, perhaps we will at last have a firm recommendation. You wouldn't give me a preview of your own recommendation, would you? Ah, but the chairperson is ready to begin. Let's listen.

Chair: Will this session come to order? Thank you. Today we approach the fulfillment of the mandate entrusted to us by our fellow citizens. May I remind you of the heartfelt need we seek to satisfy. The scroll, please. (An enormous scroll is wheeled out. It is unrolled ceremoniously by a taxpayer, a teacher, a parent, and an official.) Here are the signatures of thousands upon tens of thousands upon hundreds of thousands, upon millions of men and women who have petitioned on behalf of middle age. Children are the jewels and delights of our society; the young adults are the hope for a better future; the aged are the cherished veterans of life whose health and dignity must be preserved. But we, the neither young nor old, remain unsung. Worse, we remain *unclassified.* There is no bold legend for us, no diagnosis, no fascinating mystique. We—we are just *here* . . . taken for granted (Chair starts to weep) . . . Forgive me. I'll—I'll try to control myself. Our mandate is first to classify middle age, and then to do something about it. I will now call on each member of the Commission for his or her summary recommendation. Just a few words, please. Be right to the point. Your full reports will be made available through the superintendent of public documents.

We will hear first from the spokesperson for the Subcommittee on Psychiatric Generalities.

Spokesperson 1: The poet W. H. Auden has diagnosed our society and our times as "the age of anxiety." This is correct, but too broad. It is the middle-aged person who truly lives in the age of anxiety. The pressures and uncertainties of modern life and the ever-present dangers from forces deep within the person and from a hostile universe are at their peak during these years. Middle age should be classified as a psychiatric disaster area—we *are* the age of anxiety! And the federal government should, of course, provide proper relief, as it does for any disaster area.

Chair: Thank you so much. Now the spokesperson for the Subcommittee on Aggravation.

Spokesperson 2: Right to the point, just like you asked. Middle age is The Aggravated Generation. Do I need to say another word? Of course not! Everybody here knows what I am talking about. We are the people who are taken for granted, from whom everybody wants something, from whom everybody *expects* something! And, I ask you—no! I tell you—what do we get in return? (Speaker pauses, closes eyes, raises both arms up toward heaven). Aggravation! *That's* what we get!

Chair: So true. The spokesperson for the Subcommittee on Philosophical Reassurances, if you will.

Spokesperson 3: (Softly, and with a mild, strangely irritating smile that remains throughout) Anxiety? Aggravation? These are but the bubbles of indigestion. Life has so much to offer us, a feast set forth by nature and by the great minds of the centuries. We have only to choose wisely. The ill-considered choice produces that existential indigestion we know as *anxiety* and *aggravation*, and many other names. Middle age, dear friends, is the time of *maturity*. And the time of maturity is the time of *balance*. As my friend Aristotle was saying just the other century, here is that most precious period of life when one is neither too young nor too old. A perfect balance in which reason and virtue command the realm. We are no longer tart green fruit, nor yet the withered residue. All that is best and most becoming to a person exists in perfect balance. Let us tell the world proudly (within the proper limits of rational humility, of course) that we are the mature generation!

Chair: I couldn't agree more. And now the spokesperson for the Subcommittee on Biopolitical Cynicism.

Spokesperson 4: Maturity? Hah!

Chair: Ummm. Don't you have something more to say?

Spokesperson 4: Sure. Maturity? Hah-*hah!*

Chair: Well that certainly is to the point, but—

Spokesperson 4: Middle age is the wasteland between adolescence and senescence. It is nature's way of yawning between one important job—getting us grown-up—and another one—getting us out of the way. Yes, occasionally something does happen between youth and old age that you might call *maturity*. We come across all kinds of unusual phenomena in biology, zoology, and medicine. But for most men and women, the middle years make up a prolonged interval of deadening routine. See, look at yourselves right now: you are starting to yawn even as I remind you of the truth. Youth is the dawning generation. We, the middle-aged, are the yawning generation. Did I say enough, Chair?

Chair: (Stifling a yawn) Yes, definitely. Couldn't have said it better myself. We are privileged to hear now from the Spokesperson for the Subcommittee on Paramilitary-Industrial Benevolence.

Spokesperson 5: First, everybody up! Up, up! Out of your seats. Let's tone up those muscles, send that blood coursing through that too, too solid flesh. Run in place. One-two-one-two-one-two-one-two. Faster! Keep it up! Feels great doesn't it! One-two-one-(cough)-two-one-(wheeze)-two. That—that will do for now, I (whoosh) don't want to tire you out.

Physical fitness is all this generation needs to fulfill its destiny. We have all the other muscles already. Money. Power. Experience. Control. Who is really running things? Who should be running things? Yes, you in the back row, what do you say? Right, *us!* But say it again, with more confidence. All together. Who is running things? Us!! Us!! We are *the establishment generation.* The power, the authority, the responsibility is right where it should be, in our hands! We have worked hard to be where we are. Let's not let anybody think they can take this away. Or think they can shame us. Is it wrong to have power, is it shameful? Of course not! Our critics—and they are many—want this power for their own. But they are too young or too old to wield the power as only we can. We are *the* generation to reckon with. Raise your fists in salute, brothers and sisters of the mid-lifespan! We are the strong and proud ones, the establishment generation!

Chair: Oh, yes! Yes! Yes! Whew! Let me admit that I am exhausted from this stimulating series of reports. But it does seem to me that perhaps we still have a little—ah—adjustment of the various conclusions to form a final recommendation, and there may be still other voices to hear from. This session stands adjourned.

CONCEPTIONS OF MIDDLE AGE

We have just attended an imaginary session of The Commission To Do Something About Middle Age. But the problems and the views expressed are real enough. Before we examine them with a more analytical eye, it will be helpful to reflect a moment on the background of available knowledge. The mainstream of developmental theory is reduced to a trickle by the time it reaches the adult years. Both the theories and the studies tell us much about the process of growing up. But what is it to *be* a grown-up? The voices of theory and research are remarkably quiet here.

Activity picks up again when we jump ahead to old age. The problems experienced by many old people—and society's problems in adjusting to an increasing proportion of elders—have stimulated increased attention in recent years. Furthermore, "aging" is seen by many as a fundamental process. In old people, it is easier to detect something going on, a process that perhaps is the undoing or reversal of "development." We can at least enjoy the illusion that we know what is happening in childhood and adolescence (development) and again in later life (aging). The inbetween period does not yield itself so readily to the traditional ways of viewing human behavior and experience. This makes the middle-aged person something of a mystery (and maybe something of an embarrassment) to developmental theory.

Sometimes middle age is given the benefit of the doubt. It is treated as though developmental processes are certainly still operative. But there is little specification of precisely what is developing and what principles are involved. There is danger that development can lose its distinctive meaning with such vague usage. Instead of trying to win a quick, apparent victory by claiming that middle age is a developmental phase, we will examine some of the alternatives.

HOLDING ON: MIDDLE AGE AS REDUCTION TO THE ROUTINE

This possibility is not likely to find favor with developmentalists, but it should be considered

One possibility is that the years roughly between forty and sixty are *not* marked by any process that deserves to be called *development*. The surge of growth has run its course. Physical development is an important case in point. It is easy to find indices of continued physical development around adolescence, and some examples can even be discovered in early adulthood. But those types of systematic change that improve functioning and adaptation are difficult to see in the physical status of middle-aged men and women. Nobody even bothers to look for them (which might be a mistake). Based on present knowledge and expert opinion, what we usually call development is not part of the physiological and anatomical scene for the middle-aged adult.

If we do not develop *physically* at this time of life, then do we develop at all? Remember that through the earlier phases, there has always been an interplay between physical and psychosocial development. One could maintain that the termination of physical growth "shuts off the juices" for continued psychosocial development. The young child, for example, must develop certain cognitive abilities to make use of his or her emerging physical potentialities. There is also good reason for further intellectual development around puberty if one is to become a successful parent and help perpetuate the species. This kind of logic involves some reading of purpose into the developmental process, an approach that is deservedly controversial. But, taking this logic further along the lifespan, if there is no new physical development, what is the bio-social point of new mental or emotional development? Why should we expect it? Put the burden of proof, then, on those who assert that development does continue through middle age.

Sometimes the middle-aged person feels like a member of the "yawning generation" or is seen that way by others. (Photo by Talbot Lovering.)

The spokesperson who characterized middle adulthood as *the yawning generation* would agree with the view that is being expressed in this section. The argument is that the typical adult settles into a routine. Day follows day and year follows year in a rather predictable repetition of actions and experiences. These are punctuated by a little excitement here, a surprise there, and occasionally a stressful or disruptive event that rocks the boat. It may be extreme to classify the middle-aged as bored and boring. But it would also be extreme to refer to this thoroughly patterned, self-repeating form of existence as development!

In summary, the middle years can be regarded as a time of *holding onto* the functional abilities that emerged from the now-terminated developmental process. It is, at best, a time when the decrements are relatively small and gradual, covered nicely by "know how" and confidence gained through successful living. Life can be pleasant and rewarding, but this is not to be mistaken for development.

HOLDING SOCIETY TOGETHER: MIDDLE AGE AS A PERIOD OF SOCIAL INTEGRATION

The **biosphere** is the realm of physiological and biochemical processes

Another possibility is that an important kind of development does occur during the middle adult years, but we must look for it in the right place. Shift the spotlight away from the **biosphere.** Don't become preoccupied with purely cognitive processes either. Instead, observe the network of human relationships, observe the middle-aged adult functioning in society. This is where the developmental process now becomes most dominant.

Just a moment ago we were emphasizing the individual's tendency to hold onto his or her personal abilities after early adulthood. But a very different perspective is now being proposed. Society must be held together as well. Who is going to take the responsibility for preserving a viable society? Who will see to the day-by-day functioning of all the formal and informal subsystems that keep society going? Everybody, or almost everybody, has a role to play in society's survival. But the middle-aged person appears to have the greatest opportunity and responsibility. Well established as an individual, he or she is in a good position to serve as "the establishment" for society in general.

Social integration is an overall measure of how a person is tied to society through mutual obligations, demands, and responsibilities.

This is the time of life when the individual is at the peak of his or her **social integration.** There are important family and occupational responsibilities. Much is expected; there are many obligations and demands in various sectors of life. Another way of saying this is to refer to the middle-aged person as thoroughly "engaged" (Cumming & Henry, 1961). The person who lives up to these obligations, expectations, and opportunities often must schedule hours and days carefully. There always seems to be so much to do and so few hours in the day. The middle-aged individual is likely to feel very much a part of things. In fact, there is often an unspoken understanding that

Here is where decisions get made—and it is often the mid-life adult who has these responsibilities, as at this school board meeting. (All work not otherwise credited is from the Allyn and Bacon Photo Collection.)

the middle-aged adult "is" society. This is the person who is visualized when it is asserted that society wants this, demands that, or forbids something else. Standards are set, defended, and enforced by the powerful middle generation.

Stress, aggravation, and harassment may go along with this position. Too much may be expected of the middle-aged (by themselves as well as by others). Extremely difficult decisions may confront them (e.g., governmental officials whose judgment may swing the balance between war and peace, prosperity and hard times). The young and upcoming generation may attack them as they seek their own share of power. At the same time, the older generation may decline to relinquish the control the middle-aged echelon believes should now be in their hands. There may, in addition, be a sense of being taken too much for granted, being faulted and assaulted when things go wrong, and seldom being appreciated for their steady contribution when things go right.

Looked at in this way, we have a possible answer to the question of purpose. There *is* a need for continued development in mid-life, but not directly for the individual as such. Instead, the need is for development along those lines that will preserve and strengthen society at large. At the least, this means holding together a society that protects the young during their long and vulnerable period of development. This is a biosocial point, for humans are among the species that take a long time to reach full functional potential. Beyond this minimum, it could be argued that society must also be maintained at a sufficiently high level to nurture the aged and those who are vulnerable or impaired at any age for a variety of reasons. To go even further, it could be argued that society should advance, should evolve, should become better and better. Therefore, the middle-aged generation must make progress *possible,* even if some of the important specific innovations come from the young.

Not everybody will want to go quite this far. The idea of progress—and the idea of looking after the vulnerable and "unproductive"—has not won universal acceptance. People, within our own society and others, differ on these points. But the general point remains: whatever it may be that society is "for," some people must accept the responsibility for making it work. In our society, the middle-aged appear to be the designated operatives to keep us all more or less together and functioning.

If this view is accepted, then in middle age we could expect to find continued development of skill in interpersonal relationships, in long-range planning, and in coping with the complex realities of the practical world. Performance on a factor-analyzed test of mental functioning would be less relevant than discovering how a businessman or woman meets the payroll in difficult times. Studies of the electrical activity of the brain would be less revealing than learning how a mediator solves a tough and potentially violent dispute between parties with differing interests and backgrounds. In short, we would have to look for development elsewhere and by other methods than those ordinarily used in the study of growth processes during the earlier phases of life.

CONDITIONAL, SELECTIVE, AND PLURALISTIC DEVELOPMENT: MIDDLE AGE AS A TIME OF POSSIBILITY

Where does it say human development is simple?

Consider one more general possibility. Perhaps middle age is a constellation of possibilities that belong together. Perhaps development continues. But only for some people. And not necessarily in the same way for those who do continue to develop. And not necessarily under all circumstances.

This view has not yet been firmly introduced into developmental theory. Let's be clear about how it differs from the more customary ways of regarding human development. It is usually assumed that everybody develops. There are exceptions, but these are explicitly taken as exceptions. We look for the cause of arrested or even retarded development in the early years of life. There is no expectation that some people will normally remain at a five year old's level while others will normally move ahead. I am suggesting here that this expectation may not be in accord with the facts, especially when we reach the adult years. To assume that either *everybody* develops or *nobody* develops amounts to a self-imposed blindfolding. It is not in the scientific spirit of fresh and determined observation.

It is also assumed that development follows a single, primary path. If we are going to develop, then we are going to go *that* way, and no other. This view is expressed through the many stage theories that have come forth over the years. Piaget and Erikson have provided two of the most influential theories of this general type, but organizing observations and opinions in terms of "stages" has proven a very popular approach for other developmentalists as well.

Pluralism is the view that there is more than one "normal" pattern of development

Notice that the concept of staging would be decidedly less powerful if we had to make room for **pluralism,** for alternate pathways. This is illustrated in Figures 16–1, 16–2, and 16–3. Many theories assume there is only a single developmental pathway (*see* Figure 16–1). An individual must avoid the pitfalls along the way and not wander off the "straight and narrow." Another common type of theory holds that some people continue to develop, to move through the maze of life, while others settle into a more limited position (*see* Figure 16–2). The first type of theory distinguishes between "normal" and "deviant" (off-the-path) development; the second type distinguishes between those who progress greater and lesser distances along the essentially predetermined developmental stations.

But we might also think of humans developing in a more individual manner. The tree-of-life model (*see* Figure 16–3), as we might call it, suggests that there are *many* possible forms of development a person might take. None of these patterns are necessarily more or less "normal" than the others. This makes for a less tidy theory. It is just as plausible logically, however, and less likely to result in people becoming stereotyped if they develop in a distinctive way. The tree-of-life, pluralistic model further suggests that individuality might increase as the person moves further beyond his or her starting point. It implies that a strict stage theory might be relatively accurate for early development, but not for later development. Clinging to conventional stage theory through the adult years could then become an increasing source of distortion and error.

Notice also one important psychosocial implication of a pluralistic model of development. We no longer have to classify many people as *deviates* because they stray from the one-and-only path. The inclination to judge people as normal or deviate comes readily when we are limited to the single-path model. With a pluralistic model, we can concentrate more carefully on the individual's *style* of development without having to fit him or her into a fixed mold.

Part of the model of development proposed in this section is concerned with the *situation*. Some situations are more conducive to development than others; some situations are more conducive to certain kinds of development than other kinds. All modern theories of human development give some attention to the situation (*see* Chapter 7), but the situation is especially significant when we also recognize that whether development will continue at all, and whether it will continue in one way or another, is a conditional matter. Let's make use again of a principle suggested earlier in this section. It was said

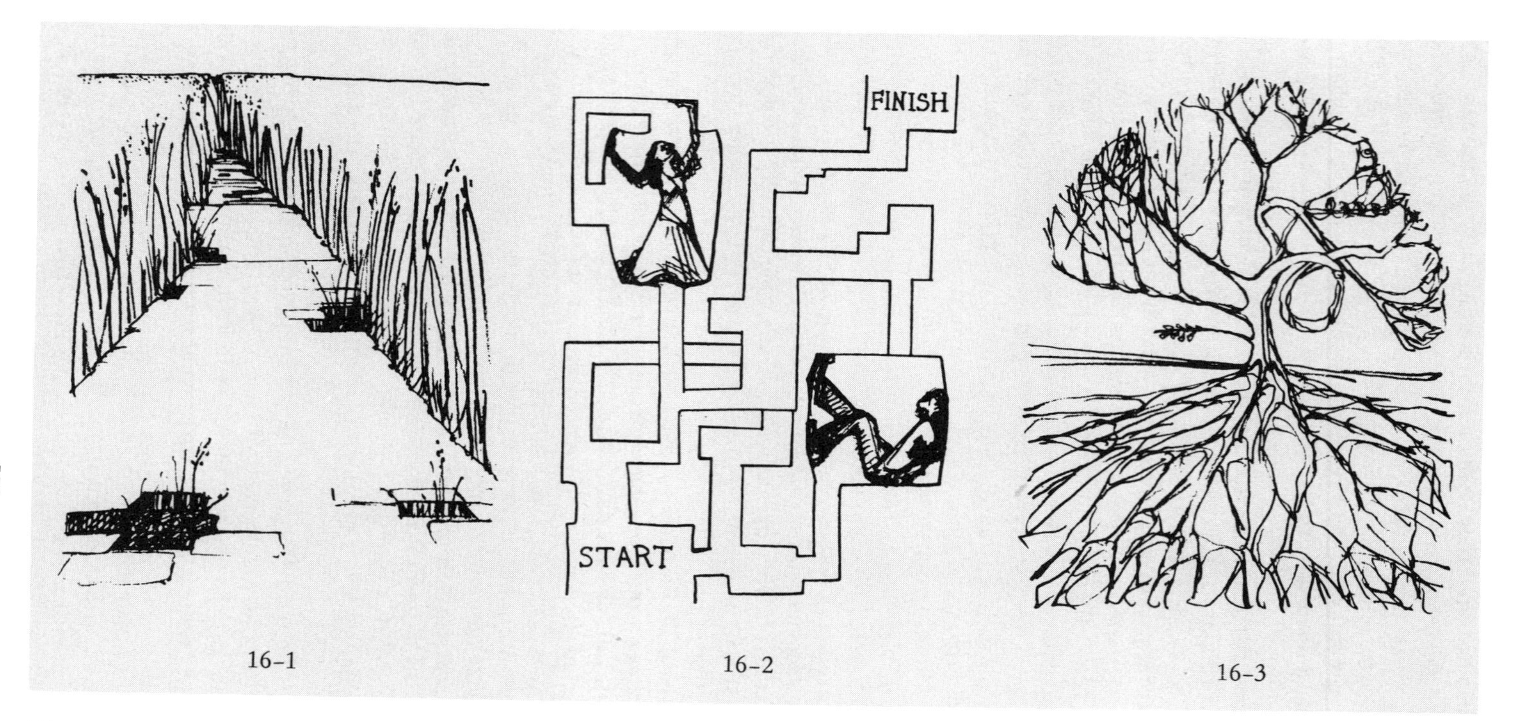

Figures 16-1, 16-2, 16-3. Three models of human development. *Many psychological theories assume there is only one straight-and-narrow path through life; one must be careful not to stumble and fall to the side (16-1). Stage theories are a more complex variant that suggest some people go farther than others in their maturation (16-2). Proposed here is a pluralistic, tree-of-life model in which a variety of individual forms of development is possible, akin to the variety of forms seen elsewhere in nature (16-3).*

that *the number of possible pathways of development increase the further we go along the lifespan.* Retain this concept and augment it with the proposition that *continued development becomes more problematic (less automatic) the further we go along the lifespan.* The odds are strongly in favor of continued development from, let's say, middle to late childhood. The odds for continued development from early to middle adulthood would not be nearly so favorable, if we actually had a way of determining them. And the situation in which the person must function will have much to do with the possibility and directionality of any further development.

What do we have, then, if we accept the possibility that development is conditional, selective, and pluralistic? We have, for one thing, a basis for appreciating the striking differences that can in fact be observed among adults at any age level. If some people seem to be continuing to develop, while others seem to have "run out of program" (or "out of steam"), then we can take these differences as serious starting points for investigation. We do not have to abolish or ignore either the developing or not-developing people in order to keep our theory propped up. We also have the impetus to improve our concepts and observations in order to take the variety of human development into account, and "permission" to find dimensions of human growth in later life that are not limited to those familiar in earlier life. We can approach more adequately a true lifespan theory, rather than burden ourselves with concepts and methods heavily indebted to studies of the young.

Moreover, the *number* of people who continue to develop into and beyond the middle years of life becomes only of secondary interest. If even *one* person actually shows significant development, then this is sufficient to establish the reality of the phenomenon. Careful study of the situation as well as of the individual's previous life experiences can perhaps help us foster continued development for more people. But the basic point is that the existence of a true developmental process that continues throughout life deserves our keen attention, even though this may be a process that takes different forms and may be restricted to only some people in some situations.

WHAT WE DO AND WHAT WE BECOME: OCCUPATION AND MID-LIFE DEVELOPMENT

Consider the general life situation of the middle-aged adult. In what ways might this situation either facilitate or impede continued development? Because so much time and energy is spent on the job, there is much that could be learned by discovering the developmental implications of occupational experience. It will also be important to bear in mind that the conditions of work do not necessarily remain constant over the decades. What is expected of a person on the job today—in terms of loyalty, hours worked, decision-making, exposure to hazards, etc.—often differs from what a person of the same age might have encountered if he or she had been living and working years ago. Similarly, occupational experience in the future may not be what it is today.

TIME ABSORBED BY WORK

Future data will have to consider the new retirement age of seventy

One of the simplest illustrations of how life situation affects continued development in middle age is the matter of how much of a person's time is absorbed by work. It has been estimated that in 1870 the "average" person started working at age fourteen. When did the working career end? With his or her death at about age sixty-one. This individual spent about 3120 hours on the job each year, in all, about 146,640 working hours during a 47 year span (Miernyk, 1975). By contrast, today it is estimated that the average person enters the work force at about age twenty and retires at age sixty-five. While the number of years as a worker has remained about the same, the hours have shrunk to about 2000 per year and about 90,000 over the entire span. This represents almost a 40 percent reduction in the number of hours a person spends in the work situation.

Notice some of the implications this has for continued development. The working adult today has considerably more time left free from on-the-job obligations. This can mean more opportunity to develop interests, skills, and a breadth of knowledge hard to come by when one is bound more extensively to the work situation. Furthermore, the later entry into the work force also suggests the opportunity for additional educational experiences, both formal and informal, that might contribute to continued personality growth. The level of formal education has, in fact, increased when compared with, say, the generations to which our grandparents and greatgrandparents belonged. Furthermore, the fact that many people today out-live their job obligations offers at least the possibility for cultivating developmental opportunities that were not available to the man or woman who died still in harness.

These considerations, although important for illustrative purposes, are only quantitative. They do not take us very far toward understanding the trade-off in qualitative experiences between a working career that begins in mid-adolescence and lasts until death, and one that begins later, makes fewer time demands along the way, and leaves the person with some time left to call his or her own after retirement. Some people in the past managed to learn much through their work experiences and cultivated distinctive personality styles that showed growth and richness. Some people today seem to grow stale or run out of developmental program early, despite the rigorous time demands of work. There is much to learn on this topic, learning that would help us to anticipate, and perhaps influence, the *future* relationship between work career and individual development through the adult years. At the present time, however, we lack some of the most basic information. For example, does a working life that absorbs a great deal of the individual's time and energy tend to encourage a *different* kind of developmental process through the middle and later years? If so, what are the differences? Are there different strategies needed to pursue individual development under varying employment conditions? Answers to questions such as these are worth vigorous pursuit.

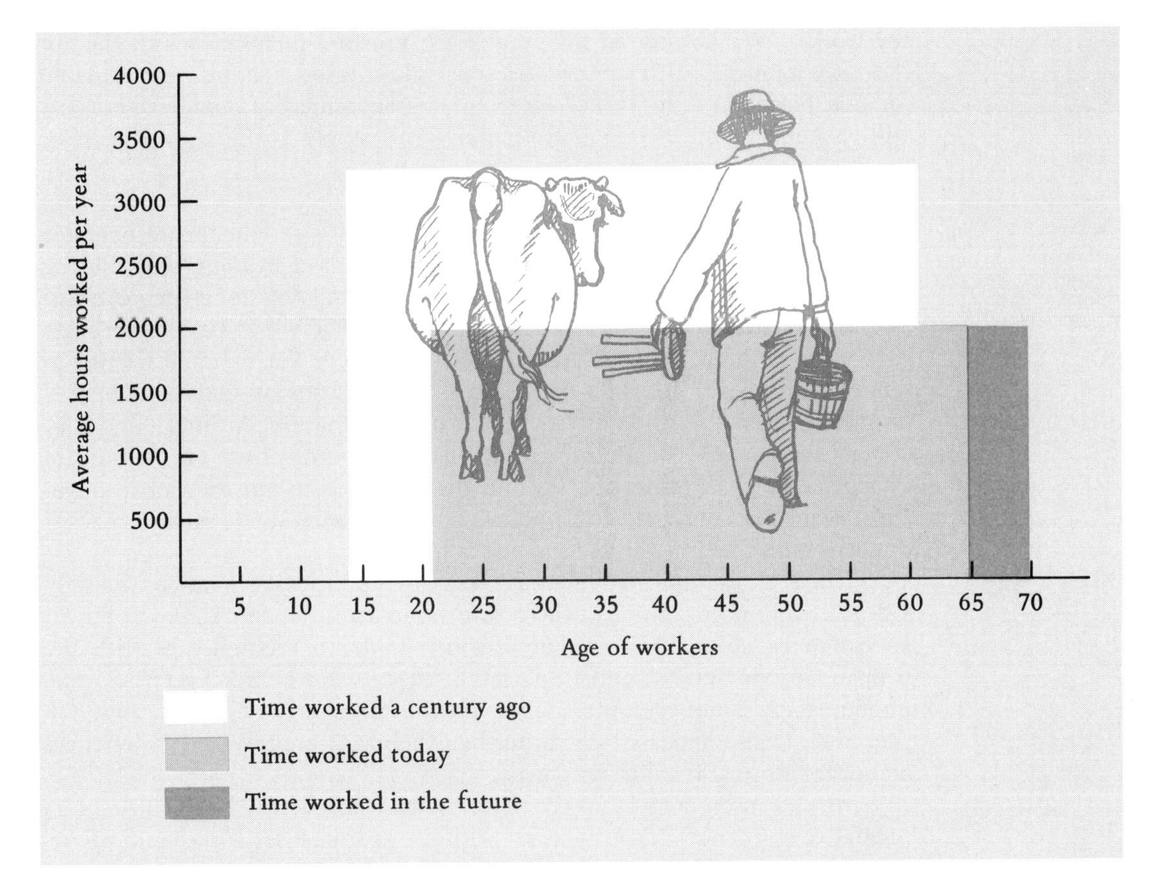

Figure 16-4. *A lifetime of labor: the worker's log today and a century ago. Compared with the worker of a century ago, a person today enters the labor force later, logs fewer hours per year, and continues until a more advanced age because of greater longevity (gray area). Recent legislation that delays mandatory retirement until age seventy will add to this trend (darkly colored area). (Adapted and updated from Miernyk, 1975 with permission.)*

ASK THE PEOPLE: WORKERS VIEW THEIR SITUATIONS

Formal research on the relationship between the work experience and adult development has not ripened to the point where it can tell us very much. Let's step more directly into the arena, then, and sample what middle-aged men and women have to say for themselves. Studs Terkel's *Working* (1972) essentially consists of descriptions and reflections about occupational experience by a wide variety of adults in the U.S.A. These are communicated in the workers' own words in the context of informal interviews with the author. It is a journalistic rather than a scientific enterprise, but no less valuable for that reason. While Terkel's respondents encompass a broad age-range, I will concentrate on commentaries from a few people in the middle years of life.

Fran at the switchboard.* Frances Swenson is a switchboard operator at a motel frequented by conventioneers. She spends eight hours a day sitting on the same chair, handling perhaps a thousand calls (by her estimate). Mrs. Swenson is literally on call all the time. To start the work day she becomes part of the switchboard apparatus herself, donning the headset and starting to remove and replace cords on the board as the calls flow in and out. There is no time to converse or do anything of a personal nature. Although she is in contact with many people throughout the day, there is little opportunity to establish any communication beyond that required to put the call through. Mrs. Swenson is expected to be polite, no matter how she is feeling or what the person on the other end of the line is saying.

This is an example of a job that takes over and restricts much of the individual's functioning. Some of the possible opportunities for self-development appear to be absent. She cannot establish in-depth relationships with the people she contacts; stimulation is limited to what comes over the headphones; there is not even the chance to turn thoughts over in her mind. Furthermore, it also appears to be a dead-end job that promises no experiences and opportunities at a different level.

Mrs. Swenson does not voice these thoughts, but she is aware of the job's restrictive influence. ("You're never without your headset. Your cords are retractable and you're talking as you get a drink of water. It's a pitcher we have about fifteen feet away. We're still plugged in and we're saying, 'Can I help you, sir?' ") Occasionally she has the impulse to break out of the situation or transform it in some way. There is the fantasy of yanking a great big handful of plugs out of the switchboard. This idea grows out of the constant tension of the job, but also may reflect the need to be more than a polite, efficient part of the system. Other ideas of a playful character occur to her, but these would not be tolerated in the job situation. Self-described as a "happy-go-lucky person," she has very little opportunity to display this side of her personality.

There is another and perhaps more subtle restriction on her developmental possibilities, particularly in the social sphere. The assistant manager is addressed as *mister.* But she is called *Fran.* Neither the bosses nor the guests offer much respect to Mrs. Swenson, although she is an adult, a mother, a widow, a useful and uncomplaining worker (she doesn't even complain about the lack of respect, but accepts it as one of the natural conditions of the job). If a person's likelihood of continued development can be enhanced by favorable expectations from others, then restricting an adult to the

* Direct quotations in this section are taken from Terkel, 1972, pp. 32–35.

So many of our adult hours are spent at work (or complaining about it) that we cannot expect to have a true lifespan psychology without understanding the satisfactions and stresses involved. (Photo by H. Armstrong Roberts.)

very limited role set for a switchboard operator might well impede continued personal growth. Mrs. Swenson may be up against the double bias of being a woman and being a person no longer young, as well as occupying a working role that allows for little personal expression. Nevertheless, she reports that she likes her job. How could this be? It is partly because she selected the job herself, and partly because even the fleeting contacts with other human voices relieves her of a sense of lonesomeness. Furthermore, she recognizes that the work she is doing is important both to the guests and to the motel, even if not much appreciation for it comes her way.

Hub at the crane.† Hub Dillard is a forty-eight-year-old construction worker in Chicago. His environment is much different from Mrs. Swenson's. While she is encapsulated within a motel office, he is "out there eating dust and dirt for eight, ten hours a day." Yet he is also very much a part of an apparatus-oriented system, operating a boom crane while other men nearby work with other tools and equipment. There is tension in his job as well. In Mr. Dillard's case, a major source of the tension is the risk of injury or death. "They're so very easy to upset if you don't know exactly what you're supposed to do. And it happens so quick. . . . It's not so much the physical, it's the mental." Mr. Dillard has himself suffered serious injury on the job. He obeyed the command to swing the stub section of the boom from the front to the back of a tractor, although it was not an appropriate maneuver in his own

† Direct quotations in this section are taken from Terkel, 1972, pp. 22–26.

judgment. The crane tipped over backwards, hurled him out, and sent a five hundred pound weight across his leg, crushing his ankle and hip. The injury itself cost him eighteen months of pain and financial concern; the residual incapacity limits the kind of work he can do now. He has also seen and known of others who have had even worse accidents on the job. He knows that a moment of carelessness or bad judgment on the part of one construction worker could result in crippling or death for several others. As he states, "it's the mental." The tensions associated with his particular kind of work, the operation of heavy equipment, lead to physiological stress and impairment. He reports that the average crane operator lives to be only about fifty-five and is often cut down by heart attack (he has had one himself).

Death related concern need not be neurotic; it may be a realistic assessment of job hazards

Mr. Dillard is struck by the contrast between the value of the machines people like himself operate and the value attached to the men. He wonders why he is not treated—financially or in terms of safety conditions—as well as the expensive machines he operates. This, by the way, is a theme voiced by other men interviewed by Terkel, the observation that they, themselves, are considered of little value, as only one of the less expensive components of the machine.

Nevertheless, Mr. Dillard also takes pride in his work. "It's food for your soul that you know you did it good." He enjoys being able to say that he helped put up this building or built that bridge.

How does working with tension hour after hour under difficult physical conditions affect a person's opportunity for continued growth? Does the challenge stimulate a certain kind of developmental process that we have not yet learned to appreciate? Or do the job demands take so much from the individual that he forgoes most of his opportunity to invest mind and energies in anything else? We really don't know the answer to questions like this. Furthermore, Mr. Dillard's work offers a type of reward whose developmental implications are also not well understood. He derives satisfaction from building something that is public, substantial, and enduring. Out of his work comes, for example, a hospital that will serve many people over the years. It represents a visible outcome of his efforts, and one that can serve as a kind of monument to him in the years ahead. He can tell his children that he helped to build that place. By contrast, Mrs. Swenson's accomplishments do not take visible and enduring form. If all else were equal, Mr. Dillard would seem to have an advantage in the form of markers to his achievements and a heritage for future generations. How important is this? The fact is we do not know very much about the influence of visible, enduring achievements on the course of individual development. This would be a useful area for future research.

Will at the wheel.‡ Will Robinson is a forty-seven-year-old cab driver, working the swing shift. He speaks of tension also. It is common among the

‡ Direct quotations in this section are taken from Terkel, 1972, pp. 201–206.

other drivers he knows, as well as himself. "Most of the drivers, they'll suffer from hemorrhoids, kidney trouble, and such as that. I had a case of ulcers." There is very little time to relax, and pressure comes from many sides: the fear of robbery, the competition for fares, the hazards of traffic, the possibility of running afoul of regulations in some way, fairly or unfairly, etc. He cannot help but take the tension home with him. The nature of the job itself has changed since he started right after World War II. Mr. Robinson was among the first blacks to be hired in what had been a predominately whites-only job in Chicago. There was a sense of satisfaction and status attached to this breakthrough. He could feel good about the job and himself. The work situation itself had a number of incentives and good touches, a place to relax, a little library. These have vanished.

In describing his own situation, specific as it is, Mr. Robinson also seems to be speaking of a number of other adults of his age. "I'm too young to get a pension and too old to be a checker. . . ." He is caught in a no-advancement, no ease-off situation between youth and age. In Mr. Robinson's case, it is given special emphasis by the hazards he sees for both the driver and the passenger. His doctor has told him to stop driving for health reasons, and he recognizes that his reflexes are not as sharp as they once were. But he has no alternative but to keep driving. He mentions a friend who died of bleeding ulcers because the company doctors insisted he could continue to work although his private doctors said otherwise. His job each day is now something just to get through. "Like a machine, that's about the only way I can feel."

A person who sees no chance for career advancement may be socially old, although young chronologically

Each of these middle-aged adults experiences tension in his or her daily work life, and each feels overly entangled in a machine-like system. None of them entertains dreams of great future advancement in the system. They take pride in doing their jobs well, but all have suggestions (the employer's never ask for these) on how the job situation might be improved for themselves and for others. Glimpses into the lives of three working adults cannot tell us very much about the overall relationship between life situations and the possibility of continued development from mid-life onward. However, we may at least put ourselves on the alert to consider development in its actual context of everyday human lives rather than in theoretical abstractions alone. It is entirely possible that Mrs. Swenson, Mr. Dillard, and Mr. Robinson have undergone significant personal development since early adulthood and may continue to do so. But developmental psychology has its own work to accomplish to explore these possibilities adequately.

WHAT WE HAVE BEEN AND WHAT WE BECOME: THE CONTINUITY OF SELF

The type of work situation an adult experiences probably influences subsequent development. And the same probably can be said for all other major aspects of the individual's total life situation. But *who* the person has been from infancy onward also remains of central importance. We will now examine several studies of developmental trajectories to improve our understanding of the relationship between who we are at one time of our lives and who we become later.

VERY BRIGHT PEOPLE: A LONG LONGITUDINAL STUDY

The intellectually gifted child. Early in this century, a young graduate student became interested in the differences between bright and dull children. This led him to introduce and standardize a French intelligence scale for use in the United States, the still-popular Stanford-Binet Test. With this assessment procedure in hand, Lewis M. Terman could and did identify children with exceptional mental ability. He recognized that a longitudinal study of their lives would provide more powerful information than investigations limited to cross-sectional approaches. In 1921 he launched such a study. Terman's study of gifted children (who became known affectionately by some as "Termanites") did indeed do much to clarify our understanding of very bright youngsters. The first book-length publication (Terman, 1925) made a big impact at the time and shattered permanently some of the popular beliefs about the mentally superior child. More significantly for our purposes, the longitudinal study continued well beyond childhood—as a matter of fact, it is still in progress, with the participants (minus a natural attrition rate) now well advanced in their adult lives. The determination and resourcefulness of Terman, his colleagues, and his successors merit the appreciation of all who would base their conceptions of lifespan human development on firm data.

Some cling to the prodigy–weakling myth despite the facts

The original sample included 857 males and 671 females. Almost all had IQ's of 140 or above, with the mean at 151. They were studied at the outset with an unusually broad spectrum of procedures, especially considering the pioneering state of personality assessment at that time. Family background, physique, health, school performance, personal interests, play habits, personality traits, and "character" were all studied.

The first wave of findings revealed that the typical gifted child was superior in many ways besides the high IQ score. It demolished a then-popular notion that the very bright child was really a sickly, weak-kneed, socially withdrawn prodigy-monster. Instead, the "Termanites" were superior physically to the average child. They were spontaneous, energetic boys and girls with unusually wide spanning interests and activities. They were quite normal in their sex roles, but more mature than most other children of the same age. As might have been expected, they excelled academically. This was represented not only by their tendency to be accelerated in grade placement, but more substantially by their mastery of subject matter about 44 percent faster than other children. On a battery of seven character tests, they placed well above average on every measure. The nine-year-old gifted child had the moral understanding and beliefs of a twelve year old. They were rated by teachers as above average (often very much above the average) in twenty-four out of twenty-five personality-related traits (e.g., originality, will power and perseverance, conscientiousness, leadership). It was only in mechanical ingenuity that they were considered to be slightly lower than the norm—and this rating was contradicted by actual tests of mechanical ingenuity.

Overall, the differences that set these intellectually gifted children apart from other children of the same age were clearly in their favor. As Terman and his colleagues have had to repeat many times, "There is no law of compensation whereby the intellectual superiority of the gifted is offset by inferiorities along nonintellectual lines" (Terman and Oden, 1959, p. 16). These children also tended to come from family backgrounds marked by superior intellectual endowment and better-than-average physical endowment.

The second wave of findings (Burks, Jensen, & Terman, 1930) found these children still quite superior about seven years later. There were no indications that they were developing in a one-sided intellectual fashion, or experiencing unusual difficulties in emotional stability or social adaptability. In fact, these children continued to adjust and participate in a full range of life activities more adequately than most children of their age.

The gifted child in adulthood. Another wave of findings was reported when the participants were around the age of thirty. They remained among the brightest people in the land, around the 99th percentile. Their vocational success generally lived up to expectations, greater not only than the population in general, but also in comparison with other college graduates. Marital adjustment was equal or superior to other groups who were less endowed intellectually, and sexual adjustment was considered as "normal" as that found for a comparison group of 792 married couples. Serious personality maladjustment, alcoholism, and other indices of major life problems were, in general, somewhat less frequent for the gifted as they moved into their thirties, as compared with others of their age. Furthermore, the inroads of mortality in this group had been somewhat below the average. It remained clear that excellent personal endowment and the combined genetic and cultural heritage of birth within a well-endowed family was associated with unusually good life situation from childhood well into the adult years (Terman & Oden, 1947).

What did mid-life hold for these gifted people? Thirty-five years after the study began, most of the participants were still alive and in contact with the project (795 men, 629 women). This itself represents an unusual persistence over time for a group of this size. The participants (and their spouses) were described as "incredibly cooperative" (Terman & Oden, 1959). The growing richness of experience and accomplishment seemed to make these people even more impressive than they had been as very bright children many years before. The differential opportunities and obligations for adult men and women led Terman and his colleagues to examine accomplishments sepa-

May Sarton discovered poetry at age six and has made her mark on the world with thirty-six books of verse and prose. The gifted and creative adult often comes from the ranks of the gifted and creative child. (Photo by Jerry Berndt, reprinted with permission of the Boston Phoenix.*)*

rately for both sexes. But according to whatever criteria used, both men and women showed remarkable records of achievement. There were many more distinguished scientists, writers, and educators, for example, than one could reasonably expect from a group of this size. Others had become leaders in business, industry, and government. The early promise had been fulfilled, from the point of view of available measures of life success.

Furthermore, the career development of the gifted group indicates that many are still approaching their years of greatest fulfillment and contribution. There is no indication that they have reached a standstill, rather, they have a creative momentum still going for them. Very few have been occupational "failures," but there are some whose vocational achievements have been short of the group as a whole. Those with relatively less success, by and large, prove to be just as intelligent as their gifted peers. A biographical study of the more and less successful members has indicated that personality factors are extremely important differentials. Those with strong ego strength, emotional stability, and social adjustment patterns have accomplished more with their high level mental endowments.

Terman and Oden have emphasized "There is almost no one who has not improved his status, even though he may still be well below the average of the group in terms of realizing his intellectual potential in his vocational accomplishments" (1959, p. 149). The investigators were aware that vocational success is not everything. It was their impression that these gifted men and women were still growing and becoming as they moved past the mid-life mark. The participants themselves felt they were also achieving other values beyond career success, such as a happy marriage and home life, bringing up their children well, having peace of mind and, in some way, helping others and trying to make the world a better place in which to live.

At least two other observations should be made. The general high level of achievement and satisfaction in this group does not mean that they are all similar people living similar lives. Individuality was apparent at the outset of the study. But by the mid-life point of the study, individual differences had emerged even more impressively. Secondly, the "fulfillment of promise" is more difficult to evaluate with women than with men (at least at this time in our society). One could argue that the gifted girls of 1921, as a group, have not "achieved" quite as much as the boys, probably because of differential opportunities and expectancies. (Their achievements when compared with women in general are quite high.) But, as Terman and Oden noted, there is no valid basis for concluding that they have either developed or contributed less than the males. Many of the women selected the mother/housewife role and brought their outstanding intellectual and personality qualities to the situation. It is a question of value judgment as to whether this represents an application of endowment that is at all inferior to activity in the vocational sphere. Whatever view one takes, we have to recognize that it is not easy to compare the developmental trajectories of men and women in a society that has had such different expectations and reward structures available for them.

We could use another study in today's climate more conducive to equal opportunity

Satisfaction in later middle age. What kind of life satisfaction will these exceptionally bright people find as they move into their later years? A more recent analysis has compared the satisfactions they report in their early sixties with what they had to say thirty years previously (Sears, 1977). Unfortunately, the data are presently available only for the men.

It was found that their feelings about themselves remained rather consistent over the decades of their adult lives. The man who felt good about his work and his own basic worth generally continued to feel good about these matters at age sixty, just as he did at thirty. Perhaps the most interesting results centered around the *type* of satisfaction that was of greatest importance to the men themselves. As Sears pointed out, despite their exceptional occupational success and their freedom and autonomy, "these men placed greater importance on achieving satisfaction in their family life than in their work" (1977, p. 128). Adult development for them did not mean just the actualization of individual abilities. Interpersonal relationships, intimate ties with spouse, children, and friends, proved to be the source of deepest satisfaction. Many chapters ago, it was noted that development takes place between as well as within people.

COPING, RELATING, AND EXPERIENCING: VETERANS MOVING THROUGH ADULT LIFE

We switch now to a different population of adults and a different research approach. These participants are nearly one thousand normally endowed men who have cooperated in a still in-progress longitudinal study being conducted by the U.S. Veterans Administration through its community-oriented facilities in Boston. While the study as a whole encompasses many dimensions of the individual's physical, social, and psychological functioning, we will concentrate on recent findings related to personality change and continuity over the years.

Putting the terms *change* and *continuity* into the same phrase is a way of reminding ourselves that the concept of development requires both phenomena. If a person shows little or no change through the adult years, then it does not make sense to speak of development. But unless a strong continuity of self can also be observed, then we have no basis for determining *who* it is that might be developing! Additionally, we would want to be careful in assuming that all change represents development. To see developmental change in the adult years requires evidence both of continuity and systematic transformation. Research that is refined enough to reveal possible differences between those aspects of a person which develop and those which do not is especially welcome.

This study by Paul T. Costa and Robert T. McRae (1977) employs a sophisticated kind of statistical logic that cannot be detailed here. Costa and McRae have compared personality characteristics, as measured by the 16 PF instrument (Cattell & Eber, 1970), in men in the age ranges: twenty-five to thirty-four, thirty-five to fifty-five, and fifty-five plus. Through this comparison, they have identified three clusters of personality characteristics. Cluster I centers around ego strength, anxiety, emotional stability, and related traits. It would not be amiss to describe this group as representing the person's ability to *cope* with life in general. Cluster II pertains more to the style of *relating* with other people (e.g., dependent or independent, outgoing or withdrawn). Both of these clusters were found at all the age levels. This suggests a continuity or invariance in certain aspects of personality structure through a large range of the adult years.

But now we come upon a third cluster that does show appreciable change associated with adult age. Cluster III represents the way in which the person *experiences* the world. Modes of experience seem to be different for youth, middle age, and old age. It is not just that a person tends to become somewhat more introverted or reflective, for example; but the *nature* of these reflections has changed. To take an example, "middle-aged persons are not simply more conservative than other persons; their conservativism, rooted as it often is in disillusionment and sustained by the depth of their commitment to an established structure, is of a qualitatively different sort" (Costa & McRae, 1977, p. 270).

Furthermore, it makes a difference whether the individual, when tested, scores high or low on Cluster III. Those who have an attitude of being open to experience tend to score higher, whatever their age, and those who could be characterized as narrow-minded or closed to new experience, tend to score lower. It is the open-to-experience person who shows an age-related pattern that looks like continued personality development. In other words, we might be able to predict which of two young men will continue to transform himself through mid-life and beyond and which will come to something resembling a standstill in personality growth.

Let's look more carefully at the kind of person who is apt to show a change in this sector of personality. As a young adult he is likely to be concerned primarily with the experience of feelings, emotions. As a middle-aged adult, he is likely to bring a more personal quality to his inner experience, but also to be more interested in ideas and values. There is a shift from feeling toward thought. The person who has had a rich and active involvement with feelings when young and with thoughts in middle-age is the one who is most likely to *integrate or balance* both feelings and thoughts in old age. Costa and McRae remarked that C. G. Jung (1933) suggested in old age a person is able to integrate contrary psychological functions, or functions that earlier in life *seemed* contrary, such as feeling versus thinking. The data from Costa and McRae's recent study (one of a series planned to pursue this lead further) are consistent with Jung's views. But there is the strong hint that not everybody continues to develop in this way. Some people remain open to experience and become psychologically richer over the years; others do not. With close attention, then, we should be able to identify those who are still growing psychologically at mid-life and those who have appeared to close themselves off from further development. How a stalled individual in mid-life might be helped to start moving forward again is a problem that deserves more attention than it has received. (But, again, one might argue that a person has the right *not* to continue his or her development. This value issue has also been neglected.)

Jung's work remains controversial, but it holds much for the adventuresome mind

SIMPLE AND COMPLEX PEOPLE IN MID-LIFE

In Chapter 15 we examined some findings on life satisfaction from a variety of studies. These included an investigation by Marjorie Fiske Lowenthal and her colleagues (1975) that compared people at four different points in their adult life trajectory. We now return to that study for what it can add to our knowledge of developmental status in middle-aged men and women. The participants were a mixture of blue collar and white collar citizens in the San Francisco area; this differs significantly from the men-only Boston VA study and highly selected, "gifted" group recruited by Terman. In general, the participants were job- and family-oriented individuals, although other value orientations were also present.

How a person fares in mid-life has much to do with how "complex" he or she is—and whether the person is a he or a she in the first place. This pair of distinctions (simple/complex; male/female) seems to account for many of the differences observed by Lowenthal and her colleagues.

Lowenthal et al. found that people of either sex and any of the four age/stage levels could be classified in terms of their overall "complexity." This was based on both subjective and objective assessment procedures. In essence, a "simple" person is one who tends to protect himself or herself from the environment, avoid stressful situations, and show a life-style that does not ask for very much. The pattern of the inner life and of social interactions is relatively circumscribed. The "simple" person does not have many psychological resources, but does not have many deficits either. The "complex" person has a high share of both resources and deficits or problems. This person is more likely to have what some have called "growth motivation," seeking to expand his or her personality, range of functioning, and life experience. Stressful situations are accepted as part of the price one must pay for an adventurous, stimulating life. Naturally, there are many "inbetweens," people at neither extreme. Yet it is possible to use the complexity/simplicity distinction as a major basis for analyzing developmental status in these adults.

From middle age onward, the "simple" person tends to be happier. The "complex" person, however, tends to achieve more and adapt better in early adulthood. What goes wrong for this person in middle age? It is probable that frustration, boredom, self-doubt, and despair are generated by the missed opportunities and blocked pathways to further growth that many encounter in middle age. Some of Terkel's interviews may have helped prepare us for this conclusion. People who have entertained higher hopes and greater expectations may find themselves in dead-end jobs. Self-fulfillment through work no longer seems realistic. The person who did not expect much from life or from himself or herself is less likely to be disappointed.

The **developmental crisis** is a threatening situation confronted at a *certain* time in one's lifespan

Although occupational dissatisfaction may be more common and central for middle-aged men, the women appear to suffer even more. Lowenthal and her colleagues suggested that there is a **developmental crisis** in the lives of many women at mid-life. This crisis is especially evident for women with complex personalities. Here is a paradoxical situation, then. Some of the most intelligent and resourceful people in the study were also those experiencing the most stress and despair. Bright middle-aged women were becoming liberated from the energy-consuming parental role. In middle age, they had the occasion to reflect on their own needs, to see what else life might hold for them. Being a mother, and a wife, was no longer a satisfactory way of expressing abilities and receiving gratification. Yet they often felt trapped. Alternative pathways were either difficult to find, or major obstacles were seen between them and their new or renewed life goals.

Women with strong self-concepts and a firm sense of competence were more apt to meet this challenge successfully. Nevertheless, middle age was, in general, the least satisfactory stage of life for women in this study. And the more complex or growth-oriented the woman, the more possibility there was for frustration and despair. Complex and talented men were often stymied as well in their groping toward a second career or other form of self-renewal.

Lowenthal and her colleagues departed from their strictly scientific reporting to deplore the way our society wastes the potential of mature women, and also that of many mature men. The patterns of both sexism and ageism

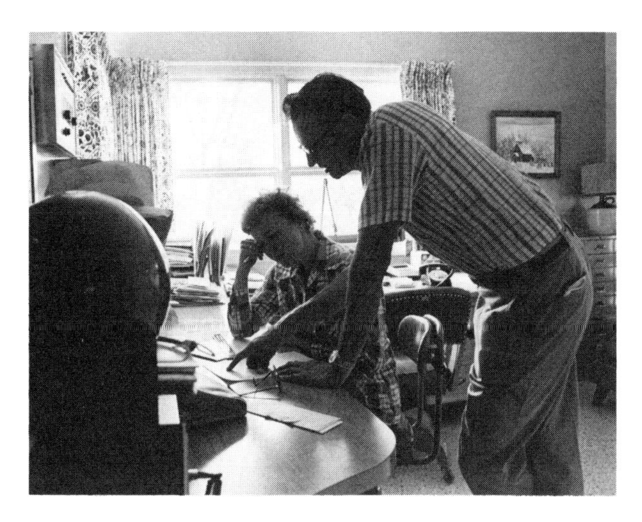

Back to work? At my age? The decision to rejoin the workforce is not always an easy one after many years of being "mother." (USDA Photo by William Kuykendall.)

that many observers have identified in our society obviously do work against the possibility of continued development. It is a tribute to the personal resources and courage of those who overcome the obstacles that they are able to overcome the barriers to growth. And their numbers appear to be increasing steadily. However, we could expect more people to renew their personal development after fulfilling parental responsibilities—and do it with less stress—if our society behaved as though it expected grown-ups to keep growing up. Making the world a better place for adult development may eventually share the more familiar value priority of creating a safe and facilitating environment for child development.

SCENARIO 11: TURNING THE CORNER

The children ran ahead, down the hill toward the beach. We couldn't keep up with them. They were jumping from rock to rock. They were laughing and excited and all. Then, all of a sudden, I just stopped and stood there. I could feel my heart pounding. With fear. No, it had nothing to do with danger for the children; they were OK. I had just realized that I couldn't keep up. Well, that's not exactly it. I mean, I didn't even *expect* to keep up. I saw myself right then for the first time as a person who was not young. *I* was the older generation! This was a very personal and a very upsetting experience.

We were all looking at the picnic pictures. My uncle, he kinda frowned. Looked like something was puzzling him. Then he said, "Hey, who's that bald old man?" I tell you, that brought us up short. We looked at the picture and we looked at him. My sister said, very quietly, "Why, that's you." That look on his face—was he ever shook! He absolutely hadn't recognized himself.

Turning the corner experiences show movement into a new developmental position

These are **turning the corner experiences.** In the two given here (taken from reports collected over the years), the people involved did not realize the corner was there until they abruptly discovered themselves in a different relationship to the world. Life would not seem quite the same any more, because the individual did not seem quite the same to himself or herself. Let's take another example in a little more detail.

It was the start of a new school year, and Mrs. Russell looked forward to meeting the new crop of teachers. At every gathering, however, the new teachers grouped themselves away from her.

"I asked Mrs. Ellis if they had something against me? Had I done something to offend them? Was there a nasty rumor about me? She said: 'Oh, no, I really don't think so. They just prefer to be with people of their own age, you know.'

"I sat there for a moment and let it sink in. She didn't seem to realize what she had said. Never before, I mean, never for a moment had I ever thought of myself as one of the old guard. I was, well, just a teacher and I suppose, in my own head, I was one of the younger ones. I know I *started* as one of the younger ones, and it hadn't occurred to me ever to think of myself any differently. So, was I supposed to start being somebody different now? Why? It was so unfair! I could have cried. . . .

"Looking back at that moment, I am grateful that I kept my feelings to myself. Mrs. Ellis wouldn't have understood, and neither would the others. Maybe some day they will."

Mrs. Russell had experienced a sense of change in her life status, at least within a certain significant setting. It did not make her "old," but it set her apart from the young. Age-status transitions of this type often seem to occur during the middle years of life, whether early or late. Sometimes the individual is the first to sense the change; sometimes the individual seems to be the last person who catches on. This teacher's outrage, "It was so unfair," may also be common, although we cannot set too much store by anecdotal reports alone. Turning the corner is a very personal experience, requiring the person to reevaluate many assumptions and expectations.

How did Mrs. Russell respond to her experience? The scenario might have taken the form of denial. She could have refused to acknowledge that she had either changed, or changed in the eyes of her peers. Using this option, Mrs. Russell might then have gone out of her way to dress and act "young." But would being aggressively youthful have solved or intensified the problem? At the least, it probably would have encouraged her to be preoccupied with her age status and taken energy from other activities.

Another possibility could have been the cave-in. She could have accepted the new instructors' implicit judgment and withdrawn from them socially and emotionally. In other words, she could have become the person she thought they thought she was—only more so. Thinking herself "too old" for the young teachers to want as a friend, she might have adopted a more reclusive life-style which, in turn, might have generated a continuing circle of rejection and isolation.

In fact, however, Mrs. Russell developed a different solution. "I felt insulted for awhile—oh, did I!" Mrs. Russell soon realized, though, that the younger teachers needed her. She did not press herself upon them, but made herself quietly available when they felt confused or anxious in learning their new jobs. The younger teachers eventually found her to be a person they could confide in on more personal matters as well. "I saw that some people had a kind of respect for me I don't remember having received before. I don't mean to flatter myself, but they really started to *look up* to me. That's something I had always wanted, I guess, but it never came my way before, and I may not have been able to handle it before." Mrs. Russell added, "Guess I gained something and lost something. There still are times when the young ones want to be off by themselves, and I hate to be left out of anything. But I have their respect, I do my own job well, and I think they really like me, too, for who I am."

HAZARD: ALCOHOLISM

Alcohol abuse is a topic that, depending on one's background, invites either moral lectures or chart after chart of doleful statistics. Neither are necessary here. There can be no question of the physical, psychological, and social harm associated with excessive drinking. Alcoholism can be cited as a hazard at any point in the lifespan (including childhood, where the health and upbringing of the young can be endangered in many ways by parental alcoholism). But singling out alcoholism as a hazard in the middle years of adult life is appropriate for several reasons. People who become heavy drinkers in adolescence or young adulthood may "get away with it" to some extent. But in middle age they are faced both with the cumulative effects of years of drinking and with the more normal physical changes that represent the transition from development to aging. The forty year old with fifteen or twenty years of heavy drinking behind him or her may have some body systems functionally aged well beyond chronological age, and may even appear deteriorated on an overall basis. Furthermore, as the more or less characteristic changes associated with continued aging occur, the damage inflicted by past and continued drinking may begin to intrude on all realms of functioning in a more obvious and severe manner.

Alcoholism in mid-life is apt to serve both as cause and effect for deepening misery. We have already seen from a number of sources that frustration, disillusionment, and other psychologically painful states are experienced by many people in their middle adult years. If a person already has some tendency to seek refuge in the bottle, then drinking may increasingly be utilized as a way of trying to cope with all of life's disappointments. In this sense, alcoholism can be the effect or outcome of the mid-life blues. But once the alcoholism pattern is well established, it tends to produce its own effects. These may include social isolation, instances of seriously poor judgment, inability to perform occupational and other responsibilities, and so on. As both cause and effect, alcoholism in mid-life contributes directly to intensification of the problems it may have been intended to solve. For some people, in fact, alcoholism appears to be a kind of prolonged suicide equivalent. The self-destructive component in alcoholism has been recognized by many suicidologists.

Yet the consequences of alcoholism in mid-life may become most crucial as the person approaches old age. An erratic life pattern may have the effect of alienating people who could be sources of strength and comfort in old age. The alcoholic may then find himself or herself a sick and lonely individual. Moreover, there is a reduced probability of surviving from one year to the next with alcoholism. The future outlook for the person who drinks heavily in the middle adult years is far from attractive.

Alcoholism justly can be regarded as a *symptom* of individual and social distress. In other words, attention must be given to what alcoholism means within its total context rather than limited only to the phenomenon of excessive drinking itself. Still, we cannot ignore the actual consequences of alcoholism, whether expressed in physical dysfunction, interpersonal relationships, or self-image. In alcoholism, the individual faces a serious menace both to continued development and continued survival. The psychosocial fall-out from alcoholism also tends to extend beyond the individual to have adverse effects on the lives of other people.

TOWARD UNDERSTANDING THE MIDDLE YEARS OF LIFE

It is not easy to integrate the many observations and viewpoints that are starting to emerge about middle age. Some of the research findings do appear to support each other. Several research studies, for example, have found that some individuals experience frustration, disillusionment, and even desperation at mid-life. Yet there are important differences as well, both within various studies and between them. People who appear to have reached a no-growth state even before getting to the middle years have been identified (e.g., Lowenthal's "simples" and Costa and McRae's "close-minded"). Yet, other people who are still on the way up, achieving, enjoying, relating well to the people in their lives have been found as well (e.g., most of the "Termanites").

The differences in the ways people cope with mid-life have something to do with the particular situations in which they find themselves at mid-life, but also with the kind of people they have been all along. *Kind of person* must be understood in terms of social opportunities and expectations as well as in terms of individual personality dynamics. Differential opportunities and expectations for men and women provide one major example here. Other examples can be provided along ethnic, racial, or even geographic (e.g., rural-urban) lines. It is difficult to arrive at simple conclusions about the nature of

"What will we do now that the children can take care of themselves?" This is not a crisis but an opportunity for many couples. (Photo by David Kelley.)

middle age as a possible stage of human development unless we are willing to force the data into questionable categories and ignore those data which do not suit our own preferences.

A clue from history? One clue worth further thought is the notion of "middle" itself. For much of human history, the average person did not survive into or beyond what we now consider to be the middle decades of life. Today there is more impetus for viewing thirties, forties, and fifties as a time when continued development should be expected—if only because it has become more realistic to project oneself into the future. What was formerly the end-point for many people needs to be reevaluated as a long stretch of time that could be used for many purposes, and also as a staging area for adaptation to old age.

Society has not yet fully caught on to the fact that middle age is changing its character because it now occupies a different place along the lifeline of the average person. Furthermore, if we are to take seriously the possibility of continued development through old age, then we cannot neglect the potentials and problems encountered in fostering development at mid-life.

Reviewing the conceptions of middle age. Let's limit ourselves to general descriptive statements for a moment. Some middle-aged people do seem to fit the conceptual model of *holding on*. Considering these people apart from the others in their lives, we might be justified in characterizing middle age as a period of routinized functioning in which the fires of youth have been banked down to provide just enough fuel for a predictable (yawn) no-growth daily life.

But suppose, instead, that we take a more socially-oriented view. It would then be possible to credit middle age as a time of binding or integration. The middle-aged person (including the man or woman who, *as an individual,* is seen as just holding on) is performing the function of *holding society together.* We can recall to mind here the family-oriented participants in Lowenthal's study. They continued to look after their parental and social responsibilities whether or not they derived full satisfaction from this activity.

Granting that there is some evidence for both of these views, we also find that the third major conception suggested earlier in this chapter has its share of observations as well. We see that *some* people continue to develop under *some* circumstances; others do not. The implication is that continued development in middle age is neither automatic nor out of the question. There is a *conditional* factor here—actually, many conditional factors that require vigorous exploration.

Furthermore, there is no evidence that compels us to accept the proposition that all individuals develop in the same way throughout the mid-life years. It remains likely that several different pathways of development will be documented as previous personality style and current situational opportunities are examined in greater depth.

More theories on the way. Until recently there was little that might be called a theory of adult development. This situation is starting to change. In addition to the ideas that have been suggested and evaluated in this chapter, a number of other theoretical frameworks are being studied that might stimulate useful research in the years to come.

Bernice Neugarten, a leading researcher in adult development, is emphasizing the inward turning of thought during the second half of life. She has found that many of her research participants show a heightened interest in taking stock of their lives and display what seems to be a more reflective orientation than they did in their earlier adult years (Neugarten, 1973, 1977). She sees the middle-aged person as one who has gained considerable proficiency through experience and as one who can apply this proficiency to achieve goals more readily than the younger individual. The increasing "interiority" of the adult in mid-life is seen as a response to the sense of diminishing physical and social power.

The cross-cultural research of David Guttman (1974, 1977) tends to support Neugarten's view that individuals move from active to "passive mastery" of the environment. The experienced adult does not have to expend so much energy and be so conspicuous in order to achieve his or her goals. As this framework for interpreting middle age continues to be elaborated, we might expect to learn more about the adult as a subtle executive. This notion is consistent with, but goes beyond, the basic theme of mentally constructing our world in childhood and adolescence, as presented by the Piagetians. If the youth learns to construct a rational and stable world, then it is the experienced adult who really knows how to live in it!

Other emerging theorists emphasize what they regard as the "crisis" of middle life. Among the most prominent observers of this type are Roger Gould (1975) and Daniel Levinson (Levinson, Darrow, Klein, Levinson, & McKee, 1974); the latter psychologist's work was drawn upon extensively for the popular book, *Passages: Predictable Crises of Adult Life* (Sheehy, 1976). In somewhat different ways, Gould and Levinson see the middle years of life as posing challenges that require at least an effort toward restructuring one's thoughts and feelings. Both approaches are consistent, in general, with the research on time perspective already presented. People do not seem to have a future perspective that encompasses all the years that lie ahead of them when they are young. At mid-life they may find themselves with "dreams" or "illusions" that are increasingly difficult to support, and which do not seem especially helpful in coping with the new realities of life.

Although these theories bring some drama to the conceptualization of the middle adult years, they have yet to be put to systematic test. Costa and McRae do not find much evidence to support the mid-life crisis theory in available data (1978), but they point out that existing studies were not especially designed to examine these hypotheses. A recent study by Madeline Cooper (1977) *was* designed with this purpose in mind. She did *not* find support for a mid-life crisis theory in her study of 315 men ranging in age from thirty-three to seventy-nine. It is too early to evaluate the adequacy of these theories, however, and, no doubt, still other conceptions of adult development will soon appear on the scene. The adult years are too interesting, complex, and significant to remain as relatively obscure as they have been in developmental theory and research.

In other words: let's keep our eyes on the new theories, but not rush to quick conclusions

Development does continue throughout all the adult years, but we have much to learn about the individual and social conditions most favorable for this process, as well as a better appreciation for the many forms that development can take.

SUMMARY

Much of adult life is spent within an age zone that is neither young nor old. Curiously, however, neither society nor the socio-behavioral sciences have established a firm conception of what middle age is or should be. Perhaps this is related to the historical fact that until recently many people did not survive into and beyond the mid-life period. We have not quite become accustomed to thinking of a forty or fifty year old as being in the middle instead of the twilight years. Furthermore, the middle generation has not been as thor-

YOUR TURN

1. Three alternative conceptions of middle age have been proposed and examined in this chapter (holding on, holding society together, pluralistic development). Can you think of any other ways of characterizing this phase of life? Draw on your own experience with middle-aged adults and with what you have been learning about human development in general. If you can come up with an alternative view, define and describe it; then muster a few observations in its support. Once your alternative view has been clearly formulated, compare it with the other conceptions.

If you cannot find any alternative views of middle-age, select whichever one of the three offered here appears most significant or convincing to you. Try to develop the best case for this conception that you can. Before concluding your argument, indicate what kinds of research or other observations we would need to demonstrate or to test the validity of your approach beyond reasonable doubt.

2. Assume that within the next decade or so the women's liberation movement achieves most of its fundamental goals. Cultural expectations and opportunities for women become on a par with those for men. How would this be likely to affect the mid-life developmental pattern for men and women? In exploring this possibility, you might draw on the rapidly increasing literature on the place of the woman in contemporary society and on differential development from infancy onward for both sexes.

3. If you consider yourself to be a young person, can you imagine what type of situation might lead you to experience a turning of the corner? Try to anticipate one or more circumstances in which you might encounter this experience. Make them directly relevant to your own life-style, and imagine them in as much detail as you can. Consider also how you are most likely to cope with these experiences if and when they occur. What options do you have? Which option do you think you will take—and why?

If you have already had a turning-the-corner experience, reflect on it. How did it come about? Who made the discovery first, you or somebody else? How did you feel and what did you do? What other ways of coping with this experience might you have tried?

oughly studied from a developmental standpoint as either the young or the old have. Although middle-aged men and women are plentiful in our society, then, the nature of this period of life has not been charted extensively and in depth.

This chapter began by eavesdropping on a session of The Commission To Do Something About Middle Age. The arguments expressed there can be reduced to three major views: that middle age is a period of *holding on* as best one can to accomplishments, security, and status; that it is a period of *holding society together* through parental, occupational, civic, and other responsible roles; or that it is a situation that yields *conditional, selective, and pluralistic development.* This latter view suggests that patterns of development become increasingly individualistic as we move from youth into mid-life; it challenges the notion that there is only one, single pathway of human development.

With these alternative conceptions as background, we explored the possible relationship between what we do and what we become. Attention was given to the occupational experiences of men and women at mid-life, drawing on the illuminating interviews conducted by Studs Terkel, and also on some data concerning the changing work careers within the past century. The fact that, in general, fewer hours are spent at work over the entire lifespan has developmental implications that require careful consideration. The people quoted in Terkel's *Working* experienced tension and a sense of being caught up in a machine-like system with little opportunity for future advancement or self-fulfillment. Yet they found ways of taking some pride and satisfaction in their work. It appears that for many middle-aged adults, the job situation does not provide an adequate opportunity for continued self-development.

We then considered a variety of research findings on the relationship between what we have been and what we become. Detailed consideration was given to the life careers of a sample of gifted children who were followed into mid-life by Terman, his colleagues, and his successors. Terman's research dispelled many of the erroneous views once held about very bright children. His studies found them to be healthy, vigorous, and adept in all spheres of life rather than puny, little, precocious monsters. Over the years, these gifted children blossomed into men and women who were among the most accomplished and mature in the nation. Middle age did not represent a dead-end, nor even a slowing down for most of them, but rather a period in which new development and accomplishment seem to be burgeoning.

Data from an in-progress longitudinal study of men who represent a broad range of intellectual functioning suggest that styles of *coping* with life's challenges and *relating* to other people tend to remain consistent from youth to middle age. Some people, however, show important changes in the way they *experience* or interpret life. This study, being conducted by Costa and McRae, 1977, indicates that men who are relatively open to new experience when they are young are more likely to continue to develop into and beyond middle age. The ability to integrate both feeling and thinking aspects of personality into one's self-system appears to gain strength in the second half of life.

From a study of men and women at four stages of life (Lowenthal, Thurnher, & Chiriboga, 1975), we have learned that people who can be classified as *complex* and those who are *simple* in their psychosocial life-styles seem to follow different pathways through the mid-life years. There are also important *differences between men and women* in middle age that could well reflect differential opportunities and expectations from society. The interaction between complexity/simplicity and sex-related role holds clues for understanding why some people experience more stress and frustration than others. This study has further suggested that a *developmental crisis* may be encountered by many of the brightest and most complex women at mid-life.

We also examined when a person comes to the conclusion that he or she has moved from the young to the no-longer-young side of the lifespan spectrum. The private experience of *turning the corner* was illustrated through a scenario. Each individual has his or her own corner to turn, and may cope with this challenge in a number of ways.

Alcoholism in mid-life was cited as a particular hazard, both for its immediate personal effects, and for the consequences it may have later in life for the individual and for others in his or her life.

Finally, we returned briefly to the three major conceptions of middle age that were identified earlier to consider them in the light of research findings; several emerging theories of middle age were also noted.

Reference List

Burks, B. S., Jensen, D. W., & Terman, L. M. *The promise of youth: Follow-up studies of a thousand gifted children.* Stanford, Calif.: Stanford University Press, 1930.

Cattell, R. B., & Eber, H. W. *The sixteen personality factor questionnaire test.* Champaign, Ill.: Institute for Personality and Ability Testing, 1957.

Cooper, M. W. An empirical investigation of the male midlife period: A descriptive, cohort study. Unpublished honors thesis, University of Massachusetts at Boston, 1977.

Costa, P. T., & McCrae, R. R. Age differences in personality structure revisited: Studies in validity, stability, and change. *International Journal of Aging & Human Development,* 1977, 8, 261–276.

Costa, P. T., & McCrae, R. R. Objective personality assessment. In M. Storandt, I. C. Siegler & M. F. Elias (Eds.), *The clinical psychology of aging.* New York: Plenum Publishing Corporation, 1978, pp. 119–144.

Cumming, E., & Henry, W. E. *Growing old.* New York: Basic Books, 1961.

Gould, R. L. Adult life stages: Growth towards self-tolerance. *Psychology Today,* 1975, 8, 74–78.

Guttman, D. Alternatives to disengagement: Aging among the highland Druze. In R. LeVine (Ed.), *Culture and personality: Contemporary readings.* Chicago: Aldine Publishing Company, 1974.

Guttman, D. The cross-cultural perspective: Notes toward a comparative psychology of aging. In J. E. Birren & K. W. Schaie (Eds.), *Handbook of the psychology of aging.* New York: Van Nostrand Reinhold Company, 1977.

Jung, C. G. *Psychological types.* New York: Harcourt, Brace & World, 1933.

Levinson, D. J., Darrow, C. M., Klein, E. B., Levinson, M. H., & McKee, B. The psychosocial development of men in early adulthood and the mid-life transition. In D. F. Ricks, A. Thomas & M. Roff (Eds.), *Life history research in psychopathology* (Vol. 3). Minneapolis: University of Minnesota Press, 1974.

Lowenthal, M. F., Thurnher, M., & Chiriboga, D. *Four stages of life.* San Francisco: Jossey-Bass, Inc., Publishers, 1975.

Miernyk, W. H. The changing life cycle of work. In N. Datan & L. Ginsberg (Eds.), *Life-span developmental psychology/normative life crises.* New York: Academic Press, 1975, pp. 279–286.

Neugarten, B. L. Personality change in late life: A developmental perspective. In C. Eisdorfer & M. P. Lawton (Eds.), *The psychology of adult development and aging.* Washington, D.C.: American Psychological Association, 1973.

Neugarten, B. L. Personality and aging. In J. E. Birren & K. W. Schaie (Eds.), *Handbook of the psychology of aging.* New York: Van Nostrand Reinhold Company, 1977, pp. 626–649.

Sears, R. R. Sources of life satisfactions of the Terman gifted men. *American Psychologist,* 1977, *32,* 119–128.

Sheehy, G. *Passages: Predictable crises of adult life.* New York: E. P. Dutton & Co., Inc., 1976.

Terkel, S. *Working.* New York: Pantheon Books, Inc., 1972.

Terman, L. M. *Mental and physical traits of a thousand gifted children.* Stanford, Calif.: Stanford University Press, 1925.

Terman, L. M., & Oden, M. H. *The gifted child grows up.* Stanford, Calif.: Stanford University Press, 1947.

Terman, L. M., & Oden, M. H. *The gifted group at mid-life.* Stanford, Calif.: Stanford University Press, 1959.

THE OLD PERSON

In Stereotype, Reality, and Potentiality

17

CHAPTER OUTLINE

I
The process of aging and the condition of being old are not identical.

Age versus ill-health / NIMH Study (Birren) / Predictors of survival / Process of increasing vulnerability

II
Theories of aging are interesting, but largely unproven.

Primary versus secondary aging / Possibility of altering the aging process(es) / Experimental approaches to aging / Biological theories versus psychosocial theories

III
The competence of a particular person is likely to be related to chronological age only in indirect and complex ways.

Interdependence of functioning / Selective impact of aging / Environmental limitations / Behavior deficits / Response force amplifiers / Wide response topographies / Psychomotor slowing / Memory functioning (primary, secondary, tertiary) / Retrieval / Experimental style / Learning and decision-making / Motivation / Mental functioning debate / Toward Sanyāsa / Victimization by crime

IV
Much of being old in our society is built on being a person with a particular life-style oriented around social sex-role, having a certain level of income, health status, education, and residing in a particular environment.

"Typical" old person / Destructive stereotypes / Degree of "oldness"

V
The best ways to understand the intimate thoughts, feelings, and relationships of an old person are to be or know an old person.

Time perspective / Retrospective modalities / Life review / Validation / Boundary setting / Perpetuation of the past / Replaying / Death, dying, and grieving / Loving and sexuality

Where do babies come from? This question has generated some fanciful replies. But long after storkish fantasies have been relinquished, we tend to be haunted by assumptions, half-truths, and stereotypes about old age. We must pick our way carefully through many assumptions if we are to see the old person clearly. Reality must be distinguished from stereotype—and also from the full potentiality of human development in the later years of a long and eventful life.

Let's begin with the obvious question: what is aging?

WHAT IS AGING?

AGING VERSUS BEING OLD

The process of aging and the condition of being old are not identical. Later in this chapter we will discuss some of the factors that either shape the characteristics of a particular generation of elders or shape our perceptions of them. Your next door neighbor may be an elderly widow who was born in another part of the world. She is growing old in a society that was poorly prepared for a population shift from the majority of people being young to the majority being older. When the child born today reaches his or her present age, both the nature of this person's experience and of society's ability to offer appropriate resources and opportunities will probably be much different. Being old represents a different status if there are few elders in a particular society, or if there are many. It is different, again, in a society that emphasizes tradition and precedent as compared with a society that is discarding previous values and technologies in favor of "progress." An eighty year old in Samoa (Holmes, 1972) or Southwest Ethiopia (Hamer, 1972) is apt to experience an age-status that is very different from the one an age peer experiences in the United States.

There is a critical distinction to be made between a process of aging that is common to all humans and the state of being old that is so heavily influenced by cultural forces. There is a parallel distinction made in relation to development. What it is like to be a two year old, a ten year old, or an adolescent also involves complex interactions between biological processes common to our species and particular cultural traditions and situational contexts for growth. Both the biological and the psychosocial influences are active throughout the entire course of development and aging. Although some important changes may occur in the biological and psychosocial spheres with advancing age, we should not look to these late-emerging influences as The

Answer. What has gone before is no less important than the changes that may occur in the later years of life.

In practice it is difficult to separate whatever might be universal about the process of aging and all that is related to an individual's particular make-up and life experiences. Let's now examine one of the more important research efforts that has been made to determine what phenomena might most securely be attributed to aging as a fundamental and intrinsic process.

HEALTHY OLD MEN: THE NIMH STUDY

Gerontologists are scientists who study aging and the aged

A select group. One of the major alternative explanations for some of the problems experienced by old people is they are suffering from specific illnesses or impairments, rather than they are in the grips of a general and inevitable process that is known as *aging*. A pioneering study conducted by **gerontologists** associated with the National Institute of Mental Health has provided important information on this topic.

Several features of this study should be noted at the outset. It was a multidisciplinary investigation, calling on the knowledge of specialists in a variety of professional and scientific fields concerned with aging. It also involved an unusually careful attempt to identify people of advanced age who were *not* afflicted with medical disorders. Many studies of adults of any age do include people with a variety of disorders; often the extent of their health problems is not fully known. The participants in this study, then, were "old" by ordinary chronological age standards, but "healthy" by ordinary medical standards. The study was also unusually thorough in its assessment of the individual's psychological and social functioning. From the standpoint of research craftmanship and range of inquiry, the NIMH study must be regarded as one of the strongest contributions that has been made to the field of gerontology. Nevertheless, all this attention was limited to a fairly small sample, 47 individuals, all of whom were men. James E. Birren and his colleagues (1963) suggested that their study be considered as a pilot rather than a definitive effort. Some of the major results seem to be supported by subsequent research.

The participants were volunteers who lived independently in the community. All were at least 65, the mean age was 71.5. They had a wide range of educational background; many had been born in foreign countries; and most had been living in an urban environment for a long time. Most were living in intact families, including 31 who were maintaining their own households with their wives. While there were many differences in background and in work experience among these men, as a group they were somewhat better off at the time of study than old men in general in the U.S.A.

Healthy as all the men were, it was, nevertheless, possible to distinguish between two levels of physical well-being. Those in Group I (27 men) met all the medical criteria for health. Those in Group II (20 men) were also healthy,

Subclinical or asymptomatic disease is an impairment that does not show up clearly in daily life

but had one or more signs of **subclinical or asymptomatic disease.** This distinction made it possible to compare the healthy with the very healthy among elderly men, as well as to compare the total sample with what was known about physical and mental functioning in young adults.

What did they find? As an overall conclusion, Birren and his colleagues found the old men: "vigorous, candid, interesting, and deeply involved in everyday living. In marked contradiction to the usual stereotypes of 'rigidity' and of 'second childhood,' these individuals generally demonstrated mental flexibility and alertness. They continued to be constructive in their living; they were resourceful and optimistic" (1963, p. 314.).

There were some men who showed depression or other signs of emotional or mental distress. Yet psychiatric-type symptomatology was not any different from that found among younger adults, nor did it show any particular relationship to chronological age, social background, or physical status. Essentially, it was the loss of significant people in their lives that was associated with depression. Other kinds of loss in their current life situation (e.g., loss of job or income) also were related to negative feeling states. From the standpoint of adjustment, then, the healthy old men in general were doing very well. When they did have problems, these were usually associated with actual deprivations or threats of deprivation in the immediate life situation rather than with a vague general process that might be called *aging*.

Physical condition. What of physical status? Some of the major findings of Birren et al. are illustrated in Figure 17–1. You will note that two sets of comparisons are presented. Here is the logic involved. Differences between Old Men I and Old Men II are best interpreted in terms of relative freedom from disease states. Although both subsamples are healthy, OM I consists of men who are as free from acquired pathology as any group we are likely to discover. OM II has mild touches of various conditions. The second comparison is between two groups of very healthy people, OM I and Young Men. This means that only differences between OM I and YM clearly point to the effects of aging. Differences that appear between OM I and OM II are more parsimoniously explained on the basis of specific physical disorders.

In examining Figure 17–1, you will see that OM II shows impairments on six of the sixteen variables when compared with OM I. There are also six differences in the OM I–YM comparison, but only three of these are on the same dimensions as in the OM I–OM II comparison. We see, for example, that one measure of blood pressure distinguishes between the two groups of old men, but the very healthy old men are *not* different from the young men on either blood pressure measure. This suggests, of course, that when an old person has a mean systolic blood pressure that departs somewhat from the normal range, we should be thinking of specific pathology rather than writing it off as a natural, intrinsic aspect of aging. You can see how findings such as this one, multiplied along many physical dimensions, could change both our conception of what aging is and what type of preventive and interventive care would be most appropriate.

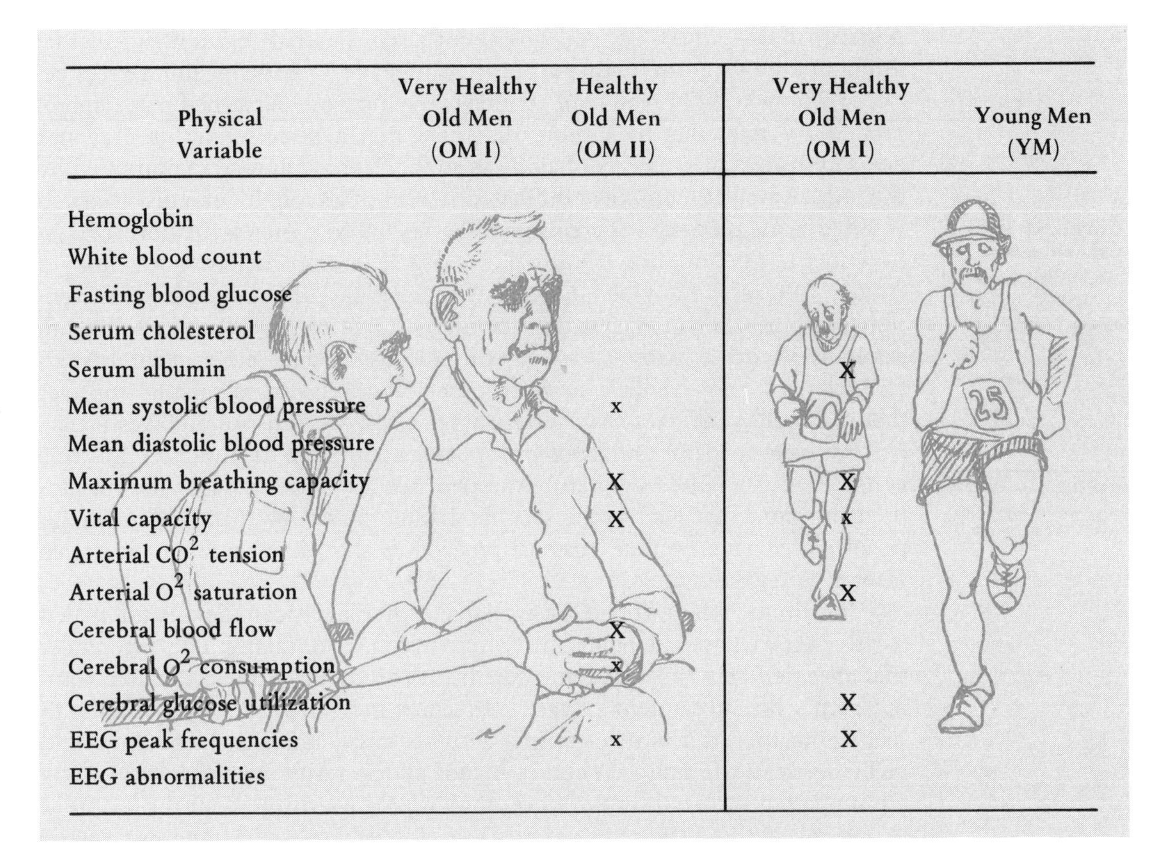

Physical Variable	Very Healthy Old Men (OM I)	Healthy Old Men (OM II)	Very Healthy Old Men (OM I)	Young Men (YM)
Hemoglobin				
White blood count				
Fasting blood glucose				
Serum cholesterol				
Serum albumin			X	
Mean systolic blood pressure		x		
Mean diastolic blood pressure				
Maximum breathing capacity		X	X	
Vital capacity		X	x	
Arterial CO^2 tension				
Arterial O^2 saturation			X	
Cerebral blood flow		X		
Cerebral O^2 consumption		x		
Cerebral glucose utilization			X	
EEG peak frequencies		x	X	
EEG abnormalities				

Figure 17-1. *Is it age or is it health? Comparing very healthy old men with healthy old men and comparing very healthy old men with young men. Capital X's indicate that this group shows a large degree of impairment in this category as compared with the comparison group. Lower case x's indicate a smaller degree of impairment. (Based on data presented by Birren, Butler, Greenhouse, Sokoloff & Yarrow, 1963.)*

Scan the overall comparison between OM I and YM. Did you expect more differences between these two groups? The assumption that a general, pervasive, and inevitable process known as *aging* is responsible for differences between young and old does not receive strong support here. Even within a particular domain of functioning, some measures indicate differences while others do not. Many important measures based on blood studies show no difference—which makes an exception such as serum albumin particularly valuable as a clue for further research. Again, we find that some indices show no differences in brain functioning, for example, cerebral blood flow is as adequate in very healthy old men as it is in young men, and the same is true of oxygen consumption in the brain. Electrical activity in the brain is as free from clinical abnormalities in OM I as in YM, yet there is a demonstrable slowing in the peak frequencies between both groups.

Aging or disease? Conclusions about age-*associated* differences must be differentiated from each other and carefully made. Observable differences between young and old people in general are sometimes explained more appropriately by reference to specific disorders that a person may or may not develop instead of speaking of aging as such. Detailed analyses conducted by the NIMH project team have indicated that even a "touch" of a disease such as **cerebral arteriosclerosis** can be related to a broader pattern of dysfunction. In one individual studied, the degree of arteriosclerosis was not yet enough to result in obvious difficulties in everyday life or even in a conventional clinical diagnosis, but the disorder seemed to be taking its toll more generally. Another man, just as old chronologically, but free from cerebral arteriosclerosis, showed little or no general functional impairment. The fact that many old people *do* have certain types of physical disorder makes it realistic for us to recognize an *association* between age and dysfunction. But this is not at all the same as equating the two. We are better advised to consider the conditions that lead to the specific disorder and how this disorder might be alleviated or prevented than to proceed as though it were an inevitable facet of the aging process.

Cerebral arteriosclerosis is a hardening of the blood vessels that serve the brain

It is only when the disease-related differences can be distinguished clearly that we have the opportunity to focus on "true" aging. The differences that appeared between OM I and YM in this study have made valuable starting points for subsequent research. It seems more fruitful, for example, to seek aging in patterns of electrical activity rather than in blood flow and metabolism in the brain. Whether or not all the results of the NIMH study are supported in detail by the work now in progress throughout the world, it is research of this kind that can bring us closer to an understanding of what just happens to us as the result of various disease processes or traumatic experiences as we move through a long life, and what might be wired in to our fundamental structure and functioning. It is only the latter type of phenomena that most gerontologists are willing to characterize as *aging*.

PREDICTORS OF SURVIVAL

Two follow-up studies to the NIMH study are worth mentioning here. Eleven years after the original study, those men who were still alive participated in a repeat study (Granick & Patterson, 1971). It was possible to determine some of the changes that had taken place in the physical and psychosocial functioning of the survivors as well as to examine factors that might have differentiated between survivors and nonsurvivors. Not surprisingly, the survival rate was higher for men in the healthier group. Those who died were more likely to have had higher blood pressure, had more arteriosclerosis, and have been heavier smokers. There were suggestions of differences in brain waves, although these are not easy to interpret. Measures of cerebral physiology in general did not seem to predict survivorship very well over the eleven-year period.

"Lack of social support" is what the scientist says; "lonely" is how it feels. People with fewer social resources to call upon are less likely to stay healthy. (Photo by R. Warren Johnson.)

Psychological measures turned out to be good predictors of survival. Performances on the Wechsler Adult Intelligence Scale at the time of the original study were related to the probability of surviving the next decade. The same was true of several other measures in the psychosocial sphere, whether assessed by formal tests or based on psychiatric interviews. In all instances, it was the stronger performance that was associated with survival. In general, then, the more able and resourceful old person—and the one with more social support in his life—was the individual most likely to stay alive. This, by the way, is not the only study that reveals a close relationship between psychological functioning and subsequent health or mortality.

Functional interdependence. One facet of the eleven-year follow-up has since had its own follow-up. Youmans and Yarrow (1971) found that as the men grew even older there was an increasing *interdependence* among different areas of their functioning. This finding is consistent with research observations made by other investigators. It suggests that as we age a problem that develops in any one aspect of our lives becomes more likely to place us in general jeopardy. Although we are whole people all our lives, to perpetuate the cliche, in old age we seem to be more whole than ever. One indication of this interdependence was mentioned earlier: the pervasive effects of even a mild cerebral arteriosclerotic affliction. But the affliction can be in any realm of functioning, the psychosocial as well as the physical, and still have a general effect.

Recently, supportive findings have come from a study of three different samples of old people (women as well as men, blacks as well as whites). Fillenbaum (1977) found that the interrelationship among various aspects of personal functioning became greater as her research participants aged. This included the relationship between physical health and the ability to perform routine tasks of daily life, as well as a variety of mental, behavioral, and social abilities. This study, along with Youmans and Yarrow's, suggests that we become more *vulnerable* with age in the specific sense of facing catastrophe from any possible direction. There is *no* aspect of an old person's functioning, then, that can be safely ignored if maintenance of personal and physical integrity is our objective.

Aging as increasing vulnerability. So far we have picked up some important clues about what aging is. The indications are that many of the adverse changes associated with living a long life are more closely related to specific disease processes rather than to an intrinsic process that could be described properly as *aging*. However, there are some changes in physical status that seem to be relatively independent of disease state. The precise extent and pattern of these physical changes is under active study in many laboratories throughout the world. We have also seen that the psychological or personal side of the individual's life is intimately related to the changes that come about with age. This suggests that any ultimate explanation of the aging process will probably have to give serious attention to all levels of functioning. For the moment, it is reasonable to think of aging as a process of increasing vulnerability. This generalization holds up whether particular phenomena are considered to be "caused" by aging, or instead to be associated with the hazards encountered in traversing the lifespan.

THEORIES OF AGING

A variety of specific theories have been suggested to explain the aging process. Frankly, an adequate presentation would take a lot of space and still leave us in an inconclusive state, while a once-over, simplistic tour does not have much to recommend itself either. For an introduction to biological theories, the recent book by Albert Rosenfeld, *Prolongevity,* (1976) is recommended; a critical overview of both biological and psychosocial theories (Kastenbaum, 1965) may be a helpful starting point as well. Articles related to various theories are published on an on-going basis in periodicals such as *International Journal of Aging and Human Development, Human Development, The Gerontologist,* and *Journal of Gerontology.*

Instead of a detailed examination of the interesting but largely unproven theories, it will be more useful here to note some of the questions, distinctions, and problems that are being addressed.

1. Many theorists distinguish between *primary* and *secondary* aging. In other words, not all changes associated with advancing age are considered

equally significant. Changes in the skin, for example, are common with aging, but are usually considered as secondary or peripheral phenomena. So far as we know, there is no theory that claims aging, in general, is "caused" by the skin changes. The more fundamental changes (those believed responsible for the overall process of aging) are less likely to be visible. Biologically oriented theorists disagree among themselves as to precisely which internal structures and events are the most fundamental, but the general distinction between primary and secondary aging is found in the work of many investigators.

Is influencing the course of childhood or remaking the aging process a good idea?

2. There is a broad range of opinion regarding the possibility of *altering* the process of aging. Some investigators believe their findings will eventually make it possible to delay, reverse, or head-off the aging process. Others are convinced that aging is so intimately built into the nature of complex creatures such as ourselves that we have no recourse but to understand and accept. Both the most extreme views and those that are more moderate have certain theories and data to organize themselves around. But it is likely that the personal philosophy and optimism-pessimism of the individuals who devise theories and conduct research play a part here as well. Controversies concerned with the possibility of modifying the basic process (or process*es*) of aging will probably be with us a long time. (Indeed, they were part of our tradition long before the scientific study of aging established itself.) What we want to remember is that even the "purest" theoretician is well aware that many hopes and expectations are riding on the results.

3. There is an increasing tendency to study the aging process by trying to bring it about or accelerate it under controlled conditions. This is based on the familiar scientific principle that we are most likely to produce convincing results when we can actually shape the course of the phenomena under study. A person who believes that nutrition is highly significant in aging and mortality will experiment with feeding patterns in laboratory animals, for example, while another researcher may be attempting to induce effects resembling old age by subjecting laboratory animals to doses of radiation, etc. You can see that this approach, sensible as it is, does nevertheless leave quite a gap to span if we are to apply the results to human beings functioning in society. Other types of research are then needed as well as experimental manipulations in the laboratory. You can also see that experimental approaches to aging have been largely within the province of the biologically oriented theorists. It is only recently, and with considerable difficulty, that psychological and social theorists have moved toward the experimental testing of hypotheses in their areas.

4. We do not have, at this time, a detailed and integrated theory of aging that accounts well for all the levels of biological functioning and also for the psychological and social dimensions. It is even difficult to find gerontologists who can "speak the language" of these different domains with expertise.

There has been a tendency to downplay the significance of psychosocial factors in aging in favor of biological explanations. Nevertheless, there is abundant evidence that the entire course of a person's life, including phenomena often taken as examples of aging, is strongly influenced by social class, personality, life experiences, etc. In the development of programs for aged people, and in the provision of clinical services (social work, psychotherapy, medical, etc.), many experienced help-givers do take a broad range of factors into account. But a strong theoretical overview of the aging *person* that adequately includes the biological, psychological, and social spheres has not yet been achieved.

5. Theoretical positions have not always been as clear as they might be on where the methodologies and facts leave off, and where personal and social values begin. A person who has strong convictions about what old people *should* be like may find it difficult to keep these sentiments from invading his or her scientific theory as such. Those of us who are consumers for scientific theories of aging, then, might do well to be aware of possible leakage between values and facts.

6. There is general agreement—from the biologists through the sociologists and anthropologists—that the phenomena of aging are best considered within a larger context. In other words, if we try to understand aging apart from general development, apart from the full range of life processes, then we will have little hope for success. This proposition works in the reverse direction as well; we will not know nearly as much as we should about the biology or psychology of development or about the structure and function of society if we neglect serious study of aging.

This is one reason why a lifespan approach to humans developing has become increasingly prominent

7. There is a significant area of *dis*agreement that should be mentioned. Biologically oriented theories usually see aging as a negative process for the individual, although a blessing for the species. Psychosocial theories are more apt to search for the positive aspects, along with the negative. This introduces a tension between the two levels of theoretical approach. On the human level, this translates into a poignant situation: here is a man or woman for whom all physical changes have a negative sign attached. The best that can be hoped for is no change or slow change. But the person himself or herself is more than the body. How is it possible to maintain worth to self and others under these conditions? A psychological or social theorist must somehow come to terms with the paradox of a person who seeks to maintain or advance as a human being while suffering the inroads of a relentlessly negative process.

The theoretical challenge of answering our questions about aging amounts to very little when it is compared with the actual challenge that faces the individual old man or old woman. It is to this situation that we now turn.

THE COMPETENT OLD PERSON

Chronological age does not tell us much about an adult's competence. This generalization still comes as a surprise to many people in our society. It certainly runs afoul of the policy to require retirement at a fixed chronological age, and of age discrimination in its many other forms. Yet both practitioners and researchers have come to recognize that they cannot place too much reliance on age as a predictive variable. It is one factor in the equation, but only one factor and not always the most important one. Sophisticated researchers no longer assume that "age" "causes" incompetence or anything else for that matter. Instead, they attempt to determine how age interacts with many other variables to result in a particular outcome.

Another generalization should be set alongside the first. There are great individual differences among people who fall into the "old age" range. There are more differences *among* elderly people than between the young and the elderly on many socially significant dimensions. Various degrees of competency can be found at all adult age levels.

Let's take one more generalization as a guide. Changes related to age do not sweep across all areas of functioning with equal power. Some psychological functions are more likely to be affected than others. Even within a specific domain, such as memory, certain abilities falter while others hold up well.

This last generalization may appear slightly at odds with one we have already encountered, the *interdependence* of functioning that contributes to increasing vulnerability with advancing age. There *is* a problem in reconciling the selective impact of aging on the individual's functioning with the in-

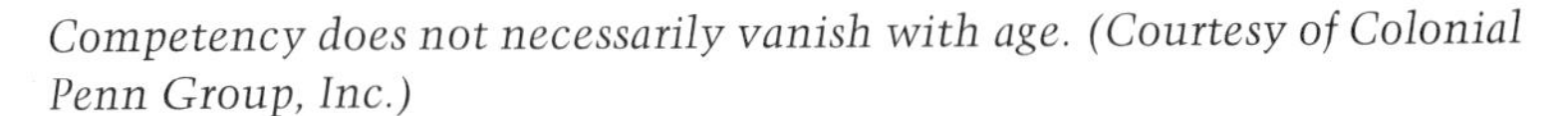

Competency does not necessarily vanish with age. (Courtesy of Colonial Penn Group, Inc.)

creasing tendency for all areas of functioning to become more dependent on each other. Research has not yet tackled this discrepancy; in fact, it appears to have gone practically unnoticed. Nevertheless, both appear to be useful guides to understanding and consistent with available knowledge.

The competency of a particular person, then, is likely to be related to chronological age only in indirect and complex ways. This becomes even more evident when we take the environment into account. As Lindsley (1964) pointed out in some detail, many of the environments in which old people function appear to make it difficult for them to exercise and maintain their skills. He proposed a number of "prosthetic" devices i.e., tools and techniques for enabling the old person to demonstrate competence and independence even though living with certain limitations or afflictions. Lindsley, one of the leading behavior modifiers, was also among the first to examine the specific situation of the old person. He proposed a careful assessment of the individual's specific *behavior deficits.* This would be followed by design of a device, system, or environmental feature that compensates for the deficit. A sample of his thinking:

> Innumerable response force amplifiers are available for normal persons. Most hand tools, for example, amplify response force. Hammers increase the force of manual pounding by extending the leverage of the arm; wrenches, the force of finger grip. In a sense, most modern machinery is designed to increase the force or accuracy of normal human action.
>
> *Response force amplifiers* [italics added] should be provided for old people with extremely weak motor responses. Geriatric environments should contain a much wider range of response force amplifiers than the fully automated factory or fully electrified home. Why, for example, must the aged open their own doors in hospitals when supermarkets and garage doors are opened electronically?
>
> For elderly people with feeble voices, the force of speech could be amplified by throat microphones and transitorized amplifiers. Such a simple device might greatly facilitate communication between older persons.
>
> *Wide response topographies** [italics added] should be provided so that palsied movements and inaccurate placement of hands and fingers would not be disabling. An individual with extreme palsy, for example, could operate a telephone with push buttons, instead of the normal dial arrangement, if the buttons were far enough apart and required enough pressure so they could not be accidentally pushed by a shaking hand. The voice-operated telephones in the Bell system design will, of course, completely prosthetize dialing deficits.†

These are but a few of the specific suggestions made by Lindsley. Since his pioneering article, many suggestions of this type have been made by a number of gerontologists, and some have in fact been introduced into practice. The general point for us here is that an environment can be indifferent

* *Wide response topographies* refer to enlarged surface areas of working spaces for people to use in performing a task.

† From "Geriatric Behavioral Prosthetics" by O. Lindsley. In R. Kastenbaum (Ed.), *New Thoughts on Old Age.* New York: Springer Publishing Co., Inc., 1964, p. 49.

or hostile to a person's specific needs, thereby decreasing his or her competence, or it can offer appropriate support. A very old person may be able to live competently in one environment, but not in another. We are accustomed to judging the individual as more or less competent. But perhaps we should also *evaluate environments and entire societies as more or less competent* in maintaining an atmosphere for good functioning on the part of individuals with specific impairments.

We will now consider several realms of functioning in old age to identify some of the factors that influence competency for the better or the worse.

SLOWING DOWN

Scientific evidence and everyday observation agree on the generalization that we tend to slow down with advancing age. A variety of studies support this conclusion. It appears to be true for other species as well.

We have already seen that the peak frequency of brain waves is slower even in very healthy old men when compared with young men (*see* Figure 17–1). But the slowdown is not limited to any one sphere of functioning. Almost all studies find that reaction time to a stimulus is not as prompt for old adults as for young adults. Furthermore, there is a generalized *psychomotor slowing*. In other words, both how long it takes to get ready and how long it takes to actually do something increases with age. Reviewing many studies on this topic, Hicks and Birren (1970) found that the decrease ranged from approximately 20 percent to 110 percent depending on the specific task-situation. That we do slow down with age must be counted as one of the most securely established facts in gerontology. There are, however, three very important and related questions for which the evidence is less conclusive.

CNS and psychomotor slowing. First, we might ask what is primarily responsible for the slowdown. There is much circumstantial evidence to suggest that it begins with alterations in the central nervous system (CNS). As you probably know, brain cells differ from most other cells in our bodies in that they do not replace themselves throughout our lifespan. When a brain cell dies or undergoes changes that make it less efficient, there is no hope for a fresh, enthusiastic replacement. If cell death and deterioration in the CNS over time were found to be the only change occurring, this might in itself be sufficient to credit the CNS with producing the slowdown. And there is no reason to limit the analysis of CNS changes to this factor alone. While the case for a CNS-instigated slowdown is plausible and can be supported at a number of points with data, it has not been demonstrated beyond reasonable doubt. Some of the necessary research is complex and time-consuming and requires linkage between the biological and behavioral spheres of functioning.

Consequences of the slowdown. Next, we might well ask about the *effect* of this slowdown on the individual's total competence. How much does it interfere with maintaining the preferred style of life? Again, there is much circumstantial evidence to suggest that the quality of life—and in some cases, the length of life as well—can be jeopardized. The elderly driver and the elderly pedestrian are at extra risk whenever age-associated slowdown penalizes them in a situation requiring fast reaction. Furthermore, people with occupations requiring unusual speed of reaction or speed of operation are also likely to experience unusual stress as the slowdown phenomenon manifests itself. You can think of many situations in daily life where that extra margin of safety or competence provided by quickness is important and reassuring. The old person tends to lose this margin. The adverse effect may show up over a period of time or it may emerge in a sudden emergency situation.

The relationship between slowdown and competence must be considered within our cultural context. We are a clock-watching, deadline-setting, time-ordered society. What some people enjoy most about a vacation is the temporary escape from time pressure, the chance to do things at a more relaxed tempo, when they want to do them, rather than when the clock proclaims they must. The aging person remains within this climate of social expectation. As long as he or she attempts to stay within the mainstream of life, there is the added pressure of remaining afloat and on course in a stream that moves faster than one's body and mind is comfortably prepared to cope with. One of the most obvious indications of "disengagement" from previous life roles, in fact, is when a person stops trying to meet all the conventional time expectations (Cumming & Henry, 1961).

Slowing down is not all bad. There is more time to observe, reflect, converse leisurely. (Photo by R. Warren Johnson.)

In our society, then, what appears to be a natural, universal slowdown in functioning becomes especially obvious for the individual. Slowing down here is much different from slowing down in a society where the general tempo is more relaxed and life is not ruled by clock and calendar. The old man or woman in our society may be put into a situation of appearing less competent because of the speed factor, whether or not there has been any other change in ability.

Psychological testing has revealed this problem in another light. Old people often fare more poorly than young people on a variety of test performances, whether these are of the paper-and-pencil or experimental type. Taking these results at face value would make it appear that old people are much less competent than the young. However, experts in evaluation and measurement know better. Much of the deficit can be attributed to the speed factor. When old people are required to race against the clock in their test performances, or when time credits figure heavily into the score, their relative competence looks most impaired. But when time is not made an important condition of performance, the discrepancy between the young adult and the old adult is reduced, sometimes to a very appreciable extent (Botwinick, 1967).

This means that we might come to a conclusion that exaggerates impairment of functioning in old age because one contributing factor—reduction of speed—can affect a wide variety of behavioral outcomes. A relatively low test score may not really signify that a person has become less competent in any fundamental sense. Rather it may mean that he or she cannot read the items, search his or her memory, make decisions, and record answers quickly enough to earn a high score.

The "real life" situation, of course, is more complex than the testing or experimental situation in many instances. It is not always easy to say that speed has one particular kind of relationship to overall functioning. A carefully designed testing procedure, for example, may indicate that an eighty-year-old person can perform adequately when the time and speed dimensions are adjusted for him or her. But when he or she steps out into the sociophysical world, these adjustments no longer exist. He or she may, in fact, experience serious difficulty and appear lacking in competence. Gerontologists know, in this instance, that the person essentially retains his or her competence, but the situations in which he or she must function may be so time-oriented that he or she is placed in the predicament of appearing less competent.

This emphasis on slowdown, by the way, should not be interpreted as the only possible source of interference with competent functioning in old age. Again, in the real life situation, any one variable interacts with many others. And in the experimental situation, even when the speed factor is taken into account, this does not always erase the total relative deficit of the older person when compared to the younger adult.

Reducing the effects of slowdown. Finally, we might ask about the possibility of reducing the adverse influence of the general neurobiological slowdown. What might be done about it? Several types of solution are worth considering. An imaginative research approach has been piloted by Diane S. Woodruff (1975). She and her colleagues are attempting to reverse the slowdown of brain waves through the biofeedback technique. Biofeedback techniques are designed to enable a person to control his or her own physiological processes, for example, blood pressure, muscular response, temperature, or level of alertness/relaxation. This is achieved in part by the use of devices that allow the individual to monitor the current functional state of the physiological process he or she wants to subject to increased control. There is now considerable literature on this topic. A useful up-to-date source is the volume edited by Schwartz and Beatty (1977). If further research has the hoped-for result, then it is possible that old people will be able to control the tempo of their own **alpha rhythms.** Presumably, this would be associated with a reversal of behavioral, as well as neurological, slowing. It is premature to judge whether or not this particular experimental approach will prove successful in the long run, but at the least it illustrates the potential for using modern scientific and clinical techniques as direct interventions in age-associated slowing.

Alpha rhythms are slow brain waves often associated with a calm, pleasant state of mind

It is known that individual adults can, themselves, influence the outcome of the slowdown, for example, older workers can often increase their attention to accuracy and quality of work. The worker's quantitative production might fall off a little, but there is a resulting gain in quality. This also suggests that a shift in work responsibility, for example, from production to inspection or quality control, might benefit both the older worker and his or her company. Additionally, Botwinick (1973) has pointed out that physical exercise tends to reduce the extent of slowdown. It is also possible that motivational factors, in general, play an important role. The person who remains active and who is intent on staying in shape is more apt to display only small slowing effects.

Individual differences are clearly recognized by researchers in this area. Some people are "quicker" than others at *any* age level. This means that a particular old man or old woman may respond and move faster than a particular young person because of life-style differences. It was not long ago that I had the privilege of seeing Artur Rubinstein in concert. His ninety-year-old fingers were obviously still much superior in speed and accuracy to most sets of young hands that launch themselves on a keyboard. More impressive yet was the mastery of music and life that he was continuing to display. Another remarkable man who remains remarkable in his nineties is Eubie Blake (*see* photograph opening this chapter). Eubie Blake's musicianship and charisma have continued to win new audiences for the composing and performing career that helped lay the foundations for the popular song and musical tradition in the U.S.A.

The instance of a great performing artist is, by its nature, exceptional. Yet we can learn from the extremes of accomplishment as well as from the annals of psychopathology and failure. To be an Artur Rubinstein at age ninety, one is well advised to have been an Artur Rubinstein at age twenty, forty, and sixty. Even the more relentless and universal effects of aging do not overcome a personality that has shaped itself forcefully right down to the fingertips.

REMEMBERING

Many of us take our memories for granted. It is only when memory fails us that we are likely to appreciate how much this ability contributes to our enjoyment and success. Furthermore, we may become apprehensive of situations in which our memories are put to the test (as, for example, on the next examination you take). Memory deficits in old age threaten the old person in at least three ways. They: (1) reduce successful functioning in specific situations, (2) contribute to other people forming an adverse opinion of his or her competence, and (3) raise doubts in the individual's own mind about his or her present and future status. These problems can interact to produce a distressful pattern, for example, the old person who withdraws from social interaction and from other activities for fear of being "betrayed" by memory lapses. An understanding of memory functioning in old age is obviously important to both the old person in making a systematic effort to maintain general competence and to the researcher in studying aging as a general process.

Types of memory. Research with people of various ages has revealed that memory is not a single ability. It is, rather, a complex system of functions. Some appreciation of this complexity is necessary to comprehend memory difficulties in old age (or in any other context). Fozard and Thomas (1975) have proposed that there are three types of memory. *Primary memory* refers to information that has just come into the person's mind. A friend has just given you his telephone number. Do you remember it? Are you sure? Primary memory has a limited scope. Only a few digits or other items can be retained on first exposure to them. *Secondary memory* encompasses both a longer time span and a larger number of possible items. Information is stored for some time and, if all goes well, is available when needed. If you learned the names and functions of the cranial nerves last week, this knowledge should be in the semi-permanent memory storage system, ready for you to dazzle your friends or impress your instructor. *Tertiary memory* is another story. In fact, it contains the oldest stories we know, the most long-ago memories that give our lives a sense of continuity. The long-term memory is apt to be very personal and to have had many opportunities for rethinking and retelling.

Old people do not experience much difficulty with primary memory. This is seldom the type of memory process that causes them trouble in daily

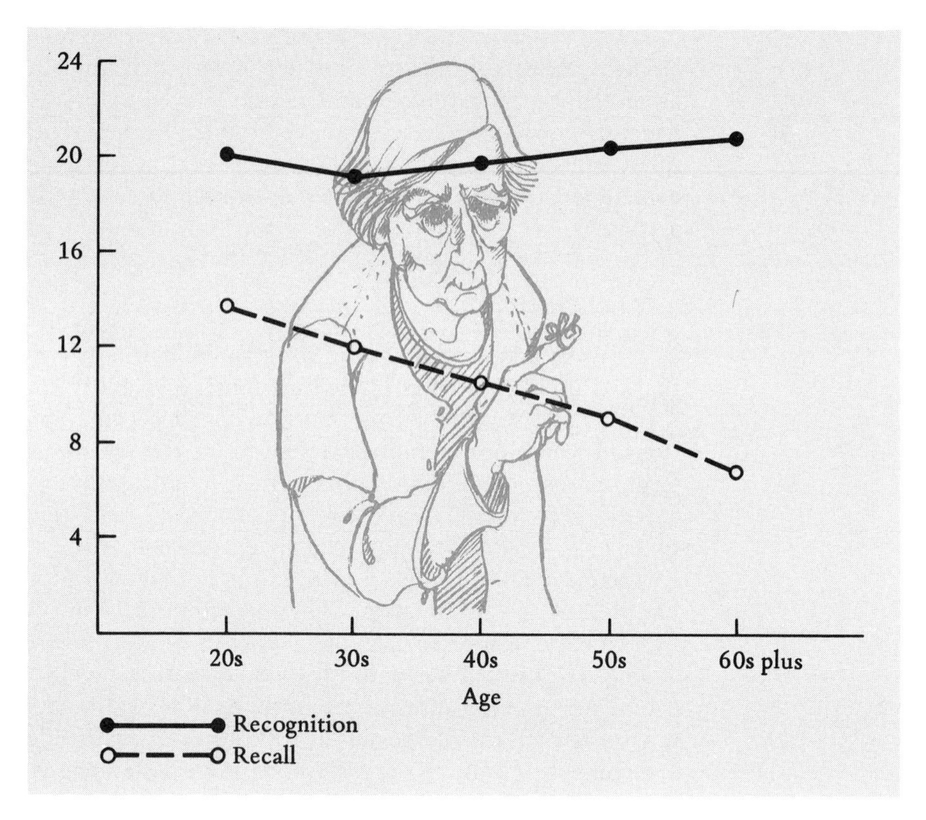

Figure 17-2. *It is easier to recognize than to recall: memory benefits from cues. When studied, older adults show no decline in memory performance as compared with younger adults when all are given the opportunity to use the recognition method. Older adults do not fare as well when using a recall method (no cue in the environment), indicating that recognition memory holds up well with age but the ability to recall without any help from the environment declines. (From "Memory Storage and Aging" by D. Schonfield and B. A. Robertson,* Canadian Journal of Psychology, *1966, 20, 228–236.)*

life or shows a significant deficit when compared with younger adults. The long-term or tertiary type of memory also holds up well for many old people. If the individual is not suffering from some form of pathology that affects the CNS or impairs general functioning, then he or she can still dip into the remote past for knowledge and skills.

This leaves us with secondary memory, and here is where most of the memory problems that plague elderly people can be located. If your CNS was made seventy-five years ago, you might experience difficulty in coming up with a telephone number (or that list of cranial nerves) that you learned two weeks ago. Much of daily life draws on knowledge that is in semi-permanent storage. When we "draw a blank," it is in this realm that the blank is most likely to appear. Many studies have attempted to determine whether the problem is mostly in the storage process itself, or in the process of *retrieval.* In other words, has the memory somehow "decayed," or is it that the old person has some special kind of difficulty in searching for the memory and bringing it out intact? The evidence is not entirely clear yet. Problems have been found both in storage and retrieval, but the conditions under which these two aspects of secondary memory create the most difficulty are still being examined.

Approaches to maintaining memory. Can old people be helped to overcome their memory difficulties, or, perhaps, can the difficulties be prevented in the first place? A wide variety of drugs and other physical treatments have been administered to people suffering from memory impairment. In general, the results have not been conclusive, but there has been the impression that enough elders experience enough improvement to make these efforts worthwhile. As Butler and Lewis (1973) have pointed out, the use of drugs with old people is a procedure that must be conducted with great care; at times, the treatment can become part of the problem. It is probable that physical approaches will continue to be developed, and some may eventually prove to be highly effective in combating memory deficit in old age. In the meantime, it makes sense to continue to explore other approaches as well.

A variety of psychological studies are converging to suggest that memory function can be maintained better through certain individual and environmental techniques. Some psychologists, for example, recommend the simple and obvious technique of talking to oneself. This is not necessarily a peculiar behavior, even if it has acquired a funny reputation. By actively repeating or rehearsing significant bits of information, we are more likely to keep that information firmly registered and available for use when needed. On a more systematic level, there are attempts to discover the best strategies for retrieving information from secondary memory. Hulicka and Grossman (1967), for example, found that young adults tended to have more effective techniques for searching their memories. The investigators reasoned that this might have something to do with the fact that the old people with whom they were compared had less formal education and, therefore, perhaps had less opportunity to develop some useful strategies for memory and learning. Based on clues in such studies, efforts are being made to help old people learn better ways to work with their memories. Indeed, just the realization that there is more than one way to use our memories, that one does not have to shrug and accept memory deficits, may become an increasing stimulant for adults of all ages to protect and enhance this important mental function.

Another line of research indicates that *experiential style* may have a close relationship to memory functioning. Experiential style is the way in which individual personality influences how knowledge is obtained and interpreted. Costa and McCrae (1976) have found that elderly men with a knack for imaginativeness and a marked openness to experience show a different pattern of memory strength and impairment from others of their age. They have a superior ability to utilize memories that were laid down years ago; curiously, however, they have more than the usual age-associated difficulty with secondary memory. Suggestive findings of this kind may eventually lead to techniques for emphasizing the particular memory strengths of individuals with certain personality characteristics and at the same time minimizing the shortcomings related to their experiential styles.

In short, there is abundant evidence that memory difficulties exist in the type of functions that are important in every day life. Instead of causing discouragement, however, these findings are stimulating researchers, educators, and clinicians to develop techniques for helping elderly people to maintain good memory function and, therefore, maintain general competence.

LEARNING AND DECISION-MAKING

Coping well with life requires more than the ability to call upon memory. It also requires the ability to learn new information and skills and to make adaptive decisions. Once again, research across the lifespan makes it clear that learning and decision-making are complex functions. The nature of the problem that confronts the individual and the nature of the situation in which the problem occurs have much to do with ease of learning and decision-making. This cautions us against sweeping generalizations; it is more useful to examine the specific situations and problems that a particular old person is facing than to indulge in generalizations. Nevertheless, a few findings appear strong enough at this time to be worth attention. And our attention must be well focused, or we will fail to make some of the distinctions vital to an understanding of mental functioning in later life.

Learning or performance? Consider first the proposition that it is common for old people to behave as though they have failed to learn new information or skills. A number of well-designed studies, such as those reviewed by Botwinick (1973), support this generalization. The extent of the deficit depends on the particular circumstances and the type of material involved, as is also the case with memory and other cognitive functions. At this point, then, it looks as though we might have to draw an important negative conclusion about what happens to mental competence with advancing age.

Yet Botwinick and other specialists in this area are not willing to stop at this point. They direct our attention to the difference between the actual learning process itself and the ways in which the learning is demonstrated. A person (at any age) may know considerably more than he or she translates into behavior or performance at a particular time. Often we know what the other person knows only by what he or she says or does under the prevailing conditions of observation. This can result in an underestimate of the actual learning that has been accomplished—and such underestimates are easy to make with old people. In daily life, part of the problem may be attitudinal: we may *expect* old people to be poor learners, therefore we see poor learning just as our prejudices tend to be self-confirmed in other spheres. Similarly, the old person may expect himself or herself to be a poor learner and then behave in ways that lead to this outcome.

One significant advantage of the experimental situation is the opportunity to distinguish between attitude and objective fact. A keenly developed study can also help to distinguish between what a person really has learned and what he or she can demonstrate under particular conditions. Here is one of the areas where current research is contributing to a revision of negative stereotypes about the capabilities of old people. Botwinick has reported: "Much of what in the past had been regarded as a deficiency in learning ability in later life has more recently been seen as a problem in the ability to express the learned information. In other words, much of what had been thought of as a deficiency in the internal cognitive process is now seen as a difficulty older people have in adapting to the task and in demonstrating what they know" (Botwinick, 1973, p. 218).

Increased emphasis is being given to "noncognitive" factors in learning. We have already touched on the role of time and speed. In some sense, speed *is* a cognitive factor, for brighter people at any age do tend to think faster. However, it is possible to separate the speed factor from other aspects of cognitive functioning. When we do so, then learning ability in old people appears in a more favorable light. One study that allowed more time both for

Every year more older Americans return to the classroom. High motivation and mature judgment help the older learner to do well. (Photo by Richard L. Maddox.)

learning and for demonstrating what had been learned indicated that old adults could do as well as the young, while under time pressure the older participants performed much less adequately (Monge and Hultsch, 1971). If these results continue to be found in subsequent studies, then we will be in better position to help older people learn under optimal conditions. A related noncognitive factor centers around the control of the task situation. Demonstration of new learning comes about more readily for old people when they have the opportunity to set their own pace rather than abide by somebody else's demands. More time, and self-paced learning and performance, then, are among the factors that minimize age-associated differences in learning competence. It would be premature, however, to claim that any one learning-performance strategy, or any known combination, can completely eradicate age-associated deficits. This still remains to be seen.

Reducing time pressure often benefits the performance of young adults as well

Motivation and mental functioning. Motivation is another factor that affects our evaluations of an old person's competence in learning and performance. A person who does not want to learn, or whose mind is preoccupied with other needs and wishes, is less likely to function well in a learning or performance situation. This may account for some of the problems experienced by old people in daily life. If they are beset by more urgent needs of close personal relevance, then they just may not have the energy and attention to invest in new learning. It follows that whatever helps an old person (or anybody else) feel at ease and function without need-generated distractions will also help him or her to learn and adapt more successfully to the now faster-paced world.

But motivation can also operate in the other direction. There is some reason to believe that old people become so aroused or activated in a learning situation that they inadvertently interfere with their own functioning (Woodruff, 1975; Botwinick, 1973). There can be both too much and too little physiological stirrings-up for optimal performance. As gerontologists learn more about this factor, it may be possible to devise strategies that enable an old person to approach learning and performance tasks with the most comfortable and effective level of activation.

Let's take just one more noncognitive factor that can result in a performance which does not do justice to the individual's knowledge and abilities. Few of us enjoy making errors, especially in public. Old people, already aware of the negative stereotypes directed toward them and of their somewhat perilous linkage with mainstream society, often are especially concerned about making mistakes. In my own clinical, research, and personal relationships with old people, I have known many who have adopted an attitude of extra caution. They seek to avoid situations in which they might fail either in their own or in other people's eyes. What the observer sees, then, is withdrawn, passive, seemingly aloof behavior. The observer might interpret this as "just old age," as indifference, or as incompetence on the old person's part. But the same old man or woman can be observed to function in a more spontaneous, decisive, and competent manner when the situation promotes com-

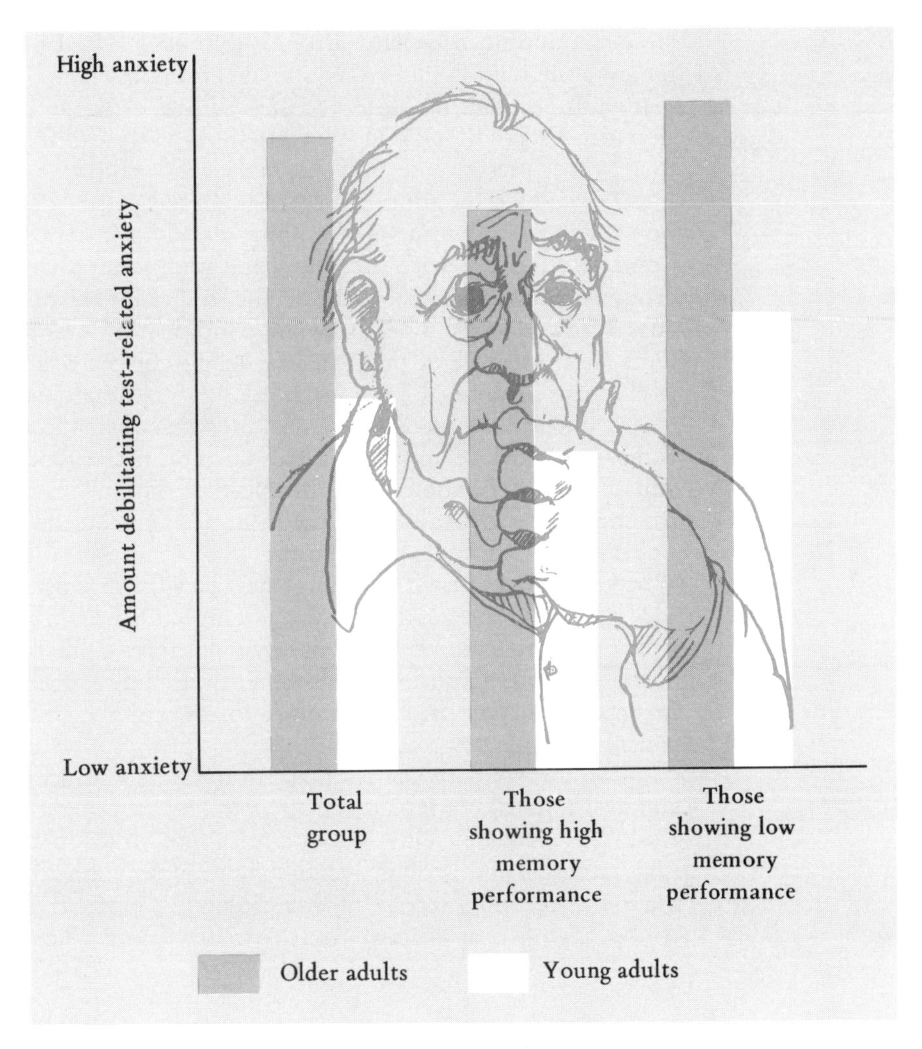

Figure 17-3. *Testing makes people anxious—especially the elderly. Elderly people show more test anxiety than younger people on a memory task. This anxiety may interfere with memory performance. (Based on data presented by Whitbourne, 1976.)*

fort and a reduction in the fear of failure. We would have no way of knowing this from a casual encounter in daily life or from the usual experimental research situation. But when we come to know the person well, or see them in situations that naturally "feel right," we find that it is the need to minimize risk rather than the loss of actual competence that is often primarily responsible for meager behavioral performances.

The risk of decision-making. All these noncognitive factors are as important in *decision-making* as in learning. Furthermore, there obviously is a close relationship between learning, performance, and making decisions. If, for any combination of reasons, an old person does not possess relevant information, it becomes much more risky for him or her to be caught in a decision-making situation. Whatever interferes with learning and performance, then, is apt to accentuate problems in decision-making.

The research findings in regard to an old person's competence in decision-making will remind us of findings already mentioned in related spheres. Compare old adults with the young and we find more difficulties in making decisions and solving problems. Yet the studies in this area are not as numerous or convincing as in learning and performance (Arenberg, 1973). Furthermore, at least some of the apparent deficit appears related not only to the noncognitive factors already discussed, but to generational differences. Some people who are elders today received their primary education in schools that emphasized rote memory rather than the more active modes of analyzing and structuring information. It is possible that the present generation of older Americans, however well equipped they are in other respects, has not been well prepared to approach new problems in varied and innovative ways. When the first approach fails, there may be a fruitless persistence rather than a shift to another modality. To the extent that the main problems are in this realm, then educational programs of various types could be an important part of the solution. The growing movement for adult education in general, and for special educational opportunities for the elderly, could make a valuable contribution here.

Despite the many factors that can interfere with learning, performance, and decision-making in later life, it would be an out-and-out mistake to conclude that old people either do not, or cannot, think competently. The question, rather, is how serious society may be in providing the opportunities for maintenance and exercise of competence throughout the entire lifespan. And, as members of society ourselves, we have some voice in what happens.

THE MIND IN OLD AGE: DECLINE, STABILITY, OR GROWTH?

But where has the mind gone? Research advances into mental functioning at all points in the lifespan have required us to examine a variety of specific processes. We must look at this or that form of memory, various types of decision-making, perceptual organization, concept formation, and so on. *The Mind* as a single entity or process does not fit into the limited dimensions of any given study. With this approach, we gain the advantages of close focus on specific processes and the heightened detail that emerges. Yet one might be forgiven if there is still a yearning to know what happens to that once-familiar standby of human experience that we learned to call *the mind.*

Researchers and theoreticians have not entirely forsaken the mind. Some of the most distinguished psychological researchers still step back from their detailed studies to offer broad conclusions. When they do so, however, their voices do not blend into perfect unison. As a matter of fact, there is a briskly bubbling controversy in progress right now concerning the trajectory of mental functioning in the later years of life.

One side of the controversy is well represented by Paul Baltes, K. W. Schaie, and their collaborators. These psychologists believe their research has substantially challenged the assumption of decline in mental functioning with age. Some of their work can be found summarized in an article prepared for general readership (Baltes & Schaie, 1974). Their concepts, methodologies, and findings have been reported in more detail in many scientific publications (e.g., Schaie, 1970, 1973; Schaie & Labouvie-Vief, 1974; Baltes & Labouvie, 1973). Their research is methodologically sophisticated and includes classic examples of cross-sequential sampling procedures with adults of various ages (*see* Chapter 2 for a discussion of this procedure). In general, their research seems to suggest that what has previously looked like intellectual decline with age is at least in part a function of limitations in the earlier studies themselves.

But there are other sophisticated researchers who are not convinced that age-related decline has proven to be a myth. Prominent among these is John L. Horn who, with Gary Donaldson, has offered a well-detailed challenge (1976). Much of their argument centers around the type of conclusion that can be drawn appropriately from particular kinds of data. I will give two illustrations of Horn and Donaldson's argument here.

Some elderly research participants were given the same tasks on repeated occasions, while others took the tasks only once. Horn and Donaldson seem to have demonstrated that the data collected under these two conditions cannot be compared too closely, thereby clouding the conclusions that can be drawn from the data. Performance on measures of **inductive reasoning,** for example, seems to keep improving through the forties (according to the repeated tasks design) *or* to start declining in the mid-twenties (according to the independent sample design), depending on the data-gathering strategy used.

Inductive reasoning is the type of logic that proceeds from specific cases to the more general

Several of the Baltes studies have shown that elderly people can improve their performances on intellectual tasks that are usually considered to show age-related decline if they are given the advantage of a training program. Horn and Donaldson praise these studies as such, however, they feel:

> . . . an interpretation commonly given to such results is that because the abilities in question are of a kind for which aging decrement has been indicated, and because the training evidence shows that these abilities are modifiable, the evidence for decline in these abilities should be discounted. In fact, such research does not speak to the question at issue in claims that there is, or is not aging decrement in intellectual abilities. (Horn & Donaldson, p. 718).

The more logical conclusion might be that there *is* decline, but its effects can be minimized by educational experiences.

No doubt this controversy will be with us for awhile and will prove instructive. Researchers on both sides of the debate recognize that mental functioning is complex at all ages, and that it is exquisitely difficult to sort out the contributions of such factors as biological change, illness, education, cultural opportunity, and individual personality.

Here I would like to emphasize two of Horn and Donaldson's observations: (1) wishful thinking can lead us to overstate or prematurely accept findings that seem to shake the age-decline stereotype, and (2) there is no reason to suppose that *all* people undergo decline with age, even if some do.

In the real life situation, there is often the expectation and assumption that elderly people do not think as well as younger adults. We have seen here that when there are actual observable differences in mental functioning between younger and older adults, this might be the result of many factors apart from intellectual decline as such (the climate of negative expectation itself being one). If research has not yet overturned all these negative assumptions as myths, it has at least made significant inroads into such careless habits of thought.

The possibility that mental functioning might do more than not decline—that it might instead reach new heights or depths of expression—should not be overlooked. Research in this area is just beginning to emerge, as can be seen in Scenario 12.

SCENARIO 12: TOWARD SANYĀSA

It was hot and dusty. Few of the usual conveniences and attractions of the tourist route were available—and it was a long, long way from home. But Michael and Tanya were pleased to be in this place unknown to travel agents. They had decided on a vacation that was not only a getting away, but a getting *to.* In making the special effort to visit this arid section of Western India, they had hoped to fathom something of a way of life that, while much different from their own, had already influenced them at a distance. Other people were visiting Nathdwara as well, but these were members of the Hindu Vaishnava sect coming as pilgrims to this small, sacred spot. Michael and Tanya both responded to Indian art. They had come to respect what they had learned about Hindu philosophy of life through readings and conversations. Now, with the help of a Hindu friend, they were having the opportunity to meet some of the finest artists still working within the ancient tradition.

"You are accepted here," their friend said. "Come, we will visit with an artist who all the others acknowledge as a true master and love as a person." Soon they were in the presence of an old man who was working with great concentration on one of the most perfect small sculptures they had ever seen. He bid them welcome and continued to work. After a few minutes, Michael and Tanya stopped feeling out of place. Through their own concentration they felt they had somehow become a small part of the creative process or, rather, of a process that was both religious and artistic, a ritualization of some impulse inside of them all. Nearby were the silent ruins of an ancient building, but also several scrubby trees crackling with the conversation of parrots.

After some time they found themselves in conversation, although it had begun so naturally they could not really recall the shift from silence to dialogue. "My life? Why, once I was young, and now I am old. I suppose that happens in your land, too, although some people here report that all Americans are young."

"Slightly exaggerated," Michael replied. "It is just that all Americans want to believe they are young."

"Young is good. When I was a young man, I was accepted as an apprentice. You know the word, *guru?* Yes, apprentice to a guru. Perhaps you can explain it better than I can to the Americans." Their Hindu friend explained the guru's role in shaping the student's life and calling, how the young person must accept instruction and correction and learn to discipline himself or herself. The artist nodded, "This is not a good piece of work! I heard that many times. At first, I would become fierce and angry inside. But then I learned to learn. I still learn."

The artist set his sculpture aside and continued. "Young is good. Not so young is good, too. I entered a second stage of life. We call the first stage *Brahmachrya.* It is necessary, but one cannot be an obedient learner forever. This second stage we call *Grihasthashrama.* I married. I seeded children. My wife and I worked to make a life together for ourselves and the children."

"You were a *householder.*"

"Yes, a man under the dominance of Kamadeva, the god of love. This guided my work and my art. Much that was new entered my life. I tried this technique and that technique in my art. I made new friends, saw more of the world. You must not be shocked, friends, when I tell you there was a strong connection between sex and creativity at this time. It was the time for pleasure. From pleasure springs art. I am a householder no longer. *Vānaprasthashrama* is the name we have for my stage of life. I am an elder; you can see that for yourself. I wish to be more alone with my thoughts and my work. I listen more carefully for messages from inside, to what the gods have to say, rather than to what the world speaks." He tilted his head toward the parrot-loud tree with the hint of a smile. "The kind of artist I am now? I am the owl who does not give a hoot—except when *he* wants to. I create to please my own mind and heart. *Āpnē masti sē*—help me with this, friend . . ."

"He does his own thing."

"Yes, that's it. An artist in my stage is as independent and strong as a water buffalo. He is ripe within himself and yet at the same time closer to the power of the gods, closer to root emotion. We do not seek pleasure as we did in the householder stage. What we need is within us—if it is any place."

The time had slipped away, even the generous, slow time of Nathdwara. In taking their leave and expressing their thanks, Michael and Tanya wished the old artist well for the future. He nodded, placed his hand on his heart and said very softly, "Toward *Sanyāsa.*"

Sanyāsa is the final and most advanced stage of the life-cycle according to Hindu belief

On their way back to the center of town, Michael and Tanya learned a little more about **Sanyāsa** from their Hindu friend. "This is the final stage of the life-cycle. But it is achieved by very few. We have been with a fine artist and a remarkable person. But his development is not yet complete. I will tell you what one sanyasi has told me":

> I am a very old man and people say I am very creative. I agree with them. I try to look at things in a different way—a new way all the time. I have lost interest in everything external to my artistic efforts, in all things except my special interests. Even though I stand before darkness and death—ready—still I try to incorporate everything that is natural in my paintings. . . . I have an awareness of the world when it concerns my paintings. My mind is always aware of the surrounding—clothes, sky, buildings, etc.—with a keen eye to incorporate these things into my paintings. Otherwise I take no notice and give no importance to other things. It is a special kind of vision which creative painters have in old age, though I have had these propensities since my childhood days. . . . The mind of the really creative painter is different from others. And I tell you clearly: after the householder stage he who is the creative painter is equal to God—he performs the highest acts of creation. (Maduro, 1974, p. 310).

Michael and Tanya also learned that recent anthropological research had confirmed the high regard Western India has for old artists. A well-designed study showed that it was the older artists who were rated as most creative by their peers (Maduro, 1974). This finding was much different from the most

Try telling this man that creativity is only for the young. The culture that expects old people to continue and deepen their creativity reaps the benefits. (United Nations/Jongen.)

frequently-cited studies of age and creativity in the United States (e.g., Lehman, 1953; Dennis, 1966). In American society, most "creative achievements" appeared to be made by the young. The old were not expected to be creative, and the old were not seen as creative. Michael and Tanya, heading back to their own society, had a few things to think about.

HAZARD: VICTIMIZATION BY CRIME

Creative development into and through old age is an ideal supported better by some societies than others. Yet it is a distinctive achievement rather than a guaranteed outcome even within a supportive framework such as the one just explored in Scenario 12. On the other side of the ledger, there are altogether too many examples of old men and women receiving insults, injuries, and mistreatments that add to whatever biological burden may be imposed by the process of aging per se. Victimization by crime is one such example. We may have to go to carefully selected surroundings to see old people encouraged and expected to function to their richest potential, but we can find the menace of crime in the life-space of many elders throughout our nation.

Goldsmith and Tomas (1974) have summarized some of the factors involved in excessive vulnerability to crime among our elderly. They noted that any loss of economic resources hits old people especially hard because many are already forced to live on reduced or low income. Furthermore, the old person is more likely to be victimized repeatedly. It is not rare for the same burgler or mugger to strike the same elderly victim over and over again. The old person is also somewhat more likely to live alone, and social isolation increases vulnerability to crime. The diminished physical strength and stamina of old people makes them attractive targets to their predators. They are less able to defend themselves or escape (although some would-be assailants have picked on the "wrong" old person and taken the consequences). Poor health and limitations in vision, hearing, or mobility can intensify those vulnerabilities more directly related to the aging process as such.

And where do old people live? Many reside in high-crime neighborhoods. Fear of becoming a victim (or becoming a victim *again*) leads some elders to adopt a very cautious life-style. This not only deprives them of opportunities for satisfaction and achievement, but may further isolate them from the community and even interfere with procuring the basic supplies for daily life. We have touched on decision-making in a somewhat abstract and experimental context already. But some old people have to face each day the decision of whether or not it is worth the risk to go shopping, visiting a friend, or outside at all. Each negative decision has the possibility of confirming a defeat by life; each positive decision opens the person up to apprehension as he or she walks slowly along a poorly-lit street, or eyes the crowd on a subway platform. Apart from the direct effects of crime, then, we must appreciate the psychophysiological impact of fear and apprehension.

This melancholy picture could be detailed much further. We might look into the reluctance of some old people to report their victimizations to the authorities—and their willingness to accept the less than satisfactory results that sometimes occur even when reports are made. We might dwell on the seemingly endless supply of fraudulent schemes to separate old people from their money. We might examine the old person as a victim of brutally violent crime. These topics cannot be pursued here, but should never be too far from our center of consciousness when we consider human development in the abstract. Growing old is not quite the same experience for the elder cherished as a still valuable, still deepening person as it is for the old person left to fend for himself or herself in a crime-ridden neighborhood.

THE OLD PERSON IN OUR SOCIETY

One of the persistent themes in this book has been the problem of establishing what in a person's total situation should be interpreted as *aging* and what should be interpreted as the result of other factors. We have examined this, for example, through the NIMH study of healthy old men and in a number of other contexts. Let's follow this same problem to still another level: the old person in our society.

THE "TYPICAL" OLD PERSON

"Typical" people exist mostly in textbooks and lecture notes. But suppose there was such a being as a typical old person. What would this person be like—and how much of the person's situation would be a direct consequence of age?

The first characteristic we might notice about the typical old person is the fact that she is a woman. This is true in most societies. In the United States today, there are approximately 3.2 million more women than men over the age of sixty-five. For every 100 men between the ages of sixty-five and seventy-four, there are 129 women. The ratio increases with age: by seventy-five and beyond, there are 160 women for every 100 men (Shanas & Hauser, 1974).

A *woman*. What is true of women in general in our society? How much of a person's experience, stress, coping style, life satisfaction, and social integration is strongly influenced by gender and sex-role? People were females or males for a long time before they became old. It is possible that we misattribute characteristics of long-lived people to old age when, in fact, much that is distinctive about them is more closely related to the life-style that has evolved around their sex-role. Our impression of old age may be colored by the fact that there are more women than men in this developmental range. Both the strengths and the difficulties that can be seen among our oldest citizens reflect something of their long-standing identities as women or men. If it

The overall picture for the elderly in our society is influenced by the fact that women clearly have the numerical advantage. (Photo by Bobbi Carrey.)

happened that there were appreciably more men than women among the aged, then we might have a rather different impression of old age. In practice, it is the *interaction* between sex and age that we have the opportunity to observe, not either one of these taken separately.

The typical old person in our society has serious financial concerns. Even with the passage of increased Social Security benefits in 1972, there remain more than 3 million Americans over the age of sixty-five who are "officially" impoverished. Robert N. Butler, director of the National Institute on Aging, does not hesitate to estimate that more than half of our elders live in a state of financial deprivation (Butler, 1975). He points out that the situation is even more grim than it might appear from statistics alone. It is not only that the old person has constant concern about money for food, housing, health maintenance, and other necessities, but that this person lives within an affluent society. This creates a sense of **relative deprivation** that intensifies the reality of objective deprivation. A person who is impoverished in a society where most other people are also hard-pressed is at least spared the gnawing frustration and discontent experienced in the U.S.A. where an elder can see many continue to enjoy the high life.

Relative deprivation is the theory that have-nots feel especially bad when comparing themselves to conspicuous haves

Pause again. What do we know about poverty? About constant concern with making ends meet? How is financial hazard and marginality likely to influence a person's moods and life-style? Take away old age from old people and leave them with their present difficulties with managing financially. We would still be likely to see some of the features that appear to characterize old age today (e.g., staying at the fringes of society, being "conservative," narrowing interests down to the details of their own daily lives). To the extent that such characteristics exist, we may be seeing the understandable correlates of financial concern rather than the manifestation of a basic process of aging.

Illness and impairment. The aging person tends to be shadowed by both the threat and reality of physical dysfunction. A touch of the flu that the individual might have shook off quickly some years ago can now become life-threatening. The accidental fall (itself more common with advancing years) is more likely to result in hospitalization, in residual impairment, even in the onset of a downward spiral ending in death. With advancing years, there is also the greater likelihood of multiple disorders that are not apt to be cured, but rather to be controlled through medication and other means. These conditions are ignored only at the risk of increased peril; and to take the conditions seriously is to make management of health a larger component of the life-style than it is to most adults.

Again we pause. Set old age itself aside for the moment. Think instead of the differences in life satisfaction, social activity, and many other variables between a healthy, vigorous person and one who must labor with sensory or motor impairments, and perhaps also with pain, fatigue, and health-related anxieties. It is probable that we often mistake some of the physical impairments experienced by old people for the nature of old age itself. There are significant differences between the person who is old and healthy, and the person who is burdened with physical distress—at any age.

Education. Formal education was less available and perhaps less important for career success when the old person of today was growing up. Although the nine years of schooling that a "typical" elder of today received

Figure 17-4. *Survival in the later years of life: a century of rising expectations. The life expectancy for older adults has increased during the twentieth century, particularly during the past three decades. Not only are more people reaching old age, but the likelihood of continued survival has been slowly improving. (Based on data from the National Center for Health Statistics.)*

	Average Life Expectancy in Years							
Period	Women				Men			
	60	65	70	75	60	65	70	75
1900	15.2	12.2	9.6	7.3	14.3	11.5	9.0	6.8
1910	14.9	12.0	9.4	7.2	14.0	11.3	8.8	6.7
1920	15.9	12.7	9.9	7.6	15.3	12.2	9.5	7.3
1930	16.0	12.8	10.0	7.6	14.7	11.8	9.2	7.0
1940	17.0	13.6	10.5	7.9	15.0	12.1	9.4	7.2
1950	18.6	15.0	11.7	8.9	15.8	12.8	10.8	7.8
1960	19.7	15.9	12.4	9.3	16.0	13.0	10.3	7.9
1970	20.8	16.9	13.4	10.2	16.1	13.0	10.4	8.1
1975	21.9	18.1	14.4	11.2	16.8	13.7	10.9	8.5

compares well with many other nations, it is clearly below the level attained by younger adults in our present population. Moreover, there has been such a proliferation of new knowledge and technology over recent decades that those long-ago days in the classroom probably did not prepare the individual for today's world as adequately as was thought at the time.

Place any individual in a complex, fast-moving situation, equipped with an out-dated and somewhat limited knowledge base. This disadvantage is likely to make it difficult to compete for employment and other opportunities, to lead to a sense of inferiority, and to withdraw from the center of social action and progress. While some individuals find a more adaptive approach, overall, the relative deficit in formal education that characterizes our older generations probably accounts for some of the difficulties they experience. Some of the problems attributed to old age, then, might be related closely to educational background instead.

Where are old people and who shares their lives? Many live alone or with people not related to them. Fewer live with spouse and family than we find in the earlier adult years. Furthermore, the quality of the physical environment often leaves much to be desired. Reference is made not just to substandard institutional settings, but also to the deteriorated neighborhoods in which some community-dwelling elders reside. The problem of social integration is often intensified by multiple loss experiences. One of the most common causes for an old woman to be living in alienated circumstances, for example, is the death of her spouse. She not only has her reduced everyday circumstances as a source of possible sorrow, then, but also her inner grief. Other significant people die in addition to the spouse. These can include siblings and grown children, lifelong friends and colleagues, and even one's own parents. The elder is also likely to have lost many of the social roles that gave meaning to life. She is no longer a mother in the way that she was when much of her life was organized around the care of young children. Mothering behavior is not regarded by society as an appropriate center for her life now. The old person may be removed from a variety of personal, occupational, and civic roles. Under these circumstances, the elder may come to feel useless, worthless.

How would *anybody* feel when cut adrift from the most important people in his or her life and from the responsibilities and opportunities that helped give purpose to existence? Some of these stressful life events befall younger adults as well—loved ones die, a job or a chance for promotion is lost, etc. It is more difficult to get up in the morning and make it through the day, let alone function energetically, optimistically, at one's best. Some of what we take as characteristic of old age proves to be characteristic, instead, of people who have losses to mourn and fewer supports (including other people) to help them cope with accumulated adversities.

Many of the people who participated in the building of America have their roots in other lands. The tradition of seeking a new life in the new world was still flourishing when today's elders were young. They—or their parents—relocated themselves here and tried to find some kind of balance between learning the new ways and preserving the most significant of the old ways. Approximately one out of every five elders was born in another country, and another one-in-five had at least one parent who immigrated here (Brotman, 1968). Nearly 40 percent of our elders have strong roots in other cultures, then. This is simply a fact, neither good nor bad in itself. In practice, it has proven sometimes good, sometimes bad. The skills and resources converging from various cultures have helped our nation in many ways. A particular individual, however, finds himself or herself disadvantaged at times because of whatever "-ism" is directed against his or her ethnic group, because of communication difficulties associated with language or behavior patterns that set him or her somewhat apart from the larger society, etc. In times of special stress and need, the old person may either receive unusually good or unusually poor support because of ethnic orientation. What we might perceive as characteristic of being old, then, is apt to draw some of its coloration from the fact that the present generation of elders happens to include many of foreign extraction.

The people who made a new life in America give the present generation of elders a flavor and style we may never see again

Up to this point we have been considering some of the characteristics and problems often associated with being an old person in our society today. But we have started to nibble away at the usual meanings that are attributed to these characteristics. Much of being old is built on the broad base of being a person with a particular life-style oriented around social sex-role, having a certain level of income, having a particular and educational background, health status, and residing in a particular environment with or without essential interpersonal support and with or without the additional complication of strong subcultural roots. These are powerful sources of influence even before we add whatever might be specific and distinctive about old age itself.

SOME DESTRUCTIVE STEREOTYPES

It is unfortunate enough when actual facts are misinterpreted. But the image of old age is also disfigured by stereotypes that lack a firm basis in reality. One of the outcomes of the stereotyping process can be what sociologists call the self-fulfilling prophecy. A negative state of affairs comes into existence because it is assumed to be inevitable.

The lack of sexuality among old people is such an example. Our society has been assuming that old men and old women have neither the interest nor the capacity for continuing sexual activity. Furthermore, the assumed loss of sexuality is generalized throughout the broader spectrum of sensuality, pleasure-giving, and attractiveness. The vibrantly alive old person is acknowledged—but as an exception. We are surprised to discover that this attractive, sexually-toned person is seventy, eighty, or ninety. As we will see later in this chapter, the expectation that old people are (or should be) asexual says rather more about our own values and knowledge than it does about reality.

Another set of assumptions centers around the relationship between old age and death. We assume it is "natural" for old people to die, but "untimely" and "tragic" for the young to do so. This attitude plays into our society's disposition toward denying death. As long as the old specialize in dying, then the rest of us seldom have to think about this unwelcome topic (Kastenbaum, 1977a).

While this orientation may bring some temporary peace of mind (at a high, if hidden, psychological cost) to society in general, the implications for old people are disturbing. The elderly may be seen as reminders of death and avoided on that count. Through our health delivery system and our total pattern of opportunities and services, we may deny first-rate attention to old people, thereby hastening their demise. This orientation is at variance with the facts in many ways. For example, it underestimates the death risk at every age level and the need to have both psychological and instrumental coping techniques available when the prospect of mortality becomes imminent. And it does the enormous injustice of blurring the distinction between the failing or acutely dying person and the old person. The needs and capacities of the dying person and the needs and capacities of the old person are far from identical, even though the important distinctions are not always made in practice. Again, as we will see a few pages from now, there is much more to understand and appreciate about an old person than his or her (estimated) distance from death.

Yet the person who accepts the "elderly mystique" (Rosenthal, 1965), essentially a caricature of what he or she really is and can be, increases his or her vulnerability to premature loss of sexuality and premature death. Here we have a reminder—as if we needed one—that it is crucial to get our facts straight about old age. Ignorance and unexamined assumptions about any part of the lifespan exact a high price.

A FEW OVERLOOKED FACTS ABOUT THE OLD PERSON TODAY

Our overview can be improved by adding a few facts that, while known to many gerontologists, have not become familiar to the public in general. One of these facts concerns degree of "oldness." The customary practice of classifying people as old at age sixty-five has many drawbacks. The most relevant drawback here is the impression it yields that all people over sixty-five are essentially the same. Actually, a wide range of research (biological, psychological, and social) suggests that there are important differences between the "young old" and the "old old." Some gerontologists believe the classification of old age should be moved to seventy-five at the least. Certainly, it has become the experience of more and more people that sixty-five does not appear to be *old*, if by this term we imply a life-style hampered by negative and irreversible changes.

The sixty-five-and-over category is so broad, it includes parents and children. The elder who speaks of mother or father may not necessarily be using the past tense. His or her parent may be alive and well. Classifying both the sixty-five year old and the ninety-five year old as *old* suggests a homogeneity that does not really exist. The thirty years difference represented here is sufficient to have provided each with life experiences—from childhood onward—that are quite dissimilar, in addition to differences in physical functioning.

We seldom recognize that the subpopulation of old people is drawn *unequally* from society in general. Take this important example: although about 12 percent of our population is black, less than 8 percent of our elders are black (Brotman, 1968). The reduced likelihood of surviving into old age is one of the most telling pieces of evidence regarding unequal opportunity for black Americans. This does not mean, by the way, that the factors creating unusual mortality risk occur especially in the later years of life. All hazards accumulate throughout the lifespan to reduce survivorship. As J. J. Jackson has pointed out, this also means that a higher proportion of black adults die before they can claim Social Security benefits for which they have paid. In effect, their premature deaths amount to a partial underwriting of Social Security benefits for the longer-lived whites (Jackson, 1971).

We also do not always recognize that whatever may be the profile of the "typical" elderly American today, it is constantly in flux and may look rather different tomorrow. Changes in the longevity of whites and blacks, for example, in patterns of immigration, in levels of education and availability of good housing and health care are among the factors that can alter this profile. Perhaps one of the most significant factors at work is the increased awareness and clout of old people themselves. On both the national and the local levels, old people have been showing greater determination in securing the full benefits due to them, and in creating the opportunities to contribute their skills and experience to society. The Grey Panthers under the charismatic leadership of Maggie Kuhn (1977) is one of several vigorous groups comprised of elderly people who are effectively combating age discrimination and stereotypes.

New roles for older people in society are being developed. Some of these have already become fairly well-known. The Foster Grandparent Program is bringing satisfaction both to children and to elderly men and women. Older people are turning up more frequently as students in a variety of traditional and innovative educational settings; indeed, the research literature in "educational gerontology" has become extensive (Bader, 1977). Although we are accustomed to thinking of old age as a time of retirement from the occupational sphere, there is a growing trend for women to re-enter the work scene in late middle age (Hendricks & Hendricks, 1977). Additionally, many men are carving out second or even third careers for themselves. Mature men and women with valuable experience and high motivation are likely to find their way into socially useful lines of work, and not to relinquish these roles just because tradition would have it so. And as more and more people have the opportunity to plan ahead for their old age, more will enter it with good uses in mind for leisure as well as work.

The Foster Grandparent Program has been one of the most successful government sponsored innovations in recent years—satisfaction for young and old alike. (ACTION Photo by Susan Biddle.)

Precisely who the "typical" old person will be in the future can be guessed at, but not fully determined at this time. It is unlikely that old age will look quite the same twenty, fifty, or one hundred years from now. The face of tomorrow's old person will look a lot like yours. And, since you have never been old before, you will be bringing something distinctive to this period of life and making your generation of elders as distinctive in its own way as the present generation is today.

INTIMATE THOUGHTS, FEELINGS, AND RELATIONSHIPS

The best way to understand the intimate thoughts, feelings, and relationships of an old person is to be an old person. The second best way is to know an old person. Every other approach, no matter how bolstered by research and expert observation, runs a distant third. Nevertheless, research can at least help us to begin to enhance our appreciation of the rich human meanings that are obscured by stereotypes of aging and the aged.

USES OF THE PAST: THE RETROSPECTIVE MODALITIES OF THOUGHT

Our relationship to time—past, present, and future—is deeply personal. Nobody else has had our own particular pattern of past experiences. And, even though we all share the present moment, we will go our own ways in the future. Furthermore, at any moment, we might be projecting ourselves into various regions of past or future. When we know something about a person's relationship to time, we also know something important about his or her total view of life, the hopes, frustrations, satisfactions, and ambiguities that cannot be read from outward behavior or chronological markings alone. Because the study of **time perspective** takes us so close to the intimate workings of heart and mind, it is one of psychology's most challenging research areas. We will concentrate here on just one facet of the old person's relationship to time: the uses of the past. This realm is complex enough to serve as introduction to the broader domain of time perspective.

Time perspective is the orientation toward past, present, and future

There is a common impression that old people "live in the past." As is often the case with such impressions, this has both a grain of truth and a chaff of assumption and careless opinion. The old person appears to have more of his or her "allotted time" on deposit in the past, while the young person is free to spend much of his or her time in the future. This is a rule that has its exceptions: witness the old woman mourning the death of her grandson. But it also is limited to an external perspective. Knowing that an old person has much "pastness" on hand tells us little about the ways in which he or she relates to or utilizes the past. The phrase, *living in the past,* is best reserved for those situations where we have clear evidence that the person is deeply engrossed in past experience to such an extent that there is virtually no psychological interchange with present or future. Otherwise, it is a careless and misleading categorization, with more than a hint of derogation. And, of course, it is not only the old person who might live in the past; individuals who are young or middle-aged by chronological standards sometimes behave as though their life ended many years ago.

How *do* old people use their pasts? It is gradually becoming recognized that there can be an *adaptive* value to thinking of the past, to reminiscing (e.g., *see* Lewis, 1973). The person can actually accomplish something through this activity; it is not necessarily a random or childish rummaging through the mental attic. It is helpful to lay aside any prejudice we might have about turning thoughts toward the past. Our culture is quite future-oriented. This is consistent with our emphasis on youth. But it is peculiar to be so indulgent about future-oriented fantasies (which may prove to be just that—fantasies) and yet be so intolerant of attention to actual past accomplishments (Kastenbaum, 1975).

Five characteristic ways of using the past have been identified, with particular relevance to the elderly. All of these phenomena can be observed in younger adults as well, but they appear to become more significant and salient with advancing age. Collectively, they have become known as **retrospective modalities.**

Retrospective modalities are ways in which the individual uses his or her past

1. *The life review.* Butler (1963) has noticed a tendency for old people to undertake critical surveys of their total life experiences as thoughts of death move into focus. This review helps the person to set his or her own

house in order and form a decisive opinion of who, in fact, he or she has become. If the individual is able to conclude that he or she has lived a good life and become a good person, then the rest of life, and the prospect of death, can be faced with equanimity. Anxiety, conflict, and even despair may result if the person concludes that he or she has wasted or ruined his or her life. The life review, then, appears to be a crucial inner process. It is certainly a long way from random, childish dabblings in the past.

2. *Validation.* This is the first of four other retrospective modalities identified through my clinical and research experiences with old people (e.g., Kastenbaum, 1975). It is not necessarily linked with anticipations of death, but is more likely to be seen in the coping behaviors of everyday life. Validation involves purposeful "visits" to the past with the hope of bolstering one's confidence in coping with the present and the future. Questioning his or her own ability to master the day's challenges, the old person searches past experience for reminders of how competent he or she has been in earlier times. If this search is successful (and the outcome cannot be guaranteed), then he or she borrows strength from his or her own past and applies it to current demands. "I used to do this well all the time! Of course, I can do it again!" The *need* for validation is heightened when the old person is caught in a social environment that itself does not remember his or her former strengths or encourage his or her current efforts. Strength must be found inside oneself, and the credentials established and on deposit in the past are a good place to look.

3. *Boundary-setting.* Psychological boundaries are what we are concerned with here, although they may have physical representation as well. Think of the old woman who is leaving her long-familiar surroundings to enter a nursing home. Is her old house still part of her life? And what else has been left "out there" as her immediate life-space shifts to the confines of an institutional setting? Does this mean the end of previous friendships? Should she forget about the neighborhood stores in which she did her shopping for so many years? Are the movie theaters, the little park around the corner, and all the other physical markings of the life she has known now off limits to her? Precisely how much of what gave her life its particular content and tone must now be filed away as over and done with? Admission to a nursing home is one of the many situations in which an old person is faced with the need to determine whether or not certain aspects of life must now be set aside as inoperative or out of bounds. "I can no longer go there . . . I should no longer do that . . . This relationship is over . . ." In such ways the old person tests out in his or her innermost thoughts what limits must be respected. It is not easy to determine that point at which a particular aspect of one's life should be treated as belonging to the past and only to the past. The decision may be premature or too-long delayed. The boundary may be readjusted firmly, or shifted again. It seems to me that much of the coping effort of old people who are in transition or crisis centers around the problem of setting appropriate boundaries between past and present/future. We understand the situation of the older

The past lives—in a family picture gallery and in the thoughts and feelings of a couple who have shared many years together. (USDA Photo by Bill Marr.)

person better when we appreciate the existence of the boundary-setting process and the intellectual and emotional demands it makes.

Some people, e.g., keep a room unchanged through the years, as if time had not moved on

4. *Perpetuation of the past.* Another way of relating to the past is to act as though it still has today's date on it. The past is not really past. Entire societies celebrate and perpetuate great events from their collective past, as anthropologists have observed. I am talking here, however, about individuals, often rather isolated individuals who must rely almost entirely on the experiences they can summon from within themselves rather than on the support and exchange from the outside world. An old person may perpetuate the past in small, subtle ways or use this technique for making it through life in general. Unfortunately this carrying along of the past can increase social isolation, even though it may be intended, in part, to compensate for social isolation already experienced. Sometimes perpetuation of the past can be seen in a particularly rich and complex form in a person who is "the last leaf on the tree"—the last of his or her family line (Kastenbaum, 1977). The person may feel a strong sense of responsibility toward all the lives that are no more. Generations are kept alive in his or her mind and no where else. How little we know about this person if we do not realize the inner responsibility he or she has taken on—and perhaps the sense of frustration in having nobody with whom to share this tapestry of lives.

5. *Replaying.* Haven't we heard this story before? Yes, perhaps many times. There he goes again, that old man, repeating an anecdote (true, con-

cocted, or somewhere inbetween) just as vividly as though it were the first time. This characteristic of some old people has been noticed for centuries (usually as the observer is trying to back off from having to listen still again). But the familiarity of this phenomenon does not mean that we understand it. Relatively little thought has been given to careful description and explanation of replaying. Here I would simply like to call a few of its distinctive features to your attention. The replayed experiences are highly selected. They still have strong emotional meanings attached. But they have torn loose from their previous attachments in the tapestry of personal time. It does not matter very much whether this incident we are hearing about occurred ten, twenty, or forty years ago. To some extent, we might compare replaying to a person reaching into a sort of personal juke-box where a few precious records or endless-reel tapes have been chosen from all the music heard over a lifetime. The date the recording was pressed hardly matters any more; it has some other quality that makes it important to the individual.

One function of replaying is to reinstate meaning and vitality by means at one's own disposal rather than depending on the chance gratifications still available in the world. But there may be another important function as well. Replaying helps the old person to step outside the rigid structure of time that most of us have lived within since late childhood, the ever pressing-forward from past and present into future. This time structure is made for achievement and ambition. It is made for mutually obligating ourselves. But some old people feel that they no longer have use for this kind of time. What is worthwhile to them developed through time, but is no longer part of time. In particular, replaying offers an alternative to a straight-out confrontation with the future. Young adults often have a strong sense of forward motion and also tend to have more anxiety about death. The old person often does not experience himself or herself as future-oriented, especially the individual who makes much use of replaying. And if the future does not count for much, then neither does death, for this is the only "place" where death is to be found.

This has been a quick sketch of the old person's uses of the past (which are themselves only one aspect of his or her relationship to time). But all these modalities have purposeful, functional roles to play in the old person's inner adaptation to life. The opportunity to share some of the accumulated riches of the past with other people can be very helpful in successful adaptation. Moreover, such sharings have much to offer the listener as well. It is a privilege to be granted entry to the unique inner worlds that old people possess; for this, we must dare to know old people as the individuals they are rather than the stereotypes we have been taught to expect.

DEATH, DYING, AND GRIEF

One of our society's odd customs is to "save" death for the old. In lifespan courses and books, for example, the topic of death is often neglected until the section on old age. This selectivity is consistent with our broader cultural orientation in which death is seen as "natural" for the aged. By implication, this means that death is somehow "unnatural" for the young. The strategy is a transparent one: keep death back, away from our own lives and thoughts, as long as we can. Leave it to old people to carry the burden of death for all the society; the rest of us do not have to think seriously about this subject.

This attitude is a luxury society could not afford in the past. Death was more of an obvious menace to people of all ages, especially to young children and to the women who bore them. It is only with the gradual reduction of the mortality rate in the younger age groups that the linkage between old age and death has been so heavily emphasized. Yet death continues to be part of life's context at every age. I have attempted to convey this message in various ways throughout this book instead of contributing to the illusion that: (1) only the old die, or (2) the old have nothing else to do except die.

The ease with which we associate aging and death can contribute to neglect, decline, and premature death. The self-fulfilling prophecy again! Priority for medical treatment of the old person remains low in many quarters. Health care personnel, like the rest of society, tend to see younger people as worth a greater investment of time, skill, and resources. The simplistic attitude, that old people are ready to die and should die, makes it that much easier to deny them the opportunity for the quality and length of life that might still be possible.

A balanced view of the old person's relationship to death, dying, and grief requires attention to facts such as the following:

1. Many, perhaps most, old people are able to acknowledge the prospect of personal death with equanimity (e.g., Munnichs, 1966).

2. The ability to accept death in a matter-of-fact, realistic manner, however, should not be mistaken for an eagerness or will to die. I have known many old men and women who are at peace with the prospect of death but who want to keep living as long as there is any quality left to life.

3. There are important individual differences among old people in their orientations toward death. One clinical study, for example, collected observations from many sources on the lives and eventual deaths of hospitalized old people. Some appeared to have accurate premonitions of their deaths, although the hospital staff did not think they were in imminent danger; others showed a variety of behaviors and attitudes ranging from unease and apprehension to serene acceptance (Weisman and Kastenbaum, 1968). It could *not* be concluded that there was any one pattern of transition from life to death. The old person's personality, ethnic background, and current social relationships, and the nature of the old person's constellation of diseases and disabilities all contributed to the individuality of his or her death. Furthermore, that difficult-to-define factor known by some as *will-to-live* often appeared to be significant. A further report from the same study indicated that the two

most common types of responding to awareness of impending death were as different from each other as could be:

> One ninety-year-old woman was a very alert, independent person who always appeared to be in supreme control of her situation. As her health began to decline, she initiated arrangements for her funeral. She also expressed a readiness for death in the most straightforward manner. She declared that she had lived her life and was now ready to see it come to an end. When she perceived that death was near, the patient refused medication and insisted that any attempt to prolong her life at this time would be a crime. (Kastenbaum & Weisman, 1972, p. 215).

She was classified among the *accepters.*

But there was another set of elders at the same hospital, also well aware of their close relationship to death, who behaved quite differently:

> One eighty-two-year-old woman typifies this approach. Upon entering Cushing Hospital she had expressed an attitude of deep resignation and a desire for death. Yet she eventually became quite involved in the social and recreational life of the institution. (Three years later), she faced death as though it were a regrettable interruption of her participation in activities and relationships. This sentiment was reciprocated by staff members and other patients. This woman, and others in her group, were classified as having been *interrupted* by death. . . . (Kastenbaum & Weisman, 1972, p. 215).

Both patterns appeared rational and appropriate to the life-styles of the individuals. Old men and women, in company with people who face death at any age, usually select a final, preterminal pattern of life if the environment makes such options possible. The popular assumption that all people die in the same way or go through a fixed sequence of stages does not have convincing research support (Kastenbaum & Costa, 1977).

4. Longings for death among old people are often expressions of depression, responses to abandonment and alienation, or fears of being a burden to others. While readiness for death deserves respect and understanding, sometimes the emotional pain that stimulates death-seeking can be identified and alleviated by people genuinely interested in helping.

Death-longings often vanish after the person's current life situation improves

5. Suicide is a risk among the aged, particularly for white men (Linden & Breed, 1975). A greater percentage of suicide attempts by the old turn into fatalities than in the younger age groups. In addition to obvious suicides, some old people place their lives at special risk through a variety of self-injurious behaviors. An institutional environment that provides little opportunity for self-expression and which seems to pay attention only when a person is clearly hurt (in the physical sense) tends to encourage self-injurious behaviors. These can lead, directly or indirectly, to premature death. Attempting to enhance the life-supporting functions of the environment can reduce self-injurious behaviors and improve morale (Mishara & Kastenbaum, 1973).

6. Living with the death of others is one of the heaviest weights borne by many old people. A secure and mature old person may be able to face the prospect of his or her own death with equanimity. But it is something else to confront every day the emptiness left by the deaths of people who had been very much a part of the survivor's life. And the longer a person lives, the more likely he or she is to see companions and intimates die. Quite possibly, it is *grief not old age* that is responsible for many of the characteristics observed in an elderly man or woman. Young adults in excellent health are often stunned, slowed down, disorganized, or drained by the death of someone close to them. The old person, forced to bear many deaths and with fewer people left to share intimate thoughts and feelings, may be suffering more from grief and grief-depletion than from phenomena intrinsic to aging. Living with loss and with the prospect of more loss in the future is one more reason for an old person's thoughts to turn to the past. Whether or not the present moment holds comfort or hope to counterbalance the inroads of grief may depend much on: (1) how well others recognize the source of the old person's distress, and (2) what, if anything, they choose to do about it.

Some of us are uncomfortable in relating to old people because we have yet to integrate the reality of death into our own thoughts and feelings. As our own awareness and security improves, we become more open with people who appear to be walking in the valley of the shadow. We can see the old person as a person, rather than as a reflection of our own anxieties.

LOVING AND SEXUALITY

It is impossible to understand loving and sexuality in old age by focusing on old age alone. Each person has a specific life history of feelings and relationships, satisfactions and disappointments. The old woman around the corner, for example, may have outlived two husbands. Although doubtful about starting any new relationship, she may feel lonely without a man in the house. Or perhaps instead, she may be a person who never married or had an enduring, intimate sexual relationship. What she wants from life now may be dignity, independence, and a measure of comfort, but not a new type of relationship that would alter her long-established life-style. Or perhaps again, she may still be married to her first love. Marital affection, including sexual relations, may remain a core part of her life. Passing this woman on the street or having a superficial conversation with her, we really are in no position to know what love and sexuality mean in her life at that moment.

And yet, our culture is brimming over with stereotyped assumptions about sexual love—or, rather, its absence—among old men and women. Lack of knowledge has not discouraged the expression of firm opinions. Old people obviously are "not interested in that sort of thing." Notice that even the Kinsey Report (1948) barely mentioned old age in its voluminous pioneering inquiry into sexual behavior. This is but one example of how the assumption that "there is nothing here worth talking about" can contribute to its own perpetuation. When we do come across an old person with sexual interests, we are likely to be shaken up, even outraged. Somehow the idea of a sexually alive old person is threatening to many of us. Only the young should be lovers. The old? Well, they have their disengagement and dying to get on with, don't they?

There is a double-bind situation here. We tend to think an old person is "proper" only if he or she has assumed a postsexual role. But we are also a culture that has such high regard for "sexiness" that this is a staple selling point for many types of merchandise, and *sexy* itself has become an all-purpose adjective expressing excitement and attractiveness. Giving up sexuality and becoming *proper*, then, also means relinquishing one of the claims a person has for being considered interesting. But retaining obvious sexual interest leads to the predictable dirty-old-man-type of humor and other fearful and unsympathetic responses from society. It is close to a no-win situation for the old person.

Fortunately, the actual facts, so far as these are known today, are more positive than the social stereotypes and attitudes. The potential for sexual intimacy is even greater than the present facts demonstrate, because some of the difficulties experienced by old people in this realm can be attributed to the negative and ignorant social climate.

Masters and Johnson (1966) have found that the capacity for a satisfying sexual life remains with many men and women well into old age. A critical factor is the type of sexual love pattern that the partners had already established in their lives. People who have had satisfying and active sexual lives

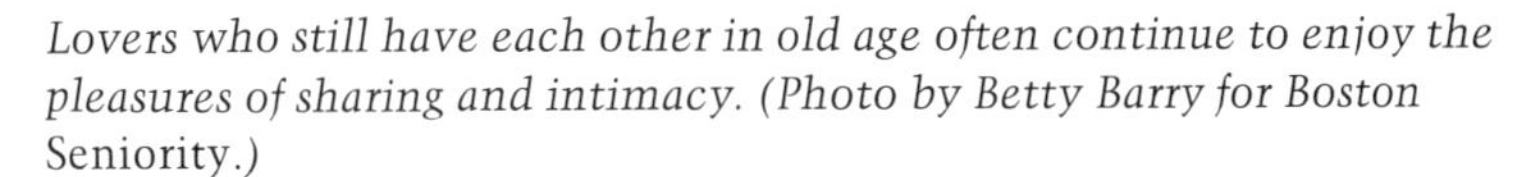
Lovers who still have each other in old age often continue to enjoy the pleasures of sharing and intimacy. (Photo by Betty Barry for Boston Seniority.*)*

are a good deal more likely to continue that life-style than those for whom sexual expression had been less satisfying and active over the years. This is also consistent with the results of longitudinal research conducted by a multidisciplinary team at Duke University (Pfeiffer, 1969). When the same person is studied over a period of time, less decline in many areas of functioning, including sexuality, is usually observed. Because cross-sectional studies are generally easier to do and quicker in yielding results, there was for awhile the impression that old people experience a marked decline in sexuality. But longitudinal studies are now finding that the decline is less marked than previously thought. The physically healthy older person who has a history of accepting and enjoying sexual love—and who has a partner still available—can and often does continue his or her sexual role into old age.

Another instance where health status proves more important than chronological age

Unfortunately, these conditions of good health and availability of a loving partner become increasingly imperiled with advancing age. More old people could remain sexually active if they received better health care, could afford better nutrition, and were better protected against environmental hazards. The differential death rate between men and women also enters forcefully here. In our society, men often marry younger women. This difference adds to the effect of the tendency for men to die at an earlier age than women. The result is a large number of women who are widowed and therefore deprived of their long-time sexual companions. Only three out of every ten women over the age of sixty-five are married, as compared with seven out of every ten men who survive to this age (Shanas & Hauser, 1974). The widow has fewer possibilities for finding a suitable new spouse because there are fewer men from whom to choose, and also because of differential social expectations. Our society appears to expect the bereaved older man to eventually find another woman who will "take care of him." But the widow must often contend with both the social and personal expectations that she should "remain faithful" to her first spouse. In any event, the remarriage rate is appreciably lower among old widows as compared with old widowers.

Furthermore, the pressure of social attitudes—old people do not and should not be sexually active—is felt by many men and women whether or not they have a partner readily available. Perhaps the core of the problem can be found in the interaction between such social expectations and certain psychobiological changes that do often occur with aging. The changes and difficulties associated with the biology of aging produce both psychological distress over sexual role and the (frequently premature) conclusion that one ought now to renounce sexual relations.

The pattern of change differs for each sex. In general, as they grow older, men start to require a longer time interval between erections. This becomes noticeable well before old age. It is often a cause of concern for men in their forties and fifties—although a concern typically kept locked in their own private thoughts. The mechanics of erection and orgasm for the older man include slower achievement of erection, less forceful ejaculation, and a quicker loss of erection after orgasm (Masters & Johnson, 1966). Worrying about these changes can increase the older man's difficulties. Afraid that he may not be able to measure up to his own or his wife's expectations, he sometimes adopts strategies that only increase the problem. Drinking, as a getting-up-courage measure or as an alternative to sex, can become a problem in its own right and damage self-esteem and overall competence. Avoidance of sexual relations in later life tends to make future sexual relations more difficult to achieve; many studies and observations are agreed on this point.

The pressure on men to *perform* is great in our society, whether it be on the job or in the bed. The twin threat of occupational retirement and a decline in sexual prowess hits many men hard. It is not just the loss of income and work role on the one hand, and the reduction and alteration in sexual activity on the other. Rather, these losses can be experienced as an assault on one's sense of oneself as a "real man." To the extent our culture equates performance with worth and emphasizes quantity and frequency, to that extent, men are set up for failure with advancing age.

The older woman does not have the potency problem with which to contend. But she may experience other changes that cause concern. During the postmenopausal years, the loss of ovarian functioning can make coitus uncomfortable or even painful. As Masters and Johnson reported, there may also be burning and aching sensations after intercourse. Physical changes associated with aging can include thinning of the vaginal walls and other signs of involution.

Whatever increases discomfort for the woman during or after intercourse can be distracting and draw attention away from the sexual interaction itself. It is difficult to feel free and spontaneous when pain intrudes. Neither partner may be fully aware of the basis for his or her own difficulties or for the problems the other is experiencing. This is a situation that can heighten self-doubt or lead to dissatisfaction with one's partner.

In a marriage blessed by good communication and strong affection, it is possible to overcome these difficulties. But the strain of poorly understood and unwelcome changes may prove too great for less substantial relationships. Furthermore, the partners may be under stress for other reasons (e.g., financial problems, concern about the health of other people) and therefore have less strength to bring to the rescue of their sex life.

As more has been learned about the basic psychobiological changes associated with sexual functioning in later life, more opportunities have developed for safeguarding the unique gratifications of physically expressed love. Certainly, a more enlightened social climate would be an important step in this direction. The general image of what an old person is has recently been studied by means of a major national survey (Harris, 1975). While there is little of direct relevance to sexuality in this survey, the overall results indicate that there is much room for the media and other major social institutions to offer a fuller, more positive conception of the old person. Recognizing that many old people are alert, competent, and deeply involved in all phases of life would also make it easier for sexuality to be respected and maintained.

More individualistic solutions are also possible. Discussions of these can be pursued in a growing number of useful books and articles. Masters and Johnson themselves made a good start with their suggestions both for hormonal and psychosocial approaches. More detailed discussions can be found in books by Rubin (1965), and Butler and Lewis (1976). All these authors rightfully give attention to the quality of the relationships between partners and the ways in which thoughts and feelings influence both the frequency and the quality of sexual functioning.

Marital happiness is not to be measured in frequency of intercourse or orgasm. The satisfaction of intimacy, of caresses and whisperings, of knowing that one can still comfort and delight the beloved, this counts for more in most relationships than any scorecard. Truly beyond measure is the shared experience of pledging continued devotion by touch as well as by word. The pleasure is not limited to keen physical sensations, but is filtered through the rich accumulation of meanings over the years.

SUMMARY

Much of what passes for knowledge about old age is a mixture of selective observations, emotionally-tinged opinions, and assumptions that have not been reflected on. A careful sifting-through process is necessary if we are to understand the relationship of old age to the total lifespan.

Many of the characteristics considered typical of old people have only indirect relationships to age. The present generation of old Americans, for example, includes many who have realistic concerns about paying for food, housing, and other essentials. Old people today also tend to have more health problems, less formal education, and less satisfactory living arrangements than younger adults. Mandatory retirement and age-discrimination in many forms adds to the pressures and deprivations that old people are likely to experience.

Each generation of old people has its own constellation of experiences, resources, and problems. The profile of our elders today, for example, includes many with roots in other nations. This appears to be a diminishing percentage. The distinctive strengths and problems associated with dual cultural affiliation will therefore differ from one generation of elders to the next. At any particular time we may have the impression that "this is what old age is like," when what we actually see is the complex interaction of biological process with a distinctive generation of people who must function within a sociocultural situation that itself is distinctive and ever-changing. This means that old age is not entirely the same phenomenon for each generation. It also means that the scientist has no easy task in differentiating changes that are closely related to an intrinsic aging process from all the individual and social factors involved. There is a difference, in other words, between *aging* and *being old.*

YOUR TURN

1. Financial concern has been identified as a major influence on the health and life satisfaction of the old person. Enter this world personally by: (1) figuring out your own budget at the moment, your income and other tangible resources as balanced against your needs and desires; (2) subtracting 60 percent from this income (a typical loss experienced by old people after retirement); and (3) analyzing in detail how this shift in your financial situation would affect your personal and social life.

2. What experiences has the "typical" person who was born around the turn of the century been through? Consider their impact on personality and life trajectory, taking into account the chronological age of the individual during such events as the major depression of the 1930s, the world wars, and other, more subtle but powerful phenomena taking place in the twentieth century. Compare these experiences and their impacts with what you have already experienced in your life. Then project yourself ahead into your own old age, if you can, and formulate the probable influence of your life experiences on the kind of person you will have become. Consider, also, some of the more significant events that may or may not occur between now and then in society as you project yourself ahead.

3. The norms of sexual expression and behavior have been changing rapidly during the past few years. Concentrating on this one realm of change, see if you can project ahead the nature of loving and sexuality for old people who have grown up amidst these new values. How is this likely to compare and contrast with the present pattern among elders today?

4. Although *slowing down* is one of the most evident phenomena of old age, the process begins gradually during earlier years. Take a personal inventory. In what ways, if any, have you already started to slow down? How has this process affected your life in general? What strategies are you developing to cope with this change? If you do not yet detect a slowing down, how do you plan to cope with it later?

An important study conducted by the National Institute of Mental Health indicated that many of the physical status differences between young and old adults disappear or are sharply reduced when health is controlled. People who are just old and not sick show relatively few changes in basic psychobiological functioning from when they were young. However, the changes that do show up become more important to study in detail (e.g., the reduced peak frequencies in electrical activity of the brain). The same study, and a number which followed, suggested that even a "touch" of physical pathology, such as cerebral arteriosclerosis or atherosclerosis, can have a significant and pervasive effect on the old person's general functioning. *Illness often masquerades as intrinsic aging.* A close relationship was also discovered between psychological functioning and subsequent health and mortality.

Theories of human aging are being actively developed and tested, although the biologically oriented approaches appear more dominant than the psychosocial at present. Many theorists distinguish between *primary* and *secondary aging,* but there is a broad range of opinion regarding the possibility of altering the process of aging. Experimental approaches to testing theory are becoming more frequent, especially in the biological domain. So far there has not been much success in integrating the psychosocial and the biological dimensions of human aging into a unitary framework, nor have theoreticians been sufficiently clear about the distinctions between facts and personal or social values. Those seeking systematic understanding of human aging recognize the need to encompass the entire developmental spectrum, rather than limiting attention to old age as such. Finally, I noted a tension between the prevailingly negative view of aging that emerges from the biological sphere and the more positive concepts and findings from the psychosocial sphere.

Much attention was given to *the competent old person.* It was argued that competence cannot be adequately explained or predicted by chronological age alone. The role of the environment in either supporting or undermining competent functioning was noted. Particular topics explored in this section were the process of *slowing down,* differential changes in *memory,* and the role of noncognitive factors in *decision-making* and *learning.* There are reasons to believe that good mental functioning in old age can be protected and enhanced by a variety of personal and socio-technical strategies.

The role of *cultural expectations* was illustrated through a research-based scenario that contrasted the low estate of *creative achievement* in old age as seen by studies in our society with the continued development and mastery of folk artists in a religious community in Western India.

Sociocultural influences were seen again in a summary of *victimization by crime,* a hazard to many old people today in our cities.

The intimate thoughts, feelings, and relationships of old people were explored within three related contexts: *uses of the past; death, dying and grief;* and *loving and sexuality.* In each of these realms, a number of negative stereotypes, based on ignorance and social bias, were set aside in favor of emerging findings from research and clinical practice. The overall findings emphasize the individual's lifelong pattern of functioning as this interacts with the specific opportunities and threats of the immediate situation. The influence of aging, as such, is understandable only within such an encompassing, interactive framework.

Reference List

Arenberg, D. Cognition and aging: Verbal learning, memory, problem solving, and aging. In C. Eisdorfer & M. P. Lawton (Eds.), *The psychology of adult development and aging.* Washington, D.C.: American Psychological Association, 1973, pp. 74–97.

Bader, J. E. Education for older adults: Selected bibliography. *International Journal of Aging & Human Development,* 1977, 8, 345–358.

Baltes, P. B., & Labouvie, G. V. Adult development of intellectual performance: Description, explanation, modification. In C. Eisdorfer & M. P. Lawton (Eds.), *The psychology of adult development and aging.* Washington, D.C.: American Psychological Association, 1973.

Baltes, P. B., & Schaie, K. W. Aging and IQ: The myth of the twilight years. *Psychology Today,* 1974, 7, 35–40.

Birren, J. E., Butler, R. N., Greenhouse, S. W., Sokoloff, L., & Yarrow, M. R. (Eds.). *Human aging: A biological and behavioral study.* Bethesda, Md.: United States Public Health Service, 1963.

Botwinick, J. *Cognitive processes in maturity and old age.* New York: Springer Publishing Co., Inc., 1967.

Botwinick, J. *Aging and behavior.* New York: Springer Publishing Co., Inc., 1973.

Brotman, H. E. *Who are the aged: A demographic view.* Ann Arbor, Mich.: The University of Michigan Institute of Gerontology, 1968.

Butler, R. N. The life review: An interpretation of reminiscence in the aged. *Psychiatry,* 1963, 26, 65–76.

Butler, R. N. *Why survive?* New York: Harper & Row, Publishers, 1975.

Butler, R. N., & Lewis, M. I. *Aging and mental health.* St. Louis: The C. V. Mosby Company, 1973.

Butler, R. N., & Lewis, M. I. *Sex after sixty.* New York: Harper & Row, Publishers, 1976.

Costa, P. T., & McCrae, R. R. Age differences in personality structure: A cluster analytic approach. *Journal of Gerontology,* 1976, 31, 564–570.

Cumming, E., & Henry, W. E. *Growing old.* New York: Basic Books, 1961.

Dennis, W. Creative productivity between the ages of 20 and 80 years. *Journal of Gerontology,* 1966, 21, 1–8.

Fillenbaum, G. G. An examination of the vulnerability hypothesis. *International Journal of Aging & Human Development,* 1977, 8, 155–160.

Fozard, J. L., & Thomas, J. C. Psychology of aging. In J. G. Howells (Ed.), *Modern perspectives in the psychiatry of old age.* New York: Brunner/Mazel, Inc., 1975, pp. 107–169.

Goldsmith, J., & Tomas, N. E. Crimes against the elderly: A continuing national crisis. *Aging,* June–July 1974, 10–12.

Granick, S., & Patterson, R. D. *Human aging. II: An eleven year followup biomedical and behavioral study.* Bethesda, Md.: United States Public Health Service, 1972.

Hamer, J. H. Aging in a gerontocratic society: The Sidamo of southwest Ethiopia. In D. O. Cowgill & L. D. Holmes (Eds.), *Aging and modernization*. New York: Appleton-Century-Crofts, 1972, pp. 15–30.

Harris, L., & Associates. *The myth and reality of aging in America*. Washington, D.C.: National Council on Aging, 1975.

Hendricks, J., & Hendricks, C. D. *Aging in mass society*. Cambridge, Mass.: Winthrop Publishers, Inc., 1977.

Hicks, L. H., & Birren, J. E. Aging, brain damage, and psychomotor slowing. *Psychological Bulletin*, 1970, *74*, 377–396.

Holmes, L. D. The role and status of the aged in a changing Samoa. In D. O. Cowgill & L. D. Holmes (Eds.), *Aging and modernization*. New York: Appleton-Century-Crofts, 1972, pp. 73–90.

Horn, J. L., & Donaldson, G. On the myth of intellectual decline in adulthood. *American Psychologist*, 1976, *31*, 701–719.

Hoyer, W. J., Labouvie, G. V. & Baltes, P. B. Modification of response speed and intellectual performance in the elderly. *Human Development*, 1973, *16*, 233–242.

Hulicks, I., & Grossman, J. Age-group comparisons for the use of mediators in paired associate learning. *Journal of Gerontology*, 1967, *22*, 46–51.

Jackson, J. J. The blacklands of gerontology. *International Journal of Aging & Human Development*, 1971, *2*, 156–172.

Kastenbaum, R. Time, death, and ritual in old age. In J. T. Fraser & N. Lawrence (Eds.), *The study of time. II*. New York: Springer-Verlag, 1975, pp. 20–38.

Kastenbaum, R. *Death, society, and human experience*. St. Louis: The C. V. Mosby Company, 1977. (a)

Kastenbaum, R. Memories of tomorrow: Interpenetrations of time in old age. In A. D. Wessman & B. Gorman (Eds.), *The personal experience of time*. New York: Plenum Publishing Corporation, 1977, pp. 194–214. (b)

Kastenbaum, R., & Costa, P. A. Psychological perspectives on death. In *Annual review of psychology* (Vol. 28). Palo Alto, Calif.: Stanford University Press, 1977, pp. 225–249.

Kinsey, A. C., Pomeroy, W., & Martin, C. E. *Sexual behavior in the human male*. Philadelphia: W. B. Saunders, 1948.

Kohlberg, L. Stages and aging in moral development: Some speculations. *The Gerontologist*, 1973, *13*, 497–502.

Kuhn, M. E. Learning by living. *International Journal of Aging & Human Development*, 1977, *8*, 359–366.

Lehman, H. C. *Age and achievement*. Princeton, N.J.: Princeton University Press, 1953.

Lewis, C. N. The adaptive value of reminiscing in old age. *Journal of Geriatric Psychiatry*, 1973, *6*, 117–121.

Linden, L. L., & Breed, W. The demographic epidemiology of suicide. In E. S. Shneidman (Ed.), *Suicidology: Contemporary developments*. New York: Grune & Stratton, Inc., 1975.

Lindsley, O. Geriatric behavioral prosthetics. In R. Kastenbaum (Ed.), *New thoughts on old age*. New York: Springer Publishing Co., Inc., 1964, pp. 41–59.

Maduro, R. Artistic creativity and aging in India. *International Journal of Aging and Human Development*, 1974, *5*, 303–330.

Masters, W. H., & Johnson, V. E. *Human sexual response*. Boston: Little, Brown & Company, 1966.

Mishara, B. L., & Kastenbaum, R. Self-injurious behavior and environmental change in the institutionalized aged. *International Journal of Aging & Human Development*, 1973, *4*, 133–146.

Monge, R., & Hultsch, D. Paired-associate learning as a function of adult age and the length of the anticipation and inspection intervals. *Journal of Gerontology*, 1971, *26*, 157–162.

Munnichs, J. M. A. *Old age and finitude.* New York: Karger, 1966.

Pfeiffer, E. Sexual behavior in old age. In E. W. Busse & E. Pfeiffer (Eds.), *Behavior and adaptation in late life.* Boston: Little, Brown & Company, 1969.

Rosenfeld, A. *Prolongevity.* New York: Alfred A. Knopf, Inc., 1976.

Rosenthal, R. H. The elderly mystique. *Journal of Social Issues*, 1965, *21*, 37–43.

Rubin, I. *Sexual life after sixty.* New York: Signet Books, 1965.

Schaie, K. W. A reinterpretation of age related changes in cognitive structure and functioning. In L. R. Goulet and P. B. Baltes (Eds.), *Life-span developmental psychology.* New York: Academic Press, Inc., 1970.

Schaie, K. W. Methodological problems in descriptive developmental research on adulthood and aging. In J. R. Nesselroad & H. W. Reese (Eds.), *Life-span developmental psychology: Methods, issues.* New York: Academic Press, Inc., 1973.

Schaie, K. W., & Labourvie-Vief, G. Generational versus ontogenetic components of change in adult cognitive behavior: A fourteen year cross-sequential study. *Developmental Psychology*, 1974, *10*, 105–120.

Schonfield, D., & Robertson, B. A. Memory storage and aging. *Canadian Journal of Psychology*, 1966, *20*, 228–236.

Schwartz, G. E., & Beatty, J. (Eds.). *Biofeedback: Theory and research.* New York: Academic Press, Inc., 1977.

Shanas, E., & Hauser, P. Zero population growth and the family life of old people. *Journal of Social Issues*, 1974, *30*, 79–92.

Weisman, A. D., & Kastenbaum, R. *The psychological autopsy: A study of the terminal phase of life.* New York: Behavioral Publications, 1968.

Whitbourne, S. K. Test anxiety in elderly and young adults. *International Journal of Aging and Human Development*, 1967, *7*, 201–210.

Woodruff, D. S. A physiological perspective of the psychology of aging. In D. S. Woodruff & J. E. Birren (Eds.), *Aging: Scientific perspectives and social issues.* New York: D. Van Nostrand Company, 1975, pp. 179–200.

Youmans, E. G., & Yarrow, M. Aging and social adaptation: A longitudinal study of healthy old men. In S. Granick & R. Patterson (Eds.), *Human aging. II: An eleven year followup study.* Bethesda, Md.: United States Public Health Service, 1971, pp. 95–105.

If There Were Another Chapter

Yes, done! We have finished this book together, and nobody can take this satisfaction away from us. But what *if* there were to be, say, one more chapter? What would there be left to explore?

Perhaps we would leave the path of the individual and range instead into the time zones before and after his or her time. It is the individual who is born, develops, flourishes, ages, and dies. But there is a fascinating kind of development that can occur between individuals as well. We might dwell on the ways each succeeding generation draws from its parental lineage and how, in turn, it shares something of the accumulated past with the generation it produces. We could examine "the family name" and how individuals either succeed or fail in measuring up to it and carrying it forward. We could look into the scattered evidence that human development across generations is changing rapidly in our times. Possibly individuals are not transmitting (or receiving) as much along family lineage routes as once was the case. And, if in fact, the traditional influence of one generation on the next is lessening, what new responsibilities are on the shoulders of grandparents today? And if social and economic power rapidly falls into the hands of the young in our society, what is the reverse effect on the elders? Do young adults either stimulate or retard the development of their elders? Many questions here; all are worth pursuing.

But instead we might concentrate on illusions. Take, for example, the illusion that things will really continue to be the same. This appears to be deeply rooted within us. No matter how well we know that development means changes, means transformations, we are often caught with looks of innocent astonishment when we see how fast baby has grown up, and how suddenly papa and mama have grown old. Much of our mental development centers around the construction of a stable universe, principles and structures that a mature mind can count on. And yet, time and change are always at work on us. Perhaps there is something very naive, very illusionary in our quest for stability. The question, therefore, might be how to reconcile the reality of change with our cognitive and emotional disposition to lock time into place.

We might also examine the related illusion that changes, when they do occur, are rational. This is not necessarily a prime illusion of the citizenry in general, but it is part of the faith of the scientist. In the sphere of most immediate interest to us, it is expressed in the assumption that all important change that occurs over the human lifespan is *developmental*. That some important changes might be accidental or irrational is rarely taken as a serious possibility by theory-makers. The result is that textbooks read like textbooks: everything is known or will be known sooner or later; it's all under control, folks. These are examples of illusions that have subtle effects on our ideas of human development.

Still again, that one more chapter might be used to warn against the loss of good, critical perspective in times ahead. The history of any science consists in part of stepping over and around the mistakes of its past. Some of the misconceptions in the field of human development have been confronted in this book. But who knows how much of our present knowledge will prove erroneous or sadly inadequate? Furthermore, in a culture as rich and active as our own, there is a continual influx of themes and advocacies. Every few years a new aspect of human development is discovered or rediscovered. Some people drop everything to accept and respond; others shake their heads and turn away. It is not easy to keep a flexible perspective. In recent years, for example, consciousness-raising has been evident in the areas of sexuality, female development, and dying and death. These are all, in my judgment, significant topics that deserve the attention they are finally receiving. But not every new cause is a significant one, and not every response to a new theme is useful. We must learn somehow to choose wisely among the various inputs that would reshape our ideas of human development.

There are also pendulum swings along the control-helplessness continuum. At times, our culture insists that we can, should, and must remake the world. We are encouraged to take one side in the ancient debate over the role of humankind in the universe. But then the fashion shifts, and we are instructed to look on ourselves as the more or less passive recipients of events we cannot master. These are broad mood swings in our culture, but they have their impact on such specific developmental issues as child-rearing practices, the privileges and responsibilities of adolescents, and the place of the aged. Keeping our balance as the culture rolls from side to side is not easy. Some knowledge of the history of thought, of current events, and of the best-established developmental principles can be helpful, however.

Or if we did have that final chapter, perhaps we would take the opportunity to fix our eyes on goals that have not yet been reached by individuals and society. What are the true potentialities of human development? Are there ways of carrying some of the freshness and spontaneity of childhood through the lifespan, while also adding the emerging dimensions of critical thought, responsibility, and competency? Is it possible for intellectual curiosity to flourish throughout the entire lifespan, instead of atrophying at an early age? Might creativity be a life-style for most people in a society, instead of for a few? Can our "prosocial" emotions develop beyond their customary point so that the spirit of love encompasses more than a few people, dogs, and cats in our lives? Old age: can this truly be a time when the person is complete, fulfilled?

To my knowledge, no society has ever expressed equally favorably attitudes toward all stations of the lifespan. Where youth was worshipped, old age was devalued. Where infants and small children were indulged, adolescents were persecuted and faced with grim challenges, and so on. A person who insists on being in a favorable life situation at all times might have to arrange to be born in one society, grow up in another, function as an adult somewhere else, and choose still another place in which to grow old. Not a very practical arrangement for most of us. A concluding chapter might, then, offer the prospect of a society in which every time of life is a good time, where personal value and social worth are not doled out here only to be taken away there.

Or, we might instead look into—but wait . . . it is, after all, your turn. . . .

Glossary

Abortion Evacuation of the embryo or fetus from the uterus (womb) before it is viable

Accommodation Response of the eyes to changes in the distance between the eyes and the object viewed

Adaptation Response of the eyes to changes in lighting condition

Affective climate The emotional tone that is characteristic of a particular group (e.g., a family) or situation

Alpha rhythm A slow brain wave that is often associated with a calm, pleasant state of mind

Altruistic behavior Actions that are intended to help others

Amniotic sac The gelatinous fluid protecting the embryo

Artificial insemination The introduction of sperm into the female through a laboratory procedure designed to increase the probability of conception

Autonomy Self-determination, independence. Erikson views autonomy as the second of eight developmental tasks to be achieved through the lifespan

Average expectable environment The situation that usually exists; what one would expect to find if no extreme or emergency factors were operating

Battered child syndrome A general term for children who have been physically assaulted by adults

Behavioral genetics The study of interrelationships between genetic endowment, personality, and behavior

Biosphere The entire realm of physiological and biochemical processes, ranging from the subcellular to large units such as the entire central nervous system

Cephalocaudal principle Development proceeds chiefly from head to tail

Cerebral cortex The topmost layer of brain cells, the most highly developed of all cells in the central nervous system

Chained stimulus-response bonds A series of psychological connections built from simple stimulus-response associations which become linked one to the other

Childhood amnesia The adult's difficulty in recalling experiences from very early childhood

Chromosome The biological unit that transmits genetic information from parent to child

Class inclusion operations The process of dividing and grouping phenomena into appropriate sets

Clinical death The condition in which vital functions have ceased, but critical and irreversible bodily damage has not occurred

Combinatorial logic The ability to analyze a situation logically and organize its components in several different ways

Concrete operations A stage of mental operations in which rules, facts, and events are understood, but truly abstract thought has not yet been achieved (Piaget)

Conditionality A situation in which one event depends on another ("if . . . then"; "if not . . . then not")

Conditioned stimulus An event whose power to elicit a response requires prior learning (*see also* unconditioned stimulus)

Conservation The ability to understand that a certain object or characteristic remains the same (invariant) even though it is seen in different ways and in different circumstances

Continuous growth A developmental pattern shown by some adolescents in which personality changes occur in a smooth and gradual manner (Offer & Offer)

Controlled experimentation A study that makes use of a formal research design to make sure valid conclusions can be drawn from the observations

Critical period A brief span of time during early development when the individual is ready to be strongly influenced by specific stimuli or circumstances

Cross-sectional research design A method of research in which people of different ages are studied at the same time

Cross-sequential research design A method of research where features of cross-sectional and longitudinal studies are combined so individuals of different ages are studied at the same time, but also studied through time

Crystallized intelligence Culturally acquired knowledge and mental skills (e.g., the use of language) and the ability to deal with the familiar (R.B. Cattell)

Cultural relativism The view that values are not absolute, but depend on the patterns and expectations of a particular society

Decentering The process of shifting from an egocentric orientation to one in which relationships and perspectives are recognized (Piaget)

Descriptive method of research Observations only; there are no experimental manipulations or interventions

Dethronement The loss of special status for an "only child" when a sibling is born (Adler)

Development Literally, an unfolding or unrolling of an organism's potential

Developmental crisis A threatening situation that confronts a person at a certain time in life, as distinguished from unpredictable problems that might occur at any time

Developmental time The tempo at which actual growth occurs, as distinguished from the standardized, never-varying tempo of clock/calendar time

Dimorphism The process or state of becoming physically distinct as male or female after an earlier period of less marked differentiation

Disequilibrium A state of temporary difficulty in adaptation which spurs efforts toward a higher level of integration and functioning (Piaget)

Ego The coping and adjusting functions by which the individual attempts to meet his or her needs in the real world (Freud)

Egocentrism Viewing the world only in relationship to one's own needs and situation (Piaget)

Ego ideal A person whom another individual has taken as a model, whom he or she wants to be like

Emergent characteristics Features (physical or mental/behavioral) that appear during development, new characteristics as distinguished from those previously existing that may continue to change

Empty organism The view that the organism is less important than the stimulus-response conditions under which it is studied

Encoding The process of registering a fact or experience into memory storage

Endocrines Ductless glands (e.g., the pituitary)

Episodic memory Recall or recognition of specific events and facts

Eros A broad instinct oriented toward life and love (Freud)

Eugenics The view that the genetic pool should be improved by various types of intervention

Evolutionary idealism The theory that distinctively human development begins in adolescence and that society should take advantage of the idealistic outlook characteristic of this time (G.S. Hall)

Exploratory behavior Actions that seem to be motivated by an interest in discovering more about the environment

Expressive behavior Gestures, sounds, and actions that reveal the individual's state of mind

External disequilibrium A lack of coordination between the individual's mode of functioning and the outer environment

Factor analysis A multi-variate statistical technique that identifies, measures and groups the commonalities within complex data

Fetus The embryo is usually considered to become a fetus about two months after conception

Fluid intelligence The ability to cope with new situations and problems, to think abstractly and creatively (R.B. Cattell)

Fontanels Growing places in the neonate's cranium that close and harden before age two

Formal operations The highest stage of mental development in which thought includes abstract, flexible, and systematic operations (Piaget)

Fraternal twins Neonates who were born at the same time of the same mother, but came from separate ova fertilized by separate sperm

Future time perspective, FTP Orientation toward the future. It is often measured by the extent of its range, as well as by themes and content

(g) A symbol sometimes used to refer to *general* intelligence, as distinguished from particular kinds of mental functioning

Gender identity That aspect of an individual's total self-image that is centered around his maleness or her femininity

Generativity The ability to encourage and enjoy other people's growth and achievement. Erikson considers it to be a major developmental task for adults

Genes Tiny protein units that make up the larger (yet still very small) chromosomes

Genocide The destruction of a group of people who have a common pool of genetic characteristics

Genotype The inherited *disposition* toward a particular characteristic

Gerontologist A scientist who studies processes associated with aging

Goal-corrected partnership One person alters his or her behavior to accord better with another person's behavior (Bowlby)

Goal correction Changing one's own behavior to achieve a desired outcome as new information (feedback) is received (Bowlby)

Gonadotropic hormone A pituitary secretion that stimulates the growth of sexual organs (gonads)

Gonads The organs directly associated with sexual reproduction (in both male and female)

Growth errors Mistakes in perception or thought that occur when a child is moving from a lower to a higher level of functioning (Bruner)

Growth hormone A secretion of the pituitary that stimulates the general growth spurt in adolescence

Habituation The process of becoming less responsive to a stimulus as it becomes more familiar

Holophrastic speech Utterances in one-word units which may mean more than their literal (adult) meanings

Homeostasis The active balance (steady state) an organism attempts to maintain for adaptation and survival

Hypothalamus A structure directly below the brain that helps regulate many vital bodily processes

Id The individual's most primitive, least developed set of strivings for pleasure and the release of tension (Freud)

Identical twins Neonates who developed from the same ovum and were fertilized by the same sperm

Identification A complex process in personality development through which the child tries to become like some admired and influential person

Identity confusion A condition in which the individual has not integrated various components of the self into a single, coherent identity (Erikson)

Implantation The process by which the zygote attaches itself to the uterine wall

Imprinting The establishment of a behavior pattern during a specific time in the organism's early development if the necessary stimulus conditions exist (Lorenz)

Incorporation The process of making a part of the outside world into an aspect of one's own personality

Indexing A marking or pointing toward other phenomena

Index variable A variable that points toward influential factors in a process or situation without having direct explanatory value itself

Innate Literally "in-born," a characteristic that does not have to be learned

Instinctual energies The tensions that are generated by organ systems and seek release or discharge (Freud)

Instrumental behavior Actions that directly affect the environment or another person

Interactionist's fallacy The error of concentrating on external and internal systems to the exclusion of the self (Kastenbaum)

Interiorization Representing and perhaps manipulating external objects and processes within the mind

Internal disequilibrium A lack of coordination within the individual's own modes of functioning (Piaget)

Invariant A characteristic or condition that does not change

Involution A process of rolling inward or withering, a loss of features and abilities achieved during development

Latency phase The reduction of instinctual pressures during late childhood which facilitates development of the ability to cope with everyday life in the real world (psychoanalytic theory)

Let-down reflex A change in nipple position during nursing usually associated with a relaxed state on the part of the mother

Longitudinal research design A method of research in which the same individuals are followed (tested, observed) repeatedly over a period of time

Lose-shift strategy Trying another approach when the first response does not have the desired effect

Maternal overprotection Expression of a mother's own anxiety through excessive control of the child (Levy, Spitz)

Maturation The process by which the young attain adult structure and function

MBD, minimal brain dysfunction A state of relatively mild central nervous system impairment that interferes with learning

Mediational deficiency The inability to use a particular memory strategy *effectively*

Menarche The first menstrual flow

Mesomorph The chunky, muscular type of physique

Metamemory Knowledge of one's own memory system (Flavell)

Metropolitan youth culture The adolescent generation serving as its own major influence and source of identity (Opler)

Mitosis The process of cell division and subdivision by which growth is accomplished

Multi-variate method of research Many aspects of a situation are studied and analyzed at the same time

Mutation A change that occurs in the genetic code from one generation to the next

Myelinization The process by which nerves become coated with an insulating sheath critical to development of the nervous system in the first two years of life

Nativistic tradition A theoretical position that emphasizes the "built-in" characteristics of the individual, as distinguished from those acquired through experience

Naturalistic method of research Studying behavior in its ordinary setting, with as little interference as possible

Natural time The standardized, public passage of time, as on clock or calendar

Nature vs. nurture controversy Which are more important in the development of an organism: constitutional (innate) factors or what is learned through experience?

Number conservation The understanding that "how many" is not changed by rearrangements, that number is an independent fact in its own right

Objective dependency The realistic need of an immature or otherwise vulnerable organism for nurture and protection

Oedipal situation The young child's desire to be like the same-sex parent and "possess" the opposite-sex parent (psychoanalytic theory)

Omnibus measures Tests of intelligence that actually assess a variety of different types of mental ability, but which may yield a total score

Open words Words that refer to people and objects (Braine)

Oral stage The first psychosexual stage, centering around the mouth and its functions (Freud)

Orthogenetic principle The concept that all development starts from a global, undifferentiated state and then becomes increasingly differentiated and organized into integrated layers or hierarchies (Werner)

Parallel play The tendency of young children to play in close proximity to each other without much interaction or collaboration

Parturition The process of giving birth

Peer-group identity That aspect of the total self-concept that is based on association with others of the same age and status

Perceptual learning The process of actively extracting information from the environment

Phenomenological realm The individual's private, inner experience

Phenotype An inherited characteristic that is actually *expressed* in a particular individual

Phonemes Sounds that resemble speech

Pitch discrimination The ability to respond differentially to adjacent (higher or lower) tones

Pituitary An endocrine gland in the head whose secretions stimulate growth and sexual maturation as well as other important processes

Pivot words Verbs, nouns, and possessives (Braine)

PKU, phenylketonuria A genetically transmitted condition that can result in a form of mental retardation if not identified and treated

Placenta The temporary biological structure that connects fetus and mother

Plasticity The tendency to be influenced or shaped by experience

Play-work The child's use of play to master problems and conflicts (Kastenbaum)

Pleasure principle The impulse toward direct and immediate gratification of bodily tensions (Freud)

Pluralism The view that there is more than one "normal" pattern of development

Postpartum period The days and weeks immediately after birth

Precognition Knowledge of events before they occur without an obvious source for such knowledge

Prehension Seizing or holding onto, whether physically or through the mind (comprehension)

Preoperational stage The period of mental development before concrete and formal thought operations become established (Piaget)

Presymbolic stage A level of mental functioning at which objects, events, and relationships cannot be inwardly represented

Probability sample A relatively small number of people, events, or other data that are selected to accurately represent the larger population from which they are drawn

Production deficiency The inability to use a particular memory strategy

Protagonist The hero or central character in a struggle, contest, or drama

Psychobiological field The organism seen as an active, integrated reality who cannot be fragmented or reduced for theoretical purposes

Psycho-historian A scholar who combines the expertise and interests of psychologist and historian and examines past events within a psychological perspective

Psychological death Although the individual is alive biologically, inner (phenomenological) experience and mental life have faded out

Psychophysical disposition The organism's innate tendency to seek need satisfaction through certain patterns of attention and action (McDougall)

Psychosexual development The maturational pattern through which tensions and gratifications focus first on one, and then another area of the body (Freud)

Psychosocial moratorium A period of time in which the individual is not required to make full commitments and firm decisions about the course of life. For Erikson, this is adolescence

Puberty The state or process of reaching sexual maturity in the physical sense

Quantifiable information Observations that can be transformed into numbers for objective and precise analysis

Recapitulation theory The developing individual goes through the same stages in his or her own life that the species went through during evolution (G.S. Hall)

Receptive language The ability to understand words and other symbols

Regression A return toward an earlier, simpler way of functioning

Rehearsal A memory strategy in which the material that is to be learned and recalled is practiced before the time of performance

Relative deprivation A sociological theory that "have-nots" feel especially disadvantaged when comparing themselves with conspicuous "haves"; It is the comparison as well as the absolute deprivation that especially disturbs

Representational stage The stage of mental development during which true symbolic functions emerge (Piaget)

Retrolental fibroplasia Detachment of the retina from the inner surface of the eye

Retrospective modalities The ways in which the individual uses his or her past (Kastenbaum)

Role discontinuity A situation in which one has left, or is leaving, his or her type of involvement with society and has not yet firmly settled into another

Sanyāsa The final and most advanced stage of the lifespan that is achieved by only a few (Hinduism)

Scenario The general blueprint or situational framework from which a story develops

Semantic memory The system and rules by which memories are stored and retrieved

Semantic properties of speech The meaning of words, as distinguished from the way they sound

Sensorimotor stage The earliest period of mental development in infancy and childhood (Piaget)

Serial monogamy A sexually-oriented relationship pattern in which each partner has only one love interest at the time, but new partners replace the previous ones

Serial ordering operations The mental process of organizing phenomena into a sequence based on one or more of their characteristics

Sibling rivalry Competition for attention and power among children of the same parents

Significant other A person whose presence and caring behavior is of special importance (e.g., mother to child)

Situational ethic "Right" and "wrong" cannot be evaluated without careful attention to the specific circumstances involved

Social contract Society, or the group, gains its legitimate power when the individual voluntarily yields his or her own power to some extent, but retains a basic freedom and individuality (Rousseau)

Social death A biologically intact individual is treated as though dead or absent by others

Social immortality A deceased person "lives" on in memory or another symbolic way within society

Social integration A measure of the individual's total pattern of relationships and obligations that link him or her with society

Spatial egocentrism Being able to see a situation only from one's own spatial point of reference

Subclinical disease A condition in which there are no symptoms or few symptoms of disorder, but in which a disease process can be detected by careful examination

Subjective dependency A need a person feels for protection and sustenance from others whether or not it is objectively required by the realities of the situation

Superego The individual's "judge," "censor," or "conscience" that tries to inhibit id functions (Freud)

Surgent growth A developmental pattern shown by some adolescents that is marked by fits and starts (Offer & Offer)

Syndrome A set or configuration of symptoms

Telegraphic speech Two- or three-word utterances with rudimentary grammatical structure

Thanatos A hypothetical instinct that is oriented toward death and cessation (Freud)

Thymus gland A small gland in the chest that is important in early development and later atrophies

Time-lag research design A method of research in which new participants are added at regular intervals

Time perspective Orientation toward past, present, and future

Trauma A physical or emotional injury

Tumultuous growth A developmental pattern shown by some adolescents that is marked by a core of conflict and distress (Offer & Offer)

Turning-the-corner experience A situation in which the individual feels he or she is moving to a new developmental position

Unconditioned stimulus An event that is followed by unlearned or reflex behavior (e.g., puff of air—eye blink)

Variant A dimension or fact that can show change

Viability The ability to survive

Vocalizing Vocal sounds that do not use words or word-like structures (e.g., crooning)

Whole-properties Characteristics of a total situation, as distinguished from its specific details

Win-stay strategy Staying with a successful mode and not changing behavior arbitrarily, just for the sake of change

Zygote The fertilized egg that combines genetic contributions from both parents

Name Index

Accardo, P.J., 110, 132
Adams, R.L., 64, 65, 94
Adler, A., 62, 63, 64
Aichorn, D., 460-465
Ainsworth, M.D., 139-140, 148-149, 167
Aisenberg, R.B., 76, 95, 159, 160, 167, 306, 336, 368, 371, 535, 538
Allamen, J.D., 364, 372
Allen, V.L., 68, 94
Amatruda, C.S., 192, 203-204
Ames, L.B., 203, 298, 336
Anderson, C., 519, 537
Annis, L.F., 89, 94
Ansbacher, H., 63, 94, 200, 203
Ansbacher, R., 63, 94, 200, 203
Anthony, A., 427, 437
Anthony, S., 305, 335
Araujo, G., 519, 537
Archer, M., 160, 168
Arenberg, D., 600, 627
Aries, P., 228, 240, 382, 414
Arnheim, R., 299, 335
Artt, S., 513, 538
Ausubel, D., 186, 203
Axline, V.M., 264, 288
Bader, J., 612, 627
Bakan, 159, 167
Baker, A.A., 126, 132
Baltes, P.B., 373, 528, 538, 601, 627-628
Bandura, A., 312, 314, 329, 335, 488-489, 492, 508
Barker, R.G., 229, 240
Barry, H., 284, 288
Bartlett, F.C., 253, 288
Bauer, D.H., 404-406, 414
Bayley, N., 80, 94, 192-193, 203
Beach, D.H., 318, 335
Beatty, J., 592, 629
Bell, S., 148-149, 167
Bellugi, U., 188-189, 203-204
Belmont, L., 65, 67, 94
Bernard, H.W., 427, 437
Biller, H.B., 152, 167
Bimmerle, J.F., 126, 133
Binet, A., 525
Birch, H.W., 68, 70, 94
Birren, J.E., 578-579, 589, 627-629
Bloom, L., 188, 203
Blum, G.S., 284, 288
Bohannon, J.N., 184, 203
Botwinick, J., 592-593, 596-598, 627
Bowlby, J., 135, 140-143, 150, 154, 165, 167, 284, 288, 305, 335
Bradford, S., 115, 133
Braine, M.D.S., 188, 202, 203
Breed, W., 619, 628
Bromberg, W., 160, 167
Bronfenbrenner, U., 327-329, 335
Bronson, G., 220, 240
Brooks, J., 192, 203
Bross, I., 520, 538
Brotman, H., 610, 612, 627
Broverman, D.M., 489, 508
Broverman, I.K., 489, 508
Brown, A.L., 255, 284, 288
Brown, A.M., 79, 94, 96
Brown, R., 171, 187-189, 202-203, 252
Bruce, R.A., 524, 530
Bruner, J.J., 344, 351, 369, 372
Brunet, O., 192, 203
Bryan, J.H., 362-372
Burnet, M., 81, 94
Bushnell, N., 406, 415
Busse, E., 530, 537, 629
Butler, N.R., 89, 94
Butler, R.N., 595, 607, 614, 624, 627
Caillois, R., 247, 290
Caldwell, B.M., 220, 240
Cameron, 515, 537-538
Cannizo, S.R., 318, 336
Cantor, P., 500, 508
Capute, A.J., 110, 132
Cassel, J., 519, 538
Cattell, P., 192, 203
Cattell, R.B., 526, 528, 530, 538, 562, 574
Chinsky, J.M., 318, 335
Chiriboga, D., 516, 538, 574-575
Chomsky, N., 190, 203
Clarkson, F.E., 489, 508
Clayton, C., 115, 133
Cohen, L.B., 109, 132
Colby, B., 256, 288
Cole, M., 256, 288, 290
Conger, J.J., 75, 95
Cooper, M., 571, 574
Corman, H.H., 199, 203
Costa, P.T., 530, 537, 562-563, 568, 571, 573-574, 595, 619, 628
Cox, R.D., 530, 537
Crandall, V.C., 364, 372
Cratty, B.J., 296, 335
Crick, F.H., 71-73, 96
Crookham, G., 386, 414
Cumming, E., 537, 546, 574, 590, 627
Cutler, D.R., 17, 26
Darrow, C.M., 571, 575
Darwin, C., 207, 209-211, 238, 240
Datan, N., 552, 575
Davison, A.N., 88, 94
Dehn, M.M., 524, 537
DeLeo, J.H., 109, 132
deMause, L., 228, 240
Dennenberg, V.H., 88, 94
Dennis, W., 605, 627
Derbin, V., 513, 538
Diers, C.J., 430, 437
Disbrow, M., 182, 204
Dixon, D., 310, 336
Dobbing, J., 88, 94
Donaldson, G., 601-602, 628
Downs, E.F., 160, 168
Droegenmueller, W., 157, 167
Dublin, L.I., 340, 372
Ducasse, C.J., 18, 26
Dudley, D., 519, 537
Durkee, N., 516, 538
Eber, H.W., 562, 574
Eckerman, C.O., 137-138, 168
Edwards, C.P., 388-389, 391-392, 415
Ehrhardt, A.A., 394-395, 397, 415
Ehrlich, A.H., 229, 240
Ehrlich, P.R., 229, 240
Eifler, D., 394, 415
Eimas, P.D., 116, 132
Eisdorfer, C., 537, 539, 575, 627
Elkind, D., 352, 372, 447-451, 462, 464-465, 474, 498-499, 508
Emde, R.N., 30, 53
Engen, T., 120, 132
Engleman, T.G., 182, 204
Erikson, E.H., 49-50, 53, 274, 288, 375, 397-403, 413, 414, 451-458, 460-465, 471, 477, 489, 530, 532, 536-538
Ervin-Tripp, S.M., 184, 203-204
Escalona, S.K., 31, 53, 199, 203
Fantz, R.L., 114, 132
Farley, H-A., 305, 336
Feifel, H., 306, 335, 538
Ferreira, A.J., 90, 94
Fillenbaum, G., 582, 627
Flavell, J.H., 258, 289, 317-320, 335-336, 352, 372
Flohil, J.M., 126, 132
Fontana, F.J., 158, 167, 519, 538
Forbes, H.B., 116, 132
Forbes, H.S., 116, 132
Fozard, J., 530, 537, 593, 595, 627
Fraiberg, S., 30-34, 37, 39, 53

Fraser, D., 187, 203
Freedman, D., 79, 94
Freud, A., 154, 167, 264, 289
Freud, S., 38, 207, 209, 214-218, 238, 240, 256, 274, 289, 458
Freyberg, J., 310, 315, 336
Friedman, S., 300, 336, 538
Friesen, D., 489, 508
Furman, R., 284, 289
Galton, F., 62, 94
Gardner, R.A., 322-323, 336
Gelman, R., 344-345, 372
Gershowitz, M., 310, 336
Gesell, A., 192-193, 204
Gewirtz, J.L., 184, 204
Gibson, E.J., 310, 312, 336
Gil, D.G., 157-160, 167
Goldberg, S., 393, 414
Goldfarb, W., 157, 167
Goldsmith, I., 605, 627
Goldstein, R.H., 89, 94
Gollin, E.S., 121-122, 132
Golomb, C., 300, 336
Goodnow, J.J., 300, 336
Gordon, J.E., 126, 133
Gordon, R.E., 127, 132
Gottesman, I.I., 79, 95
Gould, R.L., 571, 574
Granick, S., 582, 627, 629
Greenberg, M.T., 258-259, 289
Greenberger, E., 483-485, 507-508
Greenfield, P.M., 344, 351, 371
Gregory, A.J., 80, 95
Greif, E.B., 362, 373
Griffiths, R., 112, 204
Grollman, E., 306, 336
Grossman, J., 595, 628
Gussow, J.D., 68, 70, 94
Guttmalher, A.F., 89, 95
Guttman, D., 570, 575
Guyton, A.C., 421, 437
Haber, A.L., 113, 132
Haddan, W., 367, 372
Hagen, J.W., 318, 335, 336
Haith, M.W., 111, 113, 132
Hall, G.S., 442-446, 460-463, 465, 489
Hamer, J.H., 578, 628
Hanks, C., 154, 168
Harlow, H.F., 155, 165, 167
Harlow, M.K., 155, 158, 167
Harris, L., 623, 628
Harryman, S., 110, 132
Hartmann, E., 477, 478, 508
Heinicke, C.M., 142, 167
Helfer, R.E., 167, 168
Hendin, H., 500-502, 508
Hendrichs, C., 612, 628
Hendrichs, J., 612, 628
Henry, W.E., 477, 508, 537, 546, 574, 590, 627
Hershenson, H., 113, 132
Hess, R., 477, 478, 508
Hetherington, E.M., 152, 153, 167, 372
Hicks, C.H., 589, 628
Hilton, I., 64, 95
Hinkle, L., 520, 538
Hokada, E., 394, 415
Holmes, C.D., 578, 628
Holmes, T.G., 319, 336
Holmes, T.H., 519, 538
Holmes, T.S., 519, 538
Honzik, M.P., 192, 204
Horibe, F., 304, 337
Horn, J.C., 526, 528, 530, 538, 601, 602, 627
Horner, M.S., 489-490, 508
Hulicka, I., 595, 628
Hultsch, D., 598, 629
Hunt, J. McV., 199, 204
Huntington, E., 229, 240
Hurster, M.M., 160, 168
Huxley, A., 121, 224, 240
Ilg, F.L., 203, 298, 326
Inhelder, B., 199, 204, 260, 289, 297, 336, 352, 355, 372, 410, 415
Itard, J.M.G., 47
Ittelson, W.H., 229, 240
Jackson, J.J., 612, 628
Jain, A., 386-387, 414
James, W., 103-105, 132, 253, 289
Jansson, B., 126, 132
Javert, C.T., 91, 95
Jayant, K., 61, 95
Jeffers, F.C., 76, 95
Joel-Nielsen, N., 79, 95
Johnson, J., 310, 336
Johnson, R., 487, 508
Johnson, V.G., 621-622, 628
Jones, H.E., 64, 95
Jones, M.M., 269, 289
Jongeward, R.H., 318, 336
Josselson, R., 483-485, 499, 507-508
Joyce, C.S., 364, 372
Jung, C.G., 397, 414, 563, 575
Kagan, J., 75, 95, 289, 408-409, 414-415
Kail, 318, 335-336
Kalish, R.A., 7-8, 26, 487, 508
Kant, I., 221, 240
Kaplan, B., 519, 538
Kastenbaum, R., 17, 26, 57, 76, 95, 159-160, 167, 306, 336, 368, 372, 512-513, 516, 535, 538, 611, 614-615, 618-619, 627, 629
Katz, M., 252, 289
Kaye, H., 119-120, 132
Keeney, T.J., 318, 336
Kellogg, L.A., 165, 167
Kellogg, R., 264, 289, 299, 336
Kellogg, W.N., 165, 167
Kempe, C.H., 157, 167-168
Kendler, H.H., 243, 252, 287, 289
Kendler, T.S., 243, 252, 287, 289
Keniston, K., 487, 508
Kern, S., 383, 414
Kessen, W., 111, 113, 132
Kimmey, J., 91, 95
Kinsey, A., 620, 628
Klein, D., 367, 372
Klein, E.B., 575
Kleitman, N., 182, 204
Klima, E.S., 189, 204
Kline, E.L., 125, 132
Kluckhohn, C., 427, 437
Knight, G.P., 408-409, 414-415
Koch, H.L., 393, 415
Koch, R., 81, 95
Kohlberg, L., 339, 358-361, 372-373, 407, 487, 628
Konopka, G., 480, 482-483, 492, 507-508
Korn, S.J., 301, 336
Kornberg, A., 72, 95
Korner, M., 150, 167
Kozol, J., 307, 336
Krebs, D.L., 362, 373
Krolick, G., 279, 290
Kuhn, M.E., 612, 628
Kurtines, W., 362, 373
LaBouvie, G., 528, 537, 601, 627-629
Lacoursíere-Paige, F., 297, 336
Lamb, M., 146, 167
Lane, H., 47, 53
Lawton, M.P., 537, 539, 575, 627
Leasor, T., 159, 167
LeBoyer, F., 131
Lee, D., 271, 289
Lee, S., 499, 509
Lehman, H.C., 605, 628
Lennenberg, E.H., 187, 204
Lerner, R.M., 301, 336
Levinson, D.J., 571, 574
Levinson, M.H., 571, 574
Levy, D.M., 151, 163, 167
Levy, H., 115, 133
Lewin, K., 237, 240
Lewis, M., 193, 204, 392-393, 414-415
Lewis, M.I., 595, 614, 624, 627-628
Lewis, R., 294, 305, 336-337
Lezine, P.U.F., 192, 203
Lindauer, B.K., 317, 336
Linden, L.L., 619, 628
Lindsley, O.G., 619, 628
Lindzey, G., 81, 95
Lipsitt, L.P., 119-120, 132, 335
Lipton, R.C., 157, 168
Locke, J., 207, 225-226, 240
Looft, R., 499, 509
Lorence, B., 382, 415
Lorenz, K.Z., 207, 218-219, 238, 240
Lowenthal, M.F., 516, 538, 563, 568, 570, 574-575
Luce, G.G., 182, 204
Luckey, E.B., 492-496, 509
Luhan, J.A., 126, 133
Luria, Z., 122, 133, 243, 250-252, 285, 289
Lynn, D., 154, 167
McCall, R.B.M., 193, 204
McCandless, B.R., 406, 415
McCarthy, D., 187, 204
McConochie, D., 483-485, 507-508
McDougall, W., 207, 211-215, 238, 240
McGhee, P.E., 333, 336
Mack, J.E., 247, 289
McKee, B., 571, 575
McNeill, D., 189, 204
McRae, J., 530, 537, 562-563, 568, 571, 573-574
Madden, J.J., 126, 133
Maduro, R., 604, 628
Manosevitz, M., 81, 95
Markus, J., 65, 67
Marlow, D.M., 175, 204
Marquis, A.L., 184, 203
Marolla, F.A., 65, 66, 94
Maslow, A., 530, 538
Masters, W.H., 621-622, 628
Maurer, A., 305, 336
Marvin, M.S., 258-259, 289
Mehrabian, A., 199, 204
Melnyk, J.M., 81, 95
Mendel, G., 55, 74, 93
Miernyk, W.H., 552, 575
Milinaire, C., 131, 133
Millar, S., 268, 271, 289
Mischel, W., 274, 289
Mishara, B.L., 500, 509, 619, 629
Moely, B.E., 319, 336
Money, J., 394-395, 397, 415
Monge, R.A., 598, 629
Moro, E., 108, 111, 112
Mossler, D.G., 258-259, 289
Mullahey, P., 274, 289
Munnichs, J., 618, 629
Murphy, L.B., 399, 415
Murphy, M.D., 255, 288
Mussen, P.H., 75, 94-95, 132, 415
Myers, S., 160, 167
Nass, G.D., 492-496, 509
Neugarten, B., 509, 516, 538, 570, 575
Newson, E., 167, 273, 289
Newson, J., 167, 273, 289
Nichols, R.C., 79, 95
Noel, B., 519, 538
Nuckolls, K., 519, 539
O'Dell, S., 264, 289, 299, 336
Oden, M.H., 559-561, 575

Offer, D., 469–470, 472–474, 476–477, 480–481, 487–489, 492–493, 505, 509
Offer, J.B., 469–470, 476–477, 480–481, 487–489, 492, 509
Olson, G.A., 319, 336
Olver, R.R., 344, 350, 372
Opie, I., 266, 289
Opie, P., 266, 289
Opler, M., 457–463, 465
Orwell, G., 224, 242
Osborne, R.T., 80, 95
Padilla, S.G., 219, 241
Palmore, E., 76, 95
Paris, S.G., 317, 336
Parmelee, A.H., Jr., 182, 204
Patterson, R., 582, 627, 629
Pavlov, I.P., 119, 133
Paykel, E., 519, 539
Petty, T.A., 264–265, 289
Pfeiffer, E., 622, 629
Phillips, B.N., 64–65, 94
Piaget, J., 48, 50, 53, 171, 194, 196–199, 202, 204, 226, 236, 243, 251, 252, 254, 257, 259–261, 267, 287, 290, 297, 317, 336, 344–345, 352–356, 359, 369, 373, 409–410, 415, 446–448, 460–463, 465, 509
Pinneau, S., 157, 168
Pinsky, R., 520, 538
Plato, 121, 207, 222–224, 229, 238, 242, 280, 382
Plummer, N., 520, 538
Pollack, C.B., 160, 168
Proshansky, H.M., 229, 241
Provence, S., 157, 168
Provenzano, F., 122, 133
Quilliam, C., 387, 415
Rahe, R., 519, 538
Ramsay, R.H., 249, 290
Rappaport, L., 147, 168
Rebelsky, F., 153, 168
Reese, H.W., 255, 288, 290, 355
Regnemer, J.L., 406, 415
Rheingold, H.L., 137, 138, 168, 184, 204
Rhine, L.E., 249, 290
Riess, I.L., 473, 509
Rivlin, L.G., 229, 241
Rosenfeld, A., 583, 629
Rosenkrantz, P., 489, 508
Rosenzweig, S., 284, 288
Ross, C., 394, 415
Ross, D., 312, 314, 335
Ross, E.M., 89, 94
Ross, H.W., 184, 204
Ross, S.A., 312, 314, 315
Rousseau, J.J., 207, 227–229, 241
Rubenstein, J.F., 110, 132
Rubin, I., 624, 629
Rubin, J., 122, 133
Sabatini, P., 513, 538
Salapatek, P., 111, 113, 114, 132
Saltz, E., 310, 337
Santos, R., 294, 337
Scarr, S., 79, 95
Schactel, E., 244, 256–257, 287, 290
Schacter, S., 64, 95
Schaie, K.W., 21, 26, 373, 528, 539, 601, 627, 629
Scheinfeld, A., 89, 95
Schofield, M., 492, 497, 509
Schooler, C., 64, 95
Schultz, C., 304, 337
Schulz, H.R., 182, 204
Schwartz, G.E., 592, 629
Schwarz, J.C., 279, 290
Scott, J.P., 219, 241
Scott, P.M., 362, 373
Scudder, H.E., 381, 382, 415
Sears, R., 561, 575
Segal, J., 181, 204
Sellein, G., 175, 204
Shantz, C., 297–298, 337
Sheehy, G., 571, 575
Silver, H.G., 158, 167
Silverman, F.N., 157, 167
Simons, B., 160, 168
Simpson, W.J., 89, 95
Sims, J.H., 477, 508–509
Sinclair, H., 260–261, 289
Singer, D.G., 337
Singer, J.L., 309–310, 336–337
Skinner, B.F., 207, 211, 224–225, 241, 280
Smirnov, A.A., 255, 290
Smith, H.T., 399, 415
Snyder, S.S., 409, 411, 415
Sontag, L.W., 90, 392, 415
Sorenson, R.C., 491–493, 497–498, 509
Spitz, R.A., 145–147, 151, 156, 168
Staats, A.W., 190, 204
Stafford, R.E., 79, 94, 96
Stechler, G., 114, 133
Steele, B.N., 157, 160, 167–168
Steele, W.G., 392, 415
Stevenson, H.W., 203, 311–312, 337, 347–348, 373
Stone, L.J., 399, 415
Strauss, S., 410, 415
Strickland, R.G., 279, 290
Suchman, E.A., 367, 372
Sullivan, L., 186, 203
Swift, J., 308, 337
Taft, L.T., 109, 133
Tanner, J.M., 79, 95, 421, 430–431, 435, 437, 463
Tapp, J.L., 361, 373
Tenney, Y.H., 255, 290
Terkel, S., 553–556, 564, 573, 575
Terman, L.M., 62, 96, 558–561, 563, 574
Theorell, T., 520, 539
Thiessen, D., 81, 95
Thomas, C.L., 126, 133
Thomas, J.C., 593, 627
Thompson, H., 203–204
Thurnher, M., 516, 538, 574–575
Timiras, P.S., 59, 96, 106, 116, 133, 421, 437
Tinbergen, N., 220, 238, 241
Tobin, S., 516, 538
Troll, L.B., 487, 509
Tronick, E., 115, 133
Tulving, F., 253, 290
Turiel, E., 358, 361–362, 373, 410, 415
Tuteur, W., 126, 133
Tyler, L.E., 31, 53
Uzgiris, I.C., 199, 204
Van Arsel, P., 519, 537
Vandenberg, S.G., 79, 94, 96
Vining, E.P.G., 110, 132
Vislie, H., 126, 133
Vogel, S.R., 489, 508
Von Grunebaum, G.E., 247, 290
Vygotsky, L.S., 250–251, 290
Wallace, R.F., 90, 95
Walters, C.E., 86, 96
Walters, R., 329, 335, 488–489, 492, 508
Wang, C., 394, 415
Wapner, S., 337, 349–350, 373
Watson, I., 211
Watson, J.D., 71–73, 96
Watson, J.S., 297–298, 337
Waxler, C.Z., 362, 373
Wechsler, D., 525
Wedenberg, E., 132
Weinraub, M., 192, 203
Weisman, A.D., 618–619, 629
Weitzman, L.J., 394, 415
Wellman, H.M., 319, 335
Werner, H., 45–46, 53, 234, 252, 308, 337, 349–350, 369, 373, 404, 415
Westheimer, I., 142, 167
White, C.B., 406, 415
Whiting, B., 388–389, 391–392, 415, 427, 437
Williams, M., 199, 204
Winick, M., 88, 96
Wittgenstein, L., 308, 337
Wolfenstein, M., 384, 415
Woodruff, D.S., 592, 598, 629
Wright, J.W., 289
Wyden, P., 158, 168
Wylie, L., 329–330, 337
Yarrow, M.R., 362, 373, 582–583, 627, 629
Yendovitskaya, T.V., 255, 290
Youmans, E.G., 582–583, 629
Zajonc, R.B., 65–68, 93, 96
Zinchenko, P.I., 255, 290

Subject Index

Abortion, 59, 91
  spontaneous, 59
Adaptation, 211
Adolescence, 418–508
Adolescent, normal, 468–483
  boy, 468–480, 505
  girl, 480–483
Adoption, 60
Affective climate, 145–147
Age, subjective, 513–515, 535
Aged, healthy, 579–584
Ages of Me, 513–515
Aging, 13–14, 578–627
  primary, 584–585
  secondary, 584–585
  theories of, 584–585, 626
Alcoholism, 567–568, 574
Alienation, 403
Altruistic behavior, 327, 362
Amniotic sac, 83
Anal stage, 216
Artificial insemination, 60
Attachment, 136, 139–144, 148
  Bowlby's phases of, 142–144
  clinging, 140–141
  visual-motor orientation, 140
Attention, 213, 342–347
Attention span, 320
Auditory discrimination, 321
Automaton model of human beings, 232
Autonomy, 452
  vs. shame and doubt, 400–401
Average expectable environment, 154, 478
Battered child syndrome, 157–158
Behavior
  altruistic, 327, 362
  clinging, 140–141
  exploratory, 137
  expressive, 142
  innate, 107
  instrumental, 142
Behavioral engineering, 121
Behavioral genetics, 78–79
Behaviorism, 211
Bereavement, 280–285, 287
Birth order, 62–68
  Anglo-American, 65
  Black-American, 65
  Mexican-American, 65
  Netherlands, 65
Blind infant, 31, 36–40
"Blooming, buzzing confusion," 103–106
Body image, 296, 397
Body representation, 299–301
Bowel and bladder functioning, 178–179, 199
Bowlby's phases of attachment, 142–144
Boy, adolescent, 468–480, 505
Brain weight, 523
Breast feeding vs. bottle controversy, 147–149
Buddhism, 7–8
Career decision, 480–481
Central nervous system, 422, 589
Cephalocaudal principle, 85
Cerebral arteriosclerosis, 582
Cerebral cortex, 111
Chained stimulus-response bonds, 252–253
Change
  developmental, 43–46
  directional, 12
Ch'eng-jen, 533–534
Child abuse, 157–161
  characteristics of the abuser, 160
  characteristics of the victim, 159
  history of, 158
  intervention in, 161
Childhood, 374–412
  history of, 381–385
Childhood amnesia, 256, 287
Chromosomes, 71
Classical conditioning, 119
Class inclusion operations, 355
Clinging behavior, 140–141
Cognitive structures, 344, 447–448, 463
Cohort effects, 20, 42
Combinatorial logic, 449
Communication. *See* Language development
Comparison, 104–106
Competence, 587–605
Concrete operations, 48, 257, 352–356, 446–447
Conservation, 345
Continuity of self, 557–565
Controlled experimentation, 48–49
Crawling, 175–176
Creativity, 10–11, 307–315, 602–605
Creeping, 175–176
Crime, 605–606
Critical period, 219
Cross-sectional research design, 18
Cross-sequential research design, 20, 406–407
Crying, 148–151
Cultural relativism, 457–460
Dating, 472–473, 476–477, 482
Day care, 277–280, 288
Death, 17, 280–285, 287, 305–306, 617–620
  accidental, 365, 534–535
  and bereavement, 280–285, 287
  clinical, 17
  and dying, 617–620
  by motor vehicle, 534–535, 537
  psychological, 17
  social, 17
Decentering, 199, 203, 499, 507
Deferred imitation, 260
Dependency, 136, 139, 141–144, 148, 388–391
  objective, 139
  subjective, 139
Deprivation, 69, 90
  relative, 607
Descriptive research method, 47
Determinism
  placental, 87–88
  socioeconomic, 68–70
Dethronement, 63–64
Development, 6, 13, 16
  mental. *See* Mental development and functioning
  physical. *See* Physical development
  pluralistic, 549–551, 573
  precocious sexual, 421
  psychosexual, 216–217, 238
  psychosocial, 545–548
  reverse, 14
Developmental
  change, 43–46
  crisis, 564
  mission, 40–41, 52
  time, 35, 52, 58, 136, 188, 218
Differentiation, 45–46, 84–85, 234–236, 239
Dimorphism, 428
Discrimination
  according to William James, 104–106
  emotional, 141
  perceptual, 141
Disequilibrium, 361
  external, 410–412
  internal, 410–412

DNA, 71-72
Double helix, 72-73
Dying. *See* Death
Early school experiences
  French, 329-331
  USSR, 325-329
Ectoderm, 85
Ego, 217, 238
  functions, 215
  instincts, 217
Egocentrism, 197-199, 203, 257, 287, 499, 507
Ego ideal, 470
Ego integrity vs. despair, 403
Eight Ages of Man. *See* Erikson's Eight Ages of Man
Ejaculation, 428-429
Embryo, 82-85, 376
Emotion, 213
Empirical outlook, 225
Employment, 521-522, 551-557, 573
Endocrine system, 422, 435
Endoderm, 85
Erikson
  on identity confusion, 453
  on psychosocial moratorium, 456, 471
Erikson's Eight Ages of Man, 49, 398-403
  autonomy vs. shame and doubt, 400-401
  basic trust vs. basic mistrust, 451-452
  ego integrity vs. despair, 403
  generativity vs. stagnation, 403
  identity vs. role confusion, 403
  industry vs. inferiority, 402-403, 452
  initiative vs. guilt, 401-402
  intimacy vs. isolation, 403
Eros, 217
Ethical consolidation, 532
Eugenics, 77
Family intellectual climate, 65-68
Fate, 220-221, 238
Fathering, 152-154
Fears, 404-406
Feeding, 147
Femininity, 386-397
  cultural differences, 388-392
Fetus, 82, 85, 376
Fontanels, 123
Formal operations, 48, 352-356, 447
Freud
  anal stage, 216
  ego, 217
  Eros, 217
  id, 231
  oral stage, 216
  pleasure principle, 217
  superego, 231
  Thanatos, 217
Future time perspective, 515, 535
Gender identity, 397
Generation gap, 470-471, 476
Generativity vs. stagnation, 403
Genes, 71
Genocide, 76
Genotype, 73
Germ cells, 71, 73
Gifted person, 558-561
Girl, adolescent, 480-483
Global state, 105-106
Goal corrections, 150
Gonadotropic hormone, 424, 426, 428
Gonads, 422, 435
Grief, 617-620
Growth hormone, 424
Guilt, 452
Habituation, 117-122
Hazards, developmental
  abortion, 90-93
  alcoholism, 567-568
  auto accidents, 534-535
  crime, 605-606
  fatal accidents, 365-369
  parental assault, 157-161
  parental death, 280-285
  postpartum psychosis, 124-128
  slow learner, 320-325
  suicide, 500-503
Head Start program, 329
Hearing, 116, 524-525
Height, 377-381
Heredity, 71
Hinduism, 7-8
Hitching, 175-176
Hormones
  gonadotropic, 424, 426, 428
  growth, 424
Human being, 6-7
Humor, 302-305
Humpty-Dumpty, 265-266
Hyperactivity, 320
Hypothalamus, 422-425, 435
Id, 231
Idealism, evolutionary, 442-446
Identification, 274, 288
Identity, 103, 451, 463
  confusion, 453
  crisis, 477, 505
  formation, 397-403
  vs. role confusion, 403
Ideological experimentation, 532
Imitation, 185
Implantation, 83
Imprinting, 218-219, 238
Incorporation, 400
Index variables, 41
Industry vs. inferiority, 402-403, 452
Initiative, 452
  vs. guilt, 401-402
I.Q. *See* Intelligence, Quotient
Instinct, 209-215, 220, 238
Intelligence, 68
  crystallized, 527, 535, 537
  fluid, 527, 537
  Quotient, 525-530, 558
Interactionist's fallacy, 231
Interdependence, functional, 583-584, 587-588
Interiorization, 196, 202-203
Intimacy, 476-477, 482, 491-500, 505, 613, 620-624, 626
  vs. isolation, 403
Inversion, 354
Kohlberg's stages of moral development, 358-362
Language development, 183-202, 249-253, 285
  internal representation, 250
  nonverbal, 180-190
  receptive language, 249
  semantic properties of speech, 251, 287
  verbal symbols, 250
Latency phase, 341
Learning, 117-122, 596-602
  disability, 320-325
  perceptual, 311-315
Learning to learn, 347-349
Let-down reflex, 148
Life
  review, 614-615
  satisfaction, 515-520, 561
Linguistic systems, culture-specific, 185
Longitudinal research design, 17-18
Lose-shift strategy, 348
Masculinity, 386-397
  cultural differences, 388-392
Maturation, 9-12
  rate of, 378
  of small muscles, 176-177, 199
MBD. *See* Minimal brain dysfunction
Memory, 253-254, 287, 317-320, 593-596
  episodic memory, 253, 287
  impairment, 321
  intentional memory, 319
  metamemory, 293, 320
  semantic memory, 253-254, 287
  strategies, 293, 318-319
Menarche, 425-431, 435
Mendelian Laws, 74
Mental development and functioning, 180-199, 202, 301-316, 342-356, 525-530, 596-602. *See also* Erikson's Eight Ages of Man; Piaget's stages of cognitive development
Mental retardation, 80-81
Mesoderm, 85
Mesomorph, 301
Metamemory, 293, 320
Metropolitan youth culture, 457-460, 465
Middle age, 542-575
Mid-life crisis, 571
Minimal brain dysfunction (MBD), 320-325
Mistrust, basic, 452
Mitosis, 72
Modeling theory, 152
Moral
  development. *See* Kohlberg's stages of moral development; Moral reasoning
  justice, 356-357
  reasoning, 358-362, 404, 406
Motivation, 598
Müller-Lyer illusion, 350-351
Multi-variate research method, 47
Muscle strength, 525
Mutation, 78
Myelinization, 179
Natural experiment, 47
Naturalistic research method, 47-48
Nature vs. nurture controversy, 208
Neonate, 7, 9, 376
Nightmares, 246-247, 285
Number
  comprehension, 315-317
  conservation, 315
Nutrition, prenatal, 88-89
Observation, 162-165
  animal, 164-165
  direct, 162-164
  indirect, 164
Occupation, 551-557, 573
Oedipal situation, 274, 288
Old brain, 110
Old person, typical, 606-610
Operational stage of mental development, 262, 354-356
Oral stage, 216
Orthogenetic principle, 45-46, 404-405
Ova, 71, 73, 82
Overprotection, maternal, 151
Parenting, 152-154
  parent-child relationships, 487-489, 505
Parturition, 101
Peer group, 307-308, 454-455
Perceptual
  handicap, 320
  learning, 311-315
Perspective taking, 257-259, 287

Phenotype, 73
Phenylketonuria (PKU), 80-81, 323
Physical development, 82-87, 122-124, 172-182, 244-245, 294-297, 377-381, 418-431, 522-525, 580-582, 589-593
Piaget
on conservation, 345
on inversion, 354
on moral justice, 356-357
on reciprocity, 354
Piaget's stages of cognitive development
operational, 262, 354-356
class inclusion operations, 355
concrete operations, 48, 257, 352-356, 446-447
formal operations, 48, 352-356, 447
serial ordering operations, 355
preoperational, 48, 260-262, 267, 352-353
representational. *See* Piaget's stages of cognitive development, preoperational
sensorimotor, 48, 194-199, 202, 259-260, 262, 267-268, 344, 352-353
Pituitary gland, 424, 435
PKU. *See* Phenylketonuria
Placenta, 83
Plasticity, 225
Play, 262-271, 287
explorative, 267
parallel, 269
play-work, 264
Pleasure principle, 217
Postpartum psychosis, 124-128
Preadolescence, 418-419
Precognition, 249
Pregnancy, risks of, 61-62
Premature birth, 130
Prenatal scene, 56-97
Preoperational phase of human development, 48, 260-262, 267, 352-353
Prepubescence, 418-419
Prime time, 510
Protagonist, 31
Psychobiological field, 233-234, 239, 376, 421
Psychomotor slowing, 589-593
Psychophysical dispositions, 212
Psychosocial
development, 545-548. *See also* Development
maturity, 483-487
moratorium, 456, 471
Puberty, 418
Pubescence, 418, 426, 435
Reality, 342-343
Reality principle, 217
Reciprocity, 354
Reflexes, 85, 107-111, 194, 211-212, 226
Babinski, 107
crossed extensor, 108
crying, 107
danse, 174
deep tendon, 108
doll's eye, 108
Galant, 109
grasping, 107
let-down, 148
Moro, 108, 111
parachute response, 109
rooting, 107
sneezing, 107
startle, 108, 111
sucking, 107
supporting reaction, 109
Umklammerungsreflex, 108, 111
yawning, 107
Regression, 15, 37
Relationships, parent-child, 487-489, 505
Representational stage of mental development, 260-262, 267, 316
Research designs
cross-sectional, 18
cross-sequential, 20, 406-407
longitudinal, 17-18
time-lag, 407
Research methods
descriptive, 47
multi-variate, 47
naturalistic, 47-48
Retrospective modalities, 613-617
Rites of passage, 427
Role
discontinuity, 460
models, 392-394, 414
Sanyāsa, 602-605
Scenario, 30-32
Schizophrenia, 81
Sensation, 226
Sense organs, 181
Sensorimotor stage of human development, 48, 194-199, 202, 259-260, 262, 267-268, 344, 352-353
Separation, 277-285, 287
anxiety, 154
Serial monogamy, 497-498
Serial ordering operations, 355
Sex
characteristics, 425-428
differentiation, 419-431
stereotypes, 489-491
Sexual development, precocious, 421
Sexuality, 495-498, 507
Shame and doubt, 452
Sibling rivalry, 265, 287
Significant other, 218
Situational ethic, 494
Sleeping and waking, 181-182, 202
Smoking, 59, 89-90
Social
compact/contract, 228
competence, 530-534
immortality, 17
integration, 546
isolation, 155-157
learning, 185
prematurity, 58
Socioeconomic influences, 68-70, 607
Somatic cells, 71
Spacial
egocentrism, 297
objectivity, 297
orientation, 296-297
Sperm, 71-73, 82
Stereotypes
of old age, 610-612
sex, 489-491
Stimulus
conditioned, 119
control, 239
-response learning, 252, 287
unconditioned, 119
Stranger reaction, 230
Stress, 90, 518-520, 535
Suicide, 500-503, 508, 619
Superego, 231
Survival
predictors of, 582-584
statistics, 340
Talking, 183-202
Teeth, 295
Thanatos, 217
Thing language, 308
Thymus gland, 421
Time
developmental, 35, 52, 58, 136, 188, 218
natural, 35, 52
perspective, 613-617
phenomenological, 52
Time-lag research design, 407
Titchener circles, 349-351
Traits, dominant and recessive, 74
Trust
basic, 451-452
vs. mistrust, 399-400
Turning-the-corner experiences, 565-567
Twins, 79-81
Umbilical cord, 84
Verticality, 172, 199
Viability, 6, 87
Vision, 112-115, 524
Visual
acuity, 113
discrimination, 321
-motor orientation, 140
Weight, 377-381
brain, 523
low birth, 130
Win-stay strategy, 348
Working, 551-557, 573
Zygote, 72, 74, 82, 376-377